RISKS, REWARDS AND REGULATION OF UNCONVENTIONAL GAS

A Global Perspective

The global energy transition from carbon-intensive to renewable fuels has increasingly demanded a better understanding of the causes and consequences of the rapid development of unconventional oil and gas. Focusing on key countries including the US, Canada, China, Argentina, the UK and Australia, this book consists of case studies and in-depth analyses that weigh up the risks and rewards at regional, national and global scales. Explaining how and why unconventional fuels are transforming the global energy landscape, the strengths, weaknesses, opportunities and threats are explored through a political, economic and governance-based perspective. Emphasis is placed on how to regulate the industry; the discussion encompasses local issues, stakeholder engagement and the social licence to operate. The new baseline studies and standards introduced in this book provide a timely insight into the trade-offs across the social, economic and environmental domains, making it an ideal text for researchers and policymakers in energy fields and for graduate students.

R. QUENTIN GRAFTON is Professor of Economics at the Australian National University (ANU) and Director of the Centre for Water Economics, Environment and Policy (CWEEP) at the ANU and holds the UNESCO Chair in Water Economics and Transboundary Water Governance. He served as Chief Economist and Foundation Executive Director of the Australian Bureau of Resources and Energy Economics (2011–2013). He has published more than 120 scholarly articles in some of the world's leading journals in economics and the life sciences and has edited or co-authored 15 books.

IAN G. CRONSHAW worked as Division Head at the International Energy Agency (Paris) between 2005 and 2011, where he was responsible for analysing global gas, coal and power developments. He was the principal author of the Agency's Medium-Term Gas Market Outlook in that period. Since then he has worked for the IEA as a consultant on the annual World Energy Outlook and on publications, such as 'The Golden Age of Gas?' and 'Golden Rules for the Golden Age of Gas', highlighting the growing importance of unconventional gas and approaches to regulation, both viewed from a global perspective.

MICHAL C. MOORE is Professor of Energy Economics at the School of Public Policy, University of Calgary, Canada, where he teaches classes in microeconomic theory, decision analysis and energy technologies. He is also Visiting Professor of Economics and Systems Engineering at Cornell University, New York. He is a former Chief Economist for the US National Renewable Energy Laboratory in Golden, Colorado. His current research focuses on energy market regulation and a pan-North-American energy strategy.

RISKS, REWARDS AND REGULATION OF UNCONVENTIONAL GAS

A Global Perspective

Edited by

R. QUENTIN GRAFTON
The Australian National University

IAN G. CRONSHAW
The Australian National University

MICHAL C. MOORE
University of Calgary, Canada

CAMBRIDGE
UNIVERSITY PRESS

University Printing House, Cambridge CB2 8BS, United Kingdom

One Liberty Plaza, 20th Floor, New York, NY 10006, USA

477 Williamstown Road, Port Melbourne, VIC 3207, Australia

4843/24, 2nd Floor, Ansari Road, Daryaganj, Delhi – 110002, India

79 Anson Road, #06–04/06, Singapore 079906

Cambridge University Press is part of the University of Cambridge.

It furthers the University's mission by disseminating knowledge in the pursuit of education, learning and research at the highest international levels of excellence.

www.cambridge.org
Information on this title: www.cambridge.org/9781107120082

First published 2017

Printed in the United States of America by Sheridan Books, Inc.

A catalogue record for this publication is available from the British Library

ISBN 978-1-107-12008-2 Hardback

Contents

Contributors

Professor R. Quentin Grafton
Crawford School of Public Policy, Building 132, Lennox Crossing, The Australian National University, Canberra ACT 2601, Australia

Ian G. Cronshaw
Crawford School of Public Policy, Building 132, Lennox Crossing, The Australian National University, Canberra ACT 2601, Australia

Dr. Michal C. Moore
School of Public Policy, University of Calgary, Downtown Campus, 906 8th Avenue S.W., 5th Floor Calgary, AB T2P 1H9, Canada

Professor Peta Ashworth
School of Chemical Engineering, Building 47A, The University of Queensland, St Luca, QLD, 4072

Dr. Silvia Barredo
University of Buenos Aires, Buenos Aires Institute of Technology, Avellaneda 2130, CP1636 Olivos – Buenos Aires, Argentina

Professor Michael Bradshaw
Warwick Business School, University of Warwick, Coventry, UK CV4 7AL

Orlando Cabrales
Energy Consultant, Calle 87 No. 10–93, Office 302, Bogotá, Colombia

Randall Cox
Office of Groundwater Impact Assessment, Department of Natural Resources and Mines, 61 Mary Street, Brisbane QLD 4000, PO Box 15216, City East QLD 4002, Australia

Stuart Day
CSIRO Energy Flagship, PO Box 330, Newcastle NSW 2300, Australia

Martin J. Evans
Anadarko Petroleum Corporation, 1201 Lake Robbins Drive, The Woodlands, Texas 77380, United States of America

Barry Goldstein
Department of State Development, GPO Box 320, Adelaide SA 5001;
Peta Ashworth Adjunct Associate Professor, University of Queensland, Unit 2602/45 Duncan St, WEST END QLD 4101, Australia

Belinda Hayter
Department of State Development, GPO Box 320, Adelaide SA 5001, Australia

Zhang Huanzhi
CNPC Economics & Technology Research Institute, 6 Liupukang St Rm 403, Xicheng District, Beijing, 100724, China

Paul Jeakins
BC Oil and Gas Commission, PO Box 9331, Stn Prov. Govt BC, Canada V8W 9N3

Dr. Lv Jianzhong
CNPC Economics & Technology Research Institute, 6 Liupukang St Rm 532, Xicheng District, Beijing, 100724, China

Dr. Vijay Kelkar
National Institute of Public Finance and Policy, 134/4-6, Ashok Nagar, Off Range Hill Road, Bhosale Nagar, Shivaji Nagar, Pune 411 007, India

Dr. Michael Carnegie LaBelle
Central European University, Nador Utca 9, 1051 Budapest, Hungary

Juan Roberto Lozano-Maya
Asia Pacific Energy Research Centre (APERC), Inui Building Kachidoki 11F, 1-13-1 Kachidoki Chuo-ku, Tokyo, 104-0054, Japan

Michael Malavazos
Department of State Development, GPO Box 320, Adelaide SA 5001, Australia

Ann Milligan
Environment and Natural Resources in Text, PO Box 7350, Kaleen ACT 2617, Australia

Dr. William Nikolakis
Faculty of Forestry, University of British Columbia, Forest Sciences Centre, 2424 Main Mall, Vancouver, BC, Canada V6T 1Z4

Dr. Francis O'Sullivan
MIT Energy Initiative, Massachusetts Institute of Technology, 77 Massachusetts Avenue, E19-307, Cambridge, MA 02139-4307, USA

Dr. Rahool Panandiker
The Boston Consulting Group, 14th Floor, Nariman Bhavan 227, Nariman Point, Mumbai 400 021, India

Ana Cristina Sánchez-Thorin
Environmental Consultant, Calle 92, No. 14–73, Office 1004, Bogotá, Colombia

Dr. Luis Stinco
University of Buenos Aires, Oleumpetra LLC, Avellaneda 2130, CP1636 Olivos – Buenos Aires, Argentina

Tim Stubbs
Yellow and Blue Pty Ltd: Environmental and Natural Resource Consulting, 10 Jamieson Avenue, Fairlight NSW 2094, Australia

Dr. Frank Umbach
European Centre for Energy and Resource Security (EUCERS), King's College, London, UK; (non-resident) Senior Fellow, US Atlantic Council, Washington DC, USA; Senior Associate and Head of the Programme "International Energy Security" at the Centre for European Security Strategies (CESS GmbH), Heideweg 4, 53578 Windhagen, Germany

Professor John Williams
Crawford School of Public Policy, Australian National University, JG Crawford Building 132, Lennox Crossing, Canberra ACT 2601, Australia

Foreword

The unconventional oil and gas industry has already transformed global energy markets and, with the arrival of unconventional gas and oil exports from the United States and Australia, this change seems likely to become more pervasive and powerful.

The International Energy Agency (IEA) first discussed this revolution in unconventional hydrocarbon production some years ago. Worldwide resources of unconventional hydrocarbons are vast, albeit still poorly understood in most countries. Whether they are developed at scale, particularly beyond North America and Australia, will depend on a number of factors, including the deployment of technology in a safe, sustainable, way that addresses concerns over social and environmental issues but also draws on the experiences of regulators in those regions where development has been most advanced. Recognising the importance of a strong regulatory presence, the IEA has convened a number of global forums to share that experience and best practice among national and provincial level regulators.

Risks, Rewards and Regulation of Unconventional Gas: A Global Perspective explains clearly that regulatory responses to the social and environmental concerns stemming from unconventional gas development have varied widely among countries and provinces. These responses reflect the vastly varying cultural, societal and economic contexts and range from outright prohibitions to cautious, evolving, policy approaches as the industry makes wide-ranging operational improvements. Public concerns have helped regulators to focus on issues such as water management, potential induced seismicity, control of fugitive methane emissions and greater transparency on the use and type of chemicals in hydraulic fracturing.

This book, which includes contributions by experts in their field, sets out the evolution of regulation in a number of jurisdictions with the most practical experience in unconventional hydrocarbon production in the United States, Canada and Australia. The editors do not shy away from the fact that even in those countries a wide range of approaches has been taken; these include moratoria or reversals of approvals. The book summarises the potential to develop new industries in many other resource-rich countries without underestimating the practical difficulties of translating experience from a few countries to a global scale.

In sum, the editors are to be commended for putting together a balanced and fact-based overview of the last decade of global unconventional gas development. Overall, this book provides a complete perspective of the risks and opportunities of unconventional gas and clearly demonstrates that the social licence for industry to exploit gas must be gained and retained, through regulation that is properly resourced and focused on key issues. The book will allow readers to understand the drivers and implications of unconventional gas at regional, national and global level.

If the unconventional gas industry is to deliver the benefits already seen in North America on a global scale then effective, transparent, regulation will be needed in all jurisdictions. There is much to be learnt from the successes and failures to date, and this volume is an important contribution that draws together the lessons learned and how to apply them in a way appropriate to each country's circumstances.

Dr Fatih Birol
Executive Director
International Energy Agency

Preface

R. QUENTIN GRAFTON, IAN CRONSHAW AND MICHAL MOORE

Energy requirements, especially those concerned with electric power, are by definition underpinned by technology and fuel choices. System operators will always seek out the most effective combination of service capacity and cost and, more recently, fuels with optimal policy support. Preferences have gradually shifted over time to natural gas technology, substituting for coal, replacing some nuclear capacity and even displacing certain hydroelectric facilities. This in turn has strained the sources of conventional natural gas and initiated a rush to develop unconventional natural gas supplies efficiently.

The last decade has seen the deployment of a number of innovative technologies and techniques to free up the gas long known to be contained in several North American hydrocarbon producing areas, firstly in Texas but then in Canada's western sedimentary basin and most recently in the Marcellus formation, located mostly in western Pennsylvania. Shale gas now accounts for more than half the US gas output, a large rise from only a few per cent in 2005. Oil output in the same period has almost doubled to levels last seen at the beginning of the 1970s, again using the same innovative techniques, rapidly deployed and improved.

Unconventional oil and gas production has transformed the global gas and oil trade. Patterns have been completely changed, so that liquefied natural gas (LNG) originally intended for the US has been sold into other markets, improving European gas security and making a major contribution to meeting Japan's power shortfall after the Fukushima disaster. In February 2016 the US began exporting LNG; a few years may see the United States rivalling Australia or Qatar as the largest global LNG exporter. And, contrary to pundit expectations, US oil imports have fallen substantially, which in turn has shifted the geographical balance of oil trade from the Atlantic Basin to Asia and the Pacific.

The rapid developments from unconventional oil and gas production have brought large benefits at national, regional and local levels, in terms of cheaper energy, jobs, and government revenues as well as enhanced energy security and reduced carbon emissions. The rapid expansion of the industry, and the huge drilling programmes associated with that expansion, have also had negative consequences in terms of environmental damage, contaminated groundwater and a negative social impact on some communities. The speed of unconventional oil and gas expansion has also strained the capacity and capability of regulators. Some of these effects in rural Pennsylvania were documented in the 2010 film *Gasland*, which generated global concerns about unconventional gas production and hydraulic fracturing.

The cumulative impacts of unconventional oil and gas are now recognised through regional or basin-wide approaches, such as those seen in parts of the United States, Canada and Australia. Consequently, over time, governments, regulators and industry have begun to respond to these issues, with better-resourced regulatory bodies, often purpose built, and more comprehensive frameworks to manage or mitigate potential water contamination, air pollution, seismic impacts and noise and drilling disruptions.

Industry has responded to community concerns, and falling prices, by remarkable improvements in productivity and improved environmental performance. Such improvements, including the greater use of pad drilling and water recycling, have driven down costs while reducing water and use of chemicals. While regulation has generally been at the state or provincial level, federal governments in all three countries have played important roles, led by the US Environmental Protection Agency (EPA). Cooperative approaches involving industry, governments and research bodies have also played an important role in improving productivity, reducing environmental impacts and managing social outcomes better. Outright prohibitions, for example on the use of certain chemicals, and proscriptive approaches to ensure well integrity, have been employed alongside more incentive-based approaches to improve standards.

Despite this progress, support for the industry, and especially its principal tool, hydraulic fracturing, is controversial. In a number of jurisdictions, some even adjacent to producing areas, the industry's hydraulic fracturing techniques are banned outright; these jurisdictions include certain states in the United States and Australia and Canadian provinces, although in these countries as a whole unconventional gas is well established. Outright bans can also be found in some countries, such as France. There is also very strong local opposition to unconventional gas production in places such as the United Kingdom, where rural population densities are high and incentives for local landowners to allow drilling are weak. Without doubt, the unconventional gas and oil industry is struggling to win widespread acceptance; its ongoing success will be critically dependent on continuous improvement in its environmental and perceived social impacts. In sum, the unconventional gas industry must gain and retain its social licence to operate.

While resources of shale gas, coal bed methane, tight gas and other types of unconventional gas are globally widespread, the industry has been slow to achieve the same success as that seen in North America. This is due to differing geological conditions, the lack of skilled and experienced human and capital resources of established markets or the result of unpredictable above-ground conditions such as taxation, foreign investment and intellectual property issues. Despite early predictions that the unconventional oil and gas industry would rapidly spread, progress in places as diverse as Argentina, China and South Africa has been slow.

A common factor shared by all regions is concern about environmental and social issues, often compounded by water management issues, to do with its use, extraction or disposal, coupled with fears of the possible contamination of drinking or agricultural water sources. These issues exist even where oil and gas have been extracted for some time but are more pronounced where there is little experience with hydrocarbon extraction or where the population density is high.

Many independent professional scientific reviews have concluded that the industry is safe, but only if it is properly managed and effectively regulated. There is concern that the industry is still relatively opaque; transparency can and must be improved, including on location and the use of enhanced well-stimulation techniques including hydraulic fracturing. Water use needs to be minimised, recycling encouraged and the disposal and beneficial use of waste streams improved. Methane emissions from gas and oil production could negate the environmental benefits of gas use, including in the power sector, industry and homes, but effective control technologies are available with the potential to largely overcome this problem.

Our book sets out to examine how unconventional gas production has grown over the last decade and how regulations have evolved to balance the demands of a new and valuable industry with pressing social and environmental issues. Developments in key producing areas are given more detailed treatment, as these regions have the most experience in dealing with these issues. The contrasts and similarities in different regulatory approaches, including outright bans, are analysed. Prospective countries and regions or where development is just beginning are also considered, and the reasons why production may be slower to develop in those places are highlighted.

We, the editors, believe that the industry can make a positive contribution at many levels, but the asymmetry between the benefits accruing at regional or national level and the many negative impacts which are borne at local level needs to be recognised and responded to in an effective manner.

The future growth of the gas industry remains uncertain. Energy demand growth is increasingly to be found in developing Asia, where gas supplies, either as local, potentially unconventional, gas production or as pipeline or LNG imports, will be expensive even at low oil prices. Hence, conventional and unconventional natural gas may struggle to compete against very low priced coal and increasingly competitive renewables, even when the versatility, flexibility and low environmental footprint of gas are taken into account; this will be especially true in the key power sector, where competition is strongest. In developed economies in Europe, gas demand is faltering, as economic growth remains weak, industrial structural change continues and new energy technologies, especially renewable power, make major inroads.

A growing consensus that urgent climate change issues must be effectively and speedily addressed will affect fossil fuel use, although it seems likely that the growing decarbonisation of the power sector – a key route to lower carbon emissions – will have a bigger impact on coal than on oil and gas supply and demand. Into this uncertain market outlook, the growth of unconventional gas and oil output outside North America is influenced by major geological differences and widely divergent institutional, industry-capability and other factors such as higher population densities and water availability.

On one point we are convinced: effective, efficient and ongoing regulation associated with unconventional gas and oil and a 'social licence' will be a prerequisite for the global success of the industry and for unlocking its benefits to landholders and communities, in addition to gas companies and governments. This book highlights these developments and the growing breadth, depth and effectiveness of regulators, so that all jurisdictions can learn, build on and profit from the lessons of others.

RQG, IGC and MCM

Acknowledgements

This book has been made possible by the support of many people, not just the contributors and editors. Special thanks go to both Vivienne Seedsman and Nur H. A. Bahar for coordinating contact and communication with contributing authors across the globe, managing the email traffic of files, figures and details and collating everything ready for the publishers. Regarding financial support, without the efforts of the Australian National University–UNESCO Chair in Water Economics and Transboundary Water Governance many costs of putting the book together could not have been met.

1

The Rise of Unconventional Gas: The Story So Far

IAN CRONSHAW, R. QUENTIN GRAFTON AND MICHAL C. MOORE

Introduction

Natural gas has steadily increased its role in OECD countries over the last 40 years, rising from 19% of total primary energy supply (TPES) in 1973 to more than a quarter by 2013. More than half the gas is used in the residential and commercial sectors, where it is the preferred fuel for heating, cooking and hot water. Industrial activity accounts for a further quarter of consumption, notably in the chemical industry but also in non-ferrous-metal production and food processing, where its clean combustion and flexibility are valued.

While many gas markets in the OECD are well developed, notable recent growth has come from the power sector, where demand has almost trebled since 1990. Since that time, gas-fired power has become the fuel of choice for new electricity generation in many, if not most, major OECD countries, accounts for almost two thirds of new power output and in the most-developed regions is the marginal dispatch technology. Since 2000, gas-fired power has grown by 80% or around 1250 TW h, which is almost equivalent to the total power output of Japan plus Australia.

The rapid increase in gas-fired power has occurred because gas enjoys a number of advantages over alternative energy sources; these include its greater flexibility and efficiency. Furthermore, the use of gas is expected to increase dramatically, both in its role as a source of generation flexibility to complement increasing shares of intermittent renewable generation and also in its generally larger role in the energy mix of economically growing non-OECD countries such as China (where coal has been the dominant source of new power). This is likely to remain true even if aggressive greenhouse gas reduction policies are implemented, which would tend to increase the share of renewables while reducing the share of fossil fuels in the energy mix (IEA 2013).

In the United States gas has held an important position in the energy mix for many decades, supplying 28% of TPES in 2012. While markets in the residential, commercial and industrial sectors are relatively mature, gas-fired power doubled between 2000 and 2012. In Australia, gas use grew from 6% of TPES in 1973 to 26% in 2012, with gas consumption increasing sixfold in the residential sector and tenfold in the electricity generation sector.

Gas Production Looked Set to Fall in OECD Countries

Until recently the majority of natural gas was produced by 'conventional' methods, frequently associated with oil production. As recently as 2007, gas production in the OECD was predicted to decline as peak output was reached in the North Sea, the United States and elsewhere. Thus the outlook at that time was for increased imports, higher gas costs and declining energy security. As a result, in the period 1996–2010 large investments were made in importing infrastructure into Europe and the East Coast and Gulf of Mexico in the United States, with corresponding outlays in gas-liquefaction plant in Qatar and elsewhere. Liquefied natural gas (LNG) was therefore set to play a greater role, and not just in those countries where it had historically provided most, if not all, the gas supply such as Japan, Korea and Taiwan and increasingly Europe and the United States and other gas-hungry countries including China and India. Indeed, the original marketing plans for Qatar, currently the world's biggest LNG exporter, envisaged sales of roughly one third to each of the United States, Europe and Asia. Current planning in Canada relies on unconventional gas reserves for LNG exports and targets the Asia Pacific market beyond 2020 as the driver for new west-coast-port development.

Unconventional Gas in the United States has Changed the Game

While expectations for declining gas output have largely been realised in Europe, especially in the United Kingdom, the situation in the United States has taken a completely different course. Beginning in around 2005, but rapidly accelerating after 2008 and building on years of research and pioneering activity by a few medium-sized companies, the United States was able to tap hitherto uneconomic sources of gas, so called 'unconventional gas' (UCG). Starting first in Texas and then in adjacent traditional hydrocarbon provinces, the new gas extraction technologies spread rapidly to other geological basins in the United States to encompass new oil production. As shown in Figure 1.1, the Marcellus Basin, which includes West Virginia and Ohio and is centred on Pennsylvania, has seen production from shale gas rise from almost nothing in 2008. By early 2015 production levels had been reached where, had it been a country, the Marcellus Basin would have rivalled Qatar as among the largest gas producers globally. This region now accounts for nearly a fifth of the United States gas production.

A key impact of these developments in the United States is that the market price for gas has fallen below $4/Mbtu (see Glossary and Conversion Factors) for an extended period, only rising above this level as the result of a very cold winter. This price is equivalent to an oil price of around $25 per barrel, well below the global oil prices in excess of $100 per barrel prevailing over the period 2011–mid 2014. Such low energy prices have made the United States an extremely competitive location for energy-intensive industry globally, a situation likely to persist for many years. The beneficiaries are those who use gas directly and for electricity generation and include both households and industries. In early 2015, even as oil prices for US crude oil fell to $40–$50, gas prices at around $2–$3/Mbtu remained very low relative to oil.

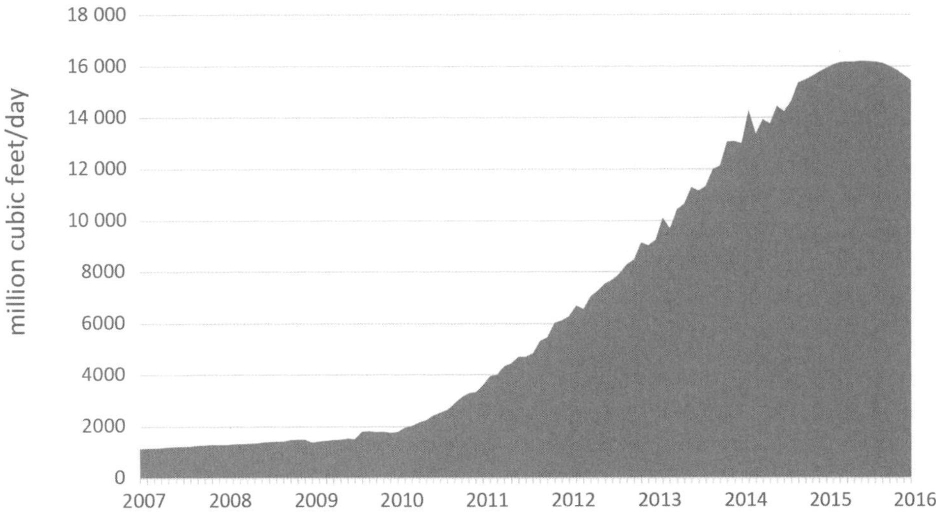

Figure 1.1 Marcellus region, natural gas production. *Source:* US Energy Information Administration (December 2015).

Unconventional Gas is Already Transforming the Global Energy Landscape

While other countries are known to have significant large resources of unconventional gas, bringing them to production is expected to take some time and will probably not happen before 2020, with exceptions in Canada, which is exploiting similar technology in its western provinces, Australia, which is rapidly growing its coal bed methane supply and possibly Argentina and China. Unconventional gas development in the United States has already had a major impact on global energy supplies and security. One notable example is the response of global gas markets to the 2011 Fukushima disaster in Japan. In its aftermath, and facing the loss of nearly 280 TW h of generation from its nuclear fleet, Japan purchased substantial extra quantities of LNG from Qatar, and elsewhere, albeit at a high price for such spot sales. This allowed it to make up some two thirds of its power shortfall. The LNG exported to Japan was effectively no longer needed in the oversupplied North American gas market. In effect, the United States had already become a virtual gas exporter.

The importance of the United States as a gas exporter will become even clearer when the first of a number of US LNG export facilities starts operation in 2016; it is based on the retrofitting of an existing LNG import terminal with a liquefaction plant. With four other such import terminals also receiving export approval, LNG exports from the United States could rival those from Qatar and Australia soon after 2020. Gas producers in Canada, previously exporting gas to the US, are also seeking new markets, generally in the Pacific region although the barriers to successful export projects there are higher largely due to limits on interprovincial pipeline approval and opposition to the construction of new port facilities.

Environmental Impacts are More Pronounced and must be Regulated

The rapid growth globally in gas reserves and production from unconventional gas is transforming the global energy landscape. As more countries join the ranks of unconventional gas producers, this transformation will become more evident. Nevertheless, the technical challenges to global unconventional gas expansion are formidable. In particular, the social, economic and environmental risks of gas extraction need to be managed if gas producers are to retain a social licence for their activities. These environmental challenges are exacerbated with unconventional gas, as a result of the higher drilling intensity required and the multiple use of hydraulic fracturing employed in some wells. These issues and the potential local or regional risks to water resources have led to growing calls for more active and specific regulation, with widely varying approaches being seen in jurisdictions worldwide. The nature of the current regulatory system, and what it should be, will be a key determinant in the longer-term future of unconventional gas development.

In this book we compare the regulatory approaches taken in a number of jurisdictions, including: Argentina, Australia, Canada, China, Colombia, India, Poland, the United Kingdom and the United States. Collectively, the volume brings together insights from these countries to provide directions for good or effective regulation in terms of unconventional gas production.

What is Unconventional Gas?

Unconventional gas is identical to natural gas, consisting essentially of methane with small concentrations of impurities; only the production methods differ from those for more conventional gas. The production methods differ because of the need to extract gas from geological formations in which the permeability is low and which may include tight gas, coal bed methane (CBM, also known as coal seam gas) and shale gas.

In the case of shale gas, economic gas extraction has been made possible by advances in the key technologies of horizontal drilling and hydraulic fracturing. The latter technique involves the injection of a fluid under pressure, typically more than 95% water with the addition of a proppant (commonly sand) to hold fractures open plus a very small proportion of certain chemicals. In the case of coal bed methane, usually the water must first be extracted and, given its often saline condition (typically 200–10 000 mg/l), needs to be treated before disposal. In coal bed methane extraction, horizontal drilling and fracturing are less widely employed, hydraulic fracturing being used in less than 5% of CBM wells, up to 2010 with that proportion increasing only to 6% (111 out of 1844 CBM wells) in 2012 and 2013. Nevertheless, hydraulic fracturing has been used in the conventional oil and gas industry for some decades in Australia, notably in the Cooper Basin straddling South Australia and Queensland. The use of hydraulic fracturing seems likely to expand, with the rapid ramp up as output in eastern Australia to meet LNG-induced demand; one LNG project is expected to use hydraulic fracturing in around 30% of its gas wells over the life of the project (i.e. 3000 to 4000 out of 10 000 wells) (APLNG 2015), with the use in other projects potentially higher.

The environmental issues, and associated regulation of the possible external costs, can differ between different methods of unconventional gas extraction, although some features are common. The focus in this book is on coal bed methane and shale gas extracted by unconventional methods.

Unconventional Gas Brings Higher Environmental Impacts

A common feature in many unconventional gas operations is higher drilling intensity relative to conventional gas developments. In unconventional fields there are often hundreds, or even thousands, of production wells being drilled in a given gas play or production area, thus increasing the actual and potential impact of drilling and associated operations on the local environment and residents. By contrast, in conventional gas fields there may be only tens or hundreds of wells. Drilling multiple wells from a single site or drilling pad, using horizontal or other drilling techniques, as is being practised more widely in US and Canada, reduces the surface impact of gas development as well as markedly reducing the costs of production (see Chapter 6 in this volume).

An aspect in the development of unconventional gas is that production wells need more complex, and sometimes ongoing, techniques to stimulate adequate gas production rates; these techniques include hydraulic fracturing for shale gas and extensive water removal or dewatering for CBM. The water extracted for CBM production may have various degrees of contamination with salt or other pollutants, which necessitates proper treatment and disposal techniques. The beneficial use, or release, of this treated water has a potential impact on existing water resources, both those on the surface and those underground. Further, the water used for fracturing can lead to a possible depletion of water supplies although newer approaches emphasise the recycling of so-called produced or formation water, which lowers the call on fresh water sources. Shale wells tend to be at a deeper level (typically 2000 metres or more) than the rather shallow CBM extraction (800–1200 metres), with potentially differing implications for water supplies.

Public concerns, and regulatory issues, while varying between regions and between gas-producing technologies, can be loosely grouped as follows:

(i) the question of land access, obviously most acute where settlement or existing land use is most intense;
(ii) water issues around the potential contamination of aquifers;
(iii) the water-treatment or disposal of the formation water and/or the fracturing or drilling liquids, which is especially important in areas of water scarcity;
(iv) conflict with other land uses or users, including loss of property value;
(v) air emissions, including fugitive methane (the oil and gas sector is a large source of methane emissions, a potent greenhouse gas and contributor to climate change);
(vi) possible seismic events triggered by high pressure hydraulic fracturing; and
(vii) surface issues such as habitat fragmentation and loss of aesthetic benefits.

The Industry is Expanding Rapidly

Many environmental issues and community concerns in the unconventional hydrocarbon industry are common to conventional hydrocarbon (oil and gas) production and have, to some extent, been addressed in existing oil and gas regulation, especially where the relevant regions have a history of such production. These include mandatory measures such as blow-out preventers and regulations to ensure well integrity through multiple cementing procedures. Where such a production history is lacking, existing regulatory coverage has, at least initially, been weak. In any case, the novel nature of production techniques and the speed with which they have been deployed in some locations have placed strains on most existing institutions, both in terms of the existing regulatory frameworks and also in terms of the resources of regulators. This is not unexpected because the acceleration of production and the rate of discovery of new locations of unconventional gas has surprised many global gas companies.

Many aspects of these gas extraction technologies have been around for decades. For example, hydraulic fracturing was first developed in the 1940s and 1950s, but the pace of its deployment has increased rapidly, especially in North America. Since first used, the hydraulic fracturing process has become almost standard and has been used on more than one million wells in the United States; it may be noted that the depth of drilling, and the need for and type of hydraulic fracturing and other components of gas extraction, vary site by site. Currently, an estimated 35 000 gas and oil wells in the United States use hydraulic fracturing annually (FracFocus 2014, EPA 2012), with an estimated 80% of shale gas wells also using it. Driven by these new techniques, the United States' shale gas output has jumped from 6% of US gas production in 2005 to more than half the US gas output in 2014, approaching 400 bcm. To date, this supply has strongly resisted the fall in gas and oil prices seen in 2014 and 2015.

Productivity has Continued to Improve Rapidly

In shale gas there has been a sharp decline in the cost of horizontal drilling and hydraulic fracturing technologies, coupled with an increase in the quality and decline in cost of advanced seismic techniques. In all these technologies, as companies have moved rapidly along the learning curve, costs have been driven down. The ability to recover natural gas liquids as co-products in the gas stream has also been an important part of the economics of gas production in North America. Low ethane prices are driving a new wave of petrochemical investment in the Gulf of Mexico region, while the United States is now the largest liquid petroleum gas (LPG) exporter. Other forms of UCG, notably CBM, rarely benefit from associated liquids, which reduces their overall profitability.

Ways Forward

The remainder of this book is devoted to examining developments in a number of countries possessing large unconventional gas resources. In most countries, the exploitation and regulation of these resources remain at an early stage, in some cases barely beyond resource identification; in others, such as China, commercial production has begun, but at a lower level of activity than anticipated.

Our attention is on those countries, provinces or regions where production is most advanced and regulatory policy, mechanisms and institutions are most evolved. Included in this volume are descriptions of the widely differing regulatory approaches adopted, often in regions adjacent to each other. We believe that the collective review and analysis in the chapters in this book provide valuable insights about the benefits, risks and opportunities of this important energy transition and about how best to manage a rapidly growing industry for the public good.

References

Australia Pacific LNG (APLNG) (2015). Fraccing. www.aplng.com.au/environment/fraccing (accessed 22 December 2015).

FracFocus: http://fracfocus.org/hydraulic-fracturing-how-it-works/history-hydraulic-fracturing (accessed 4 February 2014).

FracFocus (2014). Hydraulic fracturing: how it works. Available at: http://fracfocus.org/hydraulic-fracturing-process (accessed 10 December 2015).

International Energy Agency (IEA) (2013). *World Energy Outlook 2013*. OECD Publishing, Paris.

O'Sullivan, F. (2016). Economics of shale gas in the United States. Chapter 6 of this volume.

US Environmental Protection Agency (EPA) (2012). Study of : Progress Report. December 2012.

2

Geopolitical Dimensions of Global Unconventional Gas Perspectives

FRANK UMBACH

Introduction

The US shale gas revolution has caused unprecedented changes in the country's energy landscape and in global gas markets. In 2009, the US overtook Russia as the world's largest gas producer. It was forecast to become in 2016 a net exporter of LNG and in 2018 an overall net exporter of natural gas (IEA 2014a). The US may even overtak Russia as the world's combined largest oil and gas producer (Gold & Gilbert 2013). It has increased global oil and gas supplies as well as diversification options for energy importers and decreased oil and gas prices, as highlighted by the oil price fall in 2014 and oversupply in the markets.

In 2013, US domestic gas production met almost 94 per cent of its gas demand. Net imports declined from a peak of 107 bcm in 2007 to just 37.1 bcm or 5 per cent of total supply – the lowest level since 1989 (IEA 2014a). Consequently, US foreign and economic experts were discussing the benefits and risks of increased US LNG exports to Europe and Asia and whether to use those exports as part of a pro-active energy diplomacy (Umbach 2014a). Those discussions have increased in the light of the Russian annexation of Crimea and the energy dimensions of that event (Gonchar *et al.* 2014; Umbach 2014g) and in the light of Russia's actions in the Ukraine's eastern regions (Lenard & Sautin 2014).

The projected self-sufficiency of US natural gas and oil (at least in the North American framework) raises the question of the geo-economic and geopolitical impacts on global as well as regional energy security.

Traditionally, energy security was defined as 'the availability of energy at all times in various forms, in sufficient quantities and at affordable prices', in the 1980s and 1990s. But with the rising importance of and need for environmental and climate protection, the IEA defined energy security after 2001 as 'uninterrupted physical availability [of energy] at a price which is affordable, while respecting environment concerns'.[1] But 'sufficient quantities' and 'affordable prices' have remained rather vague terms and thus 'energy security' has still not been defined precisely. For measuring 'energy security', more and more indicators have been created and framed in new complex energy security concepts (Löschel *et al.* 2010a, 2010b).

[1] Thus the definition of 'energy security' by the International Energy Agency (IEA): http://www.iea.org/subjectqueries/keyresult.asp?KEYWORD_ID=4103.

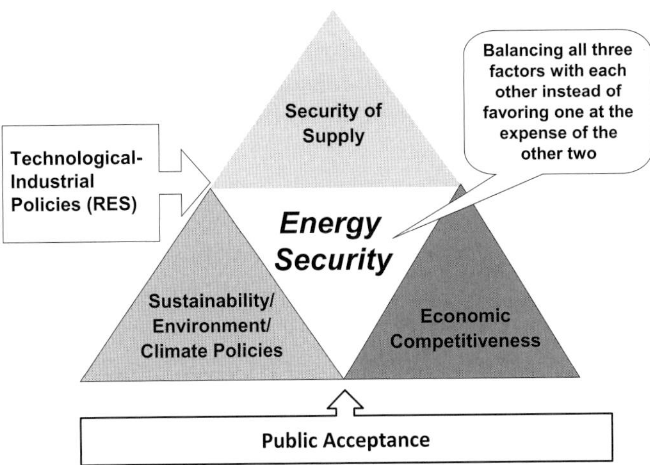

Figure 2.1 Energy triangle – objectives of energy security. *Source:* Dr Frank Umbach.

In the light of the economic–financial crisis in 2008 and the need for timely and sufficient investments in new energy sources and infrastructures to cope with the dual challenge of global energy-supply security as well as climate change, the IEA, for instance, is now differentiating between long- and short-term energy security:[2]

> Moreover, for a long time 'energy security' has had a different meaning due to the perspectives of the producer, consumer and transit states. Whereas consumer nations (like EU members) are primarily interested at security of supply, producer countries (like Russia) are more focused on security of demand from foreign markets. Transit states (like Ukraine and Turkey in the future), for their part, are often equally interested in their own national security of supply *and* security of demand from neighboring markets in order to benefit from stable and high transit fees. Furthermore the concept of 'national energy security' also depends on the individual country's geographical location and domestic policies and on the traditional state, economic and business ties it maintains with its partners.
>
> *(Umbach 2011, pp. 25–26)*

Since the end of the 1990s, international energy experts have also stressed the increasing strategic importance of energy-supply security within the 'energy triangle', whose three major objectives are economic competitiveness, environmental and climate sustainability and energy-supply security (Figure 2.1). In the view of many energy security experts, the biggest challenge is seen as maintaining a balance between the three objectives rather than favouring one at the expense of the other two. That, however, is often the case in Europe with the factor of environmental and climate sustainability, which appears often to dominate and determine all discussions of the energy–climate nexus in the EU at the expense of

[2] See the definition of 'energy security' by the IEA, at http://www.iea.org/topics/energysecurity/; downloaded on 3 June 2012.

the other two goals of the energy triangle. While the world has directed its attention to the manifold challenges of climate change and its security challenges, in Europe often the same attention is not paid to global-supply challenges and those of preserving economic competitiveness.

Maintaining the balance between all three objectives of the energy triangle has also become more difficult owing to industrial policies subsidizing renewable energy sources, as in Europe, or unconventional oil and gas exploration, as in the US, and also owing to the need to gain public acceptance in the light of NIMBY-attitudes, ideological positioning and new vested interests. This often creates 'energy trilemmas', which need to be addressed by an adequate institutional setting, above national government ministries, that can also take into account the various ministerial and vested interests, in order to obtain balanced national energy strategies and concepts (Umbach 2012).

Both US and international foreign and security policy experts are debating whether the United States will maintain its role as the 'global policeman' ('Globocop') and its stabilizing role in unstable political key regions such as the Middle East (i.e. the Persian Gulf) and the Asia–Pacific region. The Obama government, coping with a severe economic–financial crisis for years, has already redefined US foreign, security and defence policies in the light of its budget constraints and has focused its security policy more than ever on east Asia and China as its rising geopolitical rivals. These geopolitical questions center on four questions. (1) Will the United States withdraw its political and security commitments to allies in key, often unstable, political regions such as the Middle East, when in the future it will no longer be so energy dependent on this region as in the past? (2) What will be the political and security implications for global and regional stability in Europe? (3) Will the United States export its gas production surplus? (4) What will be the geo-economic and geopolitical impacts on global and regional energy security?

However, these questions are no longer related just to the US shale gas revolution but also to the geo-economic and geopolitical implications for energy security of the forthcoming worldwide shale gas development (Rühl 2014). In June 2013 the US Energy Information Administration (EIA) added nine more countries, to take the total number with technically recoverable shale gas resources to 41, in its second worldwide assessment of unconventional gas resources. This corresponds to a rise in estimated shale gas resources of 10 per cent in comparison with its first assessment, in 2011 (EIA 2013).

Indeed, the IEA expects that unconventional gas will account for around 60 per cent of the global gas demand growth by 2040 (IEA 2014b: pp. 135–170), if the industry can receive a 'social licence to operate' within stringent regulatory regimes designed to satisfy public environmental and social concerns (IEA 2012). Shale gas and other unconventional gas reserves have been identified in Argentina, Mexico, Australia, China, South Africa, northern Africa, the EU-28 (i.e. Poland, France, Germany, the United Kingdom and others), Ukraine, Turkey and other countries (EIA 2013).

However, the expansion of unconventional gas is facing grassroots opposition from environmental groups, which have concerns on ground water safety, adequate waste water management and greenhouse gas emissions (GHGE), particularly in Europe. The vested interests of other energy industries such as the Russian gas industry (i.e. Gazprom), the nuclear industry in France and the renewable energy industry in Germany also oppose shale gas projects in Europe (Umbach 2013a–d, 2014a–g) because of competitive resistance.

Shale gas developments outside the US like those in China may also decrease the forecasted global LNG growth as China is expected to become one of the largest LNG importers (Umbach 2013a–d). Also, shale gas developments in Latin America, the Middle East and Europe itself may significantly lower the anticipated future LNG import demand of these regions but will allow considerably greater exports to the rest of the world. In the short-term future, new LNG capacity is expected to come online after 2016 or 2017, which could double the world's liquefaction capacity from 288 million tons (mt) in 2013 to 575 mt in 2018 (IEA 2014b).

Beyond the geo-economic implications, the shale gas revolution could impact geopolitically in six general ways, as follows.

- The numbers of gas producers and exporters are rising and this creates more import options.
- Import dependences, risks and vulnerabilities for consuming countries will be reduced.
- Energy imports will be diversified so that they come from many more countries and regions.
- The above will lead to new bilateral (energy) relations with shale gas producing and exporting countries, with lower bilateral dependences of gas importers.
- Falling exports and fossil fuel prices are creating pressures in oil and gas producing countries (but to different degrees, as they are not equally vulnerable), since these countries are heavily dependent on stable demand-side conditions for revenues and state budget stability.
- The competitiveness of the overall economies of countries will be affected and in particular their energy-intensive industries.

Until a few years ago, the future dependence on a few countries, notably Russia, Qatar, Iran, Turkmenistan and Australia, which controlled the bulk of the future conventional gas reserves, appeared to be increasing worldwide. In contrast with the remaining conventional gas resources, unconventional gas (shale gas, tight gas and coal-bed methane) is not only abundant and more available all over the world; it challenges the market power of producer countries as well as potential export cartels, such as the Gas Exporting Countries Forum (GECF), and somewhat strengthens the position of consuming countries. Unconventional gas may even decrease economic and political stability in neighbouring producer and traditional exporter countries and regions, such as Russia and the Middle East, which are heavily dependent on high revenue income from their oil and gas exports (Hague Centre for Strategic Studies/TNO 2014).

Analysing and forecasting global and regional energy security has also become more complicated by the fall in oil and gas prices. Thus the EU's common energy policies since

2007 and the German *Energiewende* in 2011 were based on the following assumptions and core beliefs at that time:

- Fossil fuel prices would rise continuously as global demand exceeded supply.
- Europe would gain industrial and economic advantages by being the first major region in the world to develop a low-carbon economy based on renewable energy sources (RES) and other 'green technologies'.
- A gradually rising carbon price would increase the cost of externalities including air pollution and climate change, until RES becomes fully competitive.
- The negative effects of higher energy costs on competitiveness would be mitigated by a binding global agreement on climate change, with all the world's major economies making progress towards a common goal of reducing emissions.

Today, none of those core beliefs has been proven to be true:

- Fossil fuel prices have fallen dramatically, by more than 40 per cent; new drilling technologies (i.e. hydrofracking) have made unconventional gas and oil resources available for the world markets and recovery factors have increased as production costs have reduced.
- Geopolitics has brought back energy-supply security more than ever to Europe's security agenda (defined as a short-term security challenge), with the consequence that climate change is rather seen as a long-term national security challenge.
- The move to RES has given Europe no real economic advantage but proves to be extremely costly as the entire energy system needs to be changed, while the main research and development advances have been made in China (i.e. in the solar panel industry) and some other countries.
- Efforts to establish higher carbon prices have completely failed, with the result that coal, the dirtiest fossil fuel, originally supposed to be priced out, is stronger on the German and European energy markets and has phased out gas, at more climate-friendly fossil fuel.
- None of the remaining largest GHG emitters in the world (the US, China, India, Russia, Japan, Brazil etc.) has followed the EU's ambitious climate-mitigation policy. The German and EU self-images of leading by example have not produced any real major followers on the world stage. Despite a joint declaration between the US and China, and the Paris global climate summit in December 2015, there are still fundamental uncertainties and differences of opinion. Furthermore, despite having become the world leader of investments in renewable energy sources, China will only be able to decrease, but not to phase out, its coal consumption in the mid-term perspective of 2040 (Umbach 2015).

Furthermore, in the mid- and longer-term future after 2035–2040, looking at a general picture of the role of natural gas in the future global economy and geopolitics, methane extracted mainly from unconventional sources such as tight gas, shale gas, coal bed methane and, later, gas hydrates could also play a leading role.

This chapter will analyse the potential geo-economic as well as geopolitical impacts of the US shale gas revolution (it does not cover the worldwide impacts of the US shale oil revolution, however) on regional and global energy-supply security. It will focus and examine unconventional gas developments and their impact on the energy security of the US, EU, China and Russia.

The US Shale Gas Revolution and Its Geopolitical Impacts

In the view of optimistic US energy experts, the United States has already moved from being an energy-depleted country to being a newly emerging energy-rich superpower, the 'New Middle East' (Morse *et al.* 2012). American shale gas production is expected to increase from 34 per cent of total US natural gas production in 2011 to 49 per cent in 2035 and to more than 50 per cent in 2040. The share of coal use for electricity generation has already fallen by more than one third, from almost 50 per cent in 2007 to just 39 per cent in 2013 and, in the US primary energy demand mix, from 22.5 per cent in 2007 to 18.1 per cent in 2012. Over the same time, the total US energy consumption has reduced by 6.4 per cent relative to 2007 (IEA 2014a).

Between 2007 and 2012, US carbon dioxide emissions decreased by 13 per cent to the lowest levels since 1994, owing to the coal-to-gas switch, new energy-saving technologies and a doubling of renewable energy production. The greenhouse gas emission (GHGE) reduction is even more impressive if one takes into account that the real GDP in 2012 was 55 per cent higher and that the US population was 17.5 per cent larger than in 1994 (IEA 2014a).

These dramatic changes in the US energy mix and policies since 2006 are forecast to continue into the mid-term future up to about 2040. New studies suggest that the US has enough gas to last more than 100 years at present consumption rates (IEA 2014a). This optimism is also based on the fact that shale wells have higher initial returns than conventional wells. In the US initial production rates are about 85 thousand cm per day (Mcm/d) for unconventional gas as compared with just 28 Mcm/d for conventional gas. Despite the higher costs of drilling unconventional wells, the average cost of the retrieved gas was 40–50 per cent lower in 2011 than gas from conventional wells. Drilling has become increasingly faster and more cost-effective as drilling efficiency has been raised and investment costs have been repaid much faster than for conventional wells (Ryan 2014).

The development of US unconventional gas reserves has brought foreign direct investment (FDI), created more than two million new jobs and helped to strengthen the diversification of its national energy mix and to reduce unstable import dependence and GHG emissions (Gaffney, Cline & Associates 2014; Umbach 2014c). Between 2010 and the end of March 2013, the US was able to attract almost 100 chemical industry projects, valued at around US$72 bn (Anderson 2014).

For a variety of reasons concerning the preservation of its geo-economic and geopolitical interests, the US will not entirely disengage from the Middle East and other critical key regions for global economic and political stability. But even a gradual disengagement and re-balancing of US foreign policy and security commitments could have direct and indirect impacts on those regions and beyond. The ISIS security challenge in Syria and Iraq and the instabilities in North Africa demonstrate that even a gradual disengagement can rapidly create new security vacuums, to be filled by other great and regional powers or terrorist groups (supported by third parties), or it can lead to even larger instabilities or civil wars. But even then, other powers, such as the EU or Japan, may have neither the common political will nor the economic and military capabilities to replace the US security role and may not accept (fully) US requests for more 'burden-sharing' of global security commitments (Umbach 2014c).

New US LNG-supply infrastructure will allow easy access to the world's largest gas markets, those of Europe and East Asia. At present, the Cherniere Energy's Sabine Pass LNG export terminal in Louisiana is the only terminal that has had an export licence since 2010 for UK, Spain and other countries, starting in 2015, that has received both US Department of Energy and US Federal Energy Regulatory Commission approval to export to FTA and non-FTA countries. The total capacity of approved terminals that will be commissioned between 2016 and 2020 will be around 118 bcm per year. The DOE approved six export projects to non-FTA countries and a seventh with certain conditions (Jordan Cove LNG terminal in Oregon, which will supply Asia). But none of these 23 projects has finalized its financing.

In 2017 the US might be capable of producing LNG equal to one sixth of the EU consumption, but half has already been reserved by Japan, India and South Korea; the other half has been reserved by UK and Spanish companies. The presently finalized negotiations of the Transatlantic Trade and Investment Partnership (TTIP) agreement could facilitate and fasten more US LNG exports to Europe (Botzki 2014). The Senate's Energy Committee is working on a series of LNG-related bills. The 'Expedited Liquid Natural Gas for American Allies Act' of 2013 will allow easier authorization to export LNG to non-FTA partners of the US, particularly NATO members, Japan and any other foreign countries, thus promoting wider US security interests.

President Obama has made it clear, however, that even expanded LNG exports will go into the open market rather than be targeted directly to Europe (Walstad 2014). Nevertheless US geopolitical and wider security interests could change and not just hasten the approval process but also decrease LNG export prices to Europe by reducing specific tax and other costs. The present Ukraine conflict and the control of the US Senate by the Republican Party are already changing US export regulations and will push the Obama administration increasingly in the direction of adopting a proactive 'natural gas diplomacy' by increasing LNG exports and supporting new LNG terminals (as in Croatia), as well as supplying new pipelines with non-Russian gas to counteract Russia's influence in Europe and promote transatlantic energy security (Gardner 2014; Crooks 2014). Pushing the US instead towards 'energy isolationism' would not insulate it from instability in the global energy market. As Elizabeth Rosenberg argued at the beginning of 2014 in a report of the 'Unconventional Energy and US National Security Task Force', co-chaired by Ambassador Paula J. Dobriansky, Governor Bill Richardson and Senator John Warner:

Hoarding energy at home, neglecting bilateral relationships with major global energy players and forfeiting economic opportunities to export energy would leave the United States less secure. Moreover, policymakers would then be unable to use energy as a tool of economic statecraft to coerce or benefit other countries. Instead, the United States should accept the reality of energy interdependence, take steps to decrease domestic consumption and diversify supplies, facilitate broader energy exports, and more deeply and creatively integrate energy security into strategic policy and military planning'.

(Rosenberg 2014, p. 6)

Europe

In September 2014 ahead of winter and increased gas consumption in the cold period, former EU energy commissioner Guenther Oettinger warned: 'That Putin would use false information, lies and weapons was beyond my imagination. That's why I do not rule out even the worst scenarios' (Reuters 2014).

The EU–Russian gas and energy partnership has often been described as 'mutual dependence': whereas Europe benefits from Russia's stable gas supplies for its energy security, Russia benefits from European investments and technology transfers. However, it is often overlooked that their 'mutual dependence' has always had an *asymmetric nature*, in which Russia has been often the stronger actor and has steadily tested the European willingness to cooperate with Russia by taking its strategic interests (at the expense of others) into account as well as the EU's ability to oppose Russia's declared geopolitical interests. To put it in the words of a Russian defender of the Kremlin: 'To put it bluntly, Russian budget can survive without gas income, but can the fragile European economy survive a year without the supply of 25 per cent of its gas consumption?' (Mandrasecu 2013).

Against the background of three Russian–Ukrainian gas-supply crises in 2006, 2009 and 2014, the combination of the following three factors has fundamentally changed the European gas sector during the last years: (1) a drop in demand linked to the global economic recession; (2) an unexpected dramatic increase in US non-conventional shale gas production; and (3) the arrival of new LNG delivery capacity. Together, these factors created a sudden 'gas glut' of LNG in 2011–2012. The gas overcapacity made LNG in Europe less expensive and more competitive than pipeline gas (on the basis of long-term contracts) than in the past and contributed to the de-linkage of gas prices from oil prices in Europe (Kuhn & Umbach 2012).

In order to strengthen its future energy security, the European Commission's energy-demand management strategy has emphasized the broadest possible energy mix, the diversification of energy supply and imports, the promotion of renewable energies and a neutral policy towards the nuclear option. Its 20–20–20 per cent formula in its 'Energy Action Plan' (EAP) of March 2007 was aimed to reduce GHGE and to raise the share of RES as well as to improve energy efficiency and conservation (European Commission 2010). If the European Commission is able to implement and achieve its March 2007 aims by 2020 then the EU would be using 13 per cent less energy than today, which is equivalent to a saving of more than 100 billion euros and a reduction in CO_2 emissions of about 780 million tonnes per year (European Council 2007, p. 13). These policies are particularly important because

Europe is the only region in the world where no gas-production increase will take place in the foreseeable future, at a time when natural gas will become ever more important, as the cleanest fossil fuel.

Even without US LNG exports to Europe, the EU's gas-supply security, often seen as the EU's Achilles' heel, has been significantly improved since the 2009 Russian–Ukrainian gas crisis by the building of new gas interconnectors between the member states, the implementing of the Southern Gas Corridor (with the TANAP and TAP gas pipelines importing Azerbaijan gas), the expansion of LNG terminal capacities, the adoption of new gas regulations and the creation of new institutions to control and overview the EU's reform policies (the Gas Coordination Group, ENTSOG-Gas) on the way to creating a united, internal and liberalized gas market. Those reform efforts were strengthened by the Ukraine conflict, which resulted in a new EU energy security strategy favouring the further reduction in future EU gas consumption and the diversification of its gas supplies (European Commission 2014a).

Some European countries (Poland, the United Kingdom, Romania, Lithuania, Spain and Ukraine) have actively supported the test drilling and exploitation of their own unconventional gas resources (Figure 2.2), whereas others have adopted a moratorium (Germany, Bulgaria, Czech Republic) or even a ban (France) on fracking technology and the production of shale gas, owing to perceived environmental risks (Umbach 2014c, 2013a). This highlights the challenging fact that the common EU energy policies are still dependent on national energy mixes decided by national governments, despite the progress on many other fronts in developing a common EU energy policy.

Owing to high population densities, unclear regulatory conditions for social and environmental concerns and relatively less developed service industries and infrastructure, shale gas development in Europe has been slow and also has been facing problems in adopting attractive investment conditions and avoiding the over-regulation found in Poland, Lithuania, Ukraine and other countries. For all these reasons, the IEA has remained cautious and estimated last November that Europe's unconventional gas production may reach not more than 17 bcm by 2040 (IEA 2014b, p. 148). At present, for instance, even in Poland only 69 shale wells are being drilled (despite the fact that Warsaw has offered six-year tax breaks) and in the UK just 11 (Shale Gas International 2015; Vaughan 2015).

Factbox for Europe's new shale gas estimates from EIA, June 2013

- Poland's technically recoverable shale gas resource estimate has been reduced from 187 tcf to 148 tcf.
- France: its technically recoverable shale gas resource estimate has decreased by some 24% to 137 tcf, but the French reserves make up over half the total estimate for Western Europe.
- Britain: the estimate of the technically recoverable reserves has increased to 26 tcf (almost 10 times the annual British gas consumption) from 20 tcf in 2011.
- Romania: its technically recoverable shale gas estimate has increased 10 times to 57 tcf, putting Romania third in the EU after Poland and France.

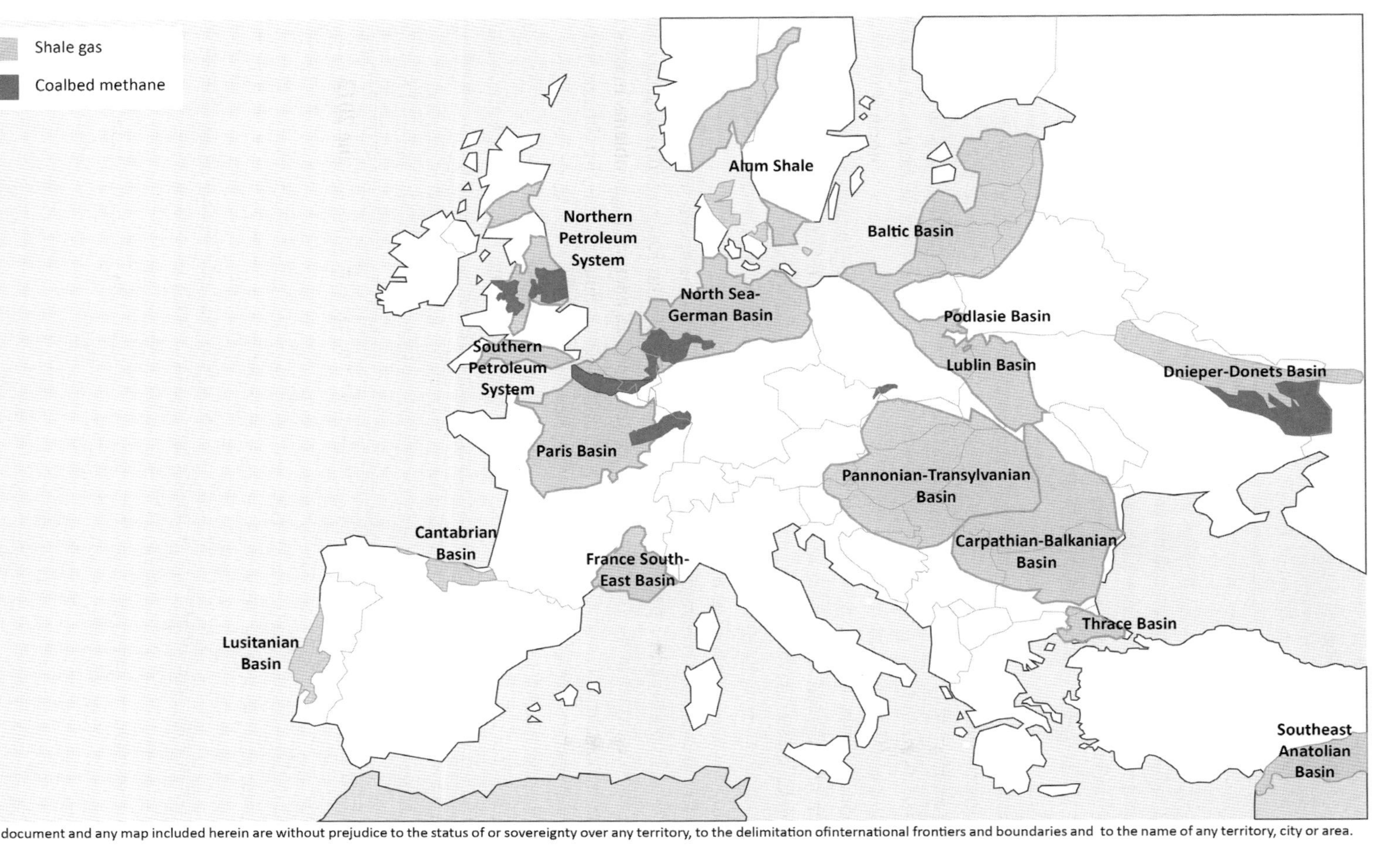

Figure 2.2 Location of main unconventional gas fields in Europe. *Source:* IEA (2012).

- Bulgaria: ranked sixth in the EU with 17 tcf, comparable with Germany.
- Ukraine's technically recoverable resource estimate has increased from 42 tcf in 2011 to 128 tcf in 2013, the third highest in Europe as a whole.

Source: Dr F. Umbach, based on EIA (2013).

In the view of the European Commission, Europe should be at least able to produce enough of its own domestic shale gas to replace its depleting conventional gas reserves, so that it avoids becoming more dependent on imports from unreliable suppliers or unstable political countries (European Commission 2014c).

As the price gap between the North American and the European oil and gas market has increasingly widened (for gas, US$2.5 per million British thermal units in the United States in comparison with US$9 in Europe and US$18 in Asia), a 're-industrialization' of energy-intensive and other industries is already underway on the US side. The future economic competitiveness of Europe and Asia in comparison with the United States is being increasingly challenged by their much higher gas prices (IEA 2013, pp. 261–300 and European Commission 2014b).

In 2013, the EU-28 spent more than €500 billion on energy imports – seven times more than in 1999 and amounting to more than 4 per cent of GDP. Increased European efforts to take greater advantage of unconventional oil and gas resources could help the EU to retain or grow industrial sector jobs and could contribute to its overall future economic competitiveness as well as overall energy-supply security.

In 2012, the EU received more than 31 per cent of its oil imports, almost 27 per cent of its coal imports and 27 per cent of its uranium imports from Russia and is paying around US$250 billion in annual energy bills to Moscow. The EU's overall gas-import dependence represents currently some 70 per cent of its gas consumption and will rise further, to more than 80 per cent by 2035. Gazprom has long-term contracts, with 'take-or-pay' clauses and with a capacity of 120 bcm, that require that Europe continues to pay at least US$50 billion for Russian gas. But meanwhile, the Southern Gas Corridor project will open around 2018, by which gas is imported from Azerbaijan, thus circumventing Russia for the first time, so Russia will lose the monopoly of gas exports from the Caspian region to Europe.

Moreover, in contrast with the 2009 gas crisis, Europe already has and will have even more alternative options for pipeline gas and LNG imports by 2020; these options include imports from Israel, the US, Africa and exploration of its own conventional and unconventional gas resources, including those in the Black Sea offshore areas (Romania, Bulgaria) and in the Adriatic Sea (Croatia, Greece). In addition Russia is eager to build the Turkish Stream gas pipeline to Turkey (this was temporarily suspended but was revived in July 2016) and announced in 2015 that it would build another two strings of North Stream with an additional capacity of 55 bcm despite the fact that it only uses its existing North Stream capacity to just 56 per cent. In 2014, Russian's pipeline gas exported to the EU-28 fell by more than 11 per cent from 2013, to just 119 bcm. Only its LNG exports to the EU increased, from 2 bcm to 4.5 bcm (Walstad 2015; Botzki 2015a, 2015b; Robinson 2015).

Like Lithuania's opening of a natural gas floating storage and regasification unit terminal (FSRU) and Poland's new LNG terminal, this reflects Gazprom's rather declining leverage of its gas supplies as a strategic instrument of foreign policy despite its present efforts to build new gas pipelines such as the Turkish Stream.

EU energy and gas consumption in 2014 compared with that in 2013

- Total energy consumption: −3.9% from 2013.
- Gas consumption: −11.6% (biggest annual decline on record) from 2013.
- Gas production: −9.8% from 2013.
- Net gas imports: −8% from 2013.
- Russian gas pipeline exports to EU-28: −11.6% from 2013.

Despite the slow and evolutionary developments of shale gas explorations in Europe, new and more positive strategic developments can be identified on the European gas market, such as the following.

- According to the latest EU gas-demand forecasts, gas imports will grow only slightly, from less than 300 billion bcm to about 340–350 bcm in 2025–2030. This might be even lower than its import demand in 2010 and is to be compared with an import demand of more than 500 bcm by 2030 of older IEA and industrial forecasts (IEA 2014b, pp. 135–170 and European Commission 2014a).
- New LNG terminals have opened in Poland, the Baltic states (Lithuania's FRSU opened in December 2014), and potentially also in Croatia and other countries, as well as new interstate gas interconnectors with reverse-flow capacities in central and southeastern Europe operating by 2020.
- The impacts of the liberalisation and unbundling processes, as part of the EU's Third Energy Package within the EU-28, should lead to much more competition.
- The discovery of new conventional gas fields off the coasts of Romania, Bulgaria, Croatia and Greece as well as the east-Mediterranean countries (Israel, Cyprus, Lebanon) with much shorter transport distances to the European consumer markets in comparison with the very expensive long-distance Russian gas pipeline (Umbach 2013c, 2013d).

The EU intends to diversify its gas supplies and expand its LNG imports, including those from the US, starting in 2015/2016. The major problem is not a lack of LNG import capacities but rather the related costs. At present, the EU has 22 LNG import terminals with a total capacity of 196 bcm per year. Six additional LNG terminals are under construction, with a capacity of 32 bcm. The EU can import much more LNG from conventional and unconventional gas reserves worldwide; its terminals were used only to 73 per cent of its re-gasification capacity in 2013. In addition, new global liquefaction capacity from Australia and southeast Asia should come online by 2017. Any additional US export projects starting in 2016 may have a disproportionately large impact on reducing global gas prices and the general tightness, in an already oversupplied LNG market in the forthcoming years (Umbach 2014e, 2014f).

China

China is the world's most populous country, with the fast-growing economy; it is the largest energy producer, consumer and (oil) importer on the globe. In 2000, China's energy demand was still only half that of the US. Today, it is also the world's largest consumer, producer and importer of coal. Despite having the third largest coal reserves in the world (after the US and Russia), it became a net importer of coal in 2009 (Umbach 2015, pp. 31–35).

Together with its continuous increase in energy demand and high GDP growth, albeit on a slower pace, China may be already the most important and influential actor in the world energy markets. During the few last years, not only has China surpassed Germany as the world's largest-export nation, but also it has surpassed the US as the world's largest economy.

By 2040, China will be consuming about 80 per cent more than the US (IEA 2014a). This brings a dual challenge: (1) an energy demand projected to rise by 44 per cent by 2040 and, at the same time, (2) a shift in its energy mix from coal to gas and non-fossil fuels like nuclear power and renewables. The country would not have the luxury just to shift its energy completely to renewables by 2050 because it would be too expensive and rather unrealistic, given its increase in energy demand and heavy dependence on fossil fuels (i.e. coal) (Umbach 2014b, 2010).

Its primary energy mix is still based on high coal consumption (i.e. for heating and power), which was at 68% in 2012 (IEA 2014: p. 178). In the short-term perspective, until 2017, China's government has planned to cut coal use to below 65 per cent, and it intends to raise the share of non-fossil fuel energy to 15 per cent by 2020 (EIA 2014). Its newest energy blueprint for its 13th Five-Year Plan (2016–2020) aims for it to be 85 per cent self-sufficient. According to this new energy strategy, non-fossil fuels will have to cover 15 per cent of the primary energy mix, gas will have to cover more than 10 per cent and coal will have to be reduced to 62 per cent. Coal imports will be banned by 2020. Conventional gas output should be expanded from 128 bcm in 2013 to 185 bcm per year by 2020 (Yiping 2014a). But China's gas consumption is planned to more than double, from 168 bcn in 2013 to 360 bcm in 2020.

Although China surpassed Japan as the third-largest natural gas consumer in 2009, the gas share in its national energy mix was just 5.9 per cent in 2013. It is projected to increase to 7.5 per cent in 2015 and 10 per cent in 2020. Its future gas-demand growth, at an annual 6 per cent on average, is expected to be far the largest in the world. Its gas consumption of 169.2 bcm in 2013 is projected to quadruple by 2035. The demand growth is due to China's efforts to diversify its energy mix, in order to reduce its large level of air pollution and the linked heavy-coal share from 68 per cent of its annual energy demand today to 53 per cent in 2035; electricity generation is the main source of its additional gas demand.

However, China's gas demand of 530 bcm by 2035 will be just 50 per cent of that of the US, which remains the world's largest single gas consuming country. China's own gas production may triple from 121 bcm in 2013 to 320 bcm in 2035. Regarding the 121 bcm demand, 117 bcm still came from conventional gas reserves. But production from conventional gas fields has struggled to keep up with demand. China has been a net natural-gas-importing country since 2007.

By 2013, China's natural gas imports had increased dramatically, to more than 30 per cent of the country's total gas demand, and they are forecast to increase further, to 50 per cent by 2020. In 2013 China needed to import 54.3 bcm via gas pipelines and as LNG. Its future gas-import dependence will rise further, to 154 bcm by 2020, and it may exceed 210 bcm by 2035 (IEA 2013, pp. 125–126). However, its gas-import demand will depend considerably on its progress in unconventional gas production, and the necessary widespread reforms, as well as on timely investment in its wholesale gas sector.

China's technically recoverable unconventional gas reserves

- Shale gas resources: 31.9 tcm
- Coal bed methane: 11 tcm
- Tight gas: 12 tcm

The potential for shale gas exploration is much larger, as, with 31.6 tcm of technically recoverable shale gas resources, China has almost as much as the US (which has 32.9 tcm). But its geology is much more complicated and its shale gas resources are in much deeper formations than in the US, raising production costs. China's water shortages are also a complicating factor in some regions (the Tarim and Ordos basins) as is a lack of US technology transfer and joint cooperation, such as joint-production-sharing contracts, and unattractive commercial conditions for foreign investors (Yiping 2014b). However, the efficiency of the drilling technology is constantly improving, and this is also reducing the water demand by multiple re-use. The official production target for China's shale gas reserves in 2015 rose from just 1.3 bcm in 2014 to 6.5 bcm in 2015, but the government originally hoped to accelerate to a mid-term production level of 120 bcm by 2035.

In August 2014 China revised its 2020 forecast for domestic shale gas production, reducing its previous target of 60–100 bcm to just 30 bcm owing to geological, technical, infrastructural, technological and topographical hurdles, to insufficient investment and to the lack of a comparative competitive system in China's upstream sector and of drilling and managing experience in hydrofracking operations (Umbach 2014b). Nonetheless, China's gas industry appears somewhat more optimistic; Sinopec, for instance, has significantly increased the efficiency and productivity of its shale gas drilling. It has decreased well costs from RMB 120 million to RMB 70–80 million, and the average full drilling time by 23 days to 60 days, compared with 2013 (Xin 2015a). However, the planned increase in production of its unconventional gas reserves also depends on an expansion of its overall gas infrastructure and timely investment as well as adequate regulation.

Until 2020, China's coal bed methane (CBM) production will play a more prominent role in China's future energy mix. In 2014, around 15.6 bcm of CBM were produced, representing around 13 per cent of China's total production. For 2015, another 12–15 bcm can be added to the production volume of the previous year. For 2020, China's government has increased the CBM production target to 40 bcm. Its total CBM resources have been estimated at 36.8 tcm alone (Xin 2015b).

China has also a very ambitious plan for expanding its coal-to-gas (CTG) production. Although this is still in the early stages of development, by using its abundant low-cost coal reserves and by building the world's largest synthetic natural gas (SNG) industry (in order to reduce its energy imports), China's first CTG plant began its production operation at the beginning of 2014. It was planned to be expanded to an annual production target of 16 bcm by 2015. China has 18 CTG projects under construction and another 50 planned, with a combined annual capacity of 225 bcm. However, the production of SNG through CTG is energy and water intensive and will ultimately also increase coal consumption, though the use of CTG for city heating could reduce air emissions and pollution (Hornby 2014). Accordingly, the National Energy Administration warned in July 2014 against 'blind development' (Xin 2014b), apparently because of technical and infrastructure problems as well as the insufficient competitiveness of many CTG projects.

In December 2014, China revised its CTG plans by excluding an additional CTG project from its next Five-Year Development Plan but it will complete the construction of approved CTG plants in order to keep its CTG production capacity to 15 bcm around 2020 (Ukraine Energy News 2014).

On 13 June 2014, China's President Xi Jinping called for an 'energy revolution' in his country with an expanded role for gas to 'restrain irrational energy consumption' and a 'revolution in supply security in diversifying into non-coal energy sources. In April 2014 the government announced an increase its gas supplies (domestic production and imports) from 174 bcm in 2013 up to 420 bcm by 2020, as it would be the most realistic energy option for achieving its 2020 carbon reduction goals. In November 2014, China's State Council released a draft of a new energy strategy that envisaged capping coal consumption at 4.2 bn tons by 2020 (after new estimates suggested that coal consumption would soon reach 5.1 bn tons) with a coal mix of no more than 62 per cent of the primary energy mix by that year (Wong 2014).

China's gas contract with Russia of May 2014 is understandable as being due to the halving of its production forecast for its shale gas reserves by 2020 (agreed by Beijing in August 2014), to the growing gap between its fast increasing gas consumption and imports and its domestic gas production, to the fact that an estimated 1.2 million Chinese die every year from coal-related air pollution (Umbach 2014d).

Despite China's expansion of gas projects and its share in the national energy mix, its government does not want to become overly dependent on an additional energy source with its rising gas-pipeline and maritime LNG import-supply risks. Beijing is rather interested in having a balanced and diversified national energy mix with a lower coal share and a rising non-fossil fuel share (I,e, nuclear and renewables) in order to reduce its heavy greenhouse gas emissions (GHGE). Furthermore, Beijing's perceived geopolitical rivalries with Japan, the US and India, alongside its rising anxiety about its dependence on maritime oil, LNG imports and coal imports, are likely to increase, when in the future China can no longer rely on its indigenous coal reserves and its own unconventional gas reserves.

The economic rise of China has raised many foreign and security policy questions for regional and global stability. China's energy policies in Africa and Iran have complicated, if not undermined, Western strategies for conflict prevention, management and sustainable development aid.

In East Asia, China has territorial maritime conflicts in the East China Sea with Taiwan, South Korea and in particular Japan, which are often directly linked to the regional offshore oil and gas reserves. Its territorial claims in the South China Sea, perceived as aggressive by its ASEAN neighbours and the US, are linked with presumed large maritime oil and, in particular, gas reserves in the region; these are often estimated as much greater than those in the US and in other East Asian countries. This drives China's security and defence policies such as its blue-water naval build-up. They are based on perceived security risks and the vulnerabilities of China's foreign trade via southeast Asia's sea lanes of communications (SLOCs) and critical choke points such as the Malacca-Straits, which are controlled by US naval forces.

As China becomes ever more dependent on increasing and stable energy imports, its energy-supply security and the diversification of its energy mix and imports from various countries as well as different forms of imports via pipelines or LNG tankers will remain on the highest political agenda of China's leadership (Umbach 2014b).

However, Beijing also has a strategic interest in the successful US shale revolution, as this could decrease US LNG imports and so decrease China's bilateral rivalry with the US in regard to many oil- and gas-exporting countries in the Middle East, Asia and Africa. A considerable increase in the geopolitical influence of China around the world could result from a US disengagement from global security commitments.

Russia – The Biggest Geopolitical Loser?

> The role of the gas industry for the future economic and political situation, now and in the future, should not be underestimated. Gas is not only the backbone of the Russian energy sector, but also one of the most powerful tools of domestic and foreign policy. It is strongly affected by the domestic economic and political processes. At the same time, the oil and gas industry itself is to a large extent defining economic performance and political stability.
>
> *(Mitrova 2014, p. 95)*

The strategic importance of hydrocarbons for Russia's wider national security, its state budget, economic growth, socio-political stability and foreign and security policies was confirmed in Russia's new 'Foreign Policy Concept of the Russian Federation' of February 2013. The Concept focused on the importance of access to energy resources globally and on the potential for military conflicts as worldwide competition for scarce energy resources increases in the years and decades ahead, for example in Central Asia (Russian Ministry of Foreign Affairs 2013). The Kremlin has used Gazprom in particular as its arm for spreading its geo-economic and geopolitical interests in Europe and Eurasia, which range far into many businesses and banks in these regions as well as beyond.

Holding the world's largest remaining conventional gas resources, Russia has no problems with a lack of gas reserves but rather with the availability of gas markets and the future competitiveness of its gas exports, owing to the much higher production and transport costs of its new, much more remote, gas fields (Mitrova 2014; Mitrova & Grigoriev 2014: pp. 140–145).

Although Russia has also huge unconventional gas reserves (i.e. CBM), it is focusing attention on its huge conventional gas reserves rather than its shale gas reserves (its technically recoverable conventional gas reserves are presently estimated at 5.7 trillion cm) due to its experience in traditional drilling technologies and a lack of new fracking technologies and operational fracking experience as well as declared higher costs (IEA 2014c, p. 95 Mitrova & Grigoriev 2014). But the latter argument has been questioned, because of the much higher production and transportation costs of the remote new gas fields in comparison with its unconventional gas reserves, some of which are partly much closer to the places of consumption, thereby saving transport costs (Shung-hwan 2014).

President Putin and Gazprom representatives have often tried to downplay the strategic importance of the US shale gas revolution and its impacts on the world gas markets by dismissing it as 'myths', shale fever' or a 'bubble that will burst very soon' (Marson & Parkinson 2013; Chazan & Buckley 2013). But the Kremlin and its gas industry have clearly become increasingly concerned as the US has become the world's largest gas producer. At the end of 2011, the largely non-subsidized US gas price (US$83) was already lower than the heavily subsidized gas from Russia (US$97). At the same time, the Kremlin has actively promoted the Rosneft–ExxonMobile cooperation for the exploration of Russia's vast unconventional oil reserves as the estimated worldwide largest (the estimate is 75 bn barrels), using almost the same fracking technology as for shale gas drilling.

Furthermore, the US shale gas revolution has threatened Russia's energy dominance in Europe – as its most lucrative market and Gazprom's central economic and geopolitical battleground – even without any US LNG exports up to now: the US shale gas revolution has significantly reduced US coal consumption, so that coal became available for export. Ultimately, cheap US coal exports and subsidized renewables have partially replaced the much more costly Russian gas in the European power sector over the last few years.

Preliminary estimates of Russia's shale and other unconventional gas resources

- 680 tcm unconventional gas reserves (shale and sandstone), which amounts to 2.5 times more than its conventional reserves, the world's largest;
- 13–84 tcm CBM resources;

- 8 tcm technically recoverable shale gas resources (4% of the world's total; the ninth largest in the world); potential shale gas resources of 48.8 tcm.

Sources: IEA (2014c) and Mitrova (2014, pp. 20–21)

In 2013, Russia exported a total of 217.7 bcm of gas, with 139 bcm to the EU-28 and 14.5 bcm from its new LNG plant in Sakhalin (opened, as its first, in 2009) to the Asia–Pacific region. But Russia's share of EU gas imports had already fallen from almost 50 per cent in 2000 to 34 per cent in 2010 and in wider Europe from 45.1 per cent in 2003 to 31.8 percent in 2010 (IEA 2011, p. 343, Bertrand 2012). Although its share of EU imports rose again to almost 30 per cent in 2013 (after it had lost 12 per cent in 2012, when Norway became Europe's largest gas importer), this was primarily the result of domestic instabilities in its rivalling exporting countries.[3] In the same year, the EU's gas consumption fell to the lowest level since 1999.

Russia's EU-28 and wider European gas exports still have not returned to the pre-crisis level of 2008. The prospect of a significantly reduced European dependence on Russian gas supplies has not just had a significant impact on Gazprom's future business and market position in Europe. It has also directly affected Russia's state budget and entire economy. All previous forecasts of Europe's gas demand have been reduced constantly downwards during the last few years. Together with the EU's liberalization efforts for its gas sector by its Third-Energy Package regulation (unbundling third-party access to pipeline and other infrastructures etc.), it has forced Gazprom to look for alternative markets and in May 2014 to sign a record-breaking deal for exporting gas to China. But its price compromise with Beijing has raised serious doubts about the profitability of that bilateral deal for Gazprom and Russia (Umbach 2014d).

Russia's energy sector is, for the Kremlin, the most important commercial asset and pillar for its domestic economic stability and foreign policy leverage. As the world's largest gas exporter, with 170–200 bcm per year, it currently provides for more than 20 per cent of the international gas trade. This status and the control of pipelines as well as other major gas infrastructures in Europe and Eurasia gives Russia a strategic influence on the world's gas prices, the 'rules of the game' in these regions, and a wider geopolitical influence on the region's economic, foreign and security policies (Mitrova 2014 p. 7).

The prospects for Russia are in fact worse, even without taking into account the present western sanctions and the fall in oil prices; the reason is that, during the Putin era, since 2000, Russia has not been able to diversify its economy away from its heavy dependence on oil and gas exports. Between 2000 and 2012, this dependence actually increased, from 47 per cent of the state budget to 50 per cent. In 2012 and 2013, the share of energy-export revenues in Russia's total exports was 70 per cent, with the EU countries alone accounting for 61 per cent (Assenova 2013; Mitrova and Grigoriev 2014, pp. 149–152). But the increasing

[3] In addition to larger infrastructure repairs in Norway in 2013, reducing its EU supplies by 5.2 per cent, exports from North Africa declined by 18.7 per cent and from Nigeria as much as 43.9 per cent (Botzki 2013).

decoupling of European gas prices from Russia's oil indexation, the inhibition of domestic price increases and the ongoing implementation of excessively expensive investment projects with uncertain prospects for payback makes future Russian gas supplies commercially and politically less attractive for Europe. This is threatening Russia's future gas export strategies and the profitability of its new expensive gas fields as well as its pipeline projects. Although Russia's energy resources will provide an important foundation of the Russian economy in the future, they will cease to be an engine of Russia's economic growth (Mitrova 2014, pp. 60–61; Mitrova & Grigoriev 2014). Therefore the state's economic underpinning and future geopolitical influence in Eurasia may be undermined (Umbach 2013c).

Russia's future gas exports, based on even more remote and super-expensive new gas fields as well as longer and more costly pipelines could become the most expensive option for Europe's future gas imports in comparison with indigenous unconventional gas production and the many future LNG imports that will be based increasingly on spot-market prices. Russia's Ministry of Economic Development warned Gazprom in August 2012 to face increased competition from North American exports of LNG which were expected to force Gazprom to lower its export prices by 2016 at the latest (Chechel 2012). The Ministry expected that after 2015 the production costs of unconventional gas in the US and other countries will decrease further and that production volumes would increase globally, leading to even more competition and lower gas prices (Godier 2013). It anticipated declining gas prices in Europe from the peak of US$411 per 1000 cm in 2014 to US$329 in 2016. But, at the same time, Russia's ageing energy sector needs a huge investment of US$2.4–2.8 trillion up to 2030 (IEA 2014c, p. 37). Russian gas producers must invest US$730 by 2035 merely to replace most of their current production of more than 600 bcm a year from Russia's ageing gas fields.

In recent years the Russian government has increasingly regarded its overdependence on the European oil and gas market as a threat to its national energy security. Its 2009 Energy Strategy for 2035 envisaged an increase in the Asian gas market share to 26–27 per cent of Russia's total energy exports and to up to 20 per cent of its gas exports by 2030. The new bilateral gas deal with China acquired an almost 'existential significance', for economic and geopolitical reasons, for the Kremlin and Gazprom. Moscow is more interested than ever in attracting Chinese capital for investments in Siberia and the far east of Russia as it faces new Western sanctions (Umbach 2014d). But it is unlikely that it will ever dominate the Chinese gas market and exercise a geopolitical influence similar to that achieved in the EU-28 and the wider Europe in the past.

China has already become the dominant actor in regional gas-price formation, whereas Russia may only be able at best to cover its investment, with little profit for Gazprom or the state budget. Russian taxpayers could end up subsidizing Chinese consumers, confirming that the two countries are not equal geopolitical partners, contrary to Russia's self-defined great-power status and geopolitical ambitions. China has a much higher GDP growth, of more than seven per cent, whilst Russia's had stagnated already before the outbreak of its Ukraine conflict in the summer of 2013. This gap in great-power status will continuously widen at the expense of Russia and could easily produce additional bilateral tensions and conflicts in their bilateral geopolitical rivalry in Eurasia and east Asia.

Furthermore, prospects for a considerable increase in Russian gas exports to China appear rather limited (even taking the newest gas deal, of another 30 bcm, into account), given China's large-scale investments in central Asia and China's own ambitious shale gas programme. This highlights China's strategic interest in diversifying its gas supplies and avoiding an overdependence on Russian gas exports, as the European example teaches (Umbach and Raszewski 2016).

Conclusions and Strategic Perspectives

The technological advance of the fracking revolution in the US has led to a collapse of national as well as international gas prices and a revival of US manufacturing as well as other energy-intensive industries. The US attracts ever more investments from Europe and the rest of the world owing to its cheap gas supplies and its overall economic competitiveness in relation to other global rivals. Geopolitically, any discussion of a present or near-future 'peak oil' situation has disappeared from the radar screen of international energy conferences and expert discussions, for at least the time being.

The historical lesson with new fossil fuels is that both availability and production levels increase – rather than decrease – in the initial phase for a longer period. But the US shale gas 'revolution' cannot be replicated in the same way, and with the same low costs of shale gas production, in Europe, Asia or many other countries owing to a variety of different reasons and conditions. Nonetheless, shale gas is expected to become a competitively priced source of energy, in particular compared with other, expensive, gas imports such as future Russian conventional gas from its new and very expensive gas fields in the remote regions of Yamal and Siberia or the Arctic offshore zones, with even longer distances for pipeline transportation.

With the share of LNG versus pipeline gas rising worldwide, the EU-28 has already benefitted in various ways from the expansion in unconventional US gas production and from cheap US coal exports to Europe, and it will do so in the future also. The EU-28 will open up new sources of LNG imports from places in the US and other newly gas exporting countries.

The US shale gas revolution, and the prospect of even limited US LNG exports to Europe, has already become a 'game changer' in regard to its manifold impacts on the global gas markets, in that it has changed the oil-indexed long-term gas contracts and successfully uses these LNG export prospects as a bargaining chip to secure lower gas prices from Russia's Gazprom in its gas contract negotiations. With the growing LNG trade in the global gas market, traditional oil-indexed gas contracts will gradually decrease further as the global gas market becomes increasingly integrated, whereas spot markets for gas will expand both in numbers and importance. Under these new conditions it will be almost impossible to maintain Russia's controversial destination and take-or-pay clauses in its long-term contracts, since these must be in compliance with EU laws and regulations. The need to renegotiate these clauses between the European Commission, the gas companies and Gazprom created increasing political frictions even before the Ukraine conflict. While long-term contracts with European companies give Moscow some assurance in regard to its demand-security concerns, those pipeline gas supplies need to respect EU competition laws (i.e. third party access for up to 50 per cent of the pipeline's capacity), which undermines the use of Moscow's pipeline policies as a geopolitical instrument.

The increasing conflicts between the EU and Russia in their bilateral 'energy partnership' even before the outbreak of the Ukraine conflict in summer 2013 (including the anti-trust investigation into Gazprom's EU business, which is still ongoing) have highlighted the increasing strength of the EU and the European Commission in that the bilateral power balance is shifting: Russia is no longer dominating the European energy landscape as in the past and the EU's energy dependence on Russia is decreasing alongside of its improving energy-supply security. Russia's energy leverage in Europe has already peaked, as its market shares and export volumes are decreasing with no prospect that they can be expanded again, owing to the increasing diversity of Europe's energy mix as well as gas imports, which will further increase with the spread of the shale gas 'revolution' beyond North America. Given the Kremlin's high dependence on energy revenues (i.e. on oil exports) and the failing diversity of its economy, Russia's strength in the past is becoming increasingly the source of its future weakness and vulnerability. For too long Europe has overlooked the fact that Russia's energy policies can only be as influential as European countries allow.

Even though a shale gas 'revolution' appears unlikely to be replicated, this does not mean that in other countries and regions no unconventional gas reserves will be explored. As it looks now, similar developments in Europe, China, the Middle East and South America (i.e. Argentina) will take a rather slower and more evolutionary path, owing to more complex geological conditions, the lack of rigs and other gas infrastructure as well as operational experience with the most advanced fracking technologies and, finally, political as well as regulatory obstacles. But the future energy and, especially, gas markets will see many more producers and exporters, which will stabilize global gas- and energy-supply security and offer considerably more diversification options as well as competition in the regional gas markets. Together with the further expansion of interregional LNG trade, this will transform the different regional gas markets into a global market in the mid- and longer-term perspective.

At the same time, the US shale gas revolution and its impacts on the regional energy markets in Europe and Asia may have far-reaching negative impacts for conventional gas producers and exporters and their state budget stability; these include Russia and the Middle Eastern and other countries and regions that are heavily dependent on high revenues from conventional oil and gas exports. A worldwide 'shale gas revolution' will create new and often more complex as well as competitive energy interdependences between consumers and importers, but it offers more positive perspectives for global energy security.

The present declines in global oil and gas prices are even more impressive when one takes into account that spreading instability as the result of the Arab Spring has led to production cuts and supply disruptions. It has affected Europe and the global oil and gas markets by reducing the gross oil production across the region to around 3.5 million barrels per day (mb/d). Until September 2015 the US shale oil and gas revolution stabilized the global oil price at around US$110–100 despite the geopolitical turmoil of falling production levels in many oil producing countries.

Saudi Arabia's unwillingness to stop the oil price fall (with related impacts on many regional gas prices) highlighted that it is more interested in maintaining its market share, particularly in the Asia–Pacific region, than in defending prices and has given Riad, with its US$ bn in reserves, the opportunity to test the robustness and break-even price for US shale oil production. The Saudi Arabia cuts in the oil price in relation to the US market indicated that Riad fights for market shares not just in Asia but also across the Atlantic to North America. In contrast with the much more capital-intensive Canadian oil sands projects, at present only 4 per cent of US shale oil output needs prices above US$80 per barrel. In order to cause greater trouble for US oil production and to have a market impact, Saudi Arabia would need to bring down oil prices, over a longer period, to less than US$50, which is in contrast with the view of some OPEC officials that larger investment would leave the US market at a higher oil price, of US$85 per barrel. Even under the financial strain of such an oil price level, some more marginal, smaller, US oil and gas producers might be able to keep their production growing for some time.

While the world has experienced dramatic oil price declines and rises in the past, such similar volatile price hikes are rather new for regional gas markets. As long as Russia and other exporters insist on oil-indexed prices in their long-term gas contracts, their gas export policies are dependent not just on the US shale gas development but also on the US shale oil revolution.

The US shale oil and gas industries thus will strengthen further its ever-increasing production efficiency and improved technologies, which then will decrease further its break-even price rate. The average cost of a US shale well is expected to decline by another 40 per cent in the next few years, simply because of better management of planning, logistics and relationships with suppliers.

Moreover, gas has become a direct competitor to oil in the US and some other countries (in the Middle East and Latin America) as an alternative in the transport sector (as CNG and LNG), and this is already slowing oil demand growth. If these energy trends in the global transport sector continue in the future, they may further constrain worldwide oil demand in the forthcoming years and thus stabilize oil prices at lower levels around US$70–80.

Now lower prices are established in the market and it seems that they would be volatile at around US$50; some corporate forecasts do not rule out prices around US$30 in the mid-term perspective. If those strategic energy trends are confirmed at a global level in the forthcoming years and decades, natural gas will contribute to global oil security as well as overall energy-supply security in the coming decades and not just be a 'bridge source' to a non-fossil-fuel age.

In this context, a future methane hydrate revolution may also influence the strategic perspectives for unconventional gas development.[4] These revolutionary changes in the global energy sector have already become a principal factor of geopolitical metamorphosis. As a result, not only will the world economy be changed through the transformation of the global energy balance, but also the world geopolitical order.

References

Anderson, R. (2014). How American energy independence could change the world. BBC News, 3 April.

Assenova, M. (2013). Russian Energy Review in 2013: consolidating state control in an uncertain market. Jamestown Foundatoin – *Eurasia Daily Monitor*, 18 January.

Bertrand, E. (2012). EU–Russia gas supply partnership. *The Voice of Russia*, 23 November.

Botzki, A. (2014). Geopolitics emphasies importance of transatlantic LNG deal. Interfax-energy.com – *Energy Policy Weekly (EPW)*, 13 March 2014.

Botzki, A. (2015a). European imports from Russia declining BP. Interfaxenergy.com – *Natural Gas Daily (NGD)*, 18 June.

Botzki, A. (2015b). European LNG imports to double by 2020 – IEA. Interfaxenergy.com – *NGD*, 15 June.

Botzki, A. (2013). Spot indexation helps Russia regain EU gas market share. Interfaxenergy.com – *EPW*, 26 June.

British Petroleum (BP 2014). Statistical review of world energy 2014.

Butler, N. (2014). European energy policy – time to start again. *Financial Times*, 27 October.

Chazan, G. and Buckley, N. (2013). A cap on Gazprom's ambitions. *Financial Times*, 5 June 2013.

[4] Methane hydrate or 'combustible ice' is a kind of unconventional gas, which is nothing other than methane dissolved in water under high pressure and low temperature. Methane hydrate deposits are mainly sedimentary rocks containing 70–80 per cent sand and clay and 20–30 per cent hydrates in ice form. The largest deposits are located along the coasts of the Atlantic, Pacific and Arctic oceans. In marine sediments, hydrate deposits may have a depth of several hundred metres. Over the world, more than 220 zones of methane hydrate deposits have been discovered on the shelves of oceans and seas. Typically, they are at sea depths where the temperature is not above +4 °C and the pressure is at least 20 atmospheres, which ensures their thermostatic balance. Reserves of methane hydrates on the planet, according to tentative estimates, are at least 250 trillion cubic metres. This is rather a pessimistic assessment but even so it exceeds the known reserves of conventional natural gas, which, according to the BP Statistical Review, amount to 187.1 trillion cubic metres. About 98 per cent of world gas hydrate reserves are concentrated in the ocean and 2 per cent on land, in the permafrost zone.

Methane hydrates remain on the agenda of systematic research in Japan, the US and Canada. Methane hydrate matters are also being actively developed in India and South Korea. China has started an active research programme to develop methane hydrates in the past decade. The list of countries interested in them expands every year: at the time of writing it includes Brazil, Chile, Australia, New Zealand and others. The development of gas hydrates will contribute in the long term to both economic development and the liberation of politicized supplies of conventional natural gas via pipelines and could lead to a revolution in the energy sector, similar to the 'shale gas revolution' in North America.

Tremendous work has been done by Conoco-Phillips on the preparation of Ignik Sikumi #1 well for the experimental testing of one of the possible, and the most advanced, technologies of methane production from gas hydrates. The essence of this technology is that, in the structure of the gas hydrate, methane molecules would be replaced by carbon dioxide molecules. This carbon dioxide hydrate would be more stable than methane hydrate. Thus, in the case of the success of such technology, three important aims could be achieved simultaneously: methane extraction, the utilization of CO_2 and the prevention of unmanageable methane-hydrate-deposits degradation. Since 2008 this technology has also been developed in Germany within the SUGAR project (submarine gas-hydrated reservoirs) by the Leibniz Institute of Marine Sciences at Kiel.

Incidentally, on 2 May 2012, the US Department of Energy announced a breakthrough in research on Ignik Sikumi #1. By the injection of a carbon dioxide and nitrogen mixture into methane hydrate deposits on Alaska's North Slope, a constant gas discharge was received in the framework of the first field test of this method.

The greatest progress has been achieved in Japan, however. On 12 March 2013, for the first time in the world, natural gas was extracted from marine gas hydrates on the bottom of the Sea of Japan, to the south of the Honshu Island. In fact, this date represents the next gas revolution – the methane hydrate revolution.

Chechel, A. (2012). Government sees trouble ahead for Gazprom 2016. Interfaxenergy.com – *NGD*, 28 August.
Crooks, E. (2014). US shakes up LNG export rule. *Financial Times*, 30 May.
Energy Information Administration of the United States (EIA) (2014). China. Analysis Briefs, 4 February (http://www.eia.gov/countries/analysisbriefs/China/china.pdf).
EIA (2013). Technically recoverable shale oil and shale gas resources: an assessment of 137 shale formations in 41 countries outside the United States. Washington DC, 10 June 2013.
European Commission (2010). Energy 2020. A strategy for competitive, sustainable and secure energy. Communication from the Commission to the European Parliament, the Council, the European Economic and Social Committee and the Committee of the Regions. Brussels, 10 November, COM(2010) p. 639 final, Brussels, 10 November.
European Commission (2014a). European energy security strategy. Communication from the Commission to the European Parliament and the Council Brussels, 28 May COM(2014) p. 330 final.
European Commission (2014b). Energy prices and costs in Europe. Communication from the Commission to the European Parliament, the Council, the European Economic and Social Committee and the Committee of the Regions. Brussels, **XXX**COM (2014) p. 21.
European Commission (2014c). On the exploration and production of hydrocarbons (such as shale gas) using high volume hydraulic fracturing in the EU. Communication from the Commission to the Council and the European Parliament. Brussels, 17 March, COM (2014), p. 23 final/2.
European Council (2007). Presidency conclusions 2007. Brussels, 14 December.
Gaffney, Cline and Associates (2014). Shale Gas: North America's economic turbocharger. can the phenomenon be exported? (http://www.gaffney-cline.com).
Gardner, T. (2014). Ukraine crisis new rallying point for US energy export backers. Reuters, 4 March.
Godier, K. (2013). Putin defends Russia's unconventional gas stance. *Newsbase*, 30 April.
Gold, R. and Gilbert, D. (2013). US is overtaking Russia as largest oil-and-gas producer. *The Wall Street Journal*, 2 October.
Gonchar, M., Chubyk, A. and Ishchuk, O. (2014). Hybrid war in Eastern Europe. Non-military dimension. Energy component. Centre for Global Studies 'Strategy XXI', Kiev, November (http://geostrategy.org.ua).
Hague Centre for Strategic Studies/TNO (2014). The geopolitics of shale gas. The Hague, HCSS. No. 17.
Hornby, L. (2014). Coal conversion plants sap China's emissions targets. *Financial Times*, 30 November.
IEA (2011). World Energy Outlook 2011. Paris: IEA/OECD.
IEA (2012). Golden Rules for a Golden Age of Gas. World Energy Outlook 2011. Special Report. Paris: IEA/OECD, June. Available online: http://www.worldenergyoutlook.org/media/weowebsite/2012/goldenrules/weo2012_goldenrulesreport.pdf (accessed 30 October 2013).
IEA (2013). World energy outlook 2013. Paris: IEA/OECD.
IEA (2014a). The United States 2014 Review. Energy policies of the IEA countries. Paris: IEA/OECD.
IEA (2014b). World energy outlook 2014. Paris: IEA/OECD.
IEA (2014c). Russia 2014. Paris: IEA/OECD.

Kuhn, M. and Umbach, F. (2011). Strategic perspectives of unconventional gas in Europe: a game CHANGER WITH implications for the EU's common energy security policies. EUCERS, *Strategy Papers*, Vol. 1, No. 1. London, May 2011, London.

Kuhn, M. and Umbach, F. (2012). Unconventional gas resources: a transatlantic shale alliance?, in: David Koranyi (ed.), *Transatlantic Energy Futures. Strategic Perspectives on Energy Security, Climate Change and New Technologies in Europe and the United States.* (Washington D . C.: Center for Transatlantic Relations, Paul H. Nitze School of Advanced International Studies, John Hopkins University), pp. 207–228.

Leanard, B. and Sautin, Y. (2014). Time for natural gas diplomacy. *The National Interest*, 5 February.

Löschel, A., Moslener, U. and Rübbelke, D.T.G. (2010a). Indicators of energy security in industrial countries. *Energy Policy*, 38, 2010, 1665–1671.

Löschel, A., Moslener, U. and Rübbelke, D.T.G. (2010b). Editorial: Energy security – concepts and indicators. *Energy Policy*, 38, 1607–1608 A.

Mandrasecu, V. (2013). The Economists' Failed Assault on Gazprom. *Russia Beyond the Headlines*, 3 April.

Marson, J. and Parkinson, J. (2013). In reversal, neighbors squeeze Russia's Gazprom over natural gas prices. *Wall Street Journal*, 1 May.

Mitrova, T. (2014). The geopolitics of Russian natural gas. Boston-Houston. Harvard Kennedy School/Belfer Center for Science and International Affairs and the Center for Energy Studies/Rice University's Baker Institute.

Mitrova, T. and Grigoriev, L.M. (eds.) (2014). *Global and Russian Energy Outlook to 2040.* Moscow: Energy Research Institute of the Russian Academy of Sciences/Analytical Centre of the Government of the Russian Federation.

Morse, E. *et al.* (2012). *Energy 2020. North America, the New Middle East.* New York: Citygroup.

Politi, J. (2014). ES energy boom fuels population and income growth. *Financial Times*, 30 March.

Reuters (2014). EU energy chief not ruling out 'worst case scenarios' on energy security. Brussels, 2 September.

Robinson, T. (2015). Russian gas exports to europe drop. Interfaxenergy.com – *NGD*, 17 February.

Rosenberg, E. (2014). Energy rush. Shale production and US national security. Unconventional Energy and US National Security Task Force, co-chaired by Ambassador Paula J. Dobriansky, Governor Bill Richardson and Senator John Warner, Washington DC, Center for a New American Security.

Rühl, C. (2014). The five global implications of shale oil and gas. *Energy Post*, 10. January.

Russian Ministry of Foreign Affairs (2013). Foreign policy concept of the Russian Federation. Moscow, 12 February.

Ryan, M. (2014). America's dash for gas. Interfaxenergy.com – *NGD*, 27 January.

Shale Gas International (2015). Is tight gas the good news Poland has been waiting for?, 18 March (http://www.shalegas.international/2015/03/18/is-tight-gas-the-good-news-poland-has-been-waiting-for/).

Shung-hwan, O. (2014). Shale gas revolution and desperate 'eastward' energy policy of Russia. *Nautilus Special Report*, 4 March.

Stirling, A. (2010). Multicriteria diversity analysis: a novel heuristic framework for appraising energy portfolios. *Energy Policy*, 38, 1622–1634.

Ukraine Energy News (2014). China plans major slowdown of new coal-to-gas projects in bid to cut emissions. 18 December.

Umbach, F. (2010). The EU–China Energy relations and geopolitics: the challenges for cooperation, in: M. Amineh and Y. Guang (eds.), *The Globalization of Energy. China and the European Union'*. Koninklijke Brill NV: Leiden-Boston: Koninklijke Brill NV, pp. 31–69.
Umbach, F. (2011). Energy security in Eurasia: clashing interests, in: A. Dellecker, A. and T. Gomart (eds.), *Russian Energy Security and Foreign Policy.* Abingdon-New York 2011: Routledge, pp. 23–38.
Umbach, F. (2012). The intersection of climate protection policies and energy security. *Journal of Transatlantic Studies*, 10 (4), 374–387.
Umbach, F. (2013a). The unconventional gas revolution and the prospects for Europe and Asia, *Asia–Europe Journal*, 11 (3), 305–322.
Umbach, F. (2013b) The risks for Europe of failing to exploit shale gas resources, Geopolitical Information Service (GIS – www.geopolitical-info.com), 19 March.
Umbach, F. (2013c). European–Russian gas partnership threatens to unravel, in: *Energy Post* (EP – http://www.energypost.eu/index.php/european-russian-gas-partnership-threatens-to-unravel/), 30 September 2013, 4 pp.
Umbach, F. (2013d). Russia seeks new strategies in the rapidly changing European gas market, Geopolitical Information Service (GIS – www.geopolitical-info.com), 24 September.
Umbach, F. (2014a). US LNG to the rescue of Europe? Does a decision on LNG exports to Europe really matter? *Energlobe*, 13 June (http://energlobe.eu/politics/us-lng-exports-to-the-rescue-of-europe).
Umbach, F. (2014b). China's growing hunger for energy resources. Geopolitical Information Service (GIS – www.geopolitical-info.com), 5 September 2014, 4 pp.
Umbach, F. (2014c). Strategic perspectives for unconventional gas in the EU. *Hazar-Caspian Report* 7, Spring 2014, 64–79. (http://hazar.org/content/yayinlar/caspian_report_sayi_07_bahar_2014_860.aspx and http://hazar.org/UserFiles/yayinlar/caspian_report/Caspian_Report7sayi.pdf)
Umbach, F. (2014d). The strategic implications of Russia's record breaking gas contract (with China), Geopolitical Information Service (GIS – www.geopolitical-info.com), 11 September.
Umbach, F. (2014e). EU can beat Russia's threat of new gas supply cuts, Geopolitical Information Service (GIS – www.geopolitical-info.com), 16 June 2014, 4 pp.
Umbach, F. (2014f). Europe faces change and competition on global LNG markets, Geopolitical Information Service (GIS – www.geopolitical-info.com), 2 May 2014, 4 pp.
Umbach, F. (2014g). Russia's gamble over Crimea could deliver major energy rewards and influence. Geopolitical Information Service (GIS – www.geopolitical-info.com), 23 June 2014.
Umbach, F. (2015). The future role of coal: international market realities vs. climate protection? EUCERS Strategy Paper Six, King's College London, May 2015.
Umbach, F. and Raszewsk, S. (2016). Strategic perspectives for bilateral energy cooperation between the Ell and Kazakhstan – geo-economic and geopolitical dimensions in competition with Russia and China's Central Asia policies. Konrad-Adenauer-Foundation/EUCERS, Berlin-Astana, EUCERS Stategy Paper No. 8, February 2016.
Vaughan, A. (2015). UK's shale gas revolution falls flat with just 11 new wells planned for 2015. *The Guardian*, 19 January.
Walstad, A. (2015). EU energy consumption returns to 1990 Level. Interfaxenergy.com – *NGD*, 11 February.

Walstad, A. (2014). Obama vows commitment to US LNG exports. Interfaxenergy.com – *Energy Policy Weekly (EPW)*, 27 March, E1–2.

Wong, E. (2014). In step to lower carbon emissions, China will place a limit on coal use in 2020. *New York Times*, 20 November.

Xin, L. (2014a). Next five years crucial for Chinese climate pact'. Interfaxenergy.com – *NGD*, 14 November, p. 4.

Xin, L (2014b). China's coal-to-gas output to disappoint next year. Interfaxenergy.com – *NGD*, 22 September, p. 6.

Xin, L. (2015a). Sinopec to march on with shale development. Interfaxenergy.com – *NGD*, 5 January.

Xin, L (2015b). China's BCM industry hopes to turn corner in 2015. Interfaxenergy.com – *NGD*, 14 January, p. 6.

Yiping, Z. (2014a). China's latest energy blueprint raises eyebrows. Interfaxenergy.com – *NGD*, 27 November, p. 8.

Yiping, Z. (2014b). China a long way from shale revolution. Interfaxenergy.com – *NGD*, 4 July, p. 6.

3

Unconventional Gas Development in Asia–Pacific: Looking for Common Ground

JUAN ROBERTO LOZANO-MAYA[1]

Introduction

In the last three decades, the development of unconventional gas resources has advanced owing to the combination of rising natural gas demand and energy security considerations. Besides the environmental constraints that have favoured the widespread consumption of natural gas, many countries face concerns that include the decline of domestic conventional gas reserves as well as the inaccessibility of, the dependence on, or the unreliability of external sources of supply. These issues have pushed several countries to strengthen their self-sufficiency in natural gas in several ways, which include the production of their own unconventional gas resources, mostly in the form of coal bed methane (also known as coal seam gas) and tight gas.

Recently, the global interest in unconventional gas has intensified again, largely owing to the addition of shale gas, which, produced on a large scale in the United States, has brought about unprecedented positive effects on the energy security of that country and on energy markets worldwide. Moreover, in parallel with the increasing unconventional gas output in the United States, the publication and subsequent update of preliminary information (EIA, 2013; 2011), which suggested that shale gas resources were better distributed across the world and in much larger volumes than conventional gas, has strengthened the impetus for more extensive development of these resources. These milestones paved the way for several countries to pursue more seriously the development of unconventional gas by enacting policies and implementing actions conducive to similarly positive outcomes.

Nevertheless, the combination of significant challenges and meagre developments outside the United States has hindered aspirations for the development of unconventional gas resources and has led many countries to manage their expectations more realistically. By 2015, many countries had slowed down or ceased their efforts to develop shale gas, and commercial production was ongoing only in Australia, Argentina, Canada and China (APERC, 2015), still at too small a magnitude to accrue any benefits to their respective economies or their energy balances when compared with the United States' experience. By

[1] The author appreciates the kind support provided by Takato Ojimi, President of the Asia Pacific Energy Research Centre.

the same token, and despite their earlier history of development, the large-scale production of coal bed methane and tight gas so far takes place in only a few countries.

These issues have stimulated a more detailed examination of the larger complexity and risks of unconventional gas development as compared with conventional gas but, interestingly, the majority of studies and discussions have remained fairly fragmented and focused at a country level. What are the common risks underlying the experiences of unconventional gas development in different countries? Can these risks be generalised and used for policy-making purposes? Motivated by these questions this chapter aims to identify critical elements shared across several countries that could be combined into a unified policy framework, in order to approach the development of unconventional gas resources from a broader regional perspective.

The scope of the chapter is set within the frame of the Asia–Pacific Economic Cooperation (APEC) mechanism and is primarily concerned with shale gas, mainly because of the recent relevance of this resource in reviving international attention to unconventional gas. Thus the first section of this chapter briefly introduces the status of natural gas in APEC, providing an overview of the unconventional gas resources potentially available in the region and their role in the collective energy agenda. The second section identifies critical areas and factors in the development of unconventional gas and condenses them into a framework proposal that is then used in the third section to review and analyse briefly, but more consistently, efforts on the development of unconventional gas undertaken in Australia, Canada, Chile, China, Indonesia and Mexico. The last section summarises these findings and highlights some insights that could be of interest for other countries pursuing these efforts that are beyond the scope of APEC.

Unconventional Gas Resources in the Asia–Pacific Region

The Role of Natural Gas in APEC's Energy Market

Established in 1989, APEC is a regional economic forum that promotes trade, integration and prosperity among 21 members located in Asia, Eurasia, North America, South America and Oceania. The APEC region has considerable weight in the global natural gas arena as it represents more than half the global production and consumption volumes.

By the end of 2013, the members of APEC consumed nearly a total of 1840 billion cubic metres of natural gas, which accounted for 54% of the worldwide demand. The United States and Russia constituted 62% of this volume, being the largest-consuming members. These countries were followed in decreasing order by China, Japan, Canada and Mexico, which altogether held another 25% (BP, 2015; APEC EGEDA, 2015; IEA, 2014). As shown in Figure 3.1, electricity generation represented 44% of the regional demand for natural gas in 2013. An additional 46% was consumed in end-uses and the remaining 10% was consumed in self-uses or lost in transmission lines. Approximately 90% of the volume of natural gas that went to end-uses was split into roughly equal shares between the industrial and the joint residential and commercial sectors, with barely 9% used as a transport fuel

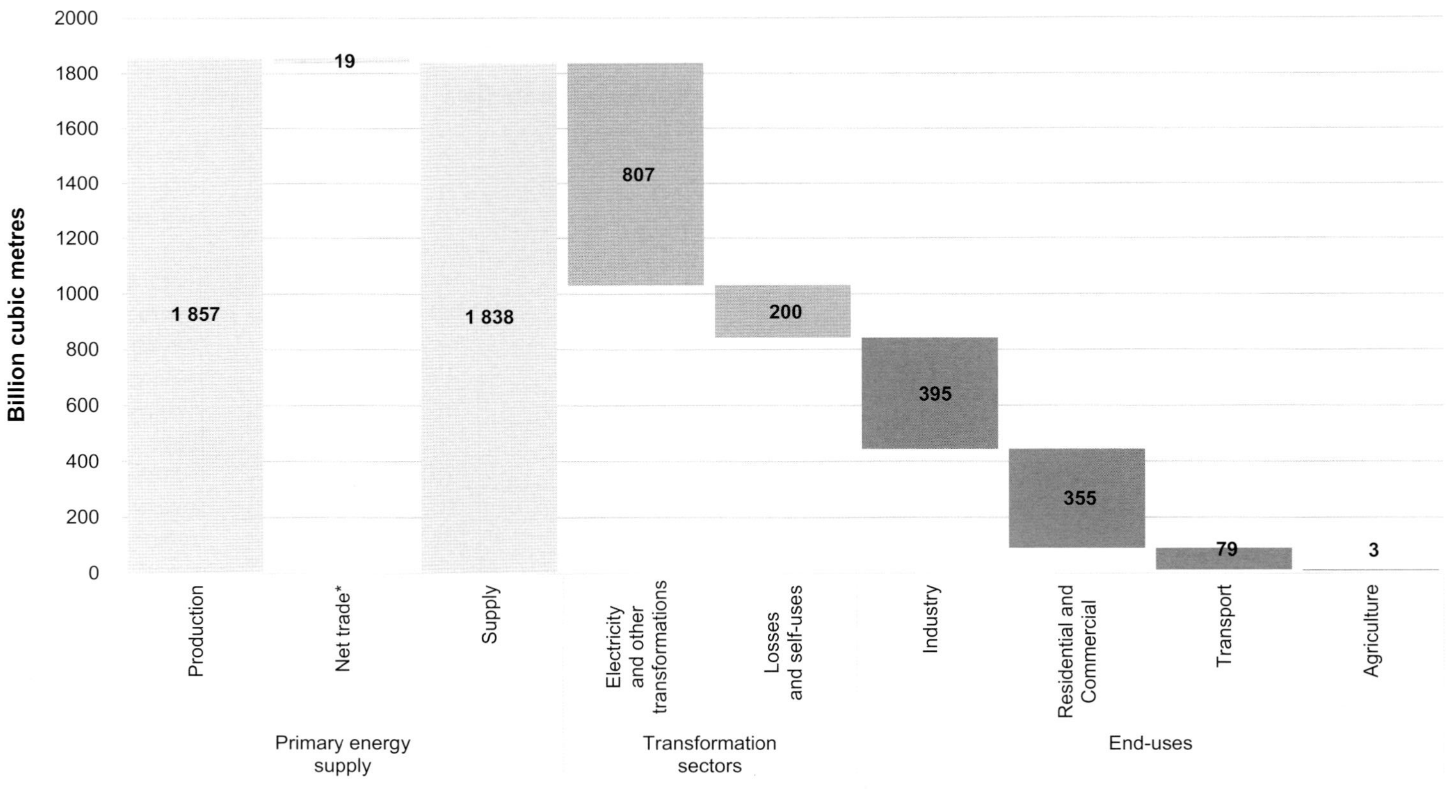

Figure 3.1 Natural gas balance in the APEC region, 2013.Original figures in tonnes of oil equivalent, converted with a factor of 1.11 billion cubic metres of natural gas per 1 million tonnes of oil equivalent. *Sources:* APEC EGEDA (2015) and IEA (2014).

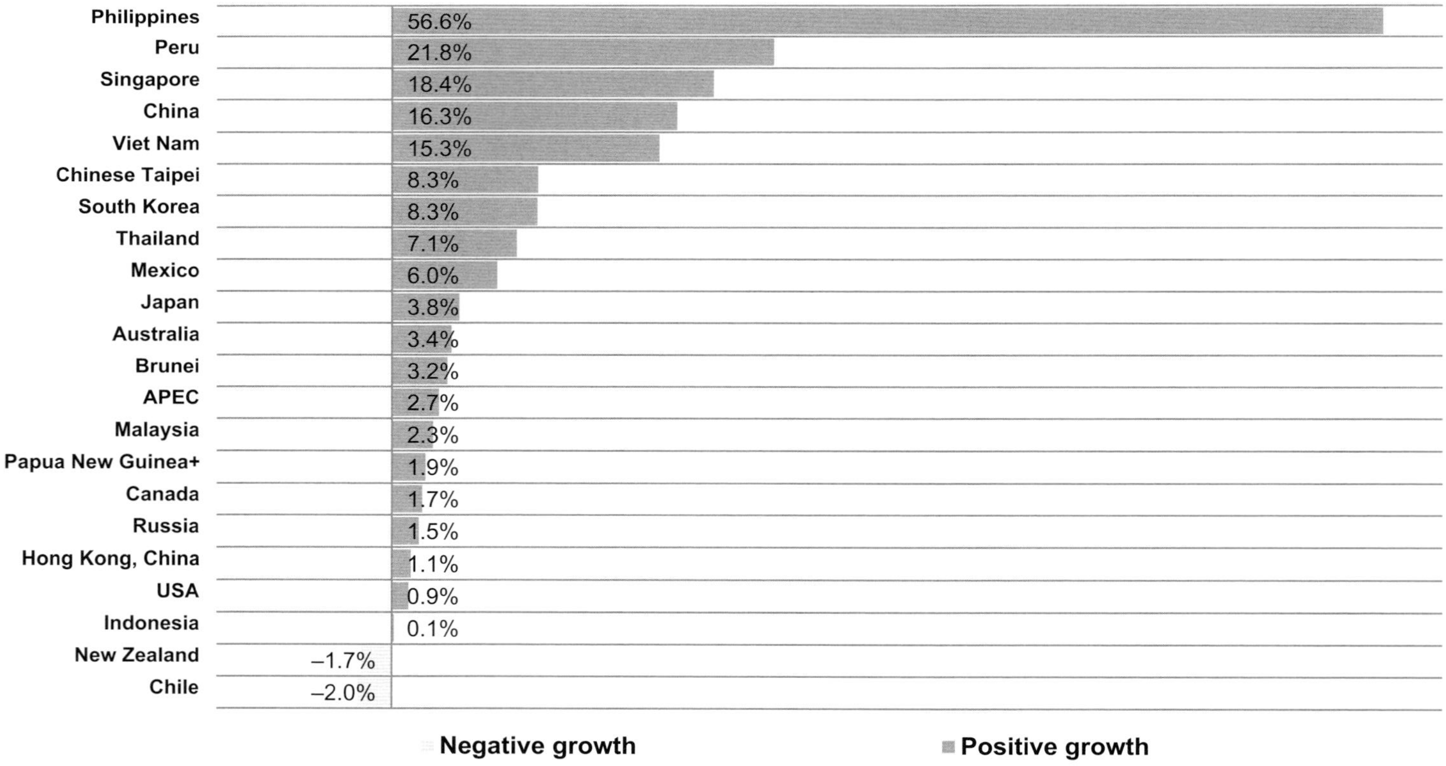

Figure 3.2 Growth in the primary demand of natural gas in APEC, showing regional and country 2000–2013 compound annual growth rates. *Sources:* APEC EGEDA (2015) and IEA (2014).+ For this country the growth rate was calculated for the 2004–2013 period.

and just a tiny fraction used in the agriculture sector. While the single largest destination of the natural gas demand in the region and in most member countries is electricity generation, in Canada, China, Indonesia and New Zealand the largest demand sector was industry.

From 2000 to 2013 the primary demand forf natural gas in APEC grew by 41%, equivalent to an annual rate of 2.7%. Many countries have seen higher growth rates, because their often underdeveloped natural gas markets have given room for a larger demand or because recently they have gained access to expanded supplies that underpin the wider consumption of natural gas. As seen in Figure 3.2, the majority of APEC members have experienced rapid growth in their primary demand for natural gas, usually above the regional average. Only in New Zealand and Chile was there a decrease between 2000 and 2013, which to a great extent is due to the diversification of their respective energy mixes.

The contribution of natural gas within the total primary energy supply stood at 21% for the region as a whole, although on a country basis it ranged from as much as 79% in Brunei and 54% in Russia to as little as 6% in China and 5% in Papua New Guinea (APEC EGEDA, 2015; BP, 2015; IEA, 2014). Business-as-usual projections indicate that the region's predominantly low share of natural gas could remain, as coal and oil will still dominate its future primary energy supply, with a combined share close to 60% by 2035; nonetheless, natural gas has shown a vigorous expansion in recent years and is estimated to become the energy source with the most rapid growth (APERC, 2013).

In terms of production, the output across APEC amounted to a total of 1857 billion cubic metres of natural gas by 2013, which is equivalent to 55% of the worldwide output, very similar to its share in worldwide demand. The United States and Russia each accounted for one third of this volume and, with Canada, China, Indonesia and Malaysia, they made up as much as 90%. With the exception of Singapore and Hong Kong, all APEC members produce natural gas; nevertheless, due to the geographical distribution patterns in their respective endowment of natural resources, there are significant volumetric mismatches between their actual production levels and their gas demand (BP, 2015; APEC EGEDA, 2015; IEA, 2014).

These asymmetries between the locations of natural gas production and demand have become sharper in recent years and have called for an increased trade of natural gas in its pipeline and liquefied forms. By 2013, gas-importing members were the largest group, consisting of Chile, China, Hong Kong, Japan, Korea, Mexico, Singapore, Chinese Taipei, Thailand and the United States. The volume of gas imported in APEC encompassed nearly three quarters of global LNG imports, with the top three LNG importers (Japan, South Korea and China) included as regional members. From Table 3.1, it can be seen that in the last ten years the individual share of gas imports has remained relatively constant only in those countries where production is absent or where it amounts to a very small share of total demand. In countries experiencing a faster expansion of their natural gas demand, their imports as a share of total demand have generally been increasing. The most notable exception is the United States, where self-sufficiency has improved substantially

Table 3.1 *Percentage natural gas self-sufficiency in APEC, showing regional and country results*

Country	2004	2005	2006	2007	2008	2009	2010	2011	2012	2013
Australia	−31	−39	−36	−31	−28	−30	−36	−37	−38	−43
Brunei	−73	−73	−74	−74	−74	−82	−74	−78	−77	−80
Canada	−51	−52	−49	−52	−50	−45	−46	−38	−37	−36
Chile	81	77	76	61	31	44	66	73	76	80
China	−6	−6	−5	2	2	5	12	21	26	29
Hong Kong, China	100	100	100	100	100	100	100	100	100	100
Indonesia	−39	−45	−44	−48	−47	−42	−41	−49	−46	−48
Japan	96	96	96	96	96	96	96	97	97	97
South Korea	100	98	99	99	99	99	99	99	99	99
Malaysia	−45	−42	−40	−38	−35	−38	−42	−40	−37	−44
Mexico	25	20	20	17	22	21	22	26	31	34
New Zealand	SS	SS	SS	SS	SS	SS	SS	SS	SS	SS
Papua New Guinea+	SS	SS	SS	SS	SS	SS	SS	SS	SS	SS
Peru	SS	SS	SS	SS	SS	SS	−28	−45	−43	−48
Philippines	SS	SS	SS	SS	SS	SS	SS	SS	SS	SS
Russia	−31	−31	−30	−29	−28	−28	−28	−29	−28	−30
Singapore	100	100	100	100	100	100	100	100	100	100
Chinese Taipei	93	95	96	97	98	98	98	98	98	98
Thailand	34	34	36	36	32	24	27	24	22	23
USA	16	17	16	16	13	12	11	8	6	5
Vietnam	SS	−22	−23	−8	−3	SS	SS	SS	SS	SS
APEC	**−6.8**	**−7.6**	**−6.2**	**−4.3**	**−4.7**	**−3.9**	**−2.9**	**−2.5**	**−0.3**	**−0.3**

Sources: APEC EGEDA (2015) and IEA (2014).

Negative values refer to net exports as a share of production; positive values refer to net imports as a share of primary demand; 'SS' denotes self-sufficiency in terms of the absence of trade, usually due to infrastructure restrictions.

+ Papua New Guinea started exporting LNG in late 2014.

in the last five years owing to its rising larger domestic production, which has been mainly the result of its robust development of shale gas.

The second largest group is formed by the gas-exporting members of Australia, Brunei, Canada, Indonesia, Malaysia, Peru and Russia. In these countries, the production of natural gas is primarily oriented to exports, with as much as 80% and 43% of their output sent abroad in Brunei and Peru, respectively; nevertheless, their domestic gas markets have been also growing in step with their production. As a whole, the APEC region has also a gas-exporting profile. In absolute terms the combined gas trade of its members amounted to a modest 5 billion cubic metres of net exports in 2013. The third and much smaller group consists of countries such as New Zealand, the Philippines and Vietnam where there are no supply options other than their domestic production, either because their own gas markets are underdeveloped and demand is stagnant or because there is no physical infrastructure such as pipelines or LNG regasification terminals to allow any gas imports. Until 2013, Papua New Guinea was also in this group but in the fall of 2014 its LNG liquefaction terminal started operations and the country is currently exporting a large share of its domestic production.

Unconventional Gas Resources on the Energy Agenda of APEC

Several APEC energy policies have stressed the potential of natural gas to support a cleaner energy transition in the region, reducing the use of oil and coal and the associated emissions of carbon dioxide and other pollutants. From 1998 to 2008, for example, APEC leaders and energy ministers launched several policies to increase investment in supply, infrastructure and cross-border transmission networks of natural gas and to facilitate a more competitive marketplace. However, starting from 2010, the natural gas agenda expanded and became more emphatic about the potential of the region's unconventional gas resources to increase regional production and trade of natural gas and ultimately to enhance the region's energy security.

In accordance with these precepts, in 2010 APEC energy ministers instructed the preparation of an Unconventional Gas Census, oriented 'to evaluate the potential of unconventional resources and to recommend cooperative actions which could increase natural gas output, boost natural gas trade and use, and moderate natural gas prices to the extent appropriate both for producers and consumers in the APEC region' (APEC, 2010). At their respective meetings in 2012, leaders and energy ministers confirmed the collective interest in the development of natural gas markets and of unconventional gas, with particular attention focused on shale resources. This resulted in another call to assess the potential production, trade and environmental impacts that the development of these resources could imply (APEC, 2012a, 2012b).

In 2013 the APEC Unconventional Gas Census was released; it highlighted that at least eight country members held significant resources of technically recoverable shale gas, coal bed methane and tight gas (Table 3.2). For reference, the estimated size of this unconventional gas resource base was approximately 5.8 times the proved reserves of natural gas of all these countries in 2013. A salient feature of this assessment was the comment that shale gas was by far the most abundant type of unconventional gas resource, accounting for 57% of those volumetric estimations.

Also, in late 2013, the Energy Information Administration of the United States released its updated assessment of shale gas resources, covering a wider geographical scope. The results of this assessment validated the widespread distribution of shale gas across the world in very large volumes, updated previous estimations and added the resource base inferred for the APEC members Chile, Russia, Thailand and Indonesia. The results from both assessments are seen in Table 3.3, together with the proved reserves of natural gas for each APEC country inferred as holding shale gas resources. Although technically recoverable resources and the proved reserves of natural gas are not directly comparable, they still provide a comparative reference to the order of magnitude of shale gas potentially

Table 3.2 *Unconventional gas resources in APEC in billion cubic metres*

Country	Shale gas	Coal bed methane	Tight gas	Country total
Australia	11 300	12 400	600	24 300
Canada	2550	1270	4830	8650
China	25 100	10 900	NA	36 000
Mexico	8410	110	NA	8520
New Zealand	NA	54	NA	54
Peru	2070	NA	NA	2070
USA	16 410	3960	14 730	35 100
Vietnam	NA	10	NA	10
Total APEC	**65 840**	**28 704**	**20 160**	**114 704**

Source: APEC EWG (2013).

Table 3.3 *Natural gas reserves and production and shale gas resources in APEC (2013) in billion cubic metres*

Country	Natural gas production	Proved reserves of natural gas	Technically recoverable resources of shale gas	
			APEC 2013	EIA 2013
Australia	58	1219	11 300	12 374
Canada	145	1889	2550	16 226
Chile	1	98	–	1359
China	125	4400	25 100	31 573
Indonesia	74	2955	–	1303
Mexico	45	484	8410	15 433
Peru	12	435	2070	–
Russia	626	47 805	–	8127
Thailand	41	256	–	142
United States	630	10 539	16 410	16 056
Total	1757	70 080	65 840	102 592
World**	**3364**	**185 696**	**65 840**	**203 910**

Sources: APEC EWG (2013) and EIA (2013); for production, APEC EGEDA (2015); for proved reserves, Oil and Gas Journal (2013).

developable in the region. These assessments along with the rising production of natural gas in the United States have led several countries in the APEC region to look forward to the development of their respective shale gas resources, as discussed in the following sections of this chapter.

The most recent initiative in APEC directed to shale gas development occurred in 2014. In line with the surge in shale gas production in the United States, but recognising at the same time the generally poor results seen in some member countries and the controversy surrounding the techniques and procedures employed in this task, APEC energy ministers convened for the pursuit of unconventional gas resources as a means to strengthen the regional gas supply, provided this is carried out following scientific solutions that deal with the environmental issues that could arise (APEC, 2014).

Critical Factors (Dimensions) for Unconventional Gas Development

Compared with conventional gas, the production of unconventional gas resources is more multifaceted, as it involves risks in several domains which in turn interlink a wide array of stakeholders and interests.

The Complexity of Addressing Unconventional Gas Development

Technically, the production of shale gas is highly variable, as the geological properties of shales can vary markedly even within the same basin or play. For this reason, technological capabilities and innovations are more relevant in the development of unconventional gas (Aguilera *et al.* 2014). In addition, the naturally faster decline of shale reservoirs calls for larger number of operating wells in order to maintain production levels.

These characteristics affect the economic profile of shale gas development as projects are typically less cost-effective and require larger amounts of capital over time to maintain production levels in comparison with conventional gas projects. Thus, producers generally deploy scalable, repeatable and modular operations reminiscent of the manufacturing sector, which requires larger land acreage, more intensive drilling and skilled human personnel and, especially, agile logistical coordination with external service suppliers. Overall, this has shaped a different business model (Binnion, 2012).

Owing to the industrial methods employed and the more intensive mobilisation of human and material resources to the well site, the extraction of shale gas increases noise, dust and road traffic, has a more visible land footprint and raises serious environmental concerns, most notably regarding the preservation of freshwater resources during the process of hydraulic fracturing. Additionally, the pervasiveness of shale gas resources and the need to exploit them more intensively have driven extraction processes to areas which are more densely populated and closer to urban centres and which had no previous contact with oil and gas activities. In consequence, shale gas development, and unconventional gas development in general, entail more controversy and conflicts, making their social acceptance and environmental sustainability more challenging and yet more crucial to their success. The novelty of all these activities also imposes learning and institutional costs on regulators, to keep abreast of the changes in the industry and the social demands.

The technical, economic, environmental, social and institutional challenges underlying shale gas development involve a diversity of perspectives that link a wider circle of interdependent stakeholders with different, and frequently, opposing interests. One of the major actors, the government, is normally the legal owner of the mineral resources, the industry regulator and an industry player through its ownership of companies. Government organisations at the federal, state and local levels pursue interests that span maximisation of the economic revenues from the extraction of natural resources, effective regulation, public health, environmental sustainability, energy security and positive externalities.

Another major stakeholder is industry, which, composed of oil and gas companies with a mixture of ownership and control structures that might include national oil companies, international oil companies and small private companies, fundamentally looks forward to achieving profitable operations. To do so, they leverage not only their technological and organisational capabilities but also networks formed with other companies, i.e. oilfield service and equipment providers, with the aim of deploying agile and cost-effective supply chains. For their part, consumers in different economic sectors, and with different sizes and needs, demand a reliable and competitively priced supply of natural gas.

In comparison with the development of conventional gas, the weight of environmental concerns, social licence and local communities in unconventional gas projects is greater. The impact of shale gas extraction involves numerous actors who are directly affected by these projects, as well as others who might not be affected but anyway feel compelled to get involved individually or through civil organisations to demand sustainable development, stricter public health measures, community development or enhanced revenue transparency for unconventional gas projects. Professional organisations, think tanks, research institutions and the communications media also influence or expand the dialogue with and interests of these stakeholders.

Adding to this multidimensionality and complexity, as much as the successful production of shale gas in the United States was the trigger for other economies to embark on the same endeavour, the unique contextual settings in that country have prevented a replication of its experience in other countries, which have accomplished modest results at best. Even in those countries where shale gas production has reached commercial status, such as Canada or China, their respective volumetric magnitude and rhythm of production seem to be nowhere close to the levels and the trend observed in the United States, largely because of different structural conditions (Tian *et al.,* 2014; Council of Canadian Academies Annual Report, 2014). Furthermore, these differences extend to other unconventional gas resources, explaining the slow and elusive results of coal bed methane development in most southeast Asia countries (Andrews-Speed and Len, 2014; Hewitt, 2014; Umbach, 2013).

Some of the main factors highlighted for the success of the United States in developing shale gas are its favourable geology, a competitive industry with wide infrastructure and skilled personnel and a multitude of small firms with the entrepreneurial spirit to produce gas under riskier conditions than larger firms in conventional gas projects. Nonetheless, the most remarkable characteristic of the United States is its predominantly private ownership of oil and gas reserves, which is very rare in other jurisdictions and is likely to be the most relevant single factor explaining why this experience will be difficult to export (Nülle, 2015).

Designing a Combined Policy Framework

In light of the issues examined above, the critical factors or *dimensions* underlying shale gas development will be identified and combined into a generic framework to assess the strengths and challenges of individual countries more homogeneously, ultimately to allow broad comparisons among them. This framework rests on the premise that unconventional gas development will not necessarily follow an identical trajectory in every place, as it is influenced by certain dimensions, as mentioned above, and by elements and stakeholders that are contingent on a country's priorities, contextual settings and experience. Thus, the framework recognises that the pace and scale in the development of unconventional gas resources will be different across the world, as it will be driven by a number of key variables that diverge on a country basis and even within the different regions of a single country.

Thus the framework, adapted from APERC (2015), is based on a holistic approach to shale gas development made up by three major dimensions that refer to (a) access to natural resources, (b) industry operations and (c) governance. The main assumption in the framework is that there is a systemic linkage between these three dimensions, whereby they are all interdependent and necessary; however, it is governance in particular which has the ability to influence the economic and institutional incentives that affect the other two dimensions. In this way, sound governance can foster the development of the industry's technology and infrastructure, which in turn will underpin improvement in the recovery factors that expand the resource base and, especially, the subset of unconventional gas resources with the potential to be commercially developed.

As shown in Table 3.4, the tripartite dimension framework is further divided into finer factors, to accommodate more accurately different countries' experiences. The first dimension refers to the access to natural resources, encompassing shale gas resources but also the water necessary to perform the hydraulic fracturing process for economic production. The argument supporting this dimension is that the existence – presumed or confirmed – of shale gas resources in a country is not a sufficient condition to pursue the development of those resources if there are legal restrictions imposed on that activity or on the methods employed to extract the gas trapped in the shale formations. This explains why countries with an allegedly vast potential of unconventional gas have declined temporarily or permanently its exploitation, not necessarily because they would not benefit from those gas resources but because of other considerations. These decisions, which are heavily influenced by the energy and political agendas in place, help to explain the different positions and legal approvals to shale gas between different State jurisdictions in a given country, as is the case between the eastern and western Canadian provinces.

The second dimension in the framework relates to the elements that allow the industry to extract shale gas economically. The elements embedded in this dimension are a combination of technological development, skilled and sufficient workforce, availability of service companies able to deploy efficient supply chains and infrastructure systems to transport the gas from the production sites up to domestic consumers, or even to foreign consumers in the form of LNG. This dimension also includes some recommended practices to make industry players harness production methods more effectively but especially to minimise several

Table 3.4 *Policy framework with critical dimensions and elements for shale gas development.*

<table>
<tr><th>Dimension</th><th>Element</th><th colspan="3">Additional points to consider</th></tr>
<tr><td rowspan="2">Access to natural resources (R)</td><td>Access to shale gas resources</td><td colspan="3" rowspan="2">• Preliminary assessments of shale gas resources
• Political position on shale gas development
• Type and proximity of water resources</td></tr>
<tr><td>Access to water</td></tr>
<tr><td rowspan="4">Industry operations (I)</td><td>Industry's technological and operational capabilities for shale gas</td><td rowspan="3">• Technological development
• Adequate workforce size and skills
• Presence of IOCs experienced in shale gas development</td><td rowspan="4">B
O
U
N
D
B
Y</td><td rowspan="4">• Market demand
• Access to capital
• Legacy of conventional (and possibly unconventional) oil and gas activities</td></tr>
<tr><td>Oil and gas field services</td></tr>
<tr><td>Gas-to-market and auxiliary infrastructure systems</td></tr>
<tr><td>Recommended industry practices</td><td>• Professional oil and gas associations</td></tr>
<tr><td rowspan="3">Governance (G)</td><td>Dedicated fiscal regime in alignment with the natural gas market structure</td><td colspan="3">• Fiscal provisions or regimes accounting for the risks and productive profile embedded in shale gas development. This is usually dependent on the market profile and the following characteristics:
○ Policies granting equal operating conditions (that is, barring monopolies or certain companies from holding a dominant industry position)
○ Open access to gas transmission infrastructure
• Deregulated natural prices and temporary subsidies</td></tr>
<tr><td>Regulatory effectiveness</td><td colspan="3">• Expertise, enforcement capacity and transparency
• Scientific- and risk-based information
• Holistic and enforceable regulation
• Adaptability to industry shifts and mutable stakeholders' expectations</td></tr>
<tr><td>Multi-stakeholder engagement</td><td colspan="3">• Proactive consultation with diverse stakeholders
• Industry's social licence to operate
• Public access to regulatory and legislative information
• Management of public expectations</td></tr>
</table>

Source: Adapted from APERC (2015, p. 60).

risks, environmental and social. In view of the variety of natural gas market structures in the world that comprise companies privately or state owned, vertically integrated or unbundled and operating under monopolistic or competitive industry settings, certain underlying fundamentals must dominate. These fundamentals are: market demand, whether domestic or external; the ability to raise capital, whether in capital markets or directly through the government's budget; the legacy of prior conventional and unconventional oil and gas activities that are likely to strengthen the industry's infrastructure, technological and operational capabilities for the production of shale gas.

The last dimension refers to the governance between the multiple stakeholders and to their relationships. Governance in this framework denotes the distributing of authority and decision-making between multiple governmental and non-governmental actors in such a way as to increase the efficiency and effectiveness of a political system (Krahmann, 2003). This dimension spans the alignment between the developers and owners of the shale resources through the acknowledgement of the risks and the productive profile of shale formations as reflected in the incumbent fiscal regime; the capabilities to adjust regulatory instruments in step with industry shifts in the pursuit of benefits for consumers and the general public alike; and finally, the search for win–win solutions in the different relationships that the government, industry and civil society have with each other. The civil society is key to companies in their quest to legitimate their operations through a social licence, and also to governments looking to manage public expectations appropriately. In fact, the mismanagement of public expectations has been cited as one of the major reasons of the recent global concern and attention around shale gas (Nülle, 2015; Lozano Maya, 2013).

Absent from this framework is any explicit consideration of environmental issues, even though the environmental footprint of shale gas, and especially its water footprint, is perhaps the most controversial and significant block to shale gas development globally. In this regard, environmental protection is implicitly responded to through recognition of the central importance of access to water resources as well as through the standard industry practices and regulations that at a minimum will include some type of environmental protection measures, notwithstanding other issues equal or more relevant such as fugitive methane emissions. In consequence, the scope and emphasis given to environmental matters will tend to vary in each country.

Country Assessments with a Unified Approach

A brief qualitative and quantitative assessment of certain countries will be given, using the elements in each of the framework's dimensions as a blueprint. The selection of these countries was driven by three criteria: their membership status in APEC; the presence of, or estimations inferring the presence of, shale gas resources in their respective territories; and indications of early development or interest in shale gas development.

The 'qualitative' assessment was made with the aid of a numerical scale that goes from 0 to 3, to grade whether a certain element was absent (0), partially in place, or not fully operational (1), partially in place, but in the process of change (2) or in place, with no major inconsistencies (3). Under this logic the higher the score, the better the conditions for a given country to develop its shale gas resources. To express qualitative assessment more effectively, a *radar graph* with the nine elements assessed was designed for each country and, for the sake of easier identification, these elements were grouped, with different shades of grey pertaining to the dimension to which they belong, visually matching the shades of grey in Figure 3.4. It must be underlined that the quantitative assessments were devised as a visual and numerical reference for comparison between countries, but the value of the framework is higher as an input for qualitative analyses oriented to highlight critical elements for every particular case.

The subsections below include the highlights for Australia, Canada, Chile, China, Indonesia and Mexico. Note that some of these countries are assessed more thoroughly in other chapters of this book.

Australia

Australia has been a long-time producer of natural gas since its first discovery in the western Australia Perth Basin in 1964. With vast conventional and unconventional gas resources that include 15.4 trillion cubic metres of technically recoverable shale gas and place it fifth in APEC and seventh in the world (EIA, 2013), the country has ramped up gas production in the last decade, including the development of coal bed methane, in response to the growing demand of overseas markets. In this way, unconventional gas resources contribute to supporting the remarkable expansion of Australia's natural gas and LNG industries.

Regarding the access to natural resources, Australia grants private companies the right to develop its oil and gas, including unconventional resources; its first commercial shale gas started in late 2012 in the Cooper Basin. Access to water exists and has not been problematic, at least where this milestone in development occurred; however, production has remained marginal and it is unclear whether water resources could support a larger scale of development. Australia's oil and gas industry is experienced in the production of coal bed methane, which began commercial production in 1996 and is currently in the process of adapting its knowledge and capabilities to the challenges of shale gas; nevertheless, the availability of oilfield services and even drilling rigs is compromised, especially in remote areas. Likewise, infrastructure is generally good in the Cooper Basin, with legacy transmission pipelines and gas processing facilities, but that is not prevalent in other potential basins. Industry practices are somewhat limited, probably due to the insignificant scale of shale operations.

Concerning the governance dimension, there are no special fiscal regimes or terms for shale gas production in Australia, although regulations have evolved rapidly, with ample studies and comprehensive rules progressing in the states of South Australia and Western Australia. Multi-stakeholder engagement by governments and the companies aiming to secure their social licence is still a weakness, albeit it has grown in importance not only in the production of conventional gas and coal bed methane but also in other energy areas,

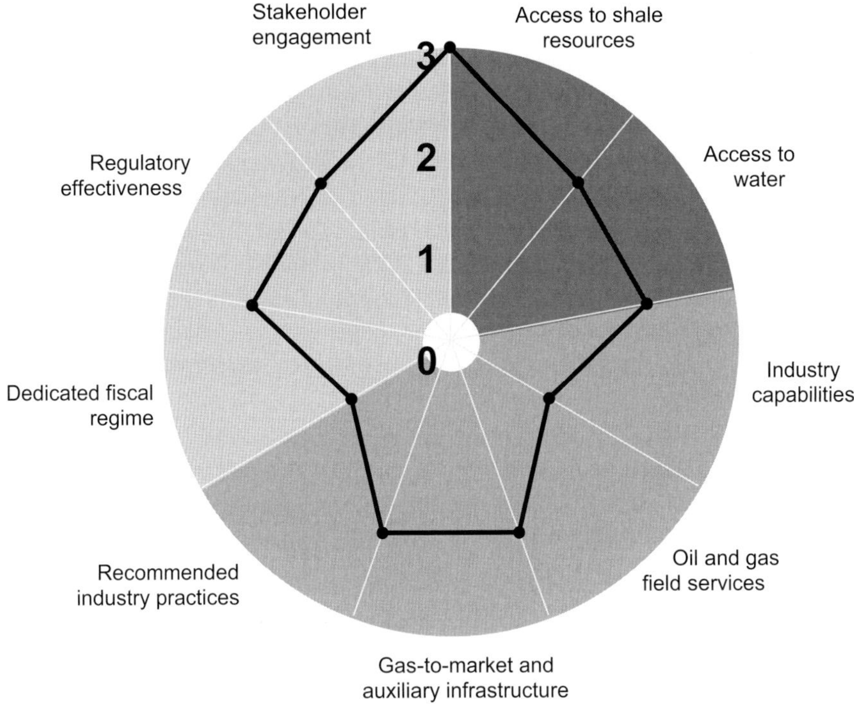

Figure 3.3 Assessment of Australia's shale gas development using the RIG policy framework. *Source:* APERC (2015, p. 71).

which include geothermal and wind energy and mining (Hall *et al.,* 2015). It is worth noting that the development of Australia's promising potential of shale gas depends greatly on the demand of external markets and on the cost-effectiveness of its production as against other conventional and unconventional gas resources. These considerations are condensed in the quantitative assessments shown in the radar in Figure 3.3.

Canada

Canada benefits from abundant energy resources. Unconventional gas resources are particularly significant, holding large amounts of shale gas that place the country at the fourth position globally and the third in APEC (EIA, 2013). Canada's potential tight-gas resources also seem to be substantial (APEC EWG, 2013).

With management and development energy resources concentrated mainly at the province level, the diversity of jurisdictions in Canada has resulted in mixed positions about shale gas development, including disapproval in some eastern provinces. Therefore, the assessment in Figure 3.4 refers only to the provinces of Alberta and British Columbia, two of the very few places beyond the United States where access to these resources exists and shale gas is currently produced. At the moment water access is granted, given support to

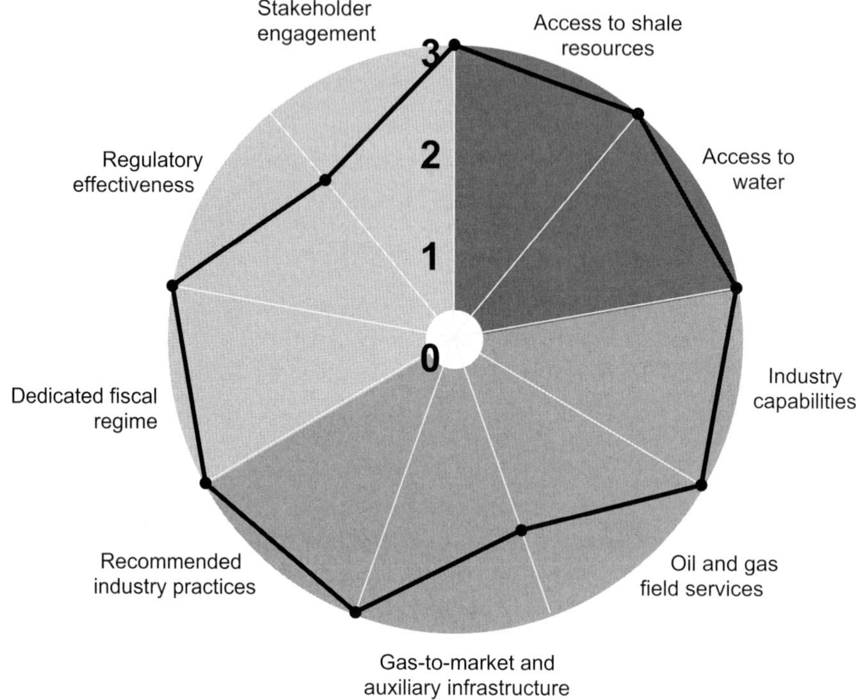

Figure 3.4 Assessment of Canada's shale gas development using the RIG policy framework. *Source:* Adapted from APERC (2015, p. 84).

a legacy oil and gas industry, although regulations in Alberta, for example, are progressively demanding the use of water from non-freshwater sources.

The industry in Canada is robust, and its infrastructure is well integrated physically and technologically with that in the United States, although gas-to-market infrastructure may still be lacking in some shale basins. In addition, while the western United States has been traditionally the main market for Canadian natural gas, the increasing self-sufficiency in the United States is pushing the western Canadian provinces to find new markets, which requires the diversification and expansion of pipelines from producing sites to consumers and, very likely, LNG terminals which at this moment do not exist but will be necessary to ship the gas production.

Alberta and British Columbia have specific fiscal regimes to address the age, type and productive profile of shale gas development; their regulations promote a non-discriminatory forward-looking approach that has introduced new notions such as play-based regulations, which, in contrast with the traditional prescriptive measures, consider the risks and effects resulting from these activities from a cumulative perspective. The engagement of different stakeholders, nonetheless, is still partial in some areas and, specifically for shale gas, the complaints of local residents and First Nation members over consultation processes and unmanaged risks have spurred social tensions regarding these projects (Shaw *et al.*, 2015).

Chile

With very scarce domestic gas resources, Chile is a net importer of natural gas; it became a gas consumer, lured by the abundance of cheap Argentinian imports, in the late 1990s but eventually this experience resulted in a severe gas shortage that taught the country the relevance of energy security the hard way. In the last years, Chile has diversified its energy supplies and it has also explored the possibility of expanding its domestic gas resources through the development of unconventional resources. By 2013, Chile was successful in the production of tight gas and has also signalled interest in possibly developing its inferred shale gas resources.

Chile has an oil gas industry in which ENAP, its National Oil Company, dominates, although increasingly private companies have been allowed to participate in the development of hydrocarbons through special operation contracts; nevertheless, access to its inferred shale gas resources has not been granted to any player, as these resources are still at an exploratory stage. Chile not only has plenty of water resources but also its hydrocarbons-rich Magallanes region is sparsely inhabited, for which water access is not likely to be compromised. Given its small fossil resources, most of them located at the southernmost part of its territory, Chile's oil and gas industry is relatively small, with little availability of oilfield services and equipment. A much more complex consideration is the unusually elongated territory of Chile, with a very long distance between the potential producing shale gas areas and its major gas-consuming centres, preventing the transmission of any future gas output unless a long – and costly – transmission line is built. In fact, the bulk of Chile's domestic gas production is consumed locally, due to the logistic near-impossibility of moving it across its territory.

In terms of governance, Chile has seen some fiscal advances with the introduction of its special operations contracts, but no special fiscal provisions exist for unconventional gas. Consistently ranked as one of the Latin American countries with the best regulations and economic efficiency, the Chilean government has aimed to streamline regulations to make the industry more efficient. Last but not least, there is some modest but active engagement by the oil and gas industry with the local community in Magallanes, which has gained it some level of social support. Because of the huge logistical and operational challenges that its own geography poses, shale gas development in Chile on a level that is significant from an economic or energy security basis is unlikely to happen. Chile's assessment is summed up in in Figure 3.5.

China

The size and energy requirements of China are huge and, despite its rising domestic production of natural gas, its self-sufficiency has been worsening in the last few years. With

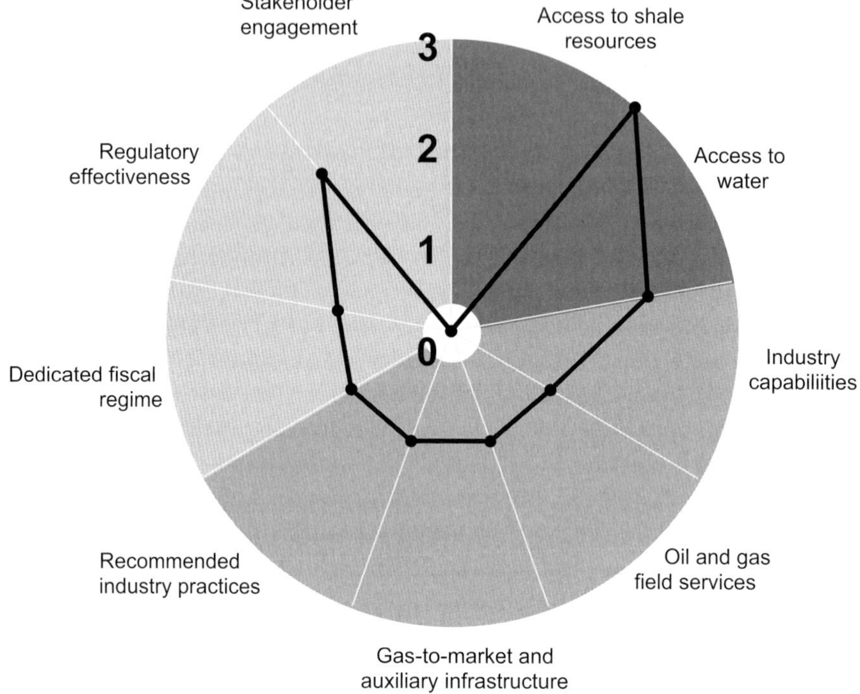

Figure 3.5 Assessment of Chile's shale gas development using the RIG policy framework. *Source:* APERC (2015, p. 93).

the largest inferred resource base of shale gas in the world (EIA, 2013), China was one of the earliest countries to pursue the development of those resources and its commercial output is rising slowly, aided by the success of its Fuling shale gas field in the southeastern part of Sichuan province.

Concerning shale gas, the Chinese government so far has conducted two tenders and signed one production-sharing contract with an international company; nevertheless, some inconsistencies prevail as to the actual legal entity that grants development rights to foreign entities, whether in the form of a production sharing contract, a joint venture or a combination of both (Deemer and Song, 2014). Water access to date has not been problematic, but China's water scarcity next to shale basins and its high population will represent hurdles to achieving a more expansive scenario.

China's oil and gas industry is dominated by three state-owned companies (CNPC, CNOOC and Sinopec), which, owing to their vertical integration, also concentrate the technology, workforce, auxiliary services and infrastructure. Moreover, nearly all the companies legally granted to develop shale gas, at the two tenders conducted to date, lack specialised expertise to undertake these activities. Therefore, China's progress to producing shale gas has occurred at the expense of significant economic losses, slow technological development and little adherence to recommended practices (Tian *et al.*, 2014). In spite of the Chinese

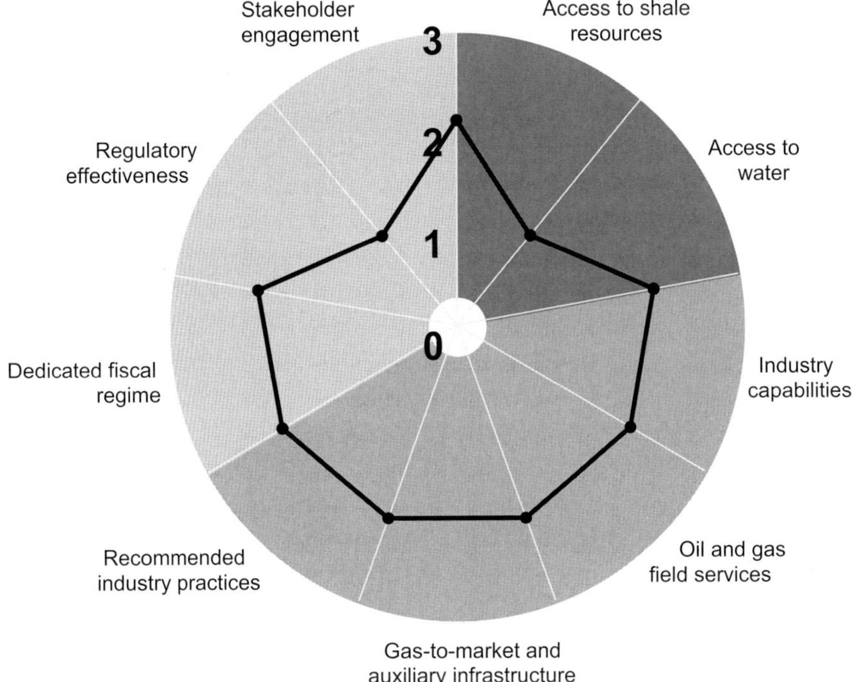

Figure 3.6 Assessment of China's shale gas development using the RIG policy framework. *Source:* Adapted from APERC (2015, p. 105).

government's strenuous efforts, governance elements are still challenging. China has given a special status to shale gas and has implemented a production subsidy; nevertheless, the scope of the existing regulations is still limited, regulatory enforcement is questionable and there is a myriad of agencies with overlapping functions. Multi-stakeholder engagement is the governance element at the lowest point, with some social issues occurring with Sichuan residents as a result of the early shale gas activities (Krupnick *et al.*, 2014). This assessment is visually conveyed in Figure 3.6.

Indonesia

Indonesia has been a major producer and exporter of conventional gas and has also looked forward to developing its coal bed methane resources and, more recently, its shale gas resources. In this sense Indonesia has already granted access rights to the exploration and eventual development of its shale gas resources, with the first production-sharing contract signed in 2013 although details for further bids are in the making. Currently, access to water is ensured but no specific guidelines have been prescribed for its use and management in shale gas development. Owing to an active interaction between the market-dominant national oil company Pertamina and numerous international oil companies, Indonesia's oil and gas industry is well developed in terms of technological and operational capabilities, oil and gas field services, gas-to-market infrastructure and industry practices. However, these capabilities need to adapt to shale gas.

As for governance issues, the fiscal regime implemented is a production-sharing regime which does not recognise the higher risks and financial differences of shale gas over conventional gas and which has also been noted (Andrews-Speed and Len, 2014) as inappropriate to fit the flexibility needed in shale gas operations. Specific unconventional gas regulations were released in 2012, but they face challenges on market access, since they give preferential access to Pertamina and open access to pipelines, which has a dubious enforcement; and, more importantly, the role of the regulator is ambiguous, as it has a transitory nature and it is still attached to the Ministry of Energy and Mineral Resources. Moreover, another regulation passed in 2013 now imposes serious restrictions on hiring foreign personnel, weakening the opportunity to build a skilled workforce that capitalises more effectively on the use of more advanced technology.

Multi-stakeholder engagement is minimal and occurs through certain social corporate responsibility principles enshrined by Pertamina and through professional associations grouping private companies. Nevertheless, this type of commitment to unconventional gas in Indonesia is not proactive, probably because even the earliest efforts directed to coal bed methane in the 1990s have not yet reached a scale that calls for it. As a result, as displayed in Figure 3.7, the governance dimension is the weakest for the development of shale gas in Indonesia.

Mexico

In the last two decades Mexico has rapidly increased its consumption of natural gas; this has been driven by policies targeted to the replacement of its oil-based electricity generation, although domestic production has not been able to keep up and has been stagnant in the last five years. Encouraged by estimates suggesting that the magnitude of its shale gas resources is the sixth largest in the world and the fourth in APEC (EIA, 2013), as well as by the fact that some shale formations are shared with the United States, the Mexican government has hastened shale gas development.

To that end, the Mexican government passed an ambitious energy reform that broke a state-owned industry monopoly of more than 75 years and now gives access to the development of conventional and unconventional oil and gas under competitive market principles. As of late 2015, however, no tenders had been conducted for shale gas and, therefore, no access rights have been granted to any company other than the national oil company Pemex, which has drilled the first exploratory wells. Water access is compromised in certain Mexican regions, especially in the north where the exploration for shale gas is at the most advanced stage.

With the legal ban imposed on private investments in the oil and gas industry since 1938, Pemex became the only player and the government's main revenue source, which over time seriously eroded Mexico's oil and gas industry capabilities at the technical, financial and operational levels. The availability of technology, skilled resources, specialised oil field

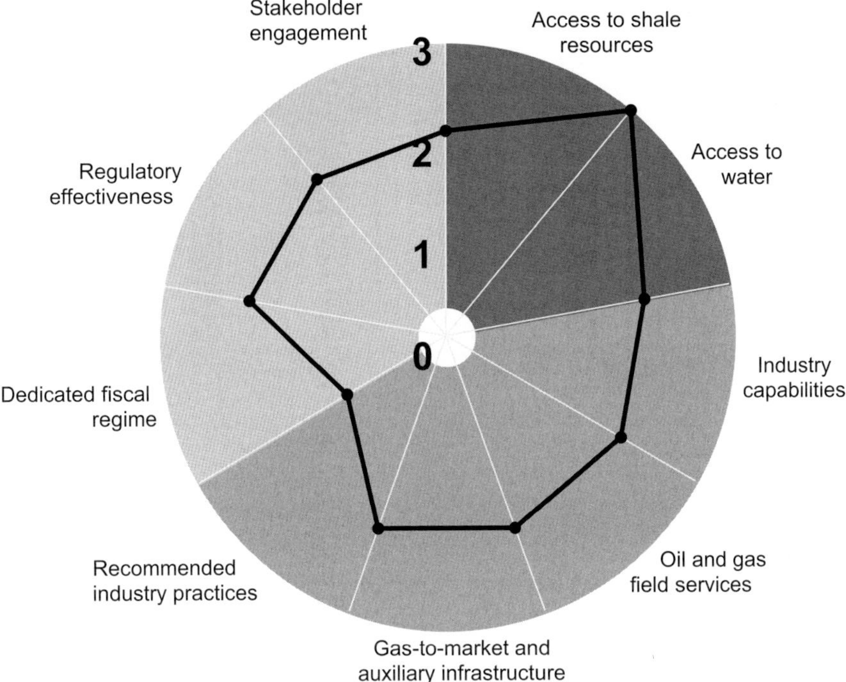

Figure 3.7 Assessment of Indonesia's shale gas development using the RIG policy framework. *Source:* Adapted from APERC (2015, p. 116).

services, infrastructure and guidelines specific for shale gas are still limited and insufficient for its development (De La Vega Navarro and Ramírez Villegas, 2015; Lozano Maya, 2013). The work under way towards full implementation of the energy reform is expected to reverse these major drawbacks but, given the lead times involved and the unfavourable timing under a global environment of low prices in 2015, many oil and gas projects, including those related to shale gas, will be delayed significantly in comparison with the government's expected timeframes (Lajous, 2015).

In the governance dimension, no fiscal regime applicable to shale gas development has been announced yet, although work is under way. In view of the pending status, to a great extent, of the reform's initiatives, the greatest underlying challenge for Mexico is its institutional capacity to enforce those regulations effectively, and avoid endemic corruption, in order to truly bring about more efficiency and benefits. Not included in the framework's dimensions or elements, but more worrisome, are considerations unique to Mexico's context concerning organised criminal activity, namely the theft of oil products in pipelines, which has risen alarmingly in the last few years. So long as this situation continues, some of Mexico's key elements that support shale gas development, especially in the industry and operations dimension, will be constrained.

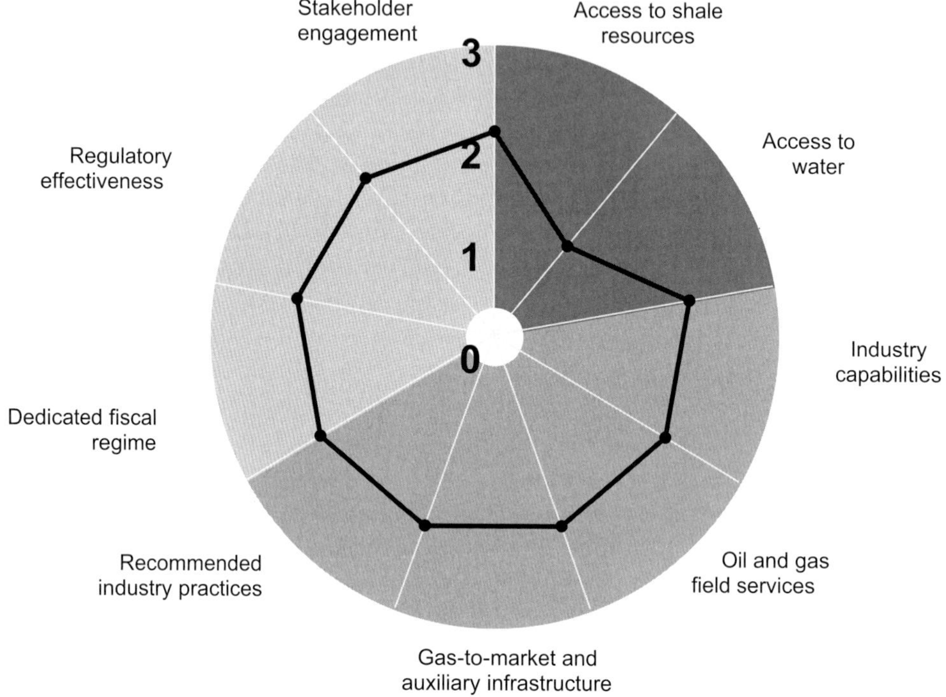

Figure 3.8 Assessment of Mexico's shale gas development using the RIG policy framework. *Source:* Adapted from APERC (2015, p. 132).

Conclusions

In many countries the development of unconventional gas, and in particular shale gas, is not an option but a necessity to meet their natural-gas and energy- security demands. Nevertheless, the development of unconventional gas resources faces more complex risks in comparison with conventional gas. These include political and economic risks from allowing the development of those geological resources and the business and financial risks in deploying enhanced technological and operational industry capabilities in order to stay cost-effective while ensuring the appropriate management of its more obvious environmental and social risks.

A tripartite framework is proposed that focuses on the combination between access to natural resources, the industry's technological and operational capabilities and governance issues. In this regard this framework underscores the major actors, interests and interdependences involved in the development of unconventional gas resources and suggests that enforcing good governance principles will improve the viability of unconventional gas and the chance that such projects will be implemented on a larger scale and with a wider geographical scope. The framework recognises that there are no 'recipes' for shale gas development but, rather, conducive pathways that will result in different time scales and volumetric magnitudes across countries and, in some cases, across regions.

By comprising only those elements which a country can influence or modify, and with the boundaries of some of the three dimensions overlapping with each other owing to their interdisciplinary nature, the framework provides a robust approximate description of the opportunities and challenges for shale gas development. It highlights the need to shift away from a piecemeal approach, paving the way to a more consistent analysis of the gaps between the actual results and the policies implemented in several countries. Thus, the approach used in this chapter should help countries design more effective and realistic energy agendas that, ultimately, foster the development of unconventional gas in a way that maximises its cost-effectiveness, environmental sustainability and social acceptance, with the higher aim of strengthening energy security and promoting mutually beneficial outcomes among diverse stakeholders.

References

Andrews-Speed, P. and Len, C. (2014). The legal and commercial determinants of unconventional gas production in East Asia. *Journal of World Energy Law and Business*, 7 (5), 408–22.

APEC (Asia–Pacific Economic Cooperation) (2010). 2010 APEC Energy Ministerial Meeting. [Online] Available at: http://apec.org/Meeting-Papers/Ministerial-Statements/Energy/2010_energy.aspx [accessed 3 November 2015].

APEC (Asia–Pacific Economic Cooperation) (2012a). 2012 APEC Energy Ministerial Meeting. [Online] Available at: http://apec.org/Meeting-Papers/Ministerial-Statements/Energy/2012_energy.aspx [accessed 3 November 2015].

APEC (Asia–Pacific Economic Cooperation) (2012b). 2012 Leaders Declaration. Annex B – Strengthening APEC Energy Security. [Online] Available at: http://apec.org/Meeting-Papers/Leaders-Declarations/2012/2012_aelm/2012_aelm_annexB.aspx [accessed 3 November 2015].

APEC (Asia–Pacific Economic Cooperation) (2014). http://www.apec.org/Meeting-Papers/Ministerial-Statements/Energy/2014_energy.aspx. [Online] Available at: http://www.apec.org/Meeting-Papers/Ministerial-Statements/Energy/2014_energy.aspx [accessed 5 November 2015].

APEC EGEDA (Asia–Pacific Economic Cooperation Expert Group on Energy Data Analysis) (2015). APEC Energy Database. [Online] Available at: http://www.ieej.or.jp/egeda/database/ [accessed 2 November 2015].

APEC EWG (Asia–Pacific Economic Cooperation Energy Working Group) (2013). *APEC Unconventional Natural Gas Census*. Singapore: Asia–Pacific Economic Cooperation.

APERC (Asia–Pacific Energy Research Centre) (2013). *APEC Energy Demand and Supply Outlook – 5th Edition*. Tokyo: Asia–Pacific Economic Cooperation.

APERC (Asia–Pacific Energy Research Centre) (2015). *Pathways to Shale Gas Development*. [Online] Tokyo: Asia Pacific Energy Research Centre. Available at: http://aperc.ieej.or.jp [accessed 6 November 2015].

Binnion, M. (2012). How the technical differences between shale gas and conventional gas projects lead to a new business model being required to be successful. *Marine and Petroleum Geology*, 31 (1), 3–7.

BP (2015). *Statistical Review of World Energy June 2015*. London: BP.

Council of Canadian Academies (2014). *Annual Report 2013/14*. [Online] Available at: http://www.scienceadvice.ca [accessed 6 Nov 2015].

De la Vega Navarro, A. and Ramírez Villegas, J. (2015). El gas de lutitas (shale gas) en México. Recursos, explotación, usos impactos. *Economía UNAM*, 12 (34), 79–105.

Deemer, P. and Song, N. (2014). China's 'Long March' to shale gas production – exciting potential and lost opportunities. *Journal of World Energy and Business*, 7 (5), 448–67.

EIA (Energy Information Administration) (2011). *World Shale Gas Resources: An Initial Assessment of 14 Regions outside the United States*. Washington: United States Energy Information Administration.

EIA (Energy Information Administration) (2013). *Technically Recoverable Shale Oil and Shale Gas Resources: An Assessment of 137 Shale Formations in 41 Countries outside the United States*. Washington: United States Energy Information Administration.

Hall, N., Lacey, J., Carr-Cornish, S. and Anne-Maree, D. (2015). Social licence to operate: understanding how a concept has been translated into practice in energy industries. *Journal of Cleaner Production*, 86, 301–10.

Hewitt, T.E.G. (2014). A progress report on Indonesian coal bed methane development. *Journal of World Energy Law and Business*, 7 (5), 468–79.

IEA (International Energy Agency) (2014). *Natural Gas Information 2014*. Paris: International Energy Agency.

Krahmann, E. (2003). National, regional and global governance: one phenomenon or many? *Global Governance*, 9, 323–46.

Krupnick, A., Wang, Z. and Wang, Y. (2014). Environmental risks of shale gas development in China. *Energy Policy*, 75, 117–25.

Lajous, A. (2015). *Mexican Oil Reform: The First Two Bidding Rounds, Farmouts and Contractual Conversions in A Lower Oil Price Environment*. New York: Columbia SIPA Center on Global Energy Policy.

Lozano Maya, J.R. (2013). The United States' experience as a reference of success for shale gas development: the case of Mexico. *Energy Policy*, 62, 70–8.

Nülle, G.M. (2015). Prospects for shale development outside the USA: evaluating nations' regulatory and fiscal regimes for unconventional hydrocarbons. *Journal of World Energy Law and Business*, 8 (3), 232–68.

Oil and Gas Journal (2013). Worldwide look at reserves and production. *Oil and Gas Journal*, 111 (12), 32–3.

Shaw, K., Hill, S. D., Boyd, A. D. *et al.* (2015). Conflicted or constructive? Exploring community responses to new energy developments in Canada. *Energy Research and Social Science*, 8, 41–51.

Tian, L., Wang, Z., Krupnick, A. and Liu, X. (2014). Stimulating shale gas development in China: a comparison with the US experience. *Energy Policy*, 75, 109–16.

Umbach, F. (2013). The unconventional gas revolution and the prospects for Europe and Asia. *Asia Europe Journal*, 11 (3), 305–22.

4

Unconventional Hydrocarbons and the US Technology Revolution

MARTIN J. EVANS

Introduction

Oil and gas exploration and development activity has traditionally focused on conventional hydrocarbon resources, with nearly all hydrocarbon production supplied from conventional fields. But in the United States, and around the world, conventional oil and gas fields are becoming more difficult to access and discover and production from existing fields is declining. The rate of discovery for new conventional fields and the size of new discoveries is decreasing. Simply put, it is an increasing challenge to discover large, low cost, conventional hydrocarbon reserves.

The American oil and gas industry has responded by adopting a technology-led approach for discovering and producing new hydrocarbon resources that were assumed until recently to be inaccessible or unrecoverable. Since the early 2000s we have seen an increasing industry focus on unconventional resources, with billions of dollars of investment deployed on exploration and development.

The rise of unconventional hydrocarbon resources has created an American energy 'renaissance' over the past decade. Exploration for unconventional resources has spread rapidly across US states and has continued its progression northwards, into Canadian sedimentary basins. Record amounts of unconventional resources and proved reserves have been found, and US production has surged. Output of both gas and liquid hydrocarbons has reached levels not seen since the 1970s and 1980s. Recoverable gas reserves have increased 25 per cent since 2010 and have doubled in the ten years since 2005. New production from unconventional reservoirs has reversed a long-term trend of declining US output and has changed the future outlook for domestic energy supply. The extraordinary growth of unconventional resource production has reshaped the balance of energy supply and demand, and the US is currently benefitting from some of the lowest energy prices available anywhere in the world. This is having a positive impact on American economic growth at a time when the global economy has struggled to fully recover from the financial crisis of 2008. Structural changes in US energy fundamentals have resulted in decreased reliance on imported oil and gas, and an increased perception of energy security. Less than ten years ago the US was planning a further expansion of liquefied natural gas (LNG) import facilities to meet domestic energy demand. Now it is moving towards becoming an energy exporter to the global markets by reengineering and reversing the flow of LNG import facilities.

The recent increases in US oil and gas production, falling energy imports and the possibility of future energy exports have created an unprecedented situation that few would have predicted at the beginning of the twenty-first century. Yet this is the new reality in which the US finds itself today, some 20 years after Mitchell Energy's first successful production tests from the Barnett Shale in Texas. Yesterday's unconventional hydrocarbons have become conventional. Developing a successful mix of drilling and completion technologies in the Barnett Shale took over a decade of investment and experimentation with horizontal well construction designs and hydraulic fracturing. These technologies are now far more widely adopted in North America than elsewhere in the world, with their sustainability based on a broad consensus of political, economic and environmental support.

The unconventional resource industry has proved to be quick-learning, inventive, adaptable and committed to achieving success. Critically, it has been capable of addressing risk and uncertainty in order to harvest the substantial rewards of unconventional hydrocarbons. The industry process of evaluating geological and engineering risk and uncertainty is built on a wealth of drilling experience, subsurface data and innovative problem solving. It has been successfully applied many times across different sedimentary basins in the US, including well known examples such as the Marcellus, the Eagle Ford and the Bakken unconventional plays.

The American unconventional energy boom has been driven by a key technology breakthrough, perhaps the most significant since rotary drilling was first combined with roller-cone drill-bit technology in the early twentieth century. Two well-established operational practices, horizontal drilling and hydraulic fracture stimulation, have been at the forefront of the technology revolution. Together with large advances in seismic imaging and reservoir characterisation, this combination of technologies has made it possible, for the first time, to find and extract the enormous hydrocarbon resources trapped in low permeability reservoirs. Independent exploration and production (E&P) companies have pioneered the technology advances and at the same time driven down their costs, making unconventional hydrocarbon exploitation commercially viable and more attractive for financing in the marketplace.

The application of this technology to unconventional reservoirs has initiated a dramatic increase in field development activity, especially in North America. Operators and service companies have learnt by experience how to drill, complete and produce unconventional reservoirs and have designed new ways of optimising well completions to boost recovery rate and volume.

The extraction of hydrocarbons from unconventional reservoirs is much more operationally intensive compared with that for conventional oil and gas fields, requiring many more wells to be drilled, completed and placed on production. To bring unconventional natural gas into production and deliver it to customers also necessitates the expansion of infrastructure in virtually all sectors of the gas value chain. Investment in new and upgraded infrastructure is required on a massive scale, and the capital expenditure for a single gas play, for example the Marcellus Shale, alone runs into billions of dollars. The infrastructure demands range from roads and railways to gas gathering pipelines, separation and processing plants and storage tanks. Long-distance gas transmission pipelines are required to transport processed gas from remote fields to regional market centres. Drilling and hydraulic fracturing operations call for large quantities of equipment, water, sand, drilling fluids and cement to be transported into areas that are often remote, necessitating the road systems to be upgraded. Extensive facilities need to be constructed in the field to support wellsite operations, including sand storage units, water storage ponds and tanks and fluid flow lines.

The optimisation of drilling and well completions technology for each unconventional play has advanced rapidly. Directional and horizontal well trajectories using advanced 'geosteering' and logging-while-drilling technology are now commonplace in field development. Using these procedures it is now possible to drill several kilometres beneath the ground surface with sufficient accuracy to penetrate a thin reservoir formation and then steer the borehole to remain within that formation. Increased daily drilling rates combined with less non-productive rig time have led to significant reductions in the number of days it takes to drill a well.

Traditional vertical production wells each require a separate pad. Today, several directional or horizontal wells can be drilled from a single, compact, land surface location (these pad sizes are often less than 1 hectare), and multi-well pad drilling has become a standard practice. This increases operational efficiency and minimises the footprint of development well sites. Typically four, six or eight wells are drilled from a pad, and there are examples of 36 wells drilled from a single multi-well drill pad. Multi-lateral wellbore architecture is also used in unconventional reservoirs and can improve drainage efficiency and hydrocarbon recovery, although it is a less common well design for shale reservoirs.

Hydraulic fracturing has evolved into the primary technology necessary for unlocking oil and gas contained within low permeability reservoirs. Hydraulic fracturing essentially 'creates' permeability in a target formation by pumping fluid into the rock at injection rates that are sufficiently high to break it by fracturing. Thus the technology places induced fractures in the hydrocarbon reservoir rock, enabling sufficient production rates and recovery efficiency on a per-well basis to make field development and production commercially viable.

Hydraulic fracturing is not a new technology. The first well to be fracture stimulated was a gas well in the giant Hugoton field of Kansas in 1947. Since then hydraulic fracturing has become a frequent operational practice to increase productivity and recovery, and this has become especially true for unconventional reservoirs over the last decade. Improvements in fracturing technology have developed rapidly in that same period. Sophisticated computer models are used to monitor and interpret fracture treatments, and multi-stage borehole fracturing has become the industry standard. There have also been significant advances in completion fluid design and in the proppant materials that keep the fractures open and facilitate the flow of hydrocarbons from the reservoir formation to the wellbore.

Three-dimensional seismic surveys have become a vital supporting source of information for exploring and exploiting unconventional plays. State-of-the-art computer technology can process large volumes of data, and analysis of the seismic information enables geoscientists to make predictions about rock depth, thickness, lithology, porosity, fluids and *in situ* stress. In addition, advanced seismic technologies are now being regularly applied to unconventional reservoirs. Seismic attribute analysis[1] can provide insight into reservoir continuity, structure and anisotropy. Seismic inversion technology[2] generates attributes that are sensitive to reservoir lithology, fluid content and the capacity of the reservoir to be hydraulically fractured.

A relatively new technique, the microseismic surveillance of hydraulic fracturing operations, is becoming more widely employed in unconventional field development. Microseismicity results from the geomechanical changes induced in the reservoir as a result of the hydraulic fracturing process, and detecting microseismic events provides a methodology to model fracture growth patterns and fracture size. The location of microseismic events is obtained by using downhole or surface receivers called geophones. Mapping the three-dimensional fracture distribution allows geoscientists and engineers to improve the subsurface earth model for an unconventional reservoir. As confidence in the subsurface model increases, drilling operations, well completions and hydraulic fracture treatments can be improved in order to obtain optimum hydrocarbon production, lower operational costs and reduced environmental impact.

Unconventional Resources

What are ‘unconventional resources’, and what exactly does the term ‘unconventional’ mean? Despite their growing importance and our increasing reliance on them, there is no generally accepted definition of unconventional resources, and no agreed framework for describing them. Geoscientists, engineers, economists and government officials all define them differently and will provide different answers to the question.

One short answer to the question is that unconventional resources are not conventional. Conventional resources are those in which hydrocarbons have accumulated in porous and permeable reservoir rocks and are concentrated in discrete structural or stratigraphic traps beneath an impermeable cap rock and above an identified hydrocarbon–water contact. Conventional oil and gas reservoirs have interconnected flow pathways in their rock matrix and natural subsurface pressure assists the flow of hydrocarbons to the wellbore, usually without the need for additional intervention such as hydraulic fracturing. Unconventional resources, however, are often geographically extensive accumulations of hydrocarbons trapped in very

[1] A seismic attribute is a quantitative measure of a characteristic of interest derived by measurement or computation from the seismic data. Many seismic attributes are calculated from the seismic data and applied to the interpretation of stratigraphy, structural geology and the properties of rocks and fluids. Common examples include amplitude, frequency, attenuation and coherence.

[2] Seismic inversion technology enables quantitative rock property information to be extracted from seismic data. This process converts the original seismic reflection data to a rock property known as the acoustic or elastic impedance.

low permeability rocks, with diffuse boundaries and no obvious hydrocarbon–water contacts. Unconventional reservoirs are characterised by a rock matrix that generally lacks connected flow pathways unless they are extensively naturally fractured. Hydraulic fracturing is required to create permeable conduits that allow hydrocarbons to flow to the well. The presence of a high subsurface pressure is a critical ingredient for improving hydrocarbon production and recovery rates from these low permeability reservoirs.

While the definition of 'unconventional' continues to evolve, a constant theme is that unconventional resources are difficult to recover from the poor-quality reservoir rocks in which they are trapped. I propose that it is useful to adopt a definition that relates unconventional resources to their method of extraction:

Unconventional resources are those that require engineering technology intervention to recover hydrocarbons from the reservoir at commercial flow rates; such intervention is outside the recovery practices that are routinely applied to develop conventional reservoirs and produce oil and gas from them.

Another useful way of defining unconventional resources employs the physical properties of the reservoir rocks and the hydrocarbons that reside in them. In this context, unconventional resources are defined as 'those petroleum reservoirs whose permeability/viscosity ratio requires the use of technology to alter either the rock permeability or the fluid viscosity in order to produce the petroleum at commercially competitive rates' (Cander, 2012).

The generic term unconventional resources includes several types of hydrocarbons, in both the liquid and gas phases. For liquids, the term covers tight oil, heavy oil and tar sands, together with oil shales and shale oil. For gas it comprises tight gas, coal bed methane (CBM) or coal seam gas, shale gas and natural gas hydrate (Figure 4.1). Natural gas hydrate is a solid crystalline compound composed of a mixture of water and methane that is stable under certain conditions of pressure and temperature found in permafrost and continental margin sediments.

There are three main types of unconventional gas resource in production today, including tight gas, CBM and shale gas (Figure 4.2). Tight gas is natural gas trapped in low porosity and low permeability clastic or carbonate reservoirs, and to establish commercial flow rates the reservoir must be hydraulically fractured. Shale[3] is a fine-grained clastic rock that is often rich in organic matter, and shale gas is the natural gas that is trapped *in situ* in petroleum source rocks. In this case shale is acting as both the source rock for hydrocarbons and the reservoir rock. Commercial extraction requires the hydraulic fracturing of the reservoir rock to create permeable conduits for hydrocarbons to flow to the well. Large volumes of gas were also generated during the coal-forming process, some of which remains trapped in the coal as CBM. It is generally produced at relatively shallow burial depths beneath the land surface.

[3] It is important to note that many unconventional reservoirs referred to as 'shale' do not conform to the formal rock-type definition of shale. They are often more complex lithology mixtures such as silty mudstones, dolomitic mudstones and siltstones, and carbonate-rich mudstones. While recognising and acknowledging this lack of rigour and consistency in terminology, the generic term shale is used herein for simplicity.

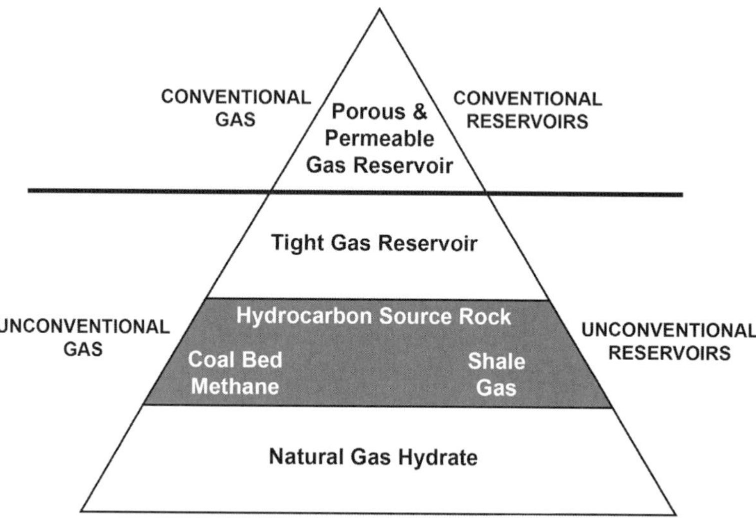

Figure 4.1 Types of conventional and unconventional natural gas. *Modified after* IFP Energies Nouvelles (2012).

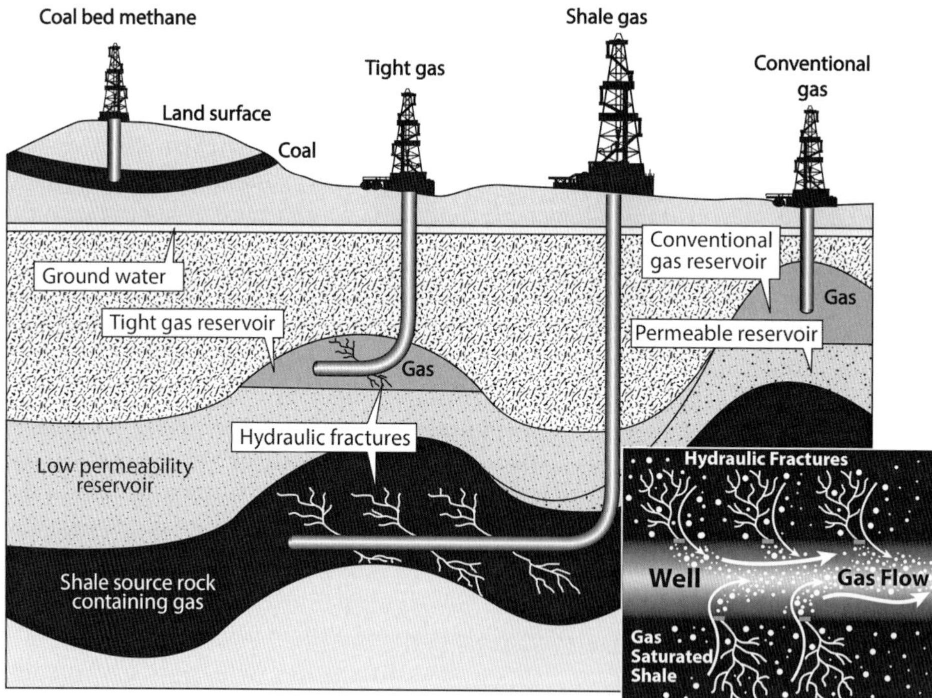

Figure 4.2 The geology and play types of conventional and unconventional natural gas. Tight gas, shale gas and coal bed methane are all considered to be unconventional resources. In low permeability tight gas and shale reservoirs, hydraulic fracturing is required to achieve commercial flow rates and the recovery of hydrocarbons.

Unconventional Resource Workflows

Large amounts, and varied sources, of information are necessary to understand unconventional reservoir rocks and hydrocarbon fluids, and over the past decade the E&P industry has collected, processed and analysed vast amounts of subsurface data. To derive the maximum benefit, and to facilitate the comparison of one unconventional play with another, it is helpful to analyse and interpret this information within a consistent framework or workflow[4] (Slatt *et al.*, 2012). An important objective of the workflow is to develop a detailed knowledge of an unconventional play within its geological context, and to make a full characterisation of the reservoir rocks and fluids. This incorporates information from a range of subsurface disciplines and technologies, including well data, seismic surveys, microseismic surveillance, and samples of both rocks and formation fluids (Figure 4.3).

The integration of subsurface information using a workflow approach allows the field development team to build static and dynamic models of the reservoir, thereby improving the field development plan. This includes identifying the best areas of the play, and optimising horizontal well placement and spacing in the most productive reservoir zones. This process is known as identifying the reservoir's 'sweet spots'. The ultimate objective of the workflow is to provide high quality information to guide decisions about the number of development wells required, optimise the field development strategy and maximise the recovery of hydrocarbons over the life of the field.

During the early stages of exploring new unconventional hydrocarbon plays it is important to recognise the primary geological controls on reservoir performance. Initially there are many geological and engineering variables and uncertainties that need to be assessed, and building an optimal development plan for the recovery of hydrocarbons is highly dependent on describing and understanding the reservoir rock properties, source rock maturity, fluid properties, and subsurface pressure regime (Figure 4.4). In the early stages of the workflow, when data availability is typically low, geological analogue models can provide valuable insight. As more data are gathered, the differences between plays become more apparent and the utility of analogue models generally decreases. Three-dimensional earth models are then constructed to describe and predict reservoir performance, as more seismic, stratigraphic, rock, fluid and pressure data become available. Eventually, when large amounts of subsurface data have been gathered, each unconventional play invariably proves to be unique in its reservoir rock and fluid characteristics.

Advanced Seismic Technology

The development of three-dimensional seismic technology was one of the most important technological breakthroughs in the oil and gas industry. Exxon shot the first three-dimensional seismic survey in 1967 over the Friendswood field near Houston, Texas, and since then tremendous increases in computational power and performance have enabled

[4] In this context, a workflow is a systematic methodology that combines a variety of geoscience and engineering analytical techniques to describe, integrate and understand the properties of unconventional reservoirs at a variety of scales.

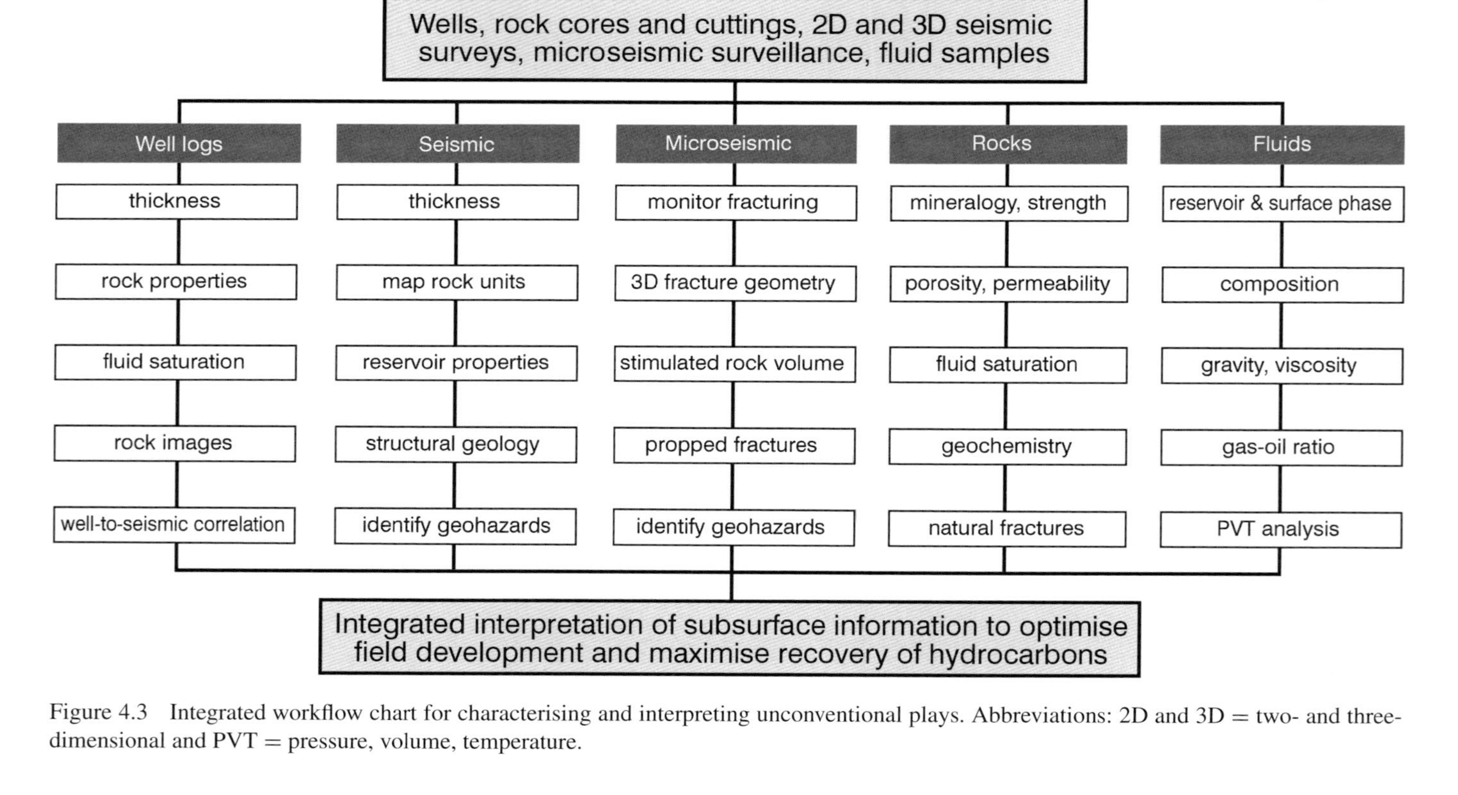

Figure 4.3 Integrated workflow chart for characterising and interpreting unconventional plays. Abbreviations: 2D and 3D = two- and three-dimensional and PVT = pressure, volume, temperature.

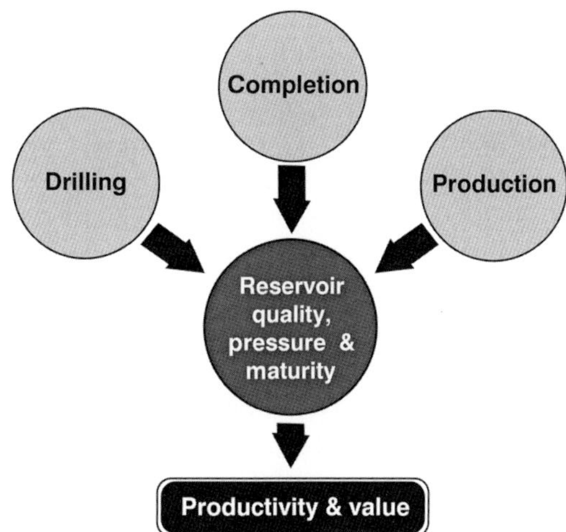

Figure 4.4 Unconventional reservoir performance drivers. Reservoir quality, subsurface pressure and organic matter maturity are primary controls on shale reservoirs. Drilling, well completion and the production of hydrocarbons are secondary drivers that can be controlled, and they may be strongly influenced by the primary controls. *Figure by Tom Miller and published with permission from Anadarko Petroleum Corporation.*

seismic acquisition and processing to advance rapidly. Seismic technology has been able to provide the high quality imaging of large volumes of the subsurface. The technology has become so successful that today it is deployed throughout the entire life cycle of an oil or gas project, from the initial imaging and mapping of exploration prospects through to development and production operations.

Three-dimensional seismic data and related attributes are becoming just as vital for reducing uncertainty and improving hydrocarbon recovery in unconventional reservoirs as they have been for conventional plays. Seismic data have a wide range of applications for improving our understanding of geology and rock properties (Figure 4.3). They enable the interpretation and mapping of structural geology and stratigraphic horizons for locating new wells and can provide valuable information for the real-time geosteering[5] of horizontal wells. When combined with geological data from wells, seismic analysis techniques can also provide information about the sedimentary environments at the time of deposition, and the three-dimensional shape and distribution of reservoir rocks. Predictions of well productivity can be improved by mapping structural geology features in the reservoir such as faults, folds and fractures. For example, mapping different fold styles in the Marcellus Shale gas reservoir of Pennsylvania has been integrated with gas production data to help define the better and poorer areas of reservoir quality. Seismic imagery can therefore be instrumental in locating the sweet spots in a challenging reservoir, and it has proven a cost-effective technology for surveying large land areas.

[5] Geosteering is the interactive steering of the drill bit while drilling is taking place, using real-time geological and directional survey information to adjust the inclination and/or azimuth and precisely position each section of the well. The accurate placement of the well within the reservoir target zone can be adjusted as more geological information becomes available.

A range of useful parameters that provide information about rock and fluid properties between well locations can be derived from three-dimensional seismic data. Collectively these are known as 'seismic attributes'. These attributes tend to be sensitive to lateral changes in the reservoir properties and can reveal relationships and patterns in the data that otherwise might go unrecognised. Attributes such as acoustic impedance and coherence can provide insight into reservoir continuity and the distribution of porosity in the reservoir. Curvature attributes can identify faults and other drilling hazards, and with sufficient resolution may resolve fracture orientation and density in the reservoir zones. Seismic inversion technology has proven to be an accurate and reliable method of quantitative reservoir characterisation, providing attributes that correspond to rock lithology and fluid content. Acoustic and elastic seismic inversion parameters, including impedance, Young's modulus and Poisson's shear modulus, can be indicators of reservoir sweet spots and provide information about the capacity of the rock to be fractured. Seismic data also respond to subsurface stress and can be transformed to give information about stress intensity and orientation.

Reservoir Rock Properties

To deepen our knowledge of unconventional hydrocarbon resources it is important to appreciate the fundamental aspects of reservoirs. These include two important rock properties: porosity and permeability. Porosity, the first essential property of a reservoir, is a measure of the space available in a rock to store fluids such as oil, natural gas and water. The open void spaces in a rock are called pores, and narrow pore throats may connect them. Mathematically, the porosity is the open pore space in a rock divided by the total rock volume (solid + pore space) and is normally expressed as a percentage of the total rock that is taken up by pore space. For example, the amount of pore space within a shale gas reservoir typically ranges between two and ten per cent. Although this is considered to be a low porosity range, it allows a large volume of natural gas to be stored in the rock matrix.

The second essential property is permeability, the capacity to transmit fluids through a porous medium such as rock. Permeability is a quantitative measure of the resistance to fluid flow, and it is a property of both the porous rock and the fluid moving through the rock. It is strongly, but not uniquely, related to the size of the pore throats that connect pore spaces.[6] The common measurement units are the darcy (D) and millidarcy (mD) (10^{-3} darcy), named after the French engineer Henry Darcy, whose experiments with water flowing through sand led him to devise the theory of permeability in 1856.

Conventional reservoir rocks are porous and permeable, typically with a porosity of more than 10 per cent and often exceeding 20 per cent. Their permeability is typically

[6] The physical dimension of permeability is the square of length, being proportional to the square of the pore throat size times a porosity factor (Nelson 2009b). The effective permeability is the ability to preferentially flow or transmit a particular fluid through a rock when other immiscible fluids are present in the reservoir, for example the ability to flow gas in a reservoir that contains both gas and water.

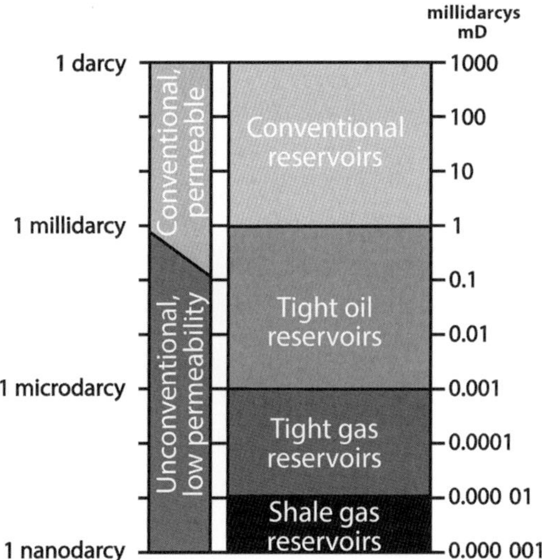

Figure 4.5 Reservoir permeability in conventional and unconventional reservoirs, expressed on a millidarcy scale. Permeability is a quantitative measure of the resistance to the flow of a particular fluid through a porous rock.

in the millidarcy to darcy range, and most conventional reservoirs produce hydrocarbons from rocks with permeability in the range of tens to hundreds of millidarcys (Figure 4.5). Reservoirs with a permeability of one darcy and above possess excellent fluid flow characteristics. As a result of their rock property attributes, conventional reservoirs readily transmit hydrocarbon fluids through a network of pores interconnected by pore throats and are generally characterised by good productivity.

Unconventional reservoirs, however, tend to be of a uniformly poor quality with low porosity and/or low to ultra-low permeability. The porosity is normally less than ten per cent and may be less than five per cent, with an upper threshold permeability of around one millidarcy although it is typically much less. As a consequence of their low permeability these reservoirs have a significant resistance to fluid movement and are described as 'tight'. Tight-gas reservoirs, by contrast, are characterised by microdarcy (μD) (10^{-3} millidarcy) permeability rock, which is less permeable than concrete; increasing amounts of gas are being produced from ultra-low-permeability, nanodarcy (nD, 10^{-6} millidarcy), shale reservoirs (Figure 4.5). For example, the measured permeability of core samples from the Barnett, Eagle Ford and Marcellus shale reservoirs is on the order of tens to hundreds of nanodarcys (Heller *et al.* 2014). This extremely low permeability makes it very challenging to release the hydrocarbons trapped in the pore spaces.

While permeability is an essential concept for characterising unconventional reservoirs, it is also instructive to consider the size range of pore throats. Pore throats are the narrow, tortuous, pathways that connect pore spaces in a rock. Measurements of the absolute size range of pore throats in the matrix of a rock provides important insight into the nature of

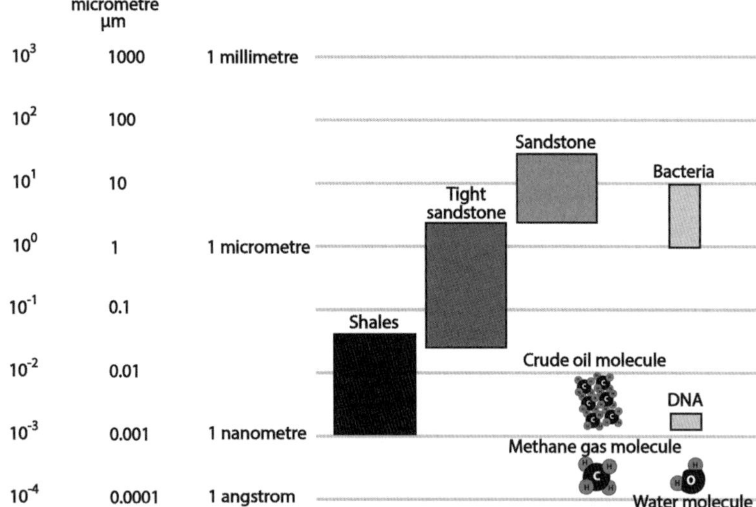

Figure 4.6 Pore throat size ranges for porous sandstone, tight gas sandstone and shale reservoirs, shown together with the molecule size for water, methane gas and crude oil. Also shown for comparison are the complex molecule deoxyribonucleic acid (DNA) and bacteria cells. The diameter, width, or size measured is in micrometres (microns, µm). *Modified after* Nelson (2009a).

unconventional reservoirs (Figure 4.6), demonstrating that there is a continuous pore throat size spectrum between conventional and unconventional reservoirs (Nelson 2009a, 2009b). In clastic rocks this continuum ranges in size from the submillimetre to the nanometre scale.

Advanced technology allows the imaging and measurement of pores and pore throats and includes field-emission scanning electron microscopy (FE-SEM), backscatter electron (BSE) imaging, micro-CT imaging, mercury injection measurements and computational chemistry. Digital rock technology is greatly improving our knowledge of pore types and pore networks in unconventional reservoirs (Loucks *et al.* 2012). These advanced techniques are providing new information about the fabric, compositional variation and pore types in very fine-grained rocks (Figure 4.7).

Reservoir Technology

Advanced reservoir characterisation methods have been developed to address the challenging geology and production mechanisms of unconventional reservoirs. Geoscientists and engineers use a wide range of technologies to describe and quantify the subsurface geology, including seismic information, well logs, core analysis and fluid property measurements (Figure 4.3). The initial exploration wells in an unconventional play satisfy a key objective, namely to gather large amounts of new rock and fluid information. To achieve this, advanced

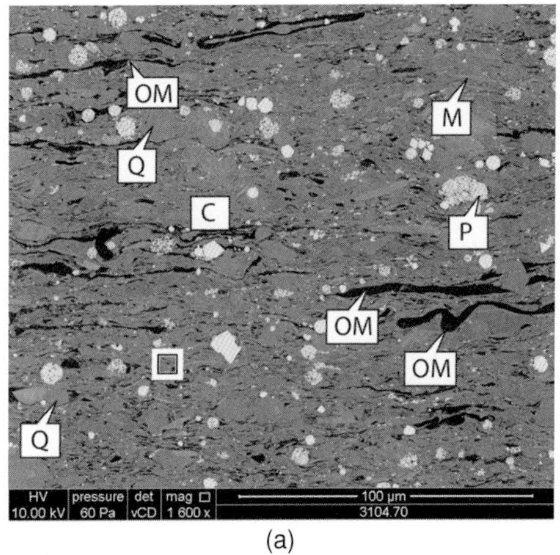

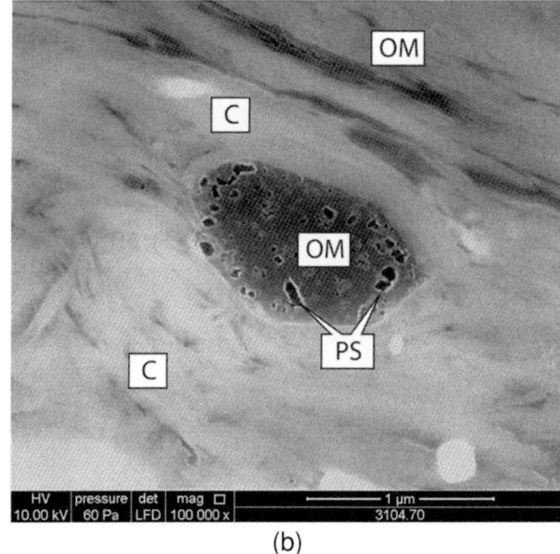

Figure 4.7 Field-emission scanning electron microscope (FE-SEM) images of a shale reservoir. The rock matrix consists of clay minerals (C), kaolinite, chlorite, illite, smectite, and mica. (a) FE-SEM backscatter electron image showing a heterogeneous texture, with laminations of organic matter (OM), quartz grains (Q), mica flakes (M), and pyrite (P). The white box highlights the area enlarged in image (b). The scale bar is 100 microns (µm). (b) FE-SEM secondary electron image showing a rounded organic-matter particle in the centre and elongate organic matter layered between detrital clay in the top right of the image. Note the nanodarcy-size pore spaces (PS) in the organic matter. The scale bar is 1 µm. *Sample provided by Anadarko Petroleum Corporation, with images courtesy of Core Laboratories.*

wireline logging technology is deployed to collect data over the formations of interest, providing measurements of parameters such as reservoir porosity and permeability, hydrocarbon type, fluid saturation and *in situ* stress (Table 4.1).

The cutting and recovery of full-diameter borehole cores from target reservoir formations is typically included in a full data acquisition programme. Logging-while-drilling technology allows accurate placement of the coring operation, and ideally a continuous core is cut across the entire reservoir interval. After the core is recovered it is transported to a laboratory facility where it is cleaned, photographed and then analysed using a suite of reservoir characterisation technologies (Table 4.2). A complete geological description is undertaken, identifying the rock types, sedimentary structures, the presence of natural fractures and diagenetic features. A basic ingredient of a full reservoir description is the analysis of core plug samples. This provides a direct measurement of rock porosity, permeability and fluid properties and helps to calibrate the wireline log information and seismic data.

Drilling

Oil and gas wells are drilled with a rotary drilling rig. There are a number of different design types, varying from traditional vertical wells to directional and horizontal wells

Table 4.1 *Well logging technology and the types of information that it can provide*

Well log technology type	Information provided by the technology
Lithology and density tool	Rock density, lithology
Compensated neutron tool	Porosity, lithology
High resolution induction resistivity tool	Hydrocarbon-bearing zones, water saturation
Dielectric tool	Water saturation
Full wave dipole sonic tool	Porosity, *in situ* stress, total organic carbon content
Natural gamma ray spectroscopy tool	Total organic carbon content, clay content
Caliper tool	Borehole size, shape, rugosity
Inclinometry tool	Borehole azimuth and angle, formation dip and angle
High resolution wellbore imaging tool	Sedimentary facies, lithology, natural fractures, thin bed studies
Nuclear magnetic resonance tool	Porosity, immoveable bound fluids, moveable fluids
Geochemical logging tool	Mineralogy, total organic carbon content, elemental analysis, lithology
Borehole seismic tool	Vertical seismic imaging, well-to-seismic-survey correlation

Table 4.2 *Rock core technology and the types of information that it can provide*

Rock core technology	Information provided by the technology
Visual core description	Lithology, sedimentary structures, fossils, diagenetic features, natural fractures
GRI crushed sample analysis	Grain density, matrix permeability, gas-filled porosity, gas saturation
Geochemistry analysis	Total organic carbon, Rock-Eval pyrolysis, thermal maturity assessment
X-ray diffraction	Mineralogy, clay type, clay abundance
Thin section petrography	Mineralogy, fabric, texture, grain size, shape and sorting
Scanning electron microscopy	Microimagery, identification of mineral grains, clay type, organic matter, pores and pore throats
Geomechanical properties	Triaxial compressive strength testing, acoustic velocity, unconfined compressive strength
Well completion testing	Fluid sensitivity assessment, hardness testing, proppant embedment, acid solubility
Nuclear magnetic resonance	Fluid identification, pore size, permeability, porosity, free fluid index, bound water volumes
Laser particle-size analysis	Grain size distribution, the sorting of sand, silt and clay-sized grains
High pressure mercury injection	Capillary pressure, porosity, pores and pore throat size distribution, matrix permeability

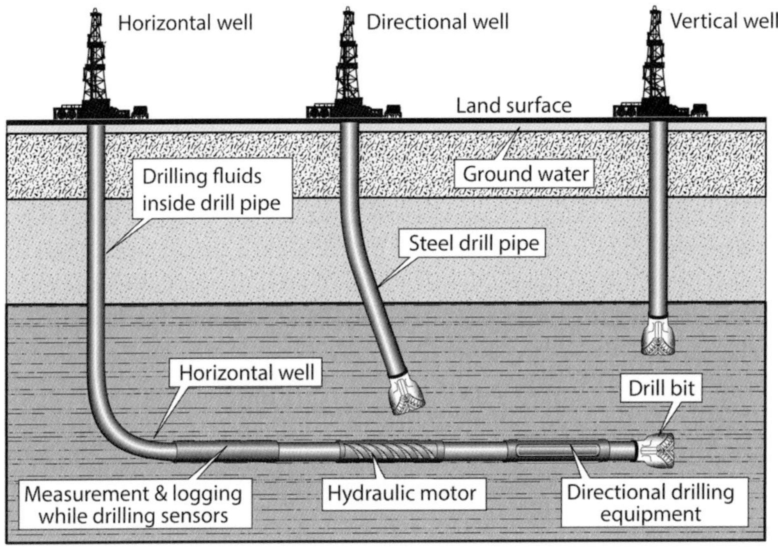

Figure 4.8 Types of well design used by the oil and gas industry for exploring, developing, and producing hydrocarbon resources.

(Figure 4.8). Boreholes are lined with steel tubing, known as casing, which serves multiple functions. It provides stability and integrity to the wellbore, it isolates and protects fresh water aquifers and it prevents small fragments of the surrounding rock formations (cavings) sloughing from the borehole wall into the well. In a typical well design there are usually three or four separate strings of casing. The first string is the surface casing, which is run to a depth beneath the deepest groundwater aquifer to protect it from contamination by the drilling, completion and production fluids and to prevent any fluids leaking to the surface. The surface casing is cemented in place by pumping cement down the casing string and forcing it out through the bottom of the casing, using pressure to circulate the cement back up the annulus space between the casing and the borehole wall. The well is then drilled deeper, and additional casing strings are run as needed and cemented in place. The casing again isolates the well fluids from native fluids in the rock formations, and prevents caving. It also facilitates the unimpeded passage of drilling assemblies and wireline logging tools.

Cementing the casing string secures it in place in the well, and the cement forms a second barrier between the formation fluids and the drilling mud. For reasons of well integrity and safety, it is critical to perform a robust cement job. The cement is pumped as a slurry, the design and properties of which vary between wells and depend on factors such as well depth, formation temperature, and pressure. It is critically important that voids are not present in the cement and that it cures rapidly, to prevent formation fluids under pressure from invading the cement and compromising its integrity. To test the cement's integrity the driller performs two pressure checks, the first ensuring that the casing string can hold pressure and the second measuring the strength of the rock formation to confirm that it can withstand high pressure without breaking down. The second formation test also verifies that the cement will remain stable under high pressure.

In 1866 one of the earliest shallow wells, drilled in east Texas, used the mechanical principles that are now embodied in the rotary drilling rig. The rig utilised an auger that was rotated by means of a large wheel driven by a steam engine. It was the first producing oil well completed in the state of Texas. It was followed by a large oil and gas discovery at Corsicana in 1894, where, four years later, the advantages of rotary drilling technology were first demonstrated. By 1900, over 600 wells had been drilled in the Corsicana field using the first rotary rigs. Thus the introduction of rotary rigs in Texas pioneered a practice that was entirely new in comparison to the cable tool drilling methods being employed in the Pennsylvania fields.

The proven effectiveness of rotary drilling was soon applied elsewhere in Texas. A Chapman rotary rig equipped with draw works to handle the drill pipe was transported from Corsicana to a site near Beaumont, and drilling was commenced in October 1900. Here, on 10 January 1901, during operations to change the bit, the drill pipe was blown out of the well and up through the derrick; it was followed by a mixture of drilling fluid, sand, and gas and then a column of black oil. The famous Spindletop oil discovery had been made (Rundell, 1977; Warner, 2007).

In the Corsicana field, borehole water circulation was employed to cool the drill bit and return rock cuttings to the surface. It was not successful at Spindletop, where a more viscous drilling fluid was required to entrain the cuttings and transport them to surface. With this in mind the drill crew ploughed the bottom of a nearby shallow water pond and drove in a herd of cows. The experiment worked: the resulting viscous mud proved a much more efficient fluid for improving the returns of cuttings. From this time forward viscous mud became the drilling fluid of choice in rotary drilling operations and, as with many innovations, it possessed more attributes than originally intended. Most importantly, the driller could control the mud density in the borehole to counteract formation fluid pressure, preventing the uncontrolled influx of hydrocarbons into the well. This led to significant improvements in well site safety. The mud served as a lubricant and coolant for the rotating drill bit and, because of its viscosity, it was efficient at entraining and transporting rock cuttings away from the bit. The mud also lined the inside of the borehole with a 'mud cake', improving its stability. As rotary drilling experience increased, artificial drilling muds were formulated to meet specific operational requirements (Rundell, 1977).

Despite gradual improvements in drilling technology, the early drag bits, consisting of rigid steel blades shaped like a fishtail, were not very efficient. In 1909 Howard Hughes invented steel roller-cone drill bits and completely revolutionised rotary drilling technology. The roller-cone bit has conical cutters or cones with spiked teeth around them, and it enabled rigs to drill more rapidly through rock formations and to reach increasingly greater depths. By 1914 rotary drill rigs using roller-cone bits had been deployed in 11 US states and 13 other countries, and it went on to become the standard oil and gas drilling technology worldwide. Today, while around 80 per cent of oil and gas wells are drilled with fixed-cutter bits, traditional roller-cone technology is still undergoing development and is valued for faster rates of formation penetration and lower drill cost per foot. Rotating polycrystalline diamond cutter bits are the preferred bit technology for shale reservoirs, and hybrid drill bits, which have characteristics of both roller-cone and fixed-cutter bits, are also being more frequently used.

Horizontal Drilling

Horizontal drilling involves the steering of a well trajectory in a planned direction or azimuth, or a combination of azimuths, to a predetermined reservoir target zone at some horizontal distance from the surface well location. The first part of the well is drilled vertically, and then a gradual curve is introduced into the well path such that by the time the well intersects the reservoir zone its trajectory is horizontal (Figure 4.8). Reservoirs that are not located directly beneath the rig footprint can thus be accessed, for example in areas where drilling is impossible owing to surface land use or prohibition by regulations.

Horizontal wells have a number of advantages over traditional vertical wells, as they extract hydrocarbons more efficiently and more economically. They increase the effective contact length of the borehole in the hydrocarbon-bearing reservoir and by doing so they improve flow rates and hydrocarbon recovery on a per well basis, reducing the number of development wells. Horizontal well recovery is usually three to five times more than that for a vertical well but is highly variable between unconventional plays, ranging from two to as much as 20 times over the life of the well. Productivity can also be enhanced by steering the well in a direction that intersects natural fractures in the reservoir, which is usually achieved by drilling orthogonally to the dominant fracture azimuth. Drilling multiple horizontal wells from a single drill pad can drain a larger area of the reservoir, thereby reducing the surface footprint of a field development area. The environmental, operational, and economic advantages today dictate that the majority of development wells are drilled horizontally.

The basic methods of directional steering evolved in the 1920s when the first borehole survey instruments were developed. The surveys indicated that in reality many wells deviated from a vertical trajectory, and to address the problem drillers invented new techniques to maintain a vertical well path. In 1929 John Eastman experimented with controlled directional drilling in the Huntington Beach field near Los Angeles, California. Tapered wedges were positioned downhole to enable deliberate deviation of the well, and by 1932 directional wells were being drilled from the Huntington beachfront to reach oil reserves located offshore.

Directional drilling became more widely utilised after 1934, when it was applied as an intervention technology to a blown-out well in the Conroe field of Texas (Rundell, 1977). Improved survey instruments and the use of a tapered wedge in the borehole allowed a directional relief well to intersect the high pressure reservoir, and water was pumped down to kill the flow of hydrocarbons into the burning well.

The Conroe relief well marked a new milestone and remained a standard in directional drilling technology until a novel approach was developed in the Talara fields of offshore Peru. Here, a tilting drilling derrick capable of starting a new well at a range of angles varying from vertical to a maximum of 30 degrees was devised and installed. The 'tilt rig' was installed on an offshore platform and the starting well trajectory was oriented directly towards its reservoir target. Using the tilting rig technology the field operator was able to drill more development wells from a single platform, allowing for a more economic field development (Storm, 1968).

The use of horizontal drilling technology became commercially viable in the late 1980s (US Energy Information Administration, 1993). It was still a relatively niche technology, confined to a limited number of productive formations, but success in the Austin Chalk of Texas and the Bakken Shale of North Dakota encouraged more widespread horizontal drilling tests. The first significant advances in the technology had been made in the early 1980s by the French company Elf Aquitaine, which drilled a horizontal test well in the Rospo Mare field oil field, in the Italian sector of the Adriatic Sea. The horizontal well showed improved productivity compared with vertical wells, and indicated for the first time that horizontal drilling technology could become commercially viable. British Petroleum drilled the first horizontal wells in Alaska's Prudhoe Bay field between 1985 and 1986 in a successful attempt to minimise unwanted water and gas invasion into the productive oil reservoir. By 1992 there were 29 horizontal wells in the field. Well performance data showed an increase in production rate (by two to three times) and accelerated oil recovery compared with vertical production wells (Broman and Schmohr 1992).

The horizontal drilling tests were facilitated by improvements in downhole drilling motors and telemetry equipment, and since then drilling motor technology has improved to a degree that enables precise location of the well trajectory. Drilling motors combined with rotary steerable systems have been introduced, providing a way of steering wells with greater efficiency and accuracy. Today, hybrid rotary steerable systems are able to pivot inside the well and navigate the drilling assembly in the desired direction, enabling the directional leg of the well to be initiated and changed at a greater depth than was previously possible and also allowing faster transitioning of the vertical well to a horizontal trajectory. The optimisation of drilling technology for each unconventional play, with increased drill penetration rates and less non-productive rig time, has led to significant reductions in the time needed to drill a development well. An extraordinary example of this drilling efficiency comes from the Eagle Ford Shale play of south Texas, where horizontal wells have been drilled to completion in less than 10 days. This results in significant cost savings for the operator.

Horizontal wells are now routinely drilled with sufficient accuracy to penetrate a thin reservoir zone several kilometres below the ground surface. Advanced well survey and logging-while-drilling technology allows the lateral leg to be geosteered and remain within the target zone within very small target tolerance limits. The height tolerance for accurately penetrating and staying within zone is often 10 m or less, and some horizontal wells have been drilled to a target tolerance of plus or minus 0.5 m, although such small height

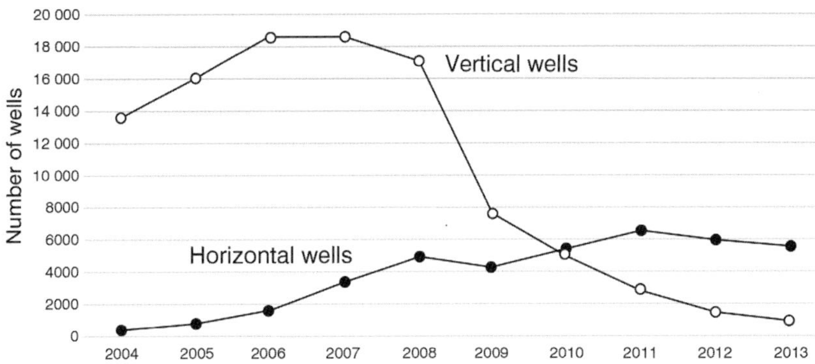

Figure 4.9 For comparison, the numbers of vertical and horizontal gas wells drilled in the US between 2004 and 2013.

tolerances can increase well costs since the rate of penetration has to be slowed in order to assure accuracy.

The clustering of wells onto a single surface location is referred to multi-well pad drilling; typically four, six or eight wells are drilled, although there are examples of 36 gas development wells completed in the Barnett Shale from a single drill pad. Multi-well pads are now the standard for unconventional field development, enabling broad areas of the reservoir to be drained from a compact land surface location. This improves operational productivity while minimising the footprint of well sites and surface environmental impact.

Today, sophisticated drilling assemblies probe unconventional reservoirs, and long-reach horizontal wells are commonplace in field operations. In 2008, Maersk Oil set a world record for the longest horizontal well section. It measured 10 900 m, and was drilled into a reservoir formation only 6 m thick. More typical horizontal well lengths range from 1000 to 3000 m, with an average of around 1600 m. Horizontal drilling technology has transformed the way in which oil gas and associated liquids are produced in the US over the last ten years. During this period the number of vertical gas wells drilled decreased by over 90 per cent, having peaked in 2006–2007 (Figure 4.9). In 2006, the ratio of vertical gas wells to horizontal was nine to one, but by 2013 this situation had reversed dramatically, with over five times more horizontal gas wells drilled than vertical wells.

Hydraulic Fracturing

Production wells completed in unconventional reservoirs typically require enhanced recovery methods to produce at sufficient rates and recovery volumes to make them economic. This is usually achieved with hydraulic fracture stimulation. The objectives of hydraulic fracturing are to establish permeable flow pathways connected to the well, maximise the stimulated reservoir rock volume, and maximise hydrocarbon production in the reservoir zone of interest. This is achieved by pumping fluid at pressures and injection rates sufficiently high to create new cracks, or open existing cracks, in the reservoir rock. Massive hydraulic fracturing treatments essentially manufacture flow pathways (fractures) in reservoirs that otherwise would not release their trapped hydrocarbons (Figure 4.2). Horizontal wells typically provide a more efficient method to drain unconventional reservoirs than vertical wells, but the majority will require the enhancement of reservoir properties by hydraulic fracturing.

After a well is drilled and the production casing is installed, holes or perforations are made in the casing. The perforations allow the fracture fluid to enter the target reservoir zone and will later be used to allow hydrocarbons to enter the wellbore when it is placed on production (Figure 4.2). The fracture fluid is mixed with a small amount of chemicals, usually less than one per cent, and fed to the truck-mounted pumping units (Figure 4.11). It is then pumped at high pressure through the casing perforations into the reservoir rock, and the hydraulic pressure is gradually increased until it exceeds the breakdown strength of the rock, causing it to fracture.

As a network of open fractures begins to form in the reservoir, the fluid pressure (known as the propagation or extension pressure) is increased to propagate the fracture network deeper into the rock formation and to keep the fractures open. Following the initial fluid fracture load, a proppant such as fine-grained sand is added to the fluid, and the resultant mixture is known as a slurry. The slurry transports the proppant particles and deposits them in the open fracture network. Once the slurry has penetrated the reservoir to the designed distance, pumping is stopped and, as the pressure decreases, the fractures start to close. The proppant holds them open, allowing hydrocarbons to flow to the wellbore.

The flow behaviour for natural gas stored within a shale reservoir is a complex process. A small volume is stored as free gas molecules in the pore space (predominantly nanopores) and natural fractures. The remaining gas molecules are bound to the walls of the pores by a mechanism called adsorption. Hydraulic fracturing increases reservoir permeability by creating fluid migration pathways that connect to the wellbore and by increasing the surface area exposure of the rock matrix permeability, for more efficient drainage (Dembicki, 2014). This enables the free gas molecules to flow to the wellbore. As gas is produced and the reservoir pressure is reduced over time, production of the adsorbed gas is enabled by the process of desorption.

Water is currently the most common and economically advantageous hydraulic fracturing fluid. The use of alternative sources of water, such as brackish and recycled water, is increasing, in for example Texas, which has large reserves of brackish to very saline groundwater that is not suitable for agriculture and other uses. The utilisation of saline waters is growing as additives are developed to be compatible with the saltier chemistry and as well operators learn how to modify completion practices.

Water is not the only type of fluid utilised for hydraulic fracturing. More viscous gels and foams can be used, but this increases the cost of the fracturing operation. Oil-based fluids can be used in water-sensitive rocks, where contact with water causes clay minerals to expand and damage the reservoir. Other types of fracture fluid include liquid propane, high pressure carbon dioxide, and foams comprising mixtures of carbon dioxide and nitrogen together with proppant.

There are two main ways of monitoring the fracture operation while it is in progress. The surface measurements include the pressure at which the slurry is injected, the number of barrels of slurry injected per minute, the proppant concentration and the pounds of proppant injected per minute. Additionally, microseismic surveillance technology can be deployed to monitor the fracturing treatment.

Hydraulic fracturing is not a new technology and the basic elements have not changed substantially since the first commercial operations were performed. However, the instrumentation, computing power and telemetry deployed to monitor the fracturing process have evolved tremendously.

The practice of stimulating wells to increase hydrocarbon flow can be traced back over 150 years. An early method of 'shooting' or 'torpedoing' a well to initiate or improve oil flow is described from shallow oil wells in the Appalachian basin in the 1860s (Bowman, 1911; Montgomery and Smith, 2010). Civil War veteran Colonel Edward Roberts invented the technique, and he received a patent in 1866 for what became known as the 'Roberts Torpedo'. It involved lowering nitroglycerine into the well and detonating an explosion to fracture the rock formation. Filling the borehole with water helped to concentrate the explosion pressure wave. Whilst this rudimentary method was sometimes successful in improving the oil flow, it frequently had the reverse effect, filling the borehole with rock detritus and impeding the flow of hydrocarbons. It was also highly dangerous, and well site accidents were commonplace.

By the 1930s, efforts were well underway to find a safer, non-explosive, well stimulation technology. Drillers experimented with pumping acid, with the objective of creating fractures by acid etching of the reservoir. Work continued into the 1940s to find a more effective technology, until Stanolind Oil and Gas Corporation demonstrated a link between well flow performance and the treatment pressures observed during acid injection. Stanolind determined that hydraulic pressure could be used to fracture reservoir rocks and improve production from oil and gas wells, and performed the first experimental hydraulic fracturing of a well in the Hugoton gas field of Kansas in 1947 (Figure 4.10). Naphthenic acid and palm oil were mixed with gasoline and injected into a gas-bearing limestone formation at an average depth of 750 m, and while gas productivity did not increase appreciably the technology showed such promising results that two years later Stanolind was granted a patent and published a paper describing the technology (Clark, 1949; Howard and Fast, 1970).

Halliburton Oil Well Cementing Company acquired the exclusive rights to this new technology, which they termed 'Hydrafrac', and performed the first two commercial fracturing treatments, in Oklahoma and Texas, in 1949 (Figure 4.10). In the first year of operations more than 300 wells were hydraulically fractured, resulting in an average production increase of 75 per cent. Such was the success of the technology that by 1955 more than 3000 well treatments per month were recorded (Montgomery and Smith, 2010).

(a)

(b)

Figure 4.10 (a) The first experimental hydraulic fracturing of a gas reservoir, conducted in 1947 on the Klepper Gas Unit number one well, in the Hugoton field, Kansas, US, by Stanolind Oil and Gas Corporation. *Photograph from Howard and Fast (1970), reproduced with the permission of the Society of Petroleum Engineers.* (b) The first commercial fracturing treatment, performed near Duncan, Oklahoma, in March 1949. *Photograph reproduced with the permission of Halliburton.*

The 1947 hydraulic fracture experiment used sand from the Arkansas River as a proppant. Over the years, many propping agents have been tried, including steel shot, aluminium pellets, plastic pellets, high-strength glass beads and nut shells. The first fracture fluids were gelled crude oil and later gelled kerosene and gasoline. When water was introduced as a fracture fluid in 1953, a number of gelling agents were developed and surfactants were added to minimise the creation of emulsions with formation fluids and to reduce the swelling effect of clay minerals. Later, when clay-stabilising additives were developed, this enabled the widespread use of water-based fracture fluids.

During the 1980s George Mitchell, together with a team of engineers and geoscientists at Mitchell Energy, headquartered in The Woodlands, Texas, experimented with ways to extract gas commercially from the Barnett Shale. Larger, rival, companies had already shut down drilling operations in such challenging shale formations, but Mitchell continued to work on finding a viable technology. In 1986 he started performing large fracture treatments using a gel fluid, but production rates were at best disappointing. The breakthrough came in 1997 when the first water-based hydraulic fracture was tried. Water was combined with a small amount of polymers to act as a lubricant. Termed a 'slick-water frac' because of the friction-reducing polymers, it created networks of microfractures in the shale reservoir, allowing natural gas to flow at economic rates. This is considered by many to be the technological breakthrough that initiated the US unconventional energy revolution. Today, the success of the unconventional resource industry owes a lot to the perseverance of George Mitchell in not giving up on his vision of achieving commercial gas production from shale.

Hydraulic fracturing is a powerful and flexible tool for recovering hydrocarbons from unconventional reservoir formations. Since the first experiment in 1947, close to 2.5 million hydraulic fracture treatments had been performed worldwide by 2012, with over one million in the US alone (King, 2012). It has become a standard industry well-completion practice. Today, approximately 60 per cent of all wells drilled are fractured, and nine out of ten onshore wells, natural gas and oil, require fracture stimulation to be economically viable (Figure 4.11).

Building on Mitchell Energy's success in the Barnett Shale, well operators and service companies experimented with various combinations of horizontal drilling and hydraulic fracturing, applying different permutations to different geological situations and learning from experience. One key industry lesson is that *each unconventional reservoir is unique*, and drilling and well completion technology must be constantly adapted and optimised. The techniques that work well in one unconventional reservoir may not necessarily succeed in another.

Hydraulic fracturing technology has advanced rapidly, with sophisticated three-dimensional computer models used in the design and monitoring of fracture treatments. Multi-stage fracturing with 20 or more fracture stages is now a standard practice, and as many as 80 fracture stages have been performed in a single horizontal well. Operators have reduced the chemical additives in the completion fluid mix to less than 1 per cent and made significant progress towards the desirable goal of achieving 100 per cent environmentally

Figure 4.11 Typical well site infrastructure for a modern hydraulic fracture operation. This example is from the Marcellus Shale of Pennsylvania, US, and shows a hydraulic fracture treatment being performed in June 2011. *Image courtesy of Anadarko Petroleum Corporation.*

friendly or 'green' completion chemicals, meaning that they must meet rigorous standards of toxicity, biodegradability, and bioaccumulation. The ultimate future goal for completion fluid composition would be to achieve water-free, chemical-free, fracture stimulations.

Microseismic Surveillance

Microseismicity occurs because of geomechanical changes induced in a reservoir as a result of the hydraulic fracturing process. The geomechanical perturbations generate microseismicity 'events', and the detection and location of these events provide a technique to monitor fracture growth patterns and overall fracture dimensions (Warpinski, 2013). When a reservoir is hydraulically fractured the brittle rock failure is typically accompanied by little 'pops' – the release of acoustic energy that is referred to as microseismicity. The microseismic events are linked to the opening or moving of fractures in the reservoir during hydraulic fracturing. They can be characterised using the same terminology developed to describe earthquakes, although the processes that produce each one are quite different and the energy released is different by orders of magnitude. On the moment magnitude

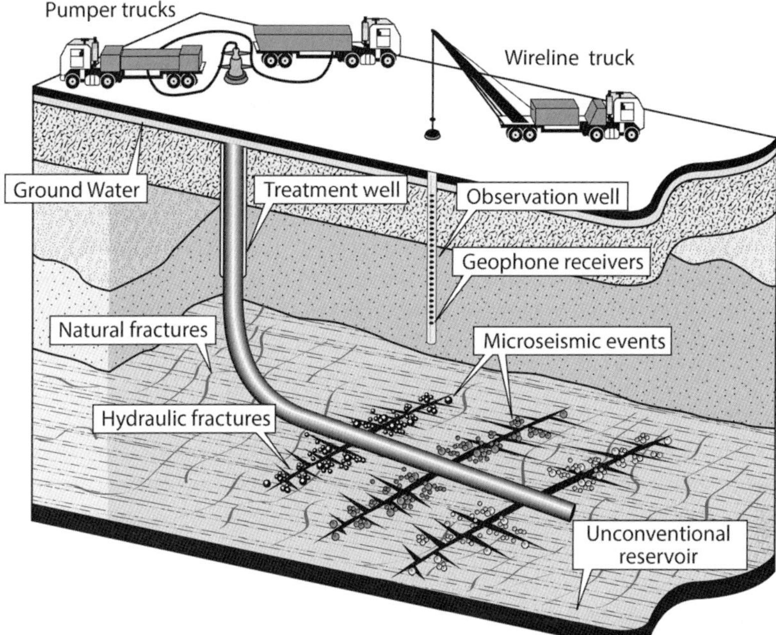

Figure 4.12 Microseismic monitoring of a horizontal well undergoing hydraulic fracture stimulation. In this example, geophone recorders are placed downhole in an observation well located near to the well that is being stimulated. The clusters of small circles represent microseismic events aligned with the hydraulic fractures. Note that the induced fractures often develop parallel to the natural fractures in the reservoir, and the use of a horizontal well is the optimum method to intersect as many open fractures as possible.

scale microseismic events have negative magnitudes, generally in the range −3 to −1. The smallest earthquake which can be felt by a person has a magnitude in the range +2 to + 3.

Microseismic monitoring technology has its origins in geothermal energy, but its application to the oil and gas industry was pioneered by Sandia National Laboratories in the late 1970s to mid-1980s. Sandia conducted a series of research projects aimed at understanding fracturing in unconventional reservoirs, and became involved in some of the earliest development of microseismic tools (Warpinski, 2013). Today, microseismic technology has become a valuable field development tool to assist the exploitation of unconventional reservoirs. The technology is increasingly being deployed to observe and measure the real-time process of fracture creation during hydraulic fracturing (Duncan, 2010).

Microseismic monitoring tools provide a measurement of the geomechanical changes in the reservoir by detecting acoustic energy waves and locating their source in time and space. This information is used to infer where fracture propagation is occurring in the reservoir, and it can help geoscientists and engineers to build a three-dimensional model of the fracture network (Figure 4.12). The acoustic 'pop' location is measured either by geophone receivers positioned downhole in an observation well (which could be horizontal or vertical) or by using surface receiver arrays. The geophones receive and record the acoustic wave and, using triangulation, the wave source point can be located. Mapping the fracture distribution in three-dimensional space enables geoscientists and engineers to improve the reservoir model, allowing well completions and fracture treatments to be optimised and operational costs to be reduced.

During early hydraulic-fracturing monitoring trials the surface-deployed geophone recorders that were then used could not detect the tiny acoustic energy events with sufficient accuracy to resolve their location. To improve detection, geophone receivers were placed downhole in a monitoring well, adjacent to the well undergoing stimulation. The first commercial deployments of microseismic technology monitored hydraulic fracturing in the Barnett Shale in the late 1990s. However, the microseismic monitoring of fracture jobs was not common practice until about a decade later, when rising levels of unconventional development activity led to more demand for this technology.

As horizontal well lengths have increased it has become more difficult to observe the entire lateral length with a single vertical monitor well because of the limited radius of investigation. The alternative to downhole monitoring has been the introduction of high-count surface receiver instrumentation. Surface arrays have made it possible to monitor entire well pads instead of single wellbores without being constrained by the locations of individual wells. Surface and near-surface receiver arrays can be as large as needed, facilitating the surveillance of wells several kilometres in length. Permanently installed surface recording systems can monitor all the development wells in a producing field, providing information over the life of the field. In response to the surface array, downhole microseismic techniques have been developed to monitor hydraulic fracturing along the full length of a horizontal well more effectively. This utilises the recently developed downhole tractor technology. An electrically driven tractor crawls along a horizontal well drilled parallel to the well that is being fractured, towing the geophone array behind it. As the treatment well is fractured from 'toe to heel' the position of the geophone array in the parallel observation well can be adjusted to maintain its field of view. This enables the full fracturing job to be monitored.

Microseismic data are currently applied to determine the orientation of stress fields in the subsurface and the extent of microseismicity both laterally and vertically within the reservoir. Microseismic data provide an important input into three-dimensional geomechanical reservoir models, and, together with geological, petrophysical, completion, and production data, they help to create an integrated model for the reservoir. Advances in computing power have made real-time processing possible, allowing data to be viewed and evaluated rapidly. Onsite telemetry enables large volumes of microseismic data to be relayed to a remote processing centre. This information helps to guide decisions about the number of fracture stages required in future wells, the optimal stage length, and the fracture proppant type and amount and ultimately leads to more accurate forecasts of well production when integrated with other data acquired during hydraulic fracturing. It can also help to inform development planning decisions about the number of wells required and the optimal well spacing.

Unconventional Resources, Reserves, and Production

It is important to understand the distinction between hydrocarbon *resources* and *proved reserves*. Resources, on the one hand, are undrilled hydrocarbons that have been identified by geological studies and are technically recoverable; they assume neither a market price nor a time schedule for discovery and production and therefore may not necessarily be commercially viable. Thus resources represent untapped potential that could be recovered by future drilling under commercially attractive circumstances and with current or foreseeable technology. Proved reserves, on the other hand, have been confirmed by exploration and appraisal wells and the flow of a significant hydrocarbon volume to the surface. Their recovery is commercially viable under the prevailing economic circumstances. Thus the concept of proved reserves provides a way of tracking the currently producible hydrocarbon inventory. While independent assessments of resources and reserves may differ in the details, there is one area of broad agreement: the United States is richly endowed in unconventional hydrocarbons.

A recent assessment of US natural gas resources is the highest in the last 50 years (Potential Gas Committee 2013). At almost 2400 trillion cubic feet (lcf), the resource size is four times larger than it was in 1966. Had the 1966 estimate remained static or decreased, the US would have run out of gas about 10 years ago. However, driven by increases in unconventional gas and especially shale gas, which are now much more accessible, total gas resources have effectively doubled during the last decade (Figure 4.13). On a global basis, it is thought that over 30 per cent of total natural gas resources are reservoired in shale formations (US Energy Information Administration, 2013).

Over the same time period the US has reversed a steadily declining trend in proved gas reserves that had persisted since the early 1970s. Proved reserves climbed to a record high 350 tcf, largely through additions made in shale reservoirs. The largest recent reserve gains have been recorded in the Marcellus Shale in the Appalachian basin and in the Barnett and Eagle Ford shales, both in Texas. With an inventory size estimated at 159 tcf, shale gas reserves account for almost half the US proved reserves portfolio (US Energy Information Administration, 2014a).

Natural gas production from unconventional reservoirs has risen rapidly, today comprising about 65 per cent of total US annual dry gas production and making it a key source of energy supply (Figure 4.14; US Energy Information Administration, 2014b). Shale gas, thought to be commercially unrecoverable as recently as a decade ago, now provides around 40 per cent of US natural gas production (US Energy Information Administration, 2014c). The rapid production increase began around 2005, an important milestone in US energy supply (Figure 4.14). It marks the first truly effective combination of horizontal drilling and hydraulic fracturing, and the beginning of large-scale commercial shale gas production.

Production data trends for recent years highlight that, alongside the burgeoning unconventional gas output, the production of crude oil, condensate, and natural gas liquids has increased sharply. The daily production of hydrocarbon liquids exceeded 11 million barrels during the first half of 2014, the highest production level seen since 1990, as unconventional

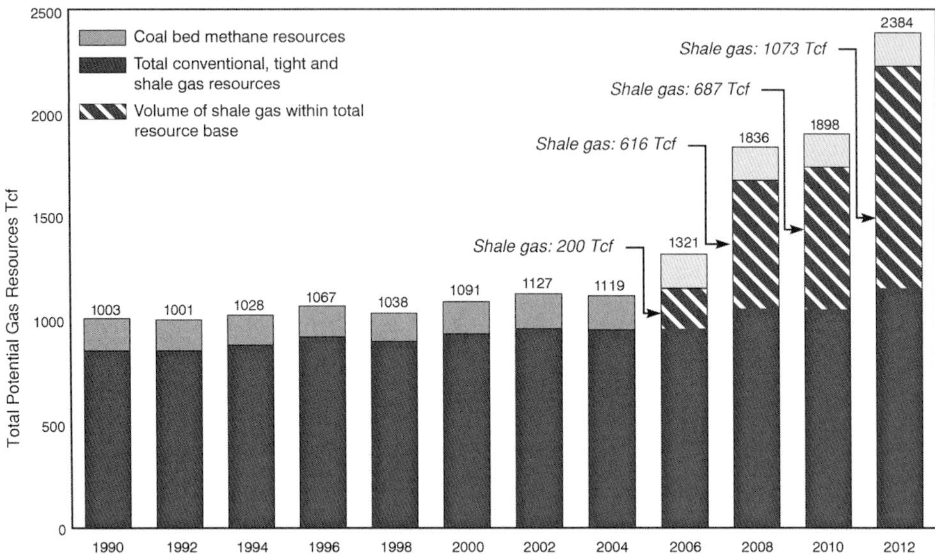

Figure 4.13 Estimate of total technically recoverable natural gas resources in the United States. The chart illustrates the huge impact that shale gas has made on US natural gas resources. Shale gas did not register in the 2004 evaluation; however, by 2012 it comprised nearly half the US gas resource base. Between 2006 and 2012 the volume of shale gas increased fivefold, from 200 tcf to over 1000 tcf. *Modified from Potential Gas Committee (2013) and published with permission.*

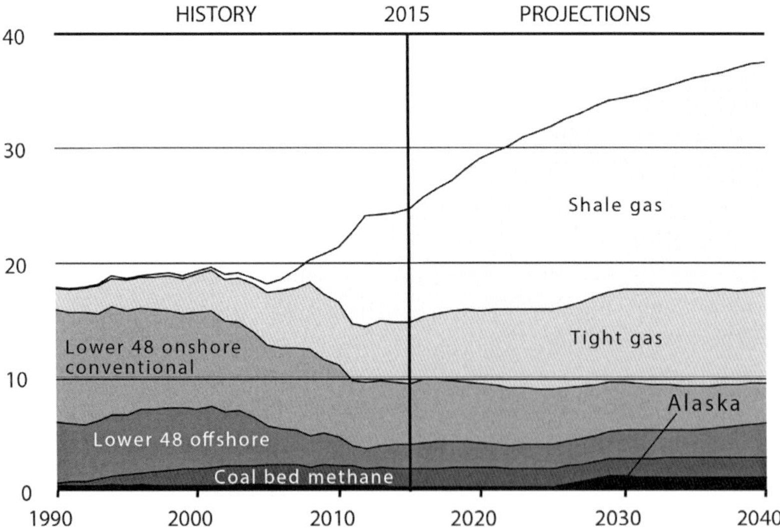

Figure 4.14 US Energy Information Administration forecast for natural gas in the United States, 1990–2040, showing the sharp increase in production driven by unconventional gas sources. Unconventional gas also drives future production growth, shale gas being the largest source followed by tight gas. Shale gas will account for more than half the total US natural gas production by 2040, highlighting its importance in the future energy mix. *Modified from the US Energy Information Administration.*

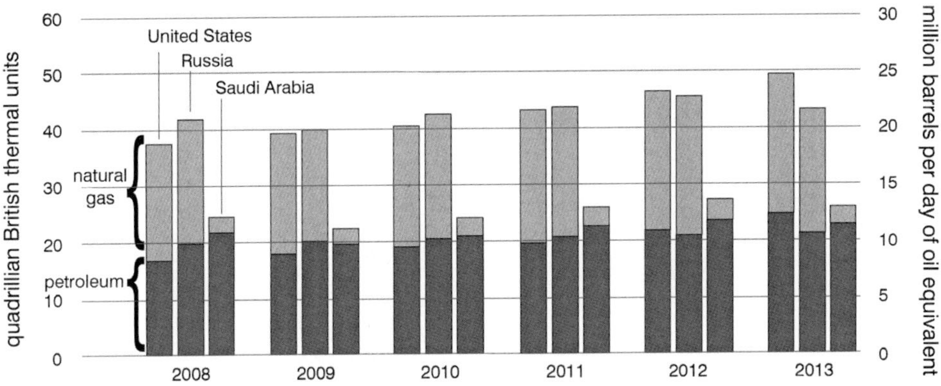

Figure 4.15 Estimate of petroleum and natural gas production in the United States, Russia, and Saudi Arabia. The United States ranks as the world's top producer of petroleum liquids (crude oil, condensate and natural gas liquids) and dry natural gas, overtaking Russia and Saudi Arabia. *Modified from the US Energy Information Administration.*

production reached record levels. Hydrocarbons extracted from unconventional reservoirs have been the driver for the production increase, and gas output has increased to such an extent that the United States is now ranked as the world's largest petroleum producer, having overtaken both Saudi Arabia and Russia (Figure 4.15).

Unconventional Hydrocarbons and the US Technology Revolution

The rise in production of unconventional hydrocarbons in the United States has ushered in a new era of energy supply and security that has been nothing short of remarkable. It has been driven by a technology revolution, perhaps the most significant since the merger of rotary drilling with roller-cone bit technology more than 100 years ago. Hydraulic fracturing and horizontal drilling have been at the forefront of this revolution and, together with multi-well pad drilling and multi-lateral drilling, they have created a quantum leap in hydrocarbon recovery efficiency from poor-quality, low permeability and ultra-low permeability reservoirs. Oil and gas resources that 20 years ago were unreachable and undevelopable with the technology available at that time have now become accessible and economic to produce.

The extraordinary growth of unconventional resources has transformed the energy sector and had a positive impact on the American economy. Increasing output has turned around a long-term trend of declining oil and gas production, initiated the expansion of infrastructure investment, and reversed trends of imports across the entire energy system and value chain. These profound changes have altered the balance between domestic production and energy imports. Upward-trending production has been mirrored by a fall in hydrocarbon imports to their lowest levels in nearly 30 years. As a consequence, regional US energy markets have expanded, creating hundreds of thousands of direct and indirect employment opportunities.

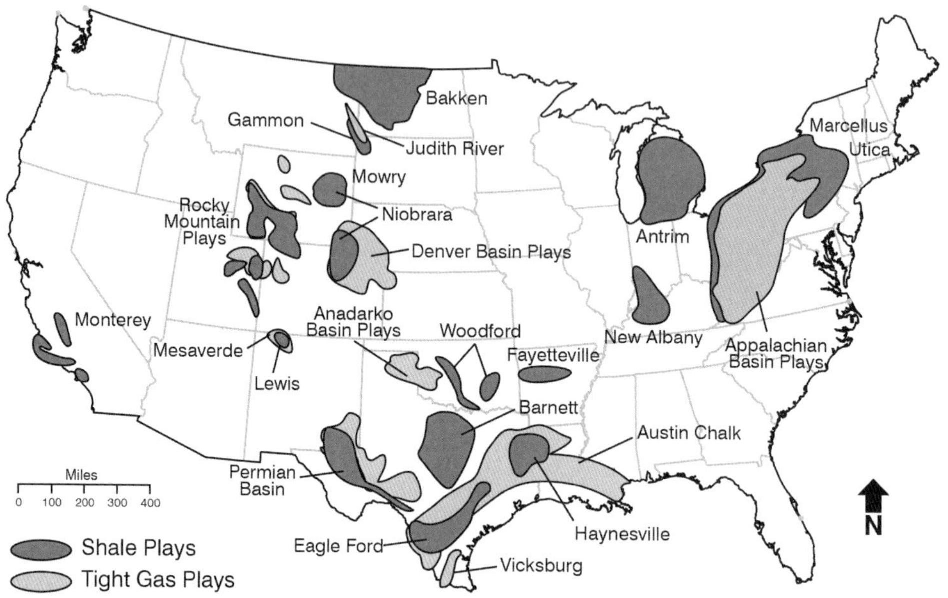

Figure 4.16 Sedimentary basins in the continental United States containing unconventional tight gas and shale gas plays. More than 35 tight gas plays and over 20 shale plays are mapped, spanning 31 states. *Modified from the US Energy Information Administration.*

The impacts of the American energy boom are spreading, directly influencing global energy markets. The production of unconventional hydrocarbons has delivered an energy abundance that has eclipsed every other country in the twenty-first century. It has launched the US into the top position as the world's largest oil and gas producer, an unprecedented situation that few would have predicted a decade ago. As recently as 2007 the US was planning to build new regasification facilities to process LNG imports. In a complete turnaround, it is now refashioning terminals on the Gulf of Mexico to reverse gas flows and export LNG cargoes to international markets. On the liquids front, the global daily output has been augmented by around 5 million barrels of US unconventional production. With total OPEC and non-OPEC supply outpacing demand, global oil prices declined in the second half of 2014 and throughout 2015, falling to a 13-year low. The rising production of unconventional hydrocarbons has been largely confined to the lower 48 contiguous states of the US. Canada is exploring and developing resources in British Columbia, Alberta, and elsewhere, although current production is relatively modest in comparison with US metrics.

It is worth asking what have been the key drivers of the unconventional energy boom, and why has this boom been largely confined to the US. Two factors emerge as the critical forces. First, the US has an exceptionally rich endowment of sedimentary basins containing viable unconventional reservoir plays (Figure 4.16). Second, the creation of a disruptive-technology mix has transformed the industry; US operators have applied technology solutions rapidly and efficiently, while incrementally driving down their costs. For the first time it has been possible to recover hydrocarbons commercially from low and ultra-low permeability rocks.

However, beyond the availability of viable unconventional plays and the technology to exploit them commercially, there is more to the answer. Several other factors have played an important role in influencing the pace at which unconventional hydrocarbons have been discovered, developed and transported to market.

The American energy renaissance has happened as a result of the confluence of a number of key enablers, some of which are currently unique to the US. For example, the private ownership of subsurface mineral rights, a circumstance that enables landowners to collect royalty payments from the field operator, has helped tremendously to align the disparate interests of landowners and E&P companies. The significance of landowner and industry alignment cannot be overstated, as it is not found anywhere else in the world. The US also benefits from ease of access to capital markets and the sheer magnitude of the potential investment pool.

The quantity, quality and density of resources and equipment, including independent drilling companies, drilling rigs, hydraulic-fracturing horsepower capacity, proppant supply, fluid separation and processing facilities, and pipeline infrastructure, is unparalleled anywhere else in the world. This factor, combined with a large, well trained and mobile workforce has resulted in 'clusters of excellence' across US sedimentary basins.

North America is a huge landmass with a generally low rural population density in the main areas of unconventional reservoir development; further, low-cost water from surface and subsurface sources is available for well completion operations. Field operators are putting significant efforts into minimising water use in well operations, recycling as much as possible to reduce consumption and increasingly using saline water for well completions.

The potential for unconventional resources is beginning to be identified elsewhere around the world, for example in Argentina, Australia and China. It can be argued that the lack of one of more key enablers may delay or even halt the commercial production of unconventional hydrocarbons in other parts of the world. But, for the countries that possess developable reserves, this is probably only a temporary situation. It is likely that the future deployment of capital, technology and human resources, coupled with sufficient energy demand, will be able to promote exploration and development and bring more unconventional oil and gas to global energy markets. Unconventional resources are set to play an increasingly important role in meeting global energy needs for the foreseeable future.

To some extent, the unconventional hydrocarbon revolution in the US can be explained by a relatively simple paradigm. The oil and gas industry has relentlessly experimented, continually learning, continually iterating, until the drilling and completion practices for an unconventional reservoir are optimised. The process has then been repeated, taking the learnings from one area, from one uncooperative and challenging reservoir, and then successfully applying them to a new reservoir play. The cumulative experience of developing a range of unconventional plays has provided a key lesson: each unconventional reservoir is unique in its character and properties, whether it is sandstone, siltstone, limestone, shale, or a mixed rock type. There is no single simple recipe for success.

References

Bowman, I. (1911). *Well-drilling Methods*, United States Geological Survey, Water-Supply Paper 257, 139 pp.

Broman, W.H. and Schmohr, D.R. (1992). Horizontal well operations at Prudhoe Bay. SPE Paper 22383, presented at the SPE International Meeting on Petroleum Engineering, 24–27 March 1992, Beijing, China, pp. 541–549. https://www.onepetro.org/download/conference-paper/SPE-22383-MS?id=conference-paper%2FSPE-22383-MS (accessed August 2015).

Cander, H. (2012). What are unconventional resources? A simple definition using viscosity and permeability. *AAPG Search and Discovery Article*, #80217. http://www.searchanddiscovery.com/documents/2012/80217cander/ndx_cander.pdf (accessed August 2015).

Clark, J.B. (1949). A hydraulic process for increasing the productivity of wells. *Transactions of the American Institute of Mining Engineers*, 186, 1–8.

Dembicki, H. (2014). Challenges to black oil production from shales. *AAPG Search and Discovery Article*, #80355. http://www.searchanddiscovery.com/documents/2014/80355dembicki/ndx_dembicki.pdf (accessed August 2015).

Duncan, P.M. (2010). Microseismic monitoring: technology state of play. SPE Paper 131777, in *Proc. SPE Unconventional Gas Conference*, 23–25 February 2010, Pittsburgh, Pennsylvania, USUSA, pp. 1–8.

Heller, R., Vermylen, J. and Zoback, M. (2014). Experimental investigation of matrix permeability of gas shales. *AAPG Bulletin*, 98 (5), 975–995.

Howard, G.C. and Fast, C.R. (1970). *Hydraulic Fracturing*. Monograph in the Henry L. Doherty Series, vol. 2, Society of Petroleum Engineers of AIME, New York.

IFP Energies Nouvelles (2012). Non-conventional hydrocarbons: evolution or revolution? http://www.ifpenergiesnouvelles.com/index.php/layout/set/print/content/download/71817/1530718/version/5/file/Panorama2012_07-VA_+Non-conventional-hydro carbons.pdf (accessed August 2015).

King, G.E. (2012). Hydraulic fracturing 101: what every representative, environmentalist, regulator, reporter, investor, university researcher, neighbor and engineer should know about estimating frac risk and improving frac performance in unconventional gas and oil wells. SPE Paper 152596, in *Proc. SPE Hydraulic Fracturing Technology Conference*, 6–8 February 2012, The Woodlands, Texas, US, pp. 1–80.

Loucks, R.G., Reed, R.M., Ruppel, S.C. and Hammes, U. (2012). Spectrum of pore types and networks in mudrocks and a descriptive classification for matrix-related mudrock pores. *AAPG Bulletin*, 96 (6), 1071–1098.

Montgomery, C.T. and Smith, M.B. (2010). Hydraulic fracturing: history of an enduring technology. *Journal of Petroleum Technology* 62 (12), 26–32.

Nelson, P.H.(2009a). It's a small world after all – the pore throat size spectrum. *AAPG Search and Discovery Article*, #50218. http://www.searchanddiscovery.com/pdfz/documents/2009/50218nelson/ndx_nelson.pdf.html (accessed August 2015).

Nelson, P.H. (2009b). Pore-throat sizes in sandstones, tight sandstones, and shales. *AAPG Bulletin*, 93 (3), 329–340.

Potential Gas Committee (2013). Potential supply of natural gas in the United States. http://potentialgas.org/download/pgc-press-release-april-2013-slides.pdf (accessed August 2015).

Rundell, W. (1977). *Early Texas Oil: A Photographic History, 1866–1936*. Texas A&M University Press.

Slatt, R.M., Philp, P.R., O'Brien, N. *et al.* (2012). Pore- to regional-scale, integrated characterization workflow for unconventional gas shales. *Shale Reservoirs – Giant Resources for the 21st Century*, in: J.A. Breyer (ed.). AAPG Memoir 97, pp. 127–150.

Storm, J.C. (1968). Slant hole drilling – offshore Peru. *Society of Petroleum Engineers of AIME*, SPE-2312, pp.1–2. https://www.onepetro.org/download/conference-paper/SPE-2312-MS?id=conference-paper%2FSPE-2312-MS (accessed August 2015).

US Energy Information Administration (1993). Drilling sideways – a review of horizontal well technology and its domestic application. http://www.eia.gov/pub/oil_gas/natural_gas/analysis_publications/drilling_sideways_well_technology/pdf/tr0565.pdf (accessed August 2015).

US Energy Information Administration (2013). Technically recoverable shale oil and shale gas resources: an assessment of 137 shale formations in 41 countries outside the United States. http://www.eia.gov/analysis/studies/worldshalegas/pdf/fullreport.pdf (accessed August 2015).

US Energy Information Administration (2014a). US natural gas reserves increase 10% in 2013 to reach a record 354 TCF. http://www.eia.gov/todayinenergy/detail.cfm?id=19051 (accessed August 2015).

US Energy Information Administration (2014b). Annual energy outlook 2014. http://www.eia.gov/forecasts/aeo/pdf/0383(2014).pdf (accessed August 2015).

US Energy Information Administration (2014c). Shale gas provides largest share of US natural gas production in 2013. http://www.eia.gov/todayinenergy/detail.cfm?id=18951 (accessed August 2015).

Warner, C.A. (2007). *Texas Oil & Gas Since 1543*. Copano Bay Press.

Warpinski, N.R. (2013). Understanding hydraulic fracture growth, effectiveness, and safety through microseismic monitoring. *Effective and Sustainable Hydraulic Fracturing*, in: A.P. Bunger, J. McLennan and R. Jeffrey (eds.). InTech, pp. 123–135. http://www.intechopen.com/books/effective-and-sustainable-hydraulic-fracturing/understanding-hydraulic-fracture-growth-effectiveness-and-safety-through-microseismic-monitoring (accessed August 2015).

5

Risks and Opportunities of Unconventional Natural Gas: Australia and the United States

IAN CRONSHAW AND R. QUENTIN GRAFTON

Introduction

Since 2008, US shale gas production has increased tenfold from around 30 bcm per year to 380 bcm in 2014, surpassing conventional gas production. While Texas remains the biggest gas supplier, the rise in production from Pennsylvania, from zero to over 80 bcm in 2013, is remarkable, and growth from the Marcellus formation has continued towards 160 bcm (see Figure 5.1). Equally remarkable is the astonishing improvement in productivity, as the more widespread use of drilling pads has increased drilling rates (especially from horizontal drilling), fracturing techniques have become more precise, and liquids recovery has improved.

Gas infrastructure in the US has struggled to keep up with increasing supply. As a result, areas such as the Bakken in North Dakota are producing substantial liquids but are flaring gas because there is an inadequate gas-gathering network. Canadian producers have also benefitted from unconventional gas (UCG) technology but, although they are integrated into the North American gas network, their netback prices are low. In turn, this has encouraged proposals for LNG export opportunities towards the higher priced east Asia gas markets.

In Australia, coal bed methane (CBM) production started in 1996 and increased modestly until 2014 but accelerated rapidly over 2014–2016. The reason is that LNG output is rapidly ramping up, as all three major LNG export facilities started production by the end of 2015, all based on CBM supplies sourced in the state of Queensland; CBM production in that state is set to rise from around 6 bcm in 2012 to more than 40 bcm before 2020. When all LNG projects are fully operational, total annual gas production in the Australian east coast gas market will be more than three times the current level, approaching 60 bcm.[1]

Hydraulic Fracturing

When geologic formation permeability is low, as in the case of shale gas, it generally takes a combination of horizontal drilling and hydraulic fracturing to achieve commercial

[1] One bcm is roughly the gas use of all households in Australia over three months.

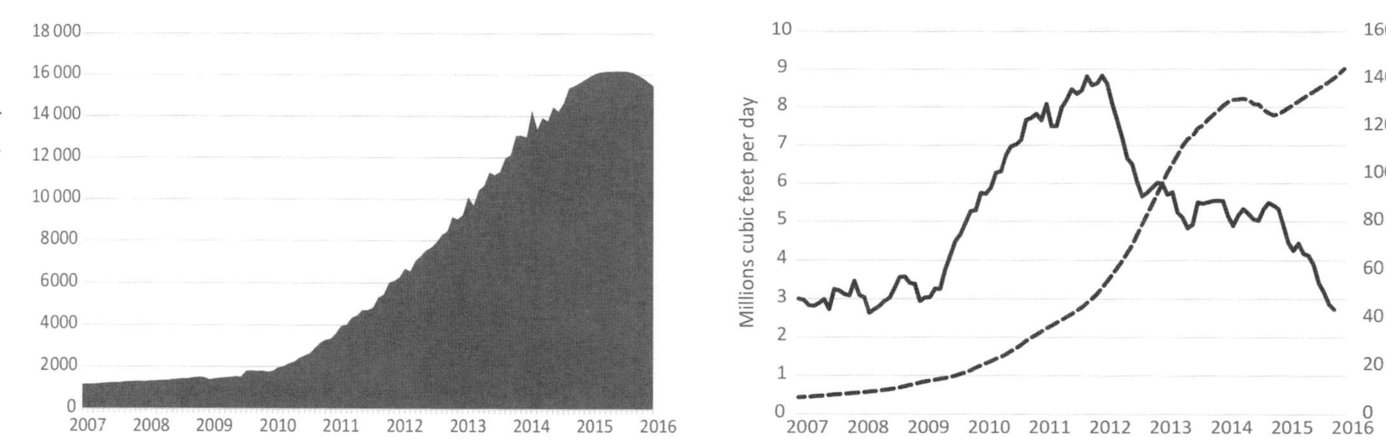

Figure 5.1 Increasing gas output from Marcellus Basin, and productivity improvements (as measured by output per rig). *Source: US Energy Information Administration (December 2015).*

levels of gas output. In non-technical terms, it involves pumping a fluid at high pressure into the well, creating tiny fractures in the rock formation, thus allowing gas to flow more readily to the surface. The fluid used in fracturing is largely water, with around 4% of fine particles, often sand, used to hold or 'prop open' the fractures created; hence this group of substances are referred to as proppants. In addition, a small amount of chemical is added, constituting typically around 1% of the total fracturing liquid. In the past, industry has been less than transparent about the composition of these chemicals, often citing commercial confidentiality. But, increasingly, the composition of fluids is being disclosed in the United States (FracFocus 2014). In many other jurisdictions, disclosure to the regulator, if not public disclosure, is mandatory.

Increasingly in the United States, the technique of hydraulic fracturing is repeated at intervals along the horizontal well bore, often ten or even twenty times, thus increasing the volume of liquid used in the fracturing operations from a few Ml to 20 Ml and the proppants from 1000 to 4000 tonnes per well. While such techniques have been responsible for rapidly increasing the production of gas, and sharp reductions in production costs, they also increase the stresses on the steel and cement well casing. This casing provides the essential barrier between the gas (plus the drilling and fracturing fluids) produced, generally at depth, and shallower aquifers. Multiple cycles of fracturing can impose greater stresses on the casing than those encountered in conventional hydrocarbon production. Thus, it is especially important that key technical parameters are designed and executed correctly (IEA 2012). Failures in the casing can lead to the contamination of surface aquifers with methane and other hydrocarbons, or cross contamination among aquifers. While such events are rare they cannot be excluded, especially where the geology is not completely understood. Well integrity is, therefore, becoming an increasing regulatory focus.

After the fracturing process has been completed, some of the fluid injected flows back up the well, as part of the produced stream, gradually being replaced by hydrocarbons. Best practice at this part of the extraction process is represented by so-called 'green completions' or reduced emissions completions, with the separation of gas and liquids. In the absence of this investment, emissions of methane can occur or the flaring of methane may need to take place. According to best regulatory practice, green completions should be obligatory. Fluids that reach the surface can be processed, recycled or disposed. Increased recycling has been observed in the United States and Canada; this is a key technique for reducing water needs and lowering demands for the safe disposal of produced liquids. Greater attention is also being given to reducing surface spills, which pose a contamination risk to groundwater aquifers.

Another issue of public concern is the seismic tremors associated with unconventional gas production, which have been recorded in Canada, the United Kingdom and the United States. All these seismic events have been small and, thus, should cause no surface damage. Hydraulic fracturing will always generate small seismic shocks but, where such operations intersect an existing fault and activate it, larger seismic shocks can be created. Best international practice requires the clear identification and characterisation of fault structures and the avoidance of fracture stimulation where such faults are identified. For deep-well fluid disposal, a reduction in injection pressure can be effective (British Columbia Oil and Gas Commission 2014). Further, the available sensing technologies allow real-time monitoring and the immediate shut down of operations if increased seismic activity is detected (Cook *et al.* 2013, IEA 2012).

Evolving Regulatory Approaches

While the existing oil and gas regulations can provide an important starting point for any regulatory approach, best practice regulation is beginning to focus on a number of newer issues (Cook *et al.* 2013, Watson 2014, SCER 2013). These regulatory approaches include:

(i) much greater levels of pre-production measurement and monitoring, especially of groundwater levels and quality, so as to establish a clear baseline for subsequent assessment;
(ii) much greater levels of transparency throughout the production process, particularly when fracturing fluids are used;
(iii) heightened scrutiny on well integrity, so as to give the greatest possible assurance that no cross contamination of water resources will occur over the lifetime of the well, including when abandoned;
(iv) water treatment of the formation water, in the case of CBM, or the fracturing fluids, for shale gas.
(v) eliminating or minimising air emissions, especially of methane by flaring or, worse, venting;
(vi) using regional approaches to measure cumulative impacts, given the large numbers of wells drilled;
(vii) ensuring high and improving performance (for example, the use of less-toxic fracturing chemicals, the greater recycling of produced water) by operating companies through rigorous and adequately resourced compliance work, funded by industry, with the monitoring of fracturing operations a critical focus;
(viii) purpose-built regulatory agencies, which appear to be a better way to address complex issues across the range of government responsibilities; such agencies might usefully regulate all unconventional gas production, bearing in mind the technical and regional differences between different unconventional gas types; and
(ix) the ongoing review of regulatory adequacy and of cost and complexity, bolstered by insights from global experience.

While additional regulation will inevitably involve higher production costs, most recent research shows such cost increases will be relatively modest, at between 7% to 10% (IEA 2012). This is a small price to pay if public confidence is to be gained and maintained. More recent work indicates such costs could be even less (EPA 2015), and the additional costs on US public lands are estimated to be smaller again. These modest extra costs from appropriately designed regulation have the payoff for gas developers in that they help to reduce the external costs from unconventional gas developments and thus help to maintain a social licence to operate.

Costs, Risks and Opportunities

Unconventional gas production is not without costs and risks, some of which are common to conventional energy production but several of which are novel. Some of these costs can be taken as a given and are due to the more industrial scale of development of unconventional gas and the generally large numbers of wells drilled. They include much greater traffic movements, higher and ongoing levels of noise from drilling, large volumes of water needed (for shale gas) or large volumes of water extracted (for CBM) and, in general, wastewater disposal issues.

Other issues are possible risks from potential local or regional water shortages, minor seismic events and the pollution of drinking or agricultural water supplies from hydraulic fracturing operations or other parts of the process, such as surface spills or other accidents. Air pollution is another possible risk, including methane leakages during well completion, production, processing and transmission (CSSD 2015). While these risks may have low probability, their local impact, or even regional impact in the case of water, may be substantial. See Table 5.1.

Benefits are Substantial, if Spread Wide

There are potentially very large rewards and opportunities in the development of unconventional gas. In the United States, municipalities benefit from more and on-going local jobs because of the need to drill large numbers of wells and re-enter them for multiple fracturing operations. In regions where there are shale oil and gas developments there can be positive total employment effects (Brown 2014, Weber *et al.* 2014). In addition, there is evidence that incomes in extraction regions are higher, by about 15% on average, in the coal bed methane locations in Queensland, Australia (Fleming and Measham 2015). These higher incomes arise from increased salaries paid as a result of the developments, as well as payments made to landholders in the form of compensation or royalties (Fleming *et al.* 2015).

Other benefits include local taxes. States and regions can also benefit from royalties or other charges. Pennsylvania introduced a well impact fee in 2012 which now raises some $400 million per year, although this is considerably below what other US States generally charge by way of severance taxes, which are similar to royalties. Current gas output could be expected to yield around three times that figure at modest rates of around 4% of wellhead value, noting that Pennsylvania is alone in the United States in not having severance taxes. Many States levy such taxes at concessional rates for non-conventional oil and gas. In terms of private royalties, Brown *et al.* (2015) estimated that, in the six most important oil and gas shale developments in the US, these amounted to US $39 billion in 2014 alone.

Table 5.1 *Strengths, weaknesses, opportunities and threats (SWOT) summary*

Strengths	Weaknesses
• Additional local employment in drilling and ongoing field operations. Increased employment in construction and operation of export facilities. • Royalty payments to resource holder (landowners in the US, mostly states or provinces elsewhere), severance payments in the US, taxation revenues to municipalities and governments. • Regionally and nationally increased GDP, export income. • Lower gas prices in some producing regions, especially North America; frequently, lower power prices as well. Lower household energy bills, clearly seen in parts of the US. • Benefits to local communities from possible Community Benefits Funds (see NSW approaches).	• Increased local traffic, accidents, noise, surface spills of chemicals. • Loss of general amenity, especially in previously quiet rural areas. • Increased air pollution from traffic, drilling equipment, hydraulic fracturing equipment. • Decreased local water availability.
Opportunities	**Threats**
• Expansion of gas exports, reductions of gas imports, improved energy security (for both producer and consumer regions or countries) • Reduced greenhouse gas emissions, to the extent that high efficiency gas use displaces higher emitting energy sources or supplements intermittent renewable power production. • Where gas prices are lowered, more competitive local gas intensive industries. • Increased property values.	• Ground water pollution, through cross contamination of aquifers, incorrect waste disposal (including waste or extracted water), the spillage or improper disposal of the chemicals used in hydraulic fracturing and the possibility that these chemicals could enter the groundwater. • Increased methane and subsequent air pollution from wellhead completions or normal operations (wellheads, processing, pipes). • Gas well blowouts, causing air, ground and water pollution. • Where linked to increased gas exports, linkage to prices (allowing for netback) from those markets could lead to higher wholesale prices for gas users in domestic markets and a consequent loss of competitiveness of gas-intensive industry and loss of the market share of gas in power (as seen, for example, in east coast Australia gas markets). • Potential seismic events.

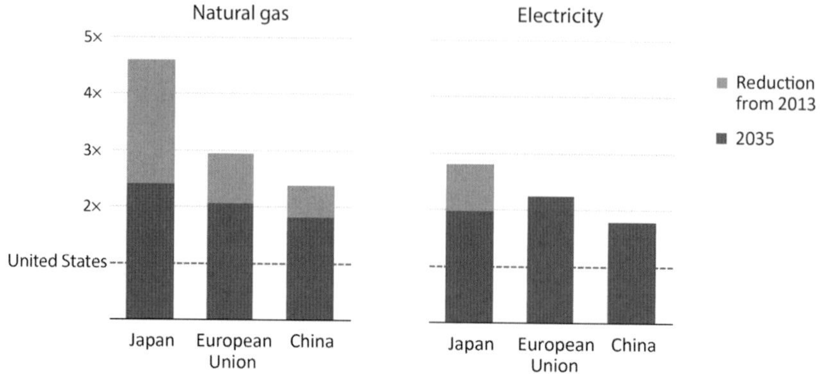

Figure 5.2 Ratios of industrial energy prices relative to the United States. Regional differences in natural gas prices are expected to narrow from today's very high levels but will remain large through to 2035; electricity price differentials will also persist. *Source:* IEA (2013).

Beyond direct payments, in the US the producing States benefit from lower gas prices for both consumers and large industrial and electricity users, even compared with neighbouring States. In the absence of this rapid growth in indigenous shale gas supplies the US could have been expected to import significant quantities of higher cost LNG, so that gas prices would have been closer to those in the United Kingdom or mainland Europe, which are at least some $5/Mbtu higher than currently. Lower gas prices are expected to save an average US household more than $900 per year, worth in aggregate more than $100 billion per year. As gas prices often strongly influence power prices, US households can expect to benefit from lower power prices too; these are estimated to have fallen by 10%, in stark contrast with almost all other IEA countries (see Figure 5.2). In terms of employment, one estimate put the number of jobs supported by unconventional gas at 1 million in the United States in 2010, rising to 1.5 million by 2015 (IHS 2012). Another estimate quoted some 175 000 new jobs in Pennsylvania over the three years to the end of 2012 (GasFields Commission 2013).

A major study in mid-2015 (Porter *et al.* 2015) assessed the economic impacts of unconventional gas and oil development as very substantial, including a value-addition of $433 billion (2012 $), direct and indirect employment of 2.7 million people, federal and State taxes and other revenues of $111 billion, split almost equally between the two levels of government, and average household savings from lower energy costs of $776 (2014 $).

United States industries benefitted from lower energy and gas feedstock (methane and ethane) prices, in some cases approaching the cheapest globally. Energy-intensive industries, including chemicals, are increasing production levels, and some $100 billion is being invested in new manufacturing plant (Obama 2014).

Replacing coal by gas in the power sector offers benefits in terms of reduced conventional pollution and lowered greenhouse gas emissions, both of which have been documented in the United States. In 2012, the US recorded a remarkable drop in greenhouse gas emissions, of nearly 4%, primarily driven by this substitution. While there was a small rebound in coal use as gas prices rose in 2013, the increase in gas fired power and the consequent greenhouse gas reductions look likely to be sustained, as older coal fired plants are retired from 2015 onwards.

The US Becomes a Major LNG Exporter

The US unconventional gas industry has spawned a major new LNG and pipeline export industry, replacing anticipated LNG imports and sharply reducing traditional Canadian pipeline imports. According to work commissioned by the US DOE, and performed by NERA (Baron *et al.* 2015), the macroeconomic benefits of LNG exports from the United States were positive to the nation across all 16 export scenarios studied. Further, the benefits increased as the level of LNG exports increased. That is, the benefits of export expansion outweighed the costs of higher domestic gas prices and subsequent slower demand growth. This study anticipated that price rises would be in the range US$0.22 to US$1.10 per Mbtu, the latter applying only in the case where gas exports ramped up very quickly. In all the scenarios, US gas prices did not rise to export netbacks but remained $6–8 below Asian gas prices, taking into account inland transport, liquefaction, shipping, and regasification.

The NERA study concluded that trade-exposed gas-intensive industry sectors that might be affected by higher prices were relatively small, with low employment and low value-addition. With five major projects now approved, having a capacity of 90 bcm/annum, the US will become a major LNG exporter by early in the 2020s. The first shipments should start in 2016.

Costs and Risks are Significant, and Often Localised

There is a fundamental mismatch between costs, risks and benefits in terms of unconventional gas. Many, if not most, of the costs and risks are borne at a local level; these include the possible loss of property values for groundwater-dependent homes close to shale gas developments (Muehlenbachs *et al.* 2015) and for houses that use private well water (Gopalakrishnan and Klaiber 2014). However, locations near but not immediately proximate to developments may benefit from higher property values (Weber *et al.* 2014).

Typically, many benefits accrue at regional or even national level. Some, such as lower gas prices, are diffuse but important. Thus, strong local opposition to unconventional gas production should not be surprising or illogical although it may still be supported at a state or national level.

Impacts can and do differ over time, with some acute effects occurring during the period when the well is completed or field development is most intense, while some benefits can be slower to appear or may even be ephemeral. Local risks may also persist over a long period of time, even after production has finished and the wells are capped. The recommendation of the NSW Chief Scientist, accepted by the NSW State Government, is that a Community Benefits Fund, designed to address these potential inequities, should be established. More concretely, the Queensland Government's 'Royalties for Regions', with co-contributions from the State and industry, is designed to target this issue.

Gasland

The 2010 film *Gasland*, by Josh Fox, raised some of the environmental issues associated with unconventional gas development, notably the possible methane contamination of water sources, especially surface aquifers, and the potential contamination of water sources by chemicals used in hydraulic fracturing fluids such as ethylbenzene and toluene. The effects of potential air pollution, including ozone formation, were also raised, along with the increased impact on local communities of population increases, truck traffic, noise and surface spills of chemicals. The difficulties of ensuring high levels of well-casing integrity over a long period, and the consequences of well failures, were also highlighted.

Subsequently, Josh Fox produced *Gasland 2* in 2013, and *The Sky is Pink*, the latter relating the potential issues associated with hydraulic fracturing to those in south central New York State and the potential adverse consequences of lifting that State's moratorium on hydraulic fracturing, at least in six counties.

The petroleum industry reacted strongly to *Gasland* by producing its own film, *Truthland*, reacting to what it claimed were factual inaccuracies in *Gasland*.

While many of the facts presented in *Gasland* were already on the public record, notably the 2005 exemption of hydraulic fracturing from the US Safe Drinking Water Act, the film did popularise actual and perceived problems and potential risks. It has been, without doubt, very influential in public debate in the United States, and indeed globally.

Responses to the film have varied from ongoing calls for open-ended moratoria on hydraulic fracturing in some States of the US to complete rebuttals. Highlighting such issues has allowed concerned groups to propose more comprehensive regulatory packages, including more effective air emission and methane controls, groundwater protection standards, greater disclosure and transparency, properly resourced compliance and enforcement, and property owner protection; (see for example Ohio Environmental Council 2013). The ability of State regulators to cope with the rapid growth of the industry, and to provide confidence that the issues raised are being addressed, has also received more scrutiny (Baizel 2012).

Well Integrity, Air Pollution

A number of local environmental issues have arisen in the United States, and are the subject of ongoing monitoring and study. For example, in North Dakota, a shale-gas well blowout in 2010 led to the uncontrolled release of hydraulic fracturing fluids and formation fluids, just outside the 2.5 mile wellhead protection zone of the water supply well of a nearby town.

Significant site clean-up was required after the well was capped, and ongoing monitoring was required.

In Washington County, in western Pennsylvania, around 1000 wells have been drilled, and large gas-processing facilities installed. In this area, air pollution levels of benzene, toluene, and formaldehyde are elevated, according to the Pennsylvania Department of Environmental Protection (DEP) data, and one gas-processing plant has been ordered to prepare an improved air pollution management plan.

In Bradford County, also in Pennsylvania, with over 2000 wells at a density of 1.8 wells per square mile, there have been numerous reports of groundwater anomalies. The DEP has issued notices of violation for infractions at wells in this area, including a gas-well blowout that released flowback and produced water. Further, methane pollution from faulty wells in Susquehanna County, Pennsylvania, led to a DEP prohibition on further drilling. Affected properties were purchased, and the residences subsequently demolished. These latter two counties are on the State's north eastern border with New York, which may explain some of the adverse reaction to shale gas production in the latter State.

Localised Water Impacts must be Managed

The US EPA study on the potential impacts of hydraulic fracturing on drinking water chose, as one of its case studies, Washington County in western Pennsylvania. Residents in several areas of the county reported negative effects on their drinking water wells; notably increased turbidity, discolouration of sinks, and transient organic odours. This study sought to determine whether any of these problems were related to oil and gas production and in particular to leaking or abandoned wells as well as to modern gas wells with enhanced output techniques. A draft report was released for public comment in mid-2015, with a final report anticipated in 2016.

Some water quality issues are being addressed by rapidly improved and tightened regulation, including for example Federal Environmental Protection Agency (EPA) rules mandating compulsory green completions, which should sharply reduce methane emissions at the completion stage of well construction. Similarly, drilling is now prohibited to a greater radius around houses and water wells. Nonetheless, it is clear that there have been, and there remain, significant potential risks for those close to drilling activity. For some people thus affected, little direct benefit from unconventional gas development may be seen.

Australia Shares Many of these Issues, with a Few of its Own

In Australia, local concerns over coal bed methane differ somewhat from those associated with shale gas. First, the requirement to remove large quantities of formation water can lead to depleted water supplies. Second, water so produced can be more saline than existing water supplies. Where hydraulic fracturing is used, some of the same water concerns as in the case of shale gas can arise. To date, there have been instances of inappropriate handling of waste water streams, and all regulatory authorities are responding to this issue. It seems likely that, at least in Queensland, area-wide reverse osmosis plants will be constructed to remove salt and allow reuse of the bulk of produced water, with careful disposal of saline waste streams, although standards and regulations remain to be developed. Such treatment seems likely at the proposed Narrabri development in New South Wales, where around 80% of the water extracted seems likely to be reusable. This water is not currently accessible or usable.

The question of the macroeconomic benefits of large-scale unconventional gas exports for Australia may not be as easily answered as for the United States, given that gas prices in Eastern Australia were set to rise rapidly in any event as long-term low-price contracts entered into, in some cases decades ago, are due to expire over the period to 2018, notably for NSW buyers. Nonetheless, it can be realistically asserted that prices at the Gladstone export hub will be at or below the export netback price, which will vary with individual contracts and with the escalators that are included, such as oil prices (these have in any event fallen sharply, as much lower oil prices flow through to LNG contracts). That is, wholesale gas prices at Gladstone should be on average around US$4/Mbtu (thus around A$6–8/GJ) below delivered prices in export markets such as Korea and Japan. However, at low LNG prices these margins may contract. As the plants ramp up production, there is a possibility that prices may rise above export netback levels for a time during this transition period (BREE 2014) or, by contrast, they may even fall to low levels as gas supply ramps up before export plants come on line.

Supplies in areas such as Narrabri in NSW can expect to be even further below export netback, given the significant pipeline transport cost savings, although future interconnections might change this situation. In particular, internal transport costs can add $2–3/GJ to wholesale prices, for example for gas delivered to Sydney from domestic sources (ACIL Tasman 2013). As in the United States, the faster exports ramp up, the more pressure is likely on prices, in this transition period, especially as the export industry will dwarf domestic demand in eastern Australia. This is very different from the US, where exports will remain a relatively small part (around 10%) of the integrated North American gas market.

One analysis (ACIL Tasman 2012) indicated that, in Australia, under current likely export levels some 20 000 to 30 000 jobs would be created over the longer term and gas and power prices would be lower over the medium to longer term. Further, Queensland government revenue from coal bed methane could be some $9 billion over the period to 2035 in net present-value terms. Despite these benefits, recent work on the costing of other sources of unconventional gas suggests that Australia will not be able to match the extremely low gas prices seen in some parts of the United States, given Australia's high costs for wages, materials and hence drilling (Cook *et al.* 2013, APPEA 2012).

Royalties Make an Important Difference in Australia

An important issue is the sharing of royalties. In the United States, generally (but not exclusively) royalties go to the landowner, typically at 12.5% but increasingly at higher levels. By contrast, in Australia the landowner is the province or the State government.

Queensland stands to make around $850 million per annum in royalty payments from CBM export development, depending on the net wellhead value of the gas, which has fallen with LNG prices. Lease payments are typically a few thousand dollars per well, a substantial sum for a large landowner who may host 100 wells but scarce compensation for a small landowner who hosts only one or two wells.

Payments to landowners highlight the mismatch between the real and potential impacts, which tend to be heavily concentrated at local levels. While some benefits, notably jobs, tend to be at local and regional levels, large-scale benefits are more at State and national level. In Queensland these benefits, according to industry and Queensland government estimates, may account for 30 000 jobs during the construction phase. Ongoing jobs are estimated to peak at 1000 to 17 000 in about ten years' time (on the basis of 45 000 wells drilled), with a boost to the state product by $3 billion per year (Energy Skills Queensland 2013)

To ensure that the costs in generating these benefits are minimised, best practice and regulation at a local level are required. This is already under way in several jurisdictions and is an absolute requirement for ensuring that the net benefits of unconventional gas development are distributed fairly across the producing state or region.

Prospects for Unconventional Gas Production

A number of uncertainties are present in any analysis of the outlook for gas markets, and for unconventional gas production and use within that outlook. On the one hand, State environmental policies, especially those directed at mitigating climate change, will lower fossil fuel demand, although the demand for gas seems likely to be much less heavily affected than that for coal. On the other hand, policies directed at improving local environmental outcomes, such as those seen in China, should increase gas demand.

On the supply side, the rate at which new players can tap unconventional gas resources has the ability to transform gas markets, in for example the Asia–Pacific region, if China could replicate North American gas-production growth rates (IEA 2015). The rate at which non-OECD governments move away from energy policies governing supply, demand and pricing, in regions or countries as diverse as the Middle East, China, India and Russia, will also impact markets profoundly. For example, Russia keeps its domestic gas prices low, thus discouraging efficiency. As a result its power sector uses slightly more gas than the US but produces barely more than 40% of the US power output.

The spectacular fall in the price of oil, beginning in mid-2014 and accelerating markedly since October 2014 has raised an even more fundamental issue: how will the change in oil prices and expectations change the global outlook for energy and gas markets? The response needs to include the consideration of supply, demand, and investment and of what will be the implications for unconventional gas production and exports in the United States and Australia.

An analysis of the supply effects of lower oil prices is taking place against an especially uncertain global and domestic economic outlook. Global economic growth remains subdued, with the US and UK slowly emerging from recession but in 2015 the euro area was still trying to avoid falling into deflation; this was further complicated by ongoing concerns over Greece. Japan was also struggling to reignite growth, while China was showing signs of a marked deceleration of growth and had dropped its official economic growth target to 7.0% from its previous 7.5% level. At the same time, a clear swing away from industrial growth to a more service-oriented economy could clearly be seen in 2015, with falling coal use.

While the fall in oil prices, with corresponding benefits for consumers, may provide a very welcome global stimulus, some countries, notably oil and gas exporters, will be negatively affected. One of the worst affected is Russia, which is already subject to sanctions associated with the war in Ukraine. While the larger Middle Eastern countries have sufficient monetary reserves to maintain spending, other smaller oil and gas producers, such as Nigeria, Venezuela, Algeria and Angola, may struggle. Many countries that have traditionally subsidised or price-controlled oil products have taken the opportunity to begin to reform their pricing policies, with beneficial impacts on public finances and energy efficiency. Such changes will also weaken any upward tendencies in demand. Oil-indexed gas prices will fall with declining oil prices, with a lag of between three and nine months. Further, gas-indexed prices will remain responsive to supply-and-demand fundamentals in energy markets, as well as to production input costs, which would seem likely to fall.

The immediate impact of the oil price fall has been seen in sharp reductions in some input costs, such as drilling supplies (bits, mud, pipe), rig costs (notably offshore rigs) and wage costs, across the hydrocarbon sector especially in North America. Drilling rig counts in oil in the US have fallen sharply for oil, but less so for gas. The transition to high productivity horizontal drilling has already happened whereas, in oil, lower productivity vertical rigs are being retired rapidly. The improvement in productivity is evident in Figure 5.1 and also the ongoing growth in total shale gas output (Figure 5.3); some change in basin activity over the period can be seen. One important factor is that there has been a decoupling of gas output from active drilling rigs, owing to the large number of wells that have been drilled, but have not yet been fully developed.

Capital spending reductions announced by most major energy companies total more than $100 billion as of early 2015. Over 2015, capital expenditure plans were reduced by 20%, with further cuts likely in 2016 (IEA 2015). These cuts were expected to have a relatively quick impact in unconventional oil output as such oil wells deplete rapidly, so that falls in oil production, or at the very least declines in growth, started to appear by the end of 2015. Hence the growth rate in US gas production seen in 2014 (15% over 2013) was likely to fall to around 5% in 2015 (according to EIA estimates).

The US gas output continued to grow, with hub prices steady or falling, despite the cold weather, to around $2/Mbtu in late 2015. With five LNG projects underway by mid-2015, and a pricing model independent of oil prices, US LNG output could rise to close to over 90 bcm by 2020, noting that such a figure is only around one tenth of the North American gas market. The taking of a final investment decision in mid 2015 on the fifth LNG project, Corpus Christi in Texas, provided an early indication of the impact of lower oil prices on US

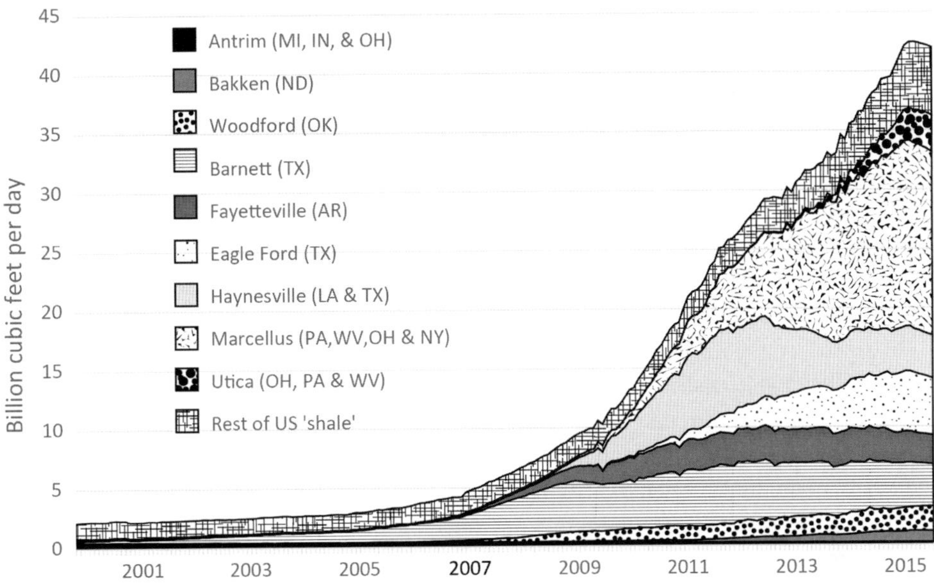

Figure 5.3 Monthly dry shale gas production. *Source:* US Energy Information Administration (December 2015).

export developments, notably that the business model appears robust, based on hub-related pricing.

Australian LNG Projects

Those of Australia's LNG projects that are under construction are generally already well advanced. Lower oil prices may result in a fall in some costs, such as imported equipment, and possibly a lesser fall in labour costs. Import cost falls will be offset by a weakening Australian dollar. In terms of realised prices for LNG, these were expected to fall from a (hypothetical) level of US$16–17/Mbtu in late 2013 to around US$8 in late 2015, again modified (but favourably) by exchange rate changes, to around A$10–11/GJ. Reductions in revenues from falling oil prices have already been priced into the equity values of affected companies, such as Santos and Origin.

Coal bed methane projects, which must continue to drill over the project life, will see a higher degree of variability in their gas production costs relative to those for large offshore developments in western and northern Australia that are based on conventional gas development. Thus, upstream gas production investment for CBM is akin to operating costs, as each project will need to drill as many as 200 to 300 wells per annum throughout the project's life and also to connect the production wells to the gas pipeline grid. Such drilling and network costs can amount to $4 million per well, or upstream operating costs of more than $1 billion per annum. Conventional gas upstream costs show more bias to initial capital cost but much lower operating costs. Thus the ongoing viability of each type of project will be affected differently, noting that in most cases the final investment decisions were taken with a generally lower oil price environment and were tested at oil prices below $100, the range varying with the companies concerned.

For major conventional LNG projects, ongoing operations seem likely to remain cash positive, although the rate of return on the project will obviously suffer and this will be compounded by any cost overruns or delays. The same reasoning applies to CBM-based projects, with added uncertainties on upstream costs. Certainly, if oil prices were to fall further, higher cost CBM projects would face much lower returns unless their drilling and other operating costs can be reduced markedly.

In any event, the developers of the seven new gas projects (conventional and unconventional) in Australia can be expected to concentrate on the rapid completion of projects and a reliable, cost effective, entry into service of LNG production. De-bottlenecking activity, targeting increasing capacity and output for minimal capital outlay, will be a priority. This behaviour could be expected to be independent of oil and LNG price movements, but it is certainly to be pursued with more urgency in a $50 oil price environment. In the eastern states, the sharp fall in oil prices, and therefore export netback gas prices, may relieve some price pressure in local markets but a degree of market turbulence as LNG projects ramp up is still to be expected.

In terms of possible new unconventional gas developments around the globe, especially those tied to export projects, it is clear that lower oil-linked gas prices will tend to stimulate demand. This gas should be more competitive in many countries and applications, notably peak and intermediate power applications in some Asian countries notwithstanding strong competition from cheap coal and rapidly growing renewable energy supplies. The question then arises whether such prices will be capable of sustaining new, relatively high cost, developments, involving UCG, or expensive offshore conventional-gas-export development.

In the case of Australia the country is a high capital cost location, in terms of construction costs for large-scale chemical engineering projects such as petrochemical plants, petroleum refineries and LNG plants. A similar observation can be made for other potential projects, for example LNG plant sites such as Canada (with remote greenfield sites in north-west British Columbia) or conventional gas projects, in for example eastern Africa, or for extreme projects such as Yamal, although some of these potential projects may be able to tap very cheap gas production. Certainly these localities are more expensive than the US Gulf of Mexico, which is arguably amongst the cheapest places in the OECD to perform such construction. The Australian record, in that the current generation of gas projects has not been constructed on time or on budget, would suggest that other locations will be favoured for future greenfield investment.

In January 2015 the shelving of the fourth LNG plant at Gladstone, planned to be a joint venture of Shell, Petrochina and Arrow Energy and to be based on Surat Basin CBM resources, confirmed that for the near future new greenfield developments are highly unlikely in Australia. At the same time as this cancellation was announced, Shell confirmed delays in proposals for LNG plants in British Columbia and in Elba Island in Georgia, as part of a marked reduction in planned capital spending. Such marked reductions in spending have been seen throughout the global petroleum industry, as noted above, and reductions in future capital spending plans have been observed and are likely to increase as the fall in oil prices is prolonged.

Despite its high construction costs, Australia still has significant advantages over some other potential LNG project locations. Although these advantages (stable policy, tax and legal frameworks) can be difficult to quantify, they do represent sources of competitive advantage, for long-lifetime high-capital-cost projects.

From a gas buyers' perspective, Australian production will comprise around one fifth of global LNG supplies in the period after 2020, raising the issue that if further plants are built in Australia then buyers may be over-exposed to one particular supplier. Australia's record as a reliable supplier is impeccable, but buyers with a high reliance on LNG may be concerned by a concentration in supply from one country. For buyers with a mix of domestic and imported pipeline and LNG (e.g. China) this is less of an issue.

Environmental regulation of Australian onshore gas production, mostly within the purview of the Australian States, needs to be mindful that technology is advancing fast, with new UCG techniques likely to be deployed in onshore basins such as the Cooper Basin, as well as in the Surat and Bowen Basins. Government activity will need to focus on robust, credible, transparent regulation plus ensuring that gas markets work efficiently to move gas around markets, particularly in the eastern States.

Elsewhere, global oil companies have continued to see cash flows fall sharply, so that commitments to new ventures, from offshore Brazil projects, to new Canadian tar sands, and new gas sources using unconventional technologies seem certain to see marked cutbacks. New greenfield developments in Mozambique and elsewhere in East Africa, which, in early 2014, appeared likely to be the next round of LNG developments, seem set to see investment decisions delayed, so gas output from them is now more probably in a 2025 timeframe.

Conclusions

International gas markets have proven extremely dynamic and unpredictable in the last decade. Large investments in LNG capacity to supply a North American market thought to be rapidly running out of gas were misplaced, as shale gas output has expanded rapidly. Global trade patterns have had to adapt rapidly and this adaptation looks set to continue as parts of the US supply chain are being speedily converted to export facilities. Any outlook projections seeking to analyse future gas markets and the role of unconventional gas need to acknowledge the many uncertainties facing oil, gas, coal, and energy markets generally.

Chinese gas demand and supply have the potential to markedly change the global gas picture or, at the very least, the Asia–Pacific balance, especially post 2020. Should Chinese efforts to develop their own shale gas and CBM resources follow a European rather than a North American trajectory, imports may need to rise. Changes to the deregulatory or liberalising path in gas pricing and transport policy have the potential to shift both demand and supply. For example, if prices rise too rapidly, demand projections may appear unrealistic. The Chinese market appears well supplied by interregional pipelines, post 2020, while the outlook for extra LNG supply looks less certain.

We have highlighted how new extraction technologies, or certainly their refined and more widely applied forms, have transformed the global gas market. Possible other new technologies could also disrupt future markets. Solar photovoltaic power costs have fallen some 40% since 2010 (on a levellised cost of electricity (LCOE) basis), as economies of manufacturing in China have driven the costs of panels down quickly. While this rate of cost decline seems unlikely to continue, costs should continue to fall and, in markets such as China and India, solar technologies have many attractions including short lead times, low operating costs, a high quality solar resource in some locations and suitability for distributed applications. Power storage technologies, if they became cheaper, lighter and more compact than those currently available, could equally transform the power and energy landscape, reducing sharply the attractions of gas-fired power as a source of flexible supply. Finally, the approach that countries take on climate change issues has the potential to radically change fossil fuel use, although gas may benefit from such policies in the short to medium term.

Notwithstanding all these uncertainties, it seems that UCG is already a permanent feature of the North American and Asia–Pacific gas markets and indeed it has already transformed global gas trade patterns. These trends seem set to continue, with the potential to spread to other countries, if perhaps not at the same rate as seen in the early-adopting regions.

This chapter draws meterial from Cronshaw and Grafton (2016).

References

ACIL Tasman (2012). Economic significance of coal seam gas in Queensland. Report prepared for the Australian Petroleum Production and Exploration Association.

ACIL Tasman (2013). Cost of gas for the 2013 to 2016 regulatory period: a report on the wholesale cost of gas for the review for standard retailers in New South Wales. Report prepared for the Independent Pricing and Regulatory Tribunal.

Australian Petroleum Production and Exploration Association (APPEA) (2012). State of the industry 2012: a status report on Platform for Prosperity – a strategy for maximising the value of Australia's oil and gas resources. www.appea.com.au/wp-content/uploads/2013/04/121130_State-of-the-Industry-2012_web.pdf (accessed 23 December 2015).

Baizel, B. (2012). Testimony before the US House Committee on Natural Resources, Subcommittee on Energy and Mineral Resources oversight hearing on 'Federal regulation: economic, job and security implications of federal hydraulic fracturing regulation. Earthworks. www.earthworksaction.org/library/detail/testimony_before_the_us_house_committee_on_natural_resources#.Vnmynnlunct (accessed 23 December 2015).

Baron, R., Bernstein, P., Montgomery, W.D. and Tuladhar, S. (2015). Macroeconomic impacts of LNG exports from the United States. *Economics of Energy & Environmental Policy*, 4 (1).

British Columbia Oil and Gas Commission (2014). Investigation of observed seismicity in the Montney trend. Technical Report.

Brown, J.P. (2014). Production of natural gas from shale in local economies: a resource blessing or curse? Federal Reserve Bank of Kansas City. Accessed from https://www.kansascityfed.org/publicat/econrev/pdf/14q1Brown.pdf on 21 December 2015.

Brown, J.P., Fitzgerald, T. and Weber, J.G. (2015). Capturing rents from natural resource abundance: private royalties from US onshore oil & gas production. Paper presented at the 2015 Agricultural & Applied Economics Association and Western Agricultural Economics Association Meeting, San Francisco, CA, 26–28 July 2015. Accessed on 21 December 2015 from http://ageconsearch.umn.edu/bitstream/205657/2/royalties_manuscript_ageconsearch2.pdf.

Bureau of Resources and Energy Economics (BREE) (2014). *Eastern Australian domestic gas market study*. Department of Industry, Canberra.

Center for Sustainable Shale Development (CSSD) (2015). Performance standards. www.sustainableshale.org/performance-standards/ (accessed 23 December 2015).

Cook, P., Beck, V., Brereton, D. *et al.* (2013). Engineering energy: unconventional gas production. Report for the Australian Council of Learned Academies, www.acola.org.au.

Cronshaw, I. and Grafton, R.Q. (2016). Economic benefits, external costs and the regulation of unconventional gas in the United States. *Energy Policy*, 98: 180–196.

Energy Skills QLD (2013). Queensland CSG to LNG Industry Workforce Plan. www.energyskillsqld.com.au/wp-content/uploads/Queensland-CSG-to-LNG-Industry-Workforce-Plan-Energy-Skills-Qld.pdf (accessed 11 Dec. 15).

Environmental Protection Agency (EPA) (2015). Assessment of the potential impacts of hydraulic fracturing for oil and gas on drinking water resources. EPA/600/R-15/047a, June 2015.

Fleming, D.A. and Measham, T.G. (2015). Local economic impacts of an unconventional energy boom: The coal seam gas industry in Australia. *Australian Journal of Agricultural and Resource Economics,* 59 (1): 78–94.

Fleming, D.A., Komarek, T., Partridge, M. and Measham, T. (2015). The booming socioeconomic impacts of shale: a review of findings and empirical methods. Working Paper available at Research Gate. DOI: 10.13140/RG.2.1.1253.7685.

FracFocus (2014). Hydraulic fracturing: how it works. Available at: http://fracfocus.org/hydraulic-fracturing-process (Accessed 10 December 2015).

GasFields Commission (2013). Shale gas regulation – an American perspective. www.gasfieldscommissionqld.org.au/whats-happening/shale-gas-regulation-an-american-perspective.html (accessed 11 December 2015)

Gopalakrishnan, S. and Klaiber, H.A. (2014). Is the shale energy boom a bust for nearby residents? Evidence from housing values in Pennsylvania. *American Journal of Agricultural Economics,* 96 (1): 43–66.

IHS (2012). The economic and employment contributions of unconventional gas development in State economies. Report prepared for America's Natural Gas Alliance. http://marcelluscoalition.org/wp-content/uploads/2012/06/State_Unconv_Gas_Economic_Contribution_Main.pdf (accessed 11 December 2015).

International Energy Agency (IEA) (2012). *Golden Rules for a Golden Age of Gas: World Energy Outlook Special Report on Unconventional Gas*. OECD Publishing, Paris.

International Energy Agency (IEA) (2013). *World Energy Outlook 2013*. OECD Publishing, Paris.

International Energy Agency (IEA) (2015). *World Energy Outlook 2015*. OECD Publishing, Paris.

Muehlenbachs, L., Spiller, E. and Timmins, C. (2015). The housing market impacts of shale gas development. *American Economic Review*, 105 (12): 3633–3659.

Obama, B. (2014). President Barack Obama's State of the Union Address. www.whitehouse.gov/the-press-office/2014/01/28/president-barack-obamas-state-union-address (accessed 11 December 2015).

Ohio Environment Council (OEC) (2013). OEC proposes safer Gas Act for oil and gas drilling. www.theoec.org/press-releases/oec-proposes-safer-gas-act-oil-gas-drilling (accessed 11 December 2015)

Porter, M.E., Gee, D.S. and Pope, G.J. (2015). *America's Unconventional Energy Opportunity: A win–win plan for the economy, the environment, and a lower-carbon, cleaner-energy future*. Harvard Business School, Boston.

Standing Council on Energy and Resources (SCER) (2013). The National Harmonised Regulatory Framework for Natural Gas from Coal Seams. http://scer.govspace.gov.au/files/2013/06/National-Harmonised-Regulatory-Framework-for-Natural-Gas-from-Coal-Seams.pdf (accessed 11 December 2015).

Watson, M. (2014). What LEED did for buildings, this could do for shale gas production. US Environmental Defense Fund blog, January 2014.

Weber, J.G. (2012). The effects of a natural gas boom on employment and income in Colorado, Texas and Wyoming. *Energy Economics* 34 (5): 1580–1588.

Weber, J.G., Burnett, J.W. and Xiarchos, I.M. (2014). Shale gas development and housing values over a decade: evidence from the Barnett Shale. Working Paper No. 14–165. US Association for Energy Economics Proceedings Paper. Accessed on 21 December 2015 at: http://works.bepress.com/wesley_burnett/9.

6

Economics of Shale Gas in the United States

FRANCIS O'SULLIVAN

Introduction

The past 15 years have been a period of tremendous change for natural gas in the US. During the first half of the 2000s, US domestic natural gas production fell and prices rose significantly. Yet, at the time, experts expressed concern that the US would have to rely increasingly on relatively expensive gas imports – in the form of Canadian pipeline gas and LNG – in order to supplement dwindling domestic output to meet demand (NPC, 2003). Remarkably, in the decade since, a very different reality has emerged. Domestic natural gas production in the US has surged to an all-time high, and the price of gas has fallen to levels not seen since the 1990s. So, what happened? Shale gas development is the explanation.

Shale gas and its rise represents one of the most significant and unexpected dynamics to impact on the global energy system in many decades, and it will certainly continue to have a profound impact on the US and international energy sectors for decades to come. The term "shale gas" is just an umbrella term used to describe natural gas produced from a variety of hydrocarbon-bearing shale and mudstone formations. "Unconventional gas" is a slightly broader term, often used to refer to gas from shale along with gas produced from coal beds and other lower-quality reservoir settings (NETL, 2013).

Although many organic-rich shale formations have long been known to contain significant volumes of hydrocarbons, owing to their being source rocks in petroleum systems, the gas resources in these rocks were historically viewed as economically unrecoverable. This is due to the very low levels of fluid permeability that are characteristic of shale-type rocks. Very low permeability makes it difficult to achieve commercially acceptable flow rates from shale wells developed using the technology typically deployed up until the early 2000s. Since the early 2000s, however, rapid innovation in the application of more complex drilling and reservoir stimulation technologies – in particular the combination of horizontal drilling and hydraulic fracturing – means that dramatically higher flow rates can now be achieved from shale wells. These much higher flow rates have, to date, more than compensated for the higher costs associated with horizontal drilling and fracturing and allowed previously unrecoverable shale gas resources to be developed.

The Rise of Shale Gas

The 1990s witnessed the full deregulation of the natural gas sector in the US. Wellhead price controls were eliminated, and competitive market forces became the driver of natural gas supply and pricing. As the market matured, gas trading increased and important trading hubs including the "Henry Hub", today's US benchmark, emerged. During this period US natural gas production increased, growing by 6.5% between 1990 and 1999,[1] while prices remained low and stable, averaging just $1.90/Mcf across the decade.[2] The dawn of the new millennium, however, brought with it changing circumstances for the US gas sector. Significant price volatility became a feature of the market in the early 2000s, as did a secular rise in the average price of natural gas.

The increasing level of volatility in US natural gas pricing witnessed during the early 2000s was the result of strong growth in gas demand during that period, coupled with a tightening of low-cost supply. At the time these dynamics were somewhat unexpected; in retrospect, however, this was simply the result of an 'overhang' of low-cost supply during the 1990s that was eliminated by demand growth by the early 2000s (Joskow, 2013). As these dynamics unfolded it was widely agreed that the US would soon become more reliant on higher cost gas resources. The role of lower quality – and thus more expensive – domestic gas resources was expected to increase, as was the scale of natural gas imports both from Canada and via LNG from the rest of the world (NPC, 2003).

At no point during the early 2000s was the shale resource that we know today broadly expected to play a significant role in supporting future US supply. Rather, it was believed that "tight" sandstone plays, particularly those in the mid-continent and Rockies, would become the dominant unconventional resource. The breakdown of gas production by resource type supported this conclusion at the time. In 2000, tight sandstone formations accounted for 21% of US output, with shale only making up 1% and much of that coming from mature plays including the Antrim Shale in Michigan (MIT, 2011). There were, however, efforts afoot elsewhere to tap the shale resource, most notably those of George P. Mitchell and his company Mitchell Energy. Mitchell's work had focused on tapping the Barnett Shale located in Texas' Fort Worth Basin, and it was there that the combination of horizontal drilling and hydraulic fracturing that is now ubiquitously applied to shale development was developed and effectively operationalized.

The production of natural gas is a techno-economic enterprise in the sense that even if a technology is available to be deployed this will not happen unless the financial return from that deployment justifies the incremental cost. A set of techno-economic circumstances evolved in the US gas market during the early 2000s that allowed Mitchell, and others, to deploy horizontal drilling and hydraulic fracturing technology in the Barnett Shale. Owing to technical progress from a decade or more of innovation, the efficiency of the overall horizontal drilling and fracturing process improved and its deployment cost fell while simultaneously the price of natural gas increased. With a greater financial return

[1] Marketed natural gas production as reported by the US Energy Information Administration. www.eia.gov
[2] Wellhead prices in nominal dollars as reported by the US Energy Information Administration. www.eia.gov

possible from the use of these new technologies, more deployment occurred and this led to a virtuous circle where experienced was gained, performance improved, costs were further reduced and even more deployment occurred. In the case of the Barnett Shale the rapidity of these developments was remarkable. Prior to 2000 only a handful of horizontal wells had been drilled in the Barnett Shale but by the end of 2005 there were almost 1200 such wells in place and, as of early 2015, almost 16 000 horizontal wells had been developed in that play.[3] The quantitative impact of this development on overall US gas supply has been very significant. In 2000 the Barnett Shale produced 0.2 bcf/day of gas from the relatively modest number of vertical wells in the play. Although not an insignificant amount of gas, such output was a vanishingly small fraction of the total US daily production that year, of over 55 bcf/day.[4] By 2005, the Barnett Shale's output had jumped to 1.3 bcf/day owing to the use of horizontal wells, and since then it has surged to beyond 5 bcf/day.[5] Put into context, this level of output is equivalent to about 8% of the total US daily gas production during the first half of the 2010s and is equal in scale to the total current output of major gas-producing nations like Malaysia, Australia and Nigeria.[6]

The Barnett Shale experience demonstrated that gas-bearing shale rock formations could be developed successfully, and it encouraged prospective plays elsewhere. Since 2005, several additional major shale gas plays have been developed. The five most prominent of these are the Eagle Ford Shale, located in the Texas Louisiana Gulf Coast Basin, the Fayetteville Shale in the Arkoma Basin, the Haynesville Shale spanning the East Texas and Arkla Basins and Oklahoma's Woodford Shale along with the largest shale play of all: the Appalachian Basin's Marcellus Shale. Combined with the Barnett, these five plays' gas output now amounts to a staggering 37 bcf/day; equal toabout half of all US gas production,[7] The growth by play of US shale gas production from 2005 to mid-2015 is shown in Figure 6.1. Along with illustrating the staggering scale of absolute production growth over the past decade, the figure also highlights how each play's contribution to this growth has evolved. The Barnett's central role in the vanguard of shale production is clear, providing as it did the majority of growth up until 2008–2009. Since 2009, however, this play's output has remained relatively flat and today development activity in the play is basically serving to maintain steady output levels rather than to grow them. Output from the Fayetteville and Woodford plays has also levelled off, while the Haynesville is in decline. The Haynesville is an interesting case in that it has a particular set of physical and economic characteristics that render its output particularly sensitive to how the price of gas evolves.

Figure 6.1 highlights that, since 2012, growth in US shale gas output has been driven by the expansion of the Marcellus and Eagle Ford. These two plays happen to be endowed with excellent shale gas resources; the growth in their output since 2012, however, is not due to

[3] Based on well completions reported by the DrillingInfo HPDI well database. www.drillinginfo.com
[4] Marketed natural gas production as reported by the US Energy Information Administration. www.eia.gov
[5] Based on well production records reported by the DrillingInfo HPDI well database. www.drillinginfo.com
[6] Based on 2014 natural gas production statistics reported by the BP Statistical Review of World Energy. www.bp.com
[7] Based on well production records reported by the DrillingInfo HPDI well database and marketed natural gas production data as reported by the US Energy Information Administration.

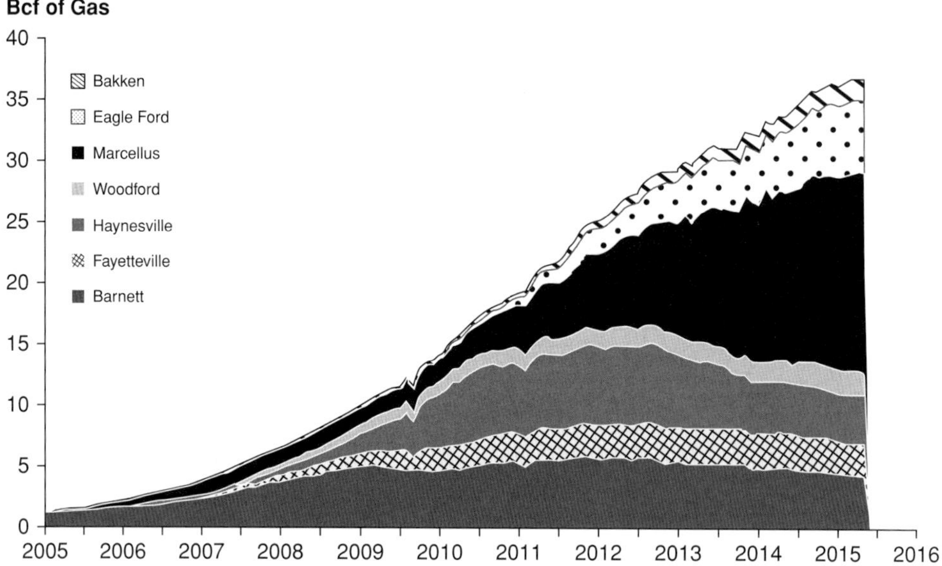

Figure 6.1 Natural gas production from major US shale plays from 2005 to 2015.

gas production alone. The Eagle Ford formations contain areas rich in oil, and these "tight oil" resources were enthusiastically developed during the period of higher oil prices that existed up until late 2014. This activity yielded significant volumes of 'oil-associated' gas production, boosting the output from the play. The Marcellus is not oil prone; however, the western portion of the play contains "wet gas", natural gas containing appreciable volumes of heavier natural gas liquids (NGLs). The monetization of these NGLs helps boost the economic attractiveness of the Marcellus development.

Along with the dramatic shorter-term impact on gas production that has resulted from the rise of shale gas, the very fact that these resources are now recoverable has had a major impact on assessments of the scale of total recoverable natural gas resources. In 2003, the US National Petroleum Council (NPC, 2003) estimated that the US endowment of recoverable natural gas resources amounted to 1438 Tcf, with 197 Tcf of that in the proved reserve category. The US consumption that year amounted to 23 Tcf, meaning that the national reserve-to-consumption (r-to-c) ratio stood at only 8.5 : 1. In that NPC assessment, the total recoverable shale gas resource was estimated to be 35 Tcf. These numbers were established before the success of Mitchell and others in the Barnett Shale and elsewhere became clear.

Since this time very significant upward revisions have been made in resource assessments in order to account for shale. One of the first such revisions by a credible assessing agency was produced by the Potential Gas Committee (PGC, 2009) in their 2008 biannual report on the supply of gas in the US. At that time the PGC reported a mean estimate of 2080 Tcf for the US total recoverable gas resource, of which it was suggested that 616 Tcf was recoverable from shale formations. In 2011, MIT estimated the mean recoverable US shale resource to be 631 Tcf, with the total resource amounting to 2156 Tcf (MIT, 2011). The most recent PGC report, published in 2015, raised their assessment of the 'most likely' recoverable shale gas resource to 1253 Tcf, which, when combined with the other resource categories, yields a total recoverable resource estimate of 2853 Tcf (PGC, 2015). Along with these enormous impacts on overall resource estimates, shale gas has also driven a large increase in proven reserves. Between 2003 and 2013, US proven reserves grew from 197 to 354 Tcf, with all the growth coming from shale gas reserve additions.[8]

Assessing the Cost of the Shale Resource

The role of shale gas in overall US natural gas production grew significantly during the latter half of the 2000s because the net returns from shale gas gained in attractiveness relative to other sources of natural gas. Prior to the early 2000s, the level of US natural gas prices meant that shale gas development was too costly. The steady increase in the price of natural gas through the early 2000s, however, coupled with improved drilling and hydraulic fracturing performance and the accompanying reduced costs, meant that shale gas became an increasingly attractive target for commercial development by the US natural gas sector. Progress in the development of the shale resource has been so significant that, since 2005, shale plays have been the focus of the vast majority of natural gas developments in the US. Perhaps the most notable illustration of this shift from "conventional" to "unconventional" gas has been in the collapse of production from the Gulf of Mexico. In 2000 the Gulf produced almost 5 Tcf or ~25% of the total US marketed natural gas. By 2014 this had fallen by over 75% to 1.2 Tcf and, owing to an overall expansion in total production during the same period, this meant that the Gulf's contribution to the total US gas output fell to just 4.5%.[9]

A variety of organizations develop, and make public assessments of the scale of, the technically recoverable shale gas resource. Few, however, integrate development economics in order to establish resource supply curves. This is unsurprising given the complexity and uncertainty associated with establishing any form of reasonably accurate supply curve for an entire resource. Of the publicly available US shale gas supply curves, one of the most widely referenced set was produced as part of the 2011 MIT Future of Natural Gas study (MIT, 2011). The study established a range of shale supply curves on the basis of varying cost and technology scenarios. Figure 6.2 illustrates two important examples of these curves, one that assumes static 2010 technology levels and one that assumes continuous technology improvement.

The curves shown in Figure 6.2 illustrate a number of salient features that characterize the shale resource. First, its scale: the resource is large, with estimates of the total recoverable volumes ranging from 600 Tcf to over 1000 Tcf of gas depending on how the technology

[8] Proved natural gas reserves as reported by the US Energy Information Administration. www.eia.gov

[9] Marketed natural gas production as reported by the US Energy Information Administration. www.eia.gov

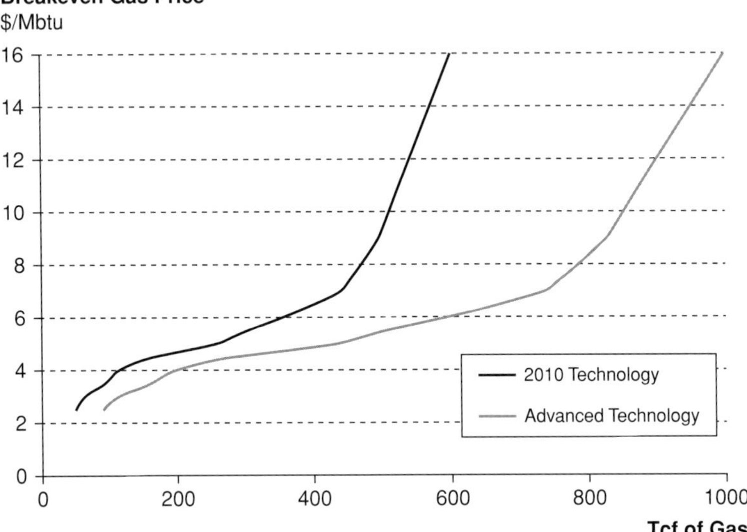

Figure 6.2 US shale gas supply curves.

evolves. Second, both curves in Figure 6.2 highlight that, although the total recoverable shale gas volumes are large, the amount of the resource that is very low cost is more modest. The MIT analysis suggests that the volume of gas recoverable at or below $4.00/Mbtu is likely to be no more than 200 Tcf and may in fact be less than 150 Tcf. This estimate leads to the important conclusion that shale gas is not a very low cost resource; rather it is a large resource of moderate cost, with somewhere between 500 Tcf and 750 Tcf of shale gas available at or below $7.00/Mbtu.

When considering the cost of the shale gas resource a few additional factors that are not captured in Figure 6.2 must be considered. The first of these is how the production of tight oil affects the economics of shale gas. Since 2010 the US has witnessed a very significant expansion in the production of tight oil, with notable examples of this occurring in the Texas Eagle Ford Shale play and in the Bakken Shale in North Dakota. The development of these and other liquids-focused plays determines shale gas economics, in the sense that the growth in oil output from these plays is accompanied by a growth in the associated gas production. In regions with an absence of gas-gathering infrastructure, this associated gas production cannot be marketed and so it is flared. By comparison, in areas where gas gathering is possible, the associated gas is sold and this output represents an additional source of very low cost shale gas that is not captured in the supply curves in Figure 6.2. The second issue that a simple review of Figure 6.2 does not reveal is how the large variability in well productivity – a salient characteristic of all shale gas development to date – means that it is not possible simply to produce gas from the lowest cost portion of the supply curve first. The reason is that operators have not found it possible to drill and complete shale gas wells that are consistently of high performance, even in the best of conditions.

Typically, the productivity distribution of any shale-gas-well ensemble is skewed and will contain more "low" than "high" performance wells. Figure 6.2 shows that this characteristic means that, to date, the development of the shale gas resource has involved sampling the entire supply curve and not simply producing the lowest cost resources first. This evidence runs counter to the belief that the production of shale gas has been very low cost given that natural gas market prices over the past number of years have been so low.

Factors Influencing the Economics of a Shale Gas Well

The economic analysis of shale gas development spans multiple dimensions. All such analysis, however, is ultimately determined by well-level economics and in particular by the breakeven gas price necessary to support development. Key factors that determine the economics of an individual shale gas well are: the costs of drilling and completion of the well and the volume of natural gas and other marketable hydrocarbons produced by the well over its lifetime. Many other factors, however, also play a role. In particular, the manner in which well production declines over time is important, as are factors such as the level of royalties paid to mineral rights owners, the severance tax levels and the well operating and maintenance costs. Further, the manner in which operators fund well development also matters, as this affects the cost of capital.

Well Drilling and Completion Costs

The cost of drilling and completing a shale gas well can vary considerably, both within individual plays and between plays. The vertical depth and horizontal length of a well, along with the scale and technical approach of the completion, are all important factors in well costs. Market factors that influence the availability of drilling rigs, the nature of fracture spreads and manpower requirements also play an important part in determining well costs. Over the course of the past decade the shale gas industry has proven remarkably adept at increasing the efficiency with which it drills and completes wells. In some circumstances these gains have translated into lower overall well costs, but often they have been used to enable operators to develop larger wells without increasing costs. The Barnett Shale provides an excellent example of this dynamic. In 2005 the cost of drilling and stimulating a typical Barnett Shale well was in the \$3–3.5M range, and involved a horizontal section length of ~2100 feet. By 2014, though, a typical new well in the plays was still in the \$3–3.5M range, but the typical horizontal section length had more than doubled to over 4300 feet.[10] Similar gains have been achieved in most shale gas plays, and these dynamics have

[10] Based on well cost data from IHS, MIT and operator disclosures along with well length records reported by the DrillingInfo HPDI well database.

Table 6.1 *Contemporary well costs in major US shale plays, in $M*

Play	Drilling	Completion	Well facilities	Total
Bakken	3.4	4.3	0.8	8.5
Barnett	1.2	1.5	0.5	3.0
Eagle Ford	3.1	3.9	0.8	7.8
Fayetteville	1.2	1.5	0.5	3.0
Haynesville	3.7	4.7	0.9	9.3
Marcellus	2.5	3.2	0.6	6.3
Woodford	2.5	3.1	0.6	6.2

been important in mitigating the effects of very low gas prices on shale gas development activity. Table 6.1 shows typical well drilling, completion and facilities costs in the major US shale oil and gas plays in early 2015.

An important aspect of shale gas development is the extent to which drilling and completion contractors pass the operational cost savings on to well operators. Unsurprisingly, this varies with market conditions. During boom times there can be very significant inflation in the cost of oil field services, including well drilling and completions, as inputs in scarce supply are bid up. By contrast, in a downturn the price of oil or gas tends to result in rapid cost deflation. Useful benchmarks exist for tracking these cost dynamics. A notable example is the IHS upstream capital cost index (UCCI). This index reports how the costs of equipment, facilities, materials and personnel across a diversified portfolio of upstream projects have varied over time relative to year 2000 levels. Figure 6.3 illustrates how the UCCI has varied since 2005 relative to changes in the price of oil and natural gas. The figure shows how upstream costs experienced significant price inflation between 2005 and 2008, a period when both oil and natural gas prices rose and shale gas development was expanding rapidly. The collapse in hydrocarbon prices that occurred in 2008 led to a brief moderation in upstream costs, but these rose again as the price of oil rebounded.

Figure 6.3 also illustrates the complex interdependency that exists between upstream costs and the oil and natural gas market conditions in North America. Since 2008 the price of natural gas in the US has remained very low, a situation that one would expect to precipitate a downward pressure on costs. The US upstream service industry, however, simply switched towards oil-focused activity. This dynamic has meant that the viability of 'dry' (without oil or appreciable NGLs) shale gas development has become very challenging over the past several years and is illustrated by a dramatic drop in dry-gas targeted drilling activity. More recently, the drop in global oil prices since mid-2014 has led to a slowdown in US oil-focused activity. In turn, this has resulted in falling upstream service costs, which has helped to improve the relative economic viability of dry shale gas development.

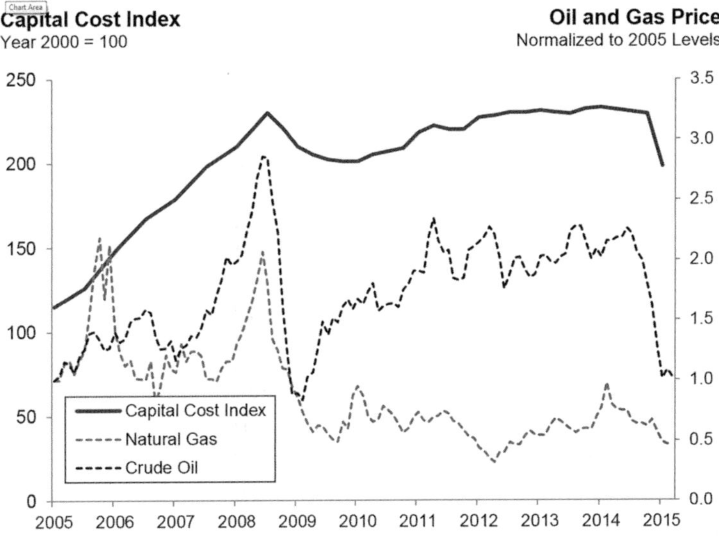

Figure 6.3 Variation in upstream capital cost index (UCCI) relative to changes in oil and natural gas prices.

Assessing Reserves and Well Productivity

Although multifactorial, the economic viability of a shale well is naturally dependent upon that well's productivity. Estimates of the amount of natural gas and other hydrocarbons that a well will produce over its lifetime are quantified by a metric known as the estimated ultimate recovery (EUR). Operators want wells with high EURs because this increases the bookable reserves and, thus, directly impacts on an operator's enterprise value, although the timing of production is also an important consideration. All else being equal, wells that produce more of their EUR earlier in their producing life are more valuable, owing to the discount rate that an operator applies to future cash flows. For example, the combination of steep production rate declines, which are a salient feature of shale gas wells, and a reasonable enterprise discount rate of 8%–12% for a shale gas producer means that contemporary shale wells yield 60% or more of their economic value within the first five years of production, even though it could take those wells more than two decades to fully produce their EURs.

The estimation of shale well EURs is an on-going challenge. One approach used involves bottom-up analysis of geological parameters and the modeling of reservoir behavior (McGlade, 2013). Unfortunately, the relatively nascent nature of shale gas development means that acquiring sufficiently reliable data to reduce uncertainties related to this is very difficult (McGlade, 2013). In addition to data limitations, the extremely low aggregate permeability of these shale-type petroleum systems means that the application of conventional reservoir engineering models and analysis is likely to be inappropriate and, therefore, is unlikely to yield reasonable EUR estimates (Strickland, 2011). For instance, the application of a straightforward material balance approach to reservoir modeling will not be reliable because the low permeability of shale means an "average" reservoir pressure parameter cannot be assumed (Drake, 1983; Strickland, 2011).

Numerical simulations of flow, which are used prodigiously with conventional resources, may prove advantageous for shale gas EUR estimation in the future, but only if large advances can be made in understanding fluid storage and transport mechanisms in these types of rock formations (Clarkson, 2012; Mustafiz, 2008). There have also been efforts to adapt rate-transient analysis, by which flow well regimes are inferred from production history, to unconventional reservoirs and completion geometries. Unfortunately, this is an area of ongoing work and suffers from many of the same limitations as reservoir modeling because of the limited understanding of flow behavior in shale formations (Clarkson, 2012; Strickland, 2011). Further, it is also difficult to estimate properties reliably at the early stages of well production.

An alternative approach to forecasting EUR involves so-called "decline curve analysis." This involves fitting an existing production decline curve model to data from the early months of a well's production (Lee, 2010; McGlade *et al.*, 2013). On the basis of this fitting, the well's EUR is then established by projecting the empirical production into the future. The application of the decline curve method is straightforward because it requires little more than data on the initial or peak production rate (IP) of a well in order to establish an EUR. Because of its relative simplicity decline curve analysis is widely used, but it does have serious limitations. In particular, the method can yield erroneous EUR estimates if used when only limited early-life production data are available. Nevertheless, despite the real issues that can emerge from the use of decline curves to establish EURs, the technique is very useful for financial modeling purposes because it typically provides accurate estimates of early-life production levels which are crucial when undertaking discounted cash flow (DCF) type analysis.

Different decline curves exist in the literature, including the traditional Arps decline curve (Arps, 1945), which tends to overestimate shale well EURs especially when applied early in a well's production life (Lee, 2010; McGlade *et al.*, 2013). Newer methods proposed by Valkó, and by Ilk *et al.* may be better suited for use with shale gas (Valkó, 2009; Ilk *et al.*, 2008). Even with these newer models, however, it remains unclear how valid the assumption is that future decline will follow early decline behavior.

Shale Well Productivity and Production Decline Characteristics

The productivity of a shale gas well can be characterized in several ways. The EUR provides an important quantification of long-term productivity. The challenge is that it can take decades for a well to produce its EUR, and so investment decisions regarding the development of shale wells must also consider shorter-term productivity metrics that characterize the scale and temporal evolution of a well's output over a timescale relevant to investors. As a result, well IP rate and early-life production decline characteristics are

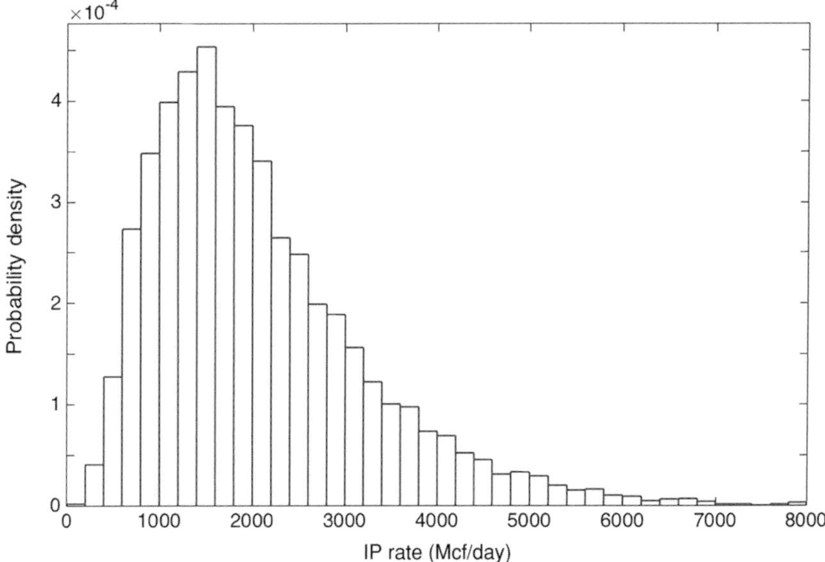

Figure 6.4 Initial production (IP) rate probability distribution of Barnett shale gas wells drilled between 2005 and 2013.

important since they determine the vast majority of a well's topline cash flow and associated present value (MIT, 2011).

Trends in Shale Gas Well Initial Productivity

A review of the IP rate data from shale gas wells brought online over the past decade in the US reveals a number of interesting characteristics that are germane to an assessment of the resource's economic profile. First, well IP rates vary significantly from play to play. For example, in 2014 the mean IP rate for a new Marcellus shale well in Pennsylvania was ~7400 Mcf/day. By contrast, the mean IP rate for a 2014 vintage Barnett well was less than 2000 Mcf/day.[11] Such a considerable variation in average IP rate is to be expected, given the geological heterogeneity of the formations that combine to form the overall shale gas resource. Within individual plays, well IP rates tend to vary significantly and are distributed in a lognormal manner (O'Sullivan, 2013; Montgomery, 2015). Given the multiplicity of factors that determine well productivity, a lognormal distribution of well productivity rates is not unreasonable; however, the broadness of the distribution, even over very limited spatial ranges, right down to individual well pads is of significance.

Figure 6.4 illustrates the empirical probability distribution of IP rates from over 9000 horizontal wells drilled in the Barnett Shale between 2005 and 2013.[12] The data highlight

[11] Based on well production records reported by the DrillingInfo HPDI well database. www.drillinginfo.com
[12] Based on well production records reported by the DrillingInfo HPDI well database. www.drillinginfo.com

both the skewedness of well productivity and the scale of variability between higher and lower performing wells. The data have a P90 : P10 IP-ratio spread of over 4 : 1, meaning that the IP rate of the top 10% of wells is over 400% higher than the bottom 10%. Some of this spread in performance reflects the macrovariations in resource quality across the play along with the fact that the data aggregate wells from a range of vintages during which significant technology evolution has occurred. However, when the IP data are disaggregated by area and vintage, this very broad variance in well IP rates remains. In fact, as discussed in O'Sullivan (2015, 2013), this significant well-to-well variation in productivity remains salient even at individual pad level in all the major plays. Important economic implications for the development of shale resources flow from these results.

First and foremost, there exists a strong stochastic component to shale well productivity. In effect, during a drilling campaign operators are sampling from a broad productivity distribution. Some wells will have very high IP rates, and their economics will look very attractive, but a greater proportion of wells will have relatively lower performance. As a result, assessments of shale gas development economics must involve a broad portfolio analysis rather than being based on individual wells or small sample sets of wells.

Interpreting Changes in Shale Well Productivity

Patterns in the shape and breadth of the distribution of shale gas well IP rates have tended to be very consistent from vintage to vintage in all the major shale plays even though mean performance levels have changed. For example, in 2010 the mean IP rate of a Marcellus shale gas well was ~3600 Mcf/day. By 2013 this had risen to ~5400 Mcf/day, and in 2014 it jumped again to 7400 Mcf/day.[13] These, and similar gains in the average performance of wells in all the major shale plays, have been attributed to many factors including difficulties in quantifying metrics such as "learning-by-doing". The most important explanation in terms of these gains is that, over time, wells have been getting bigger. Horizontal well sections have been steadily increasing in length and the sizes of hydraulic fracturing stimulations have also increased. For example, between 2010 and 2013 the average length of Barnett well horizontal sections increased from ~3100 feet to ~4100 feet while in the Eagle Ford the average length increased from ~4,300 feet to ~5,100 feet.[14] Properly accounting for changes like these is important when one is assessing the actual productivity progress being made in recovering the resource and how the economic viability of extraction is changing over time. A useful but less than perfect approach to assessing trends in shale resource productivity over time involves normalizing the IP data to the length of a well's horizontal section. This "specific" metric provides a better understanding of whether productivity is actually increasing or whether an apparent increase is simply a scale effect.

[13] Based on well production records reported by the DrillingInfo HPDI well database. www.drillinginfo.com
[14] Based on well completion records reported by the DrillingInfo HPDI well database. www.drillinginfo.com

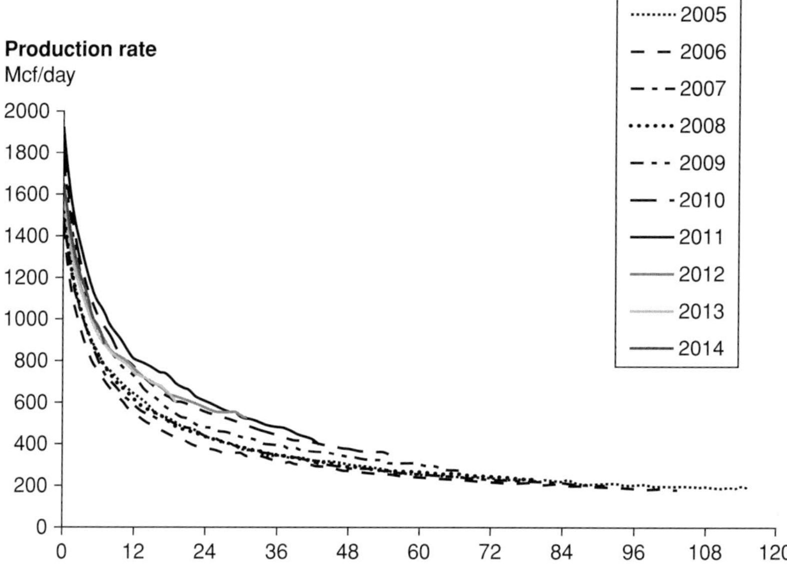

Figure 6.5 Specific initial production (IP) rate cumulative distribution functions (CDFs) for the 2005, 2010 and 2013 Barnett Shale gas well vintages.

Reviewing shale well performance using a specific-type metric such as the specific IP rate yields interesting insights. First, consider Figure 6.5, where the cumulative distribution functions (CDF) of the specific IP rates of the 2005, 2010 and 2013 Barnett shale gas well vintages are plotted.[13] The data show that the specific IP rates of Barnett wells drilled in 2005 are much higher than those of the 2010 or 2013 vintages. In fact, the complete dataset shows a secular decline in specific IP rate levels from 2005 onwards. This result and similar patterns from other plays indicate that, although absolute productivity levels might increase from year to year, the actual productivity per unit length of stimulated reservoir tends to fall over time. Some of this drop in unit length productivity is to be expected owing to losses that arise from longer horizontal well sections. A much more important factor relates to the quality of the particular resource at a well's location. Typically, the initial development of shale plays focuses on locations that predevelopment analysis indicates are of highest quality. After these locations are developed, operators are obliged to move to lower quality rock. Thus, even if they are gaining expertise in drilling and completion, the underlying quality of the resource is lower.

The fact that specific IP rates tend to fall as a play is developed is undesirable but is not necessarily a problem, because what matters is the relative rate of decline between productivity and well costs. If the costs can be reduced faster than the specific productivity falls then the economic viability of a well actually improves over time. Over the longer term, however, the gains in cost reduction are likely to slow, so that the declining quality

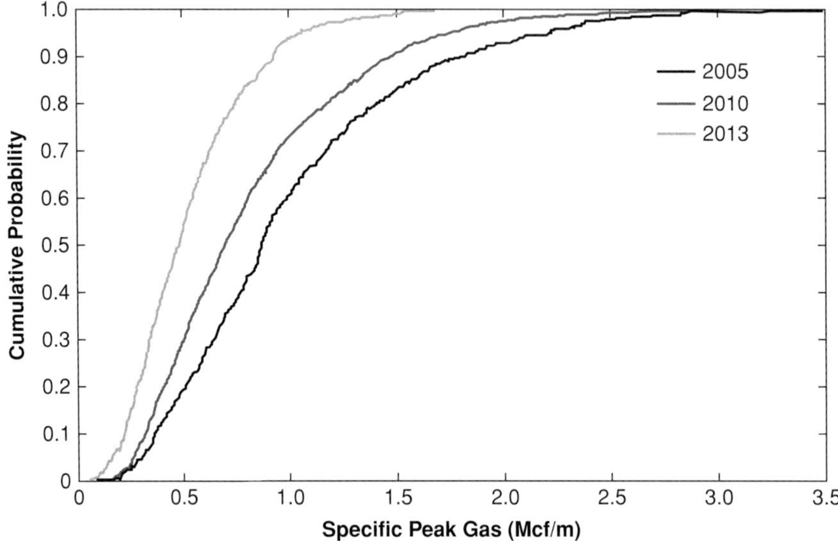

Figure 6.6 Barnett Shale well decline curves by vintage.

of the resource under development will require the well breakeven price eventually to rise in order for development of the resource to remain attractive.

Well Production Decline Characteristics

Along with the IP rate, the manner in which production declines over time plays a major role in determining a well's economic viability. While the modeling of long-term production from shale wells remains a major challenge, empirical data have become available on the evolution of production during early well life, the period of greatest relevance to a well's economic viability. Figure 6.6 shows the empirical production decline curves for each well vintage of Barnett shale from 2005 to 2014.[15] In all cases production falls significantly over the course of the first year – typically by about 60%. This contrasts sharply with conventional gas wells such as those located offshore in the Gulf of Mexico, which have much more moderate early life decline rates, often in the 10%–20% range, and even onshore conventional wells such as those in the Permian Basin, which experience year-one declines of under 40%.[16] The decline rates then tend to moderate considerably over subsequent years to a point where the annual decline in well output is less than 10% after year five or six. While decline trends are not identical in other major plays they usually show a similar

[15] Based on well production records reported by the DrillingInfo HPDI well database. www.drillinginfo.com

[16] Based on the average of 1990–2000 vintages for Federal Offshore Gulf of Mexico and Permian Basin gas wells as reported in production records in the DrillingInfo HPDI well database. www.drillinginfo.com

pattern; an exception is the Marcellus Shale where wells have a more moderate year-one decline, which tends to be in the 35%–40% range.[17]

A rapid initial production decline means that a shale well's economic performance is entirely dependent upon the first four to five years of production. Thus, the vast bulk of a well's EUR is entrained in the slowly declining tail of the production decline curve. As a result, much has been made about the rapid initial decline of production from shale gas wells. In particular, some critics have suggested that these declines mean that assessments of the scale of a shale resource are overly optimistic. The reality, however, is rather different. While it is true that well output does fall significantly during the first few years, the longer-term decline rates of shale wells appear to be more moderate than in the case of conventional gas wells. As a result, once a sufficient number of shale wells have been drilled, the output from the resource tends to be very stable even if overall drilling levels fall. Evidence of this "production stickiness" can be seen in Figure 6.1, which illustrates that output levels from plays like the Barnett, Fayetteville and Woodford have remained relatively flat since 2010 despite the fact that drilling levels in each of those plays have fallen significantly during that period as a result of low natural gas prices. In the case of the Barnett, for example, the active rig count has fallen from over 100 rigs in 2010 to less than 20 in 2015, and yet during that period the play's output has seen little if any decline.

Shale Gas Economics – Today and Tomorrow

Over the past decade shale gas has gone from playing a bit role to being the headline act in the North American natural gas market, and many interrelated factors have driven this dynamic. Early in its development higher natural gas prices supported the higher costs involved in developing shale gas resources. Subsequently, and over the past five years also, gas prices have fallen and yet shale gas output has continued to increase. Much of this continuing growth in shale gas production has been made possible by advances in technology and operational expertise, which have helped to reduce costs. Additionally, the rapid development of tight oil has yielded an additional and very low cost source of associated gas production, which has augmented the supply from dry shale plays.

An illustration of the impact that 'liquids' production has had on shale gas economics is shown in Figure 6.7. The figure shows the first 12 months of production from those new shale wells drilled during 2006, 2009 and 2012 across the major US shale plays along with the corresponding "full cycle" breakeven price for that gas.[18] In 2006 the minimum breakeven price was greater than $2.00/Mcf and most of that vintage's production had full cycle breakeven prices well above contemporary market gas prices. By contrast, in 2012 an appreciable amount of shale gas was being produced at a breakeven price of $0.00/Mcf, and a considerable fraction of the vintage's total production was at or below $4.00/Mcf. The

[17] Issues regarding the accuracy and completeness in the reporting of PA production data mean that the Marcellus analysis tends to be incomplete.

[18] "Full cycle" economic analysis includes the sunk cost of acquiring acreage in assessing a well's breakeven price. This is in contrast with "half-cycle" or "forward-looking" economics, which does not consider sunk costs.

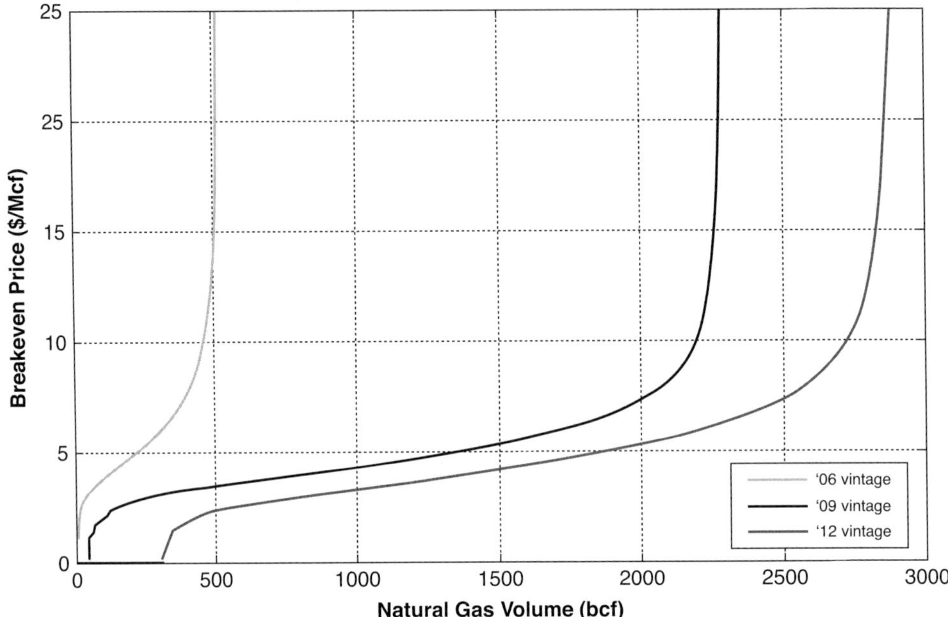

Figure 6.7 Assessment of first-12-months gas production volumes and the associated breakeven prices for the new shale gas wells drilled in the major US shale gas plays during 2006, 2009 and 2012.

"free gas" in this instance was produced from wells in NGL-rich and tight oil producing plays like the Eagle Ford, while the overall flattening and lowering of the total supply curve arose from improving technology and gains in development efficiency. Since 2012 there has been very significant further expansion in the production of ultra-low cost shale gas associated with tight oil, and this has been the major factor keeping US natural gas prices low through 2013, 2014 and 2015.

The fall in global oil prices that began in 2014 will certainly have some implications for shale gas over the medium term, although the impacts may not be dramatic. The US tight oil development activity levels have dropped as global oil prices have softened, and this will mean a lower associated shale gas output. As illustrated in Figure 6.3, the fall in the oil price has also pushed down upstream costs, and this will help to improve the economics of dry shale gas development. Over the medium to longer term the range of possible opportunities could further bolster the economic attractiveness of shale gas. One opportunity, in particular, is to develop a better physical model for the resource that could allow for a reduction in the well performance variability that is a salient feature of today's shale gas business.

Shale gas development outside North America and its economic profile remains very different. To date, the development activity that has taken place has been very limited and the results have been underwhelming. A number of salient differences between the North

American oil and natural gas sector and those elsewhere means that this situation is likely to persist. First, many countries have extensive low cost conventional resources available, and thus they have little need to target higher cost unconventional resources like shale. Second, most countries with potential shale resources have industry and regulatory structures that differ considerably from the North American paradigm.

Third, in terms of industry structure, the US and Canada have a very large and dynamic independent producer sector, and these operators have proven to be much more innovative with regards to developing shale than the large international oil and gas companies. In particular, the independents have been in the vanguard of cost reduction and operational efficiency gains. A second characteristic of the industry structure is that many of these independent entities have a different capital structure from that ofthe larger oil companies, and the investment scales associated with shale development activities align better with their business models. A third important characteristic of the industry is that the North American market benefits greatly from having a very large service industry that has developed a considerable number of shale capacities in the recent past. For example, the US has over 1200 horizontal drilling rigs, an order of magnitude more than the rest of the world. A similar comparison exists for hydraulic fracturing equipment. As a result, shale development costs are dramatically lower in North America than elsewhere. By comparison, anecdotal reports exist suggesting that shale gas wells drilled in international markets like Australia, Poland and China have cost 3–4 times more than an equivalent well in the US. A fourth industry characteristic of the US shale gas development is that it has benefitted from having access to an existing large physical midstream infrastructure that has allowed new plays to be rapidly opened up and brought to market. No international market has an equivalent infrastructure, and putting it in place adds considerably to the cost of development.

The fifth key advantage of the US is that individual landowners hold the mineral rights to their properties. This provides an important financial incentive to support development, which, in turn, also helps to reduce costs as the scale of operations can be larger as a result. By contrast, in most international markets the mineral rights are owned by the state and there is less economic incentive to support development. For instance, in the UK public sentiment has been decidedly negative towards shale development and has impaired the social licence to operate, and this has added to costs.

References

Arps, J.J. (1945). Analysis of decline curves. *Transactions of the American Institute of Mining Engineers*, 160: 228–247.

Clarkson, C.R., Jensen, J.L. and Chipperfield, S. (2012). Unconventional gas reservoir evaluation: what do we have to consider? *Journal of Natural Gas Science and Engineering*, 8: 9–33.

Drake, L.P. (1983). *Fundamentals of Reservoir Engineering*. Elsevier, New York.

Ilk, D., Rushing, J.A., Perego A.D. and Blasingame, T.A. (2008). Exponential vs. hyperbolic decline in tight gas sands – understanding the origin and implications for reserve estimates using Arps' decline curves. In *Proc. Society of Petroleum Engineers Annual Technical Conference and Exhibition*.

Joskow, P.L. (2013). Natural gas: from shortages to abundance in the United States. *American Economic Review*, 103 (3): 338–343.

Lee, J. and Sidle, R. (2010). Gas-reserve estimation in resource plays. *Society of Petroleum Engineers Economics and Management*, 2: 86–91.

McGlade, C., Speirs, J. and Sorrell, S. (2013). Methods of estimating shale gas resources – comparison, evaluation and implications. *Energy*, 59: 116–125.

MIT (2011). *The Future of Natural Gas: An Interdisciplinary Study*. Massachusetts Institute of Technology, June 2011.

Montgomery, J. (2015). Characterizing shale gas and tight oil drilling and production performance variability, MS thesis – Engineering Systems Division, Massachusetts Institute of Technology.

Mustafiz, S. (2008). State-of-the-art petroleum reservoir simulation. *Petroleum Science and Technology*, 26 (10–11): 1303–1329.

NETL (2013). *Modern Shale Gas Development in the United States: An Update*. National Energy Technology Laboratory, September 2013.

NPC (2003). *Balancing Natural Gas Policy – Fueling the Demands of a Growing Economy*. National Petroleum Council, September 2003.

O'Sullivan, F. and Ejaz, Q. (2013). The North American shale resource: characterization of spatial and temporal variation in productivity. In *Proc. International Association of Mathematical Geology Annual Meeting*, Madrid.

O'Sullivan, F. and Montgomery, J. (2015). The salient distribution of unconventional oil and gas well productivity. In review.

PGC (2009). *Potential Supply of Natural Gas in the United States*. Colorado School of Mines, Golden, CO.

PGC (2015). *Potential Supply of Natural Gas in the United States*.Colorado School of Mines, Golden, CO.

Strickland, R., Purvis, D. and Blasingame, T. (2011). Practical aspects of reserves determinations for shale gas. In *Proc. Society of Petroleum Engineers North American Unconventional Gas Conference and Exhibition*.

Valkó P.P. (2009). Assigning value to stimulation in the Barnett shale: a simultaneous analysis of 7000 plus production histories and well completion records. In *Proc. Society of Petroleum Engineers Hydraulic Fracturing Technology Conference*.

7

Unconventional Natural Gas in China

LV JIANZHONG AND ZHANG HUANZHI

As global petroleum exploration and development evolve, unconventional gas resources of for example tight gas, shale gas and coal bed methane (CBM) exhibit a great potential. In China, where unconventional gas resources are abundant with significant development potential, tight gas is being widely produced, the development and utilization of CBM and shale gas have begun and fundamental research on gas hydrate has progressively increased. However, as China started relatively late in unconventional gas exploration and development, it has to face some risks and challenges related to the resource quality, technology suitability, cost affordability, infrastructure connectivity and environmental protection (Zhou *et al.*, 2012).

Resources

China has abundant unconventional gas in the Sichuan, Ordos, Songliao, Tarim and Qinshui basins. Generally, in comparison with conventional gas, the amount of unconventional gas is six times greater in terms of resources or twice as great in terms of recoverable resources (Figure 7.1).

Shale gas

China is endowed with marine, continental and transitional mud shale. The most prospective Paleozoic marine mud shale is mainly distributed in the south China region and the Tarim Basin, covering an area of 60×10^4 to 90×10^4 km^2. The Paleozoic marine organic-rich shale in the Yangtze Platform, which has good hydrocarbon generation and shale gas storage conditions, is currently the most realistic shale gas production zone in China. In March 2012, the MLR (Ministry of Land and Resources of the People's Republic of China) published the *China Shale Gas Resource Potential Assessment and Prospects Selection*, which gave an assessment of five regions, 41 basins or blocks, 87 evaluation units and 57 gas-bearing shale beds on the basis of province, surface environment, geologic unit, stratigraphic bed series, sedimentary environment and buried depth. This assessment suggests that China's geological resources and technically recoverable resources of shale gas

Table 7.1 *Assessment of shale gas resources in China, in 10^{12} m^3*

Assessment unit	Geological resources	Technically recoverable resources	Percentage (%)
Upper Yangtze, Yunnan, Guizhou, and Guangxi	62.56	9.94	39.63
North China and northeast China	26.79	6.70	26.7
Middle-lower Yangtze and southeast China	24.16	4.64	18.49
Northwest China	19.90	3.81	15.19
Total	134.42	25.08	100

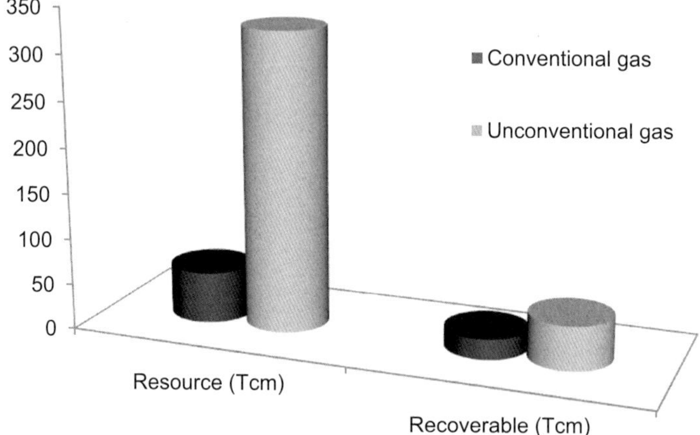

Figure 7.1 Conventional gas vs. unconventional gas resources in China.

are estimated to be 134 × 10^{12} m^3 and 25 × 10^{12} m^3 respectively. The Upper Yangtze, Yunnan, Guizhou, and Guangxi contribute 39.6% to the resources (Table 7.1).

This assessment also indicates that the assessment units, with total area of 88 × 10^4 km^2, have 93 × 10^{12} m^3 of geological resources and 16 × 10^{12} m^3 of recoverable resources. They are the most important shale gas prospects in China. So far 180 shale gas prospects have been selected – 60 prospects in Upper Yangtze, Yunnan, Guizhou, and Guangxi (33%), 57 prospects in north China and northeast China (32%), 38 prospects in northwest China (21%), and 25 prospects in middle–lower Yangtze and southeast China (14%).

As indicated by the distribution of shale gas in the provinces of China, the marine facies area in South China and northwest China contribute most to the total shale gas resources, and the top five provinces include Sichuan, Xinjiang, Chongqing, Guizhou and Hubei (Figure 7.2).

Table 7.2 *Distribution and predicted resources of tight gas in China (Jia and Zheng, 2012)*

Basin	Area/10^4 km^2	Geological resources/10^{12} m^3	Technically recoverable resources/10^{12} m^3
Ordos	25.0	6–8	3–4
Sichuan	18.0	3–4	1.5–2.0
Songliao	26.0	2.0–2.5	1.0–1.2
Tarim	3.5	4–7	2–3
Turpan-Hami	5.5	0.6–0.9	0.4–0.5
Bohai Bay	8.9	1.0–1.5	0.5–0.8
Junggar	13.4	0.8–1.2	0.4–0.6

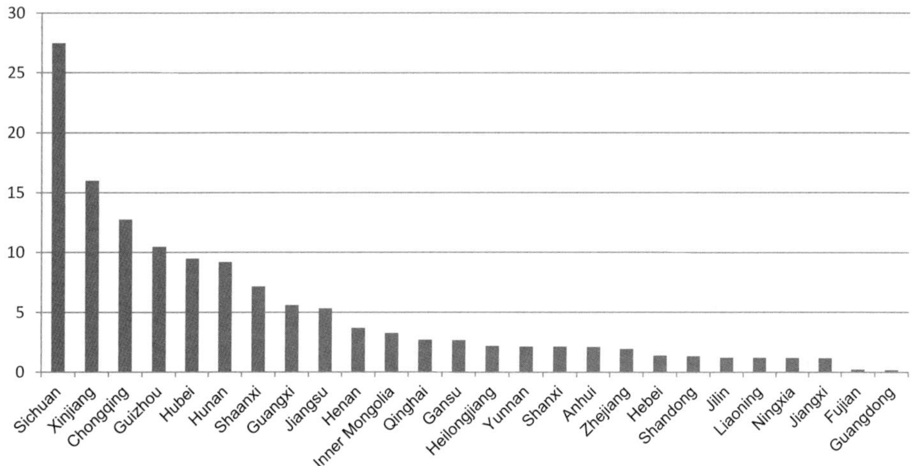

Figure 7.2 Distribution of shale gas in 10^{12} m^3 in the provinces of China (Ministry of Land and Resources, 2010).

Tight Gas

Tight sandstone gas is widely distributed over China onshore, with an overall prospect area of 32×10^4 km^2 and considerable exploration potential. According to the available exploration data, China has $(17.4–25.1) \times 10^{12}$ m^3 of tight gas geological resources (overlapping partially with conventional resources) and $(8.8–12.1) \times 10^{12}$ m^3 of technically recoverable resources. These resources are distributed over more than 10 basins, e.g. Ordos, Sichuan, Turpan-Hami, Bohai Bay, Songliao, and Junggar, of which the Ordos Basin, the Sichuan Basin and the Tarim Basin together account for 81% of total resources (Table 7.2).

Table 7.3 *Distribution and predicted resources of CBM in China (Jia and Zheng, 2012)*

Basin	Assessment area/10^4 km^2	Geological resources/10^{12} m^3	Technically recoverable resources/10^{12} m^3
Ordos	10.88	9.86	1.79
Qinshui	2.71	3.95	1.12
Junggar	3.46	3.83	0.81
Eastern Yunnan and western Guizhou	1.61	3.47	1.29
Erlian	3.48	2.58	2.10
Turpan-Hami	0.94	2.12	0.41
Tarim	4.06	1.93	0.69
Tianshan	1.05	1.63	0.67
Hailaer	1.3	1.60	0.45

Geologically, the tight sandstone reservoirs in China are commonly characterized by low porosity and low permeability, highly fractured locally ultra-low water saturation, high capillary pressure and abnormally high formation pressure. Gas reservoirs are usually buried deeper than 2000 m. Mesozoic and Upper Paleozoic sandstones contribute most to the proved reserves (Zhou *et al.*, 2012).

Coal Bed Methane

In China the coal basins are well developed, and coal seams contain abundant high-content gas resources. The Carboniferous-Permian coal basins in north China, the Permian coal basins in south China and the Jurassic coal-bearing fault depression in northeast China are the major zones where coal-bearing strata have been developed. The most important prospective areas in the eastern margin of the Ordos Basin, and Qingshui zone in the Qinshui Basin, have been the primary targets for CBM exploration and development in recent years. The result of the latest national resource assessment indicates that China's CBM prospect area is estimated to be 37.5×10^4 km^2; 40 coal basins are being developed in 5 CBM zones, coal beds buried no deeper than 2000 m have 36.8×10^{12} m^3 of CBM geological resources and coal beds buried no deeper than 1500 m have 10.9×10^{12} m^3 of technically recoverable CBM geological resources (Table 7.3). In the Ordos Basin, the Qinshui Basin, the Junggar Basin, the Eastern Yunnan and Western Guizhou Basin, the Erlian Basin, the Turpan-Hami Basin, the Tarim Basin, the Tianshan Basin and the Hailaer Basin (with respective geological resources exceeding 1×10^{12} m^3), the total geological resources are 31.0×10^{12} m^3 and the technically recoverable resources are 9.3×10^{12} m^3, corresponding to 84% and 85% of total CMB resources in China respectively. By the end of 2013, China's cumulative proved CBM reserves were estimated to be 0.57×10^{12} m^3; the proved ratio, however, was 1.5% (Qi, 2013).

Natural Gas Hydrate

Because of a weak basis and late initiation, the investigation and assessment of natural gas hydrate resources in China went through a difficult process. The preliminary study on natural gas hydrate commenced in 1995. An investigation of natural gas resources initiated by MLR in 1999, with the support of the National Development and Reform Commission (NDRC) and the Ministry of Finance (MF), by integrating the technical forces of various parties provided a vast number of fundamental exploratory works. So far, the China Geological Survey of MLR has carried out 40 voyages of natural gas hydrate investigation, accomplishing 45 800 km high-resolution multi-channel seismic surveys, 36 800 km multi-beam surveys, 7100 km sub-bottom profile surveys, 1480 stations of subsea geological sampling and 222 stations of subsea heat flow measurement. Investigation and study during the past decade have: (1) formed an integrated surveying system of natural gas hydrate which is particularly suitable for the sea areas in China; (2) independently developed some key technologies such as high-resolution multi-channel subsea seismic target detection, subsea microtopography and heat flow measurement, and subsea *in situ* pore water sampling; (3) systematically summarized the control factors and reservoir-forming modes of natural gas hydrate; and (4) created a theory of complex gas hydrate reservoir formation at epistemic passive continental margins. In the South China Sea alone, the natural gas hydrate resources are estimated to be 64 to 77 billion tons of oil equivalent (Xie *et al.*, 2012).

Current Status of Exploration and Production

Over the past decade, tight gas and CBM have been commercially developed, and production has increased rapidly in China. Shale gas, as the most attractive unconventional resource to be developed in future, is still at its exploratory stage; two major Chinese oil companies, CNPC and Sinopec, have just completed the preliminary development plan for their shale gas exploration project in the Sichuan Basin.

Shale Gas

The exploration and development of shale gas in China commenced in 2009. At present, the Sichuan Basin and surrounding areas are being explored (Figure 7.3), with a great breakthrough achieved in the Fuling Block, operated by Sinopec, in the Changning-Weiyuan and Zhaotong blocks, operated by CNPC, and in the Fushun-Yongchuan block, operated jointly by CNPC and Shell. As of July 2014, in China over RMB20 billion had been invested to develop 130 billion m^3 of reserves, drill 400 wells, and acquire 19 139 km of two-dimensional (2D) seismic data and 1451 km^2 of three-dimensional seismic data. In the same period, 54 licences were issued, with a total exploration area of 170 000 km^2. The marketable shale gas was expected to increase from 0.2 billion m^3 in 2013 to nearly 4.5 billion m^3 in 2015.

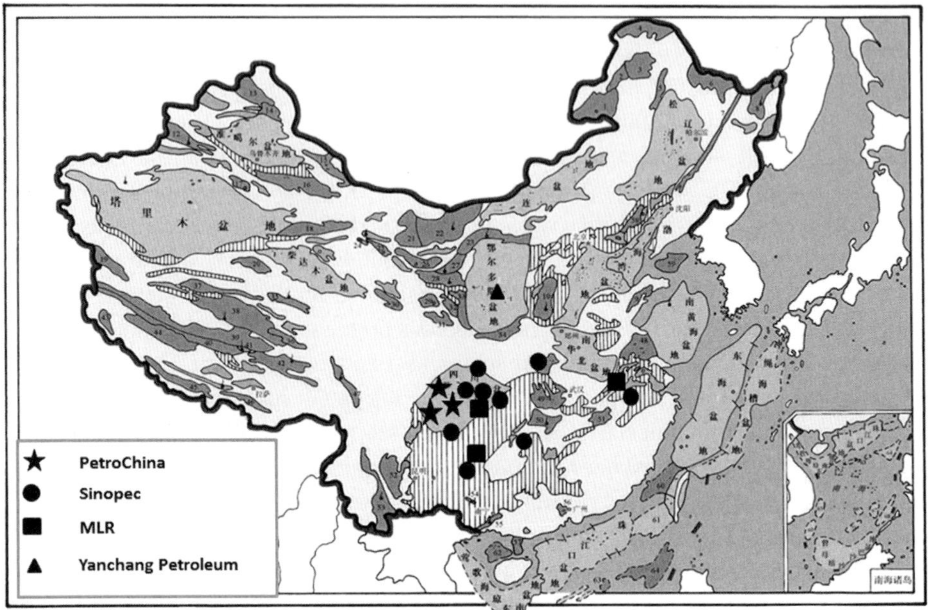

Figure 7.3 Major shale gas test zones in China.

The shale gas blocks operated by CNPC include two national demonstration blocks (Changning-Weiyuan and Zhaotong) and a foreign cooperation block (Fushun-Yongchuan). By the end of 2013, a total of RMB 5.6 billion (RMB 2.9 billion from CNPC and RMB 2.7 billion from Shell) had been spent in acquiring 4411 km of 2D seismic data and 259 km^2 of 3D seismic data and in drilling 50 wells (23 vertical appraisal wells and 27 horizontal wells) in the above blocks, of which 30 wells (16 vertical appraisal wells and 14 horizontal wells) were tested for gas production and 10 wells were put into trial production, with a cumulative marketable volume of shale gas of 79 million m^3.

Sinopec has obtained breakthroughs in exploration and development in the Lower Silurian Longmaxi Formation marine facies blocks, i.e., Fuling block and Pengshui block. The Fuling shale gas field is located within the administrative region of Chongqing. With an exploration area of 7308 km^2, it is located in the eastern Sichuan steep dip structural belt and lies within a hilly, sometimes mountainous, region, with surface evaluations ranging from 300 m to 1000 m. With an area of 200 km^2, the Jiaoshiba block is being tested for development; here the gentle-dip weakly deformed marine shale layer in Longmaxi Formation is free from the influence of major fault and thus is favourable for preserving shale gas. In 2009, 595 km^2 of 3D seismic data was acquired by Jianghan Oilfield, Sinopec, over Jiaoshiba block, and four exploration wells with a footage of 11 800 m were drilled. The gas flow from Well Jiaoye-1 on 28 July 2012 marks the beginning of shale gas development test

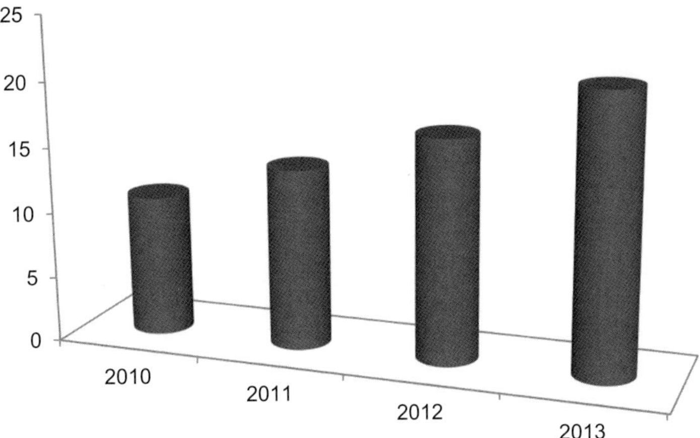

Figure 7.4 Tight gas production in bcm of the Sulige gas field.

over the Fuling block. By the end of October 2014, 102 shale gas wells were completed, with a cumulative marketable shale gas of 869 million m^3.

Tight Gas

Tight gas has become an important domain for increasing reserves and production in China. So far, the Upper Paleozoic in the Ordos Basin and the Upper Triassic Xujiahe Formation in the Sichuan Basin are two major tight gas zones. Tight gas accounts for 50% of the increase in China's gas geological reserves, which rose by 700 billion m^3 per year over the period 2009–2013. The increase per year in tight gas production accounts for over 40% of the total gas production in China. In 2013 the tight gas production was 40 billion m^3, accounting for 34% of the total gas production in China. The technologies relevant to the development of tight gas are basically mature, and the technical workflow of sweet spot prediction, quick horizontal drilling and completion, and large-scale fracturing stimulation for tight gas development has been established. Innovative technologies and management, coupled with research on and application of horizontal well technology, now enable the commercial development of sweet spots.

Sulige Gas Field, the most successfully developed tight gas field in China, is located in Suligemiao, within the administrative region of Ordos, Inner Mongolia. The surface of this area is covered by desert. The gas layers consist of multiple superimposed sand bodies, forming a typical lithologic trap reservoir, which is very large and is low in porosity, permeability and the production and abundance of reserves. Wide-range exploration over this gas field commenced in 1999, and by 2003 it had become the first giant gas field, with 533.6 billion m^3 of cumulative proved geological reserves. This gas field recorded 21.1 billion m^3 of gas production in 2013 (Figure 7.4), and contributed greatly to the

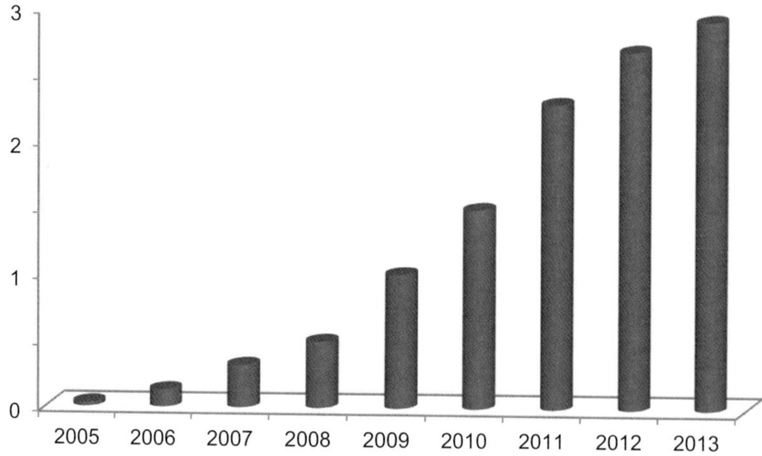

Figure 7.5 Coal bed methane production in China.

50 million tons of oil and gas equivalent achieved by Changqing Oil Field in the same year.

Coal Bed Methane

The CBM industry in China is growing rapidly. For example, the industrial production of CBM is under way in China, especially in Qinshui (Shanxi), Hancheng (Shaanxi), and Tiefa, and a trial test has been performed in Turpan-Hami (Xinjiang), in the eastern Junggar Basin, and in eastern Heilongjiang. By the end of 2013, a total of 14 042 wells had been drilled in China, of which 33 wells are pinnate horizontal wells, producing 2.93 billion m^3 of CBM (Figure 7.5). An initial cost-effective technological system, consisting of various technologies such as CBM geophysical recognition, pinnate horizontal well, gas recovery by water drainage and low-pressure transportation, was established (Sun *et al.*, 2011).

Natural Gas Hydrate

China started investigating natural gas hydrate rather late. In the 1980s, very few researchers paid attention to the progress of international natural gas hydrate research. From the late 1990s, research into natural gas hydrate started to become popular. Since 1999, the investigation and study of natural gas hydrate in the South China Sea, in the East China Sea, in onshore tundra, and in international sea bed areas have enabled China to discover a series of geological, geophysical and geochemical abnormal markers and to acquire natural gas hydrate samples in Shenhu in the northern South China Sea and Southern Qilianshan margin tundra in Qinghai. This has been a major breakthrough in prospecting operations and indicates good exploration potential. Overall, China is still at the stage of resource assessment, because natural gas hydrates have been relatively unresearched, with no advanced technologies or equipment.

Policies and Regulations

The Chinese government pays close attention to unconventional gas exploration and development and has continuously introduced policies and regulations regarding tax revenues, subsidies, resource management and environmental protection, which have greatly promoted exploration and development.

Policies

For CBM, the production companies are subsidized at the rate of RMB 0.2/m^3. They will initially be free of resource tax but will be allowed to apply for a reduction in royalty fee until 2020. Moreover, they benefit from a refund-after-collection policy and are free of import duty on equipment and of import VAT. In coal prospects, gas production comes first and priority is given to the surface development of CBM. In planned coal production zones, the practices of "recovery after extraction" and the "integrated production of coal and gas" are adopted; the recovery of CBM resources is encouraged from both surface and well-bore, but coal resources may not be recovered until the CBM content decreases below the relevant standards. For CBM construction projects, an environmental impact assessment must be conducted in strict accordance with the relevant laws, and the site should avoid ecologically sensitive regions such as natural conservation areas and drinking water sources. The standards regarding CBM emission should be strictly followed, and direct emission is prohibited. The waste water and gas generated during CBM production should be discharged only after being treated according to the applicable standards, and solid waste should be treated appropriately in order to protect underground water from pollution (National Energy Administration, 2013b).

Regarding shale gas, the Chinese government considers it as a national strategic emerging industry and is continually increasing financial support to the shale gas effort. In China, shale gas is classified as an independent mineral and receives special investment and management. The Ministry of Land and Resources has twice opened the exploration licence of shale gas for public bidding. Research on technologies related to shale gas exploration and development are considered as national major special projects and receive funding at national level. The subsidy rate for shale gas was RMB 0.4/m^3 from 2012 to 2015. Shale gas production companies receive a reduction in their mineral resource compensation fee and mineral rights royalty and also enjoy incentive policies for resource tax, VAT, and income tax. Regarding shale gas exploration and development projects, self-use equipment which was not produced domestically is exempt from custom duties according to the relevant regulations. Companies that engage in shale gas exploration and development should possess the capacity for investment appropriate to the project, have good financial conditions and sound financial and accounting rules, and be able to bear civil liability independently.

Shale gas exploration and development companies are encouraged to improve the success rate of shale gas exploration, the utilization rate of development and the economic benefit by using internationally innovative state-of-the-art technologies (National Energy Administration, 2013b).

Regulations

The Chinese government attaches great importance to environmental protection. On 24 April 2014, the Standing Committee of the National People's Congress (NPC) of China passed the revised Environmental Protection Law (EPL). If an offender's behaviour constitutes a crime, they will be held criminally liable. Owing to the early stage of unconventional gas development in China, the current environmental regulations are mainly adopted from conventional oil and gas regulations.

In 2007, the report "Technical guidelines for environmental impact assessment – constructional project of petroleum and natural gas development on land", issued by the Ministry of Environmental Protection (MEP), set clear standards for wastewater discharge and stated the leading practices to prevent pollution caused by wastewater in the oil and gas industry. In April 2008, the MEP issued the report "Emission standard of coal bed methane/coal mine gas (on trial)". The report regulates that the venting of high concentration CMM (methane concentration $\geq$ 30%) is not allowed for coal mines established after 1 July 2008 and for other existing coal mines will not be allowed from 1 January 2010. In March 2012, the MEP issued the report "Technical guidelines for oil and gas industry pollution prevention". This document gave guidance on the fracking fluids to be used in the oil and gas industry. It also required the creation of underground water monitoring schemes to confirm that the current industrial practices in underground injection did not cause any water pollution.

Although there are no specific environmental regulations on shale gas, the NDRC shale gas development plan (2011–2015) covers environmental issues. The plan states procedures for fracking water recycling, strict rules on drilling activities, and an enhanced monitoring scheme for wastewater discharge, in order that shale gas operators reduce negative impacts on the local environment.

Water Management

Water availability may prove to be one of the biggest obstacles to unconventional gas development in China, particularly in the north and west where water supplies are scarce and may be already strained by agricultural or urban needs. Water policies, regulations and plans are determined nationally, though responsibilities for management and enforcement are delegated locally. Many different entities are involved at the national, regional, and local levels, which risks limiting the co-ordination of water resources at the river basin level. National standards establish maximum discharge concentrations for pollutants into water sources and the Circular Water Law promotes the reuse and recycling of waste and produced water (OECD/EIA, 2012).

Future Risks and Challenges

Unconventional natural gas resources show great potential in China. However, their exploration and development has to face some risks and challenges related to the resource quality, technology suitability, cost affordability, infrastructure connectivity, and environmental protection.

Resource Quality

China differs greatly from North America in the quality of its unconventional gas resources. In contrast with the coal bed methane sources in North America, the Late Paleozoic and Meso-Cenozoic are major coal-forming phases in China, and thus the coal seams formed in and after the Carboniferous-Permian period were affected by Meso-Cenozoic tectonic evolution. As a result, coal seams and CBM and CBM reservoirs suffered different levels of reforming and damage, leading to the formation of a great amount of coal dust, the damage of primary structure coal, and many negative influences on CBM development activities. Compared with tight gas in North America, the tight gas reservoirs in the Ordos Basin, China, have the advantages of lateral continuity, thin sand layers and low pressure coefficient and thus can be classified as low pressure unfractured gas reservoirs. For shale gas, the geological conditions in China are much more complex than those in North America. Generally, gas-bearing shales in China are characterized by high organic abundance, thermal evolution and late-stage variation. Shale gas that occurs in marine, continental and transitional facies is characterized by being distributed in multiple bed series, by multiple origins and by complex late-stage reforming. Some shale reservoirs were buried deeper than 5000 m and distributed unevenly, with the subsurface structures severely broken. China's shale gas resources are mostly concentrated in hilly and mountainous areas in central and western China, where complex land forms and poor surface conditions are unfavorable not only for the transfer and transportation of materials and equipments for exploration and development but also for the establishment of large-scale "factory" drilling sites similar to those in North America, thereby increasing the difficulty and cost of development significantly.

Technology Suitability

China has initially formed a set of core technologies, through unconventional gas exploration and development tests, which consist mainly of unconventional gas resource assessment, horizontal well drilling and staged fracturing in horizontal wells. The core indexes of these technologies, however, still have gaps compared with those in North America. Specifically, for horizontal well drilling technology, the lengths of the horizontal intervals need to be extended, the drilling time is twice that in North America, and the drilling operation is less automated. For horizontal staged fracturing technology, although international mainstream fracturing technology has been adopted, the performance of the domestically manufactured fracturing tools needs to be improved, some core tools (e.g.

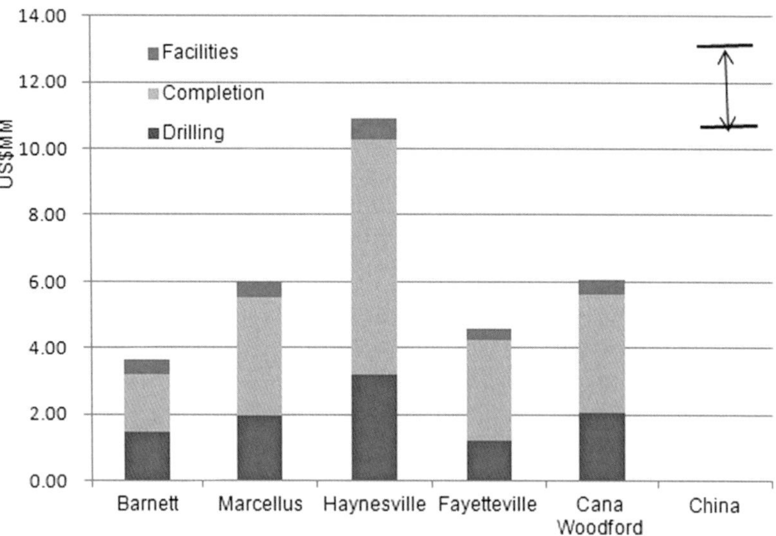

Figure 7.6 Comparison of well cost in China and the US. The arrows represent the cost range of shale gas wells in China.

mechanical bridge plugs) rely on imports, and the cyclic utilization of fracturing fluids requires the support of more advanced technologies. For factory drilling, a platform can only host 6–8 wells; mobile rigs are still relatively few, although the design and manufacture technologies are available domestically.

Cost Affordability

Cost is one of the major factors that influence the economic benefit of unconventional gas and the enthusiasm of development companies. China is just beginning to explore unconventional gas, particularly shale gas. Given the complex geological conditions, the depth and the mountainous landforms, cost is a major challenge. For example, the cost per well for the first batch of horizontal shale gas wells drilled in 2011 and 2012 ranged from RMB 80 million to RMB 100 million. Since the commencement of the domestic production of technology and equipment and the application of factory drilling mode, the current cost for horizontal well drilling and completion has been reduced to RMB 60 million, which is, however, still higher than that in North America (Figure 7.6).

Infrastructure Connectivity

China has fewer gas pipeline network and infrastructures. The total length of pipelines is 106 000 km, including 50 000 km of gas pipeline, which is only one tenth of the US's total gas pipelines; the gas transmission capacity totals 100 billion m^3 per year, the third party access to gas infrastructures is limited, and the peak load regulation capacity of gas storage is 1.5%. Furthermore, the prospective areas of unconventional gas resources are mostly located in mountainous areas in central and western China, where pipeline network construction is difficult and costly, therefore posing an obstacle to the transmission of natural gas and the development of downstream markets.

Environmental Protection

At present, most shale gas activities are at an exploration stage, where logistics requirements remain low. Some operators currently tap water from the Wujiang River, where there is no lack of water, to avoid competing with drinking water resources in Sichuan. However, given that shale gas is predominantly producing using low cost slickwater fracturing technology, China as a whole is short of water. In the future, water shortage will be a major challenge to shale gas exploration, particularly in certain blocks located in the northern and northwestern parts with no river. In addition, poor surface conditions and dense populations in shale-gas-rich regions will increase the costs for operators to build infrastructure such as roads and pipelines and will result in significant health safety, security and environmental (HSSE) exposure for any road-transport-related activity. The Sichuan Basin, for example, is one of the most densely populated regions in China, and most of the landscape in Chongqing and Guizhou provinces is highly mountainous. Shale gas development plans must address rough surface conditions and poor infrastructure in many parts of the country.

References

Jia Chengzao, Zheng Min and Zhang Yongfeng, 2012. Unconventional hydrocarbon resources in China and the prospect of exploration and development, *Petroleum Exploration and Development*, 39 (2) 129, 2012.

Ministry of Land and Resources, Strategic Research Center of Oil and Gas Resources, 2010. *National Oil and Gas Resource Assessment*. Beijing: China Land Press.

National Energy Administration, 2013a. Policies for the CBM industry.

National Energy Administration, 2013b. Policies for the shale gas industry.

OECD/EIA, 2012. World energy outlook.

Qi Haiying, 2013. *Natural Gas Resource and Its Development and Utilization*, China Petrochemical Press, pp. 186–188.

Sun Zandong, Jia Chengzao, *et al.*, 2011. *Unconventional Oil & Gas Exploration and Development*. Beijing: Petroleum Industry Press, pp. 624–625.

Xie Kechang, Huang Qili, *et al.*, 2012. Clean, efficient and sustainable development and utilization of fossil energy. In *Proc. Second CAE/NEA Energy Forum*. Beijing: China Coal Industry Publishing House, pp. 1244–1249.

Zhou Caineng, *et al.*, 2012. *Unconventional Oil & Gas Geology*. Beijing: Geology Publishing House, p. 106.

8

The Argentinian Approach for Developing Unconventional Gas Resources

LUIS STINCO AND SILVIA BARREDO

Introduction

Unconventional shale resources of 802 Tcf and 27 bbo have generated great hopes in Argentina. This paper presents the most relevant geological characteristics of the unconventional shale reservoirs and describes the exploration, drilling, completion and production activities for the Vaca Muerta, Los Molles and Agrio formations in the Neuquén Basin and the Pozo D-129 formation in the Golfo San Jorge Basin. Current and future activities are also presented, together with statistics associated with oil and gas production. Furthermore, the hydrogeology and surface and subsurface water usage are explained and related to fracking. Vital environmental issues, including concerns, procedures and regulations, are also considered together with significant aspects of the laws and decrees concerning upstream activities in the country. Finally, and in order to have a complete overview, the key economic aspects of the unconventional shale reservoirs in Argentina are discussed.

Geological Setting of Argentina's Unconventional Gas Reserves

There are six main hydrocarbon-producing and 15 frontier basins in Argentina (Figure 8.1).

Argentina's basins display a complex evolutionary history that demands an integrated surface and subsurface tectosedimentary study to model the petroleum systems. Most of the basins were located within the Gondwanides (Keidel, 1916), a wide belt of older accreted Eopaleozoic terrains sutured along the southwestern Gondwana margin. Alternating extensional and compressional events produced various basin geometries, which led to notable thickness variations that resulted in different burial histories. The infill comprises clastic, carbonatic and pyroclastic rocks derived from continental and marine realms. Tectonic evolution and sea-level variations were the main controls on the marine sedimentation of some of the most prolific troughs, whereas the non-marine record seems to have been controlled by the occurrence of large lacustrine bodies (presently significant hydrocarbon source rock) under the influence of tectonic and/or climatic fluctuations. Integrated surface and subsurface regional analysis carried out in these basins reveals that tectonic

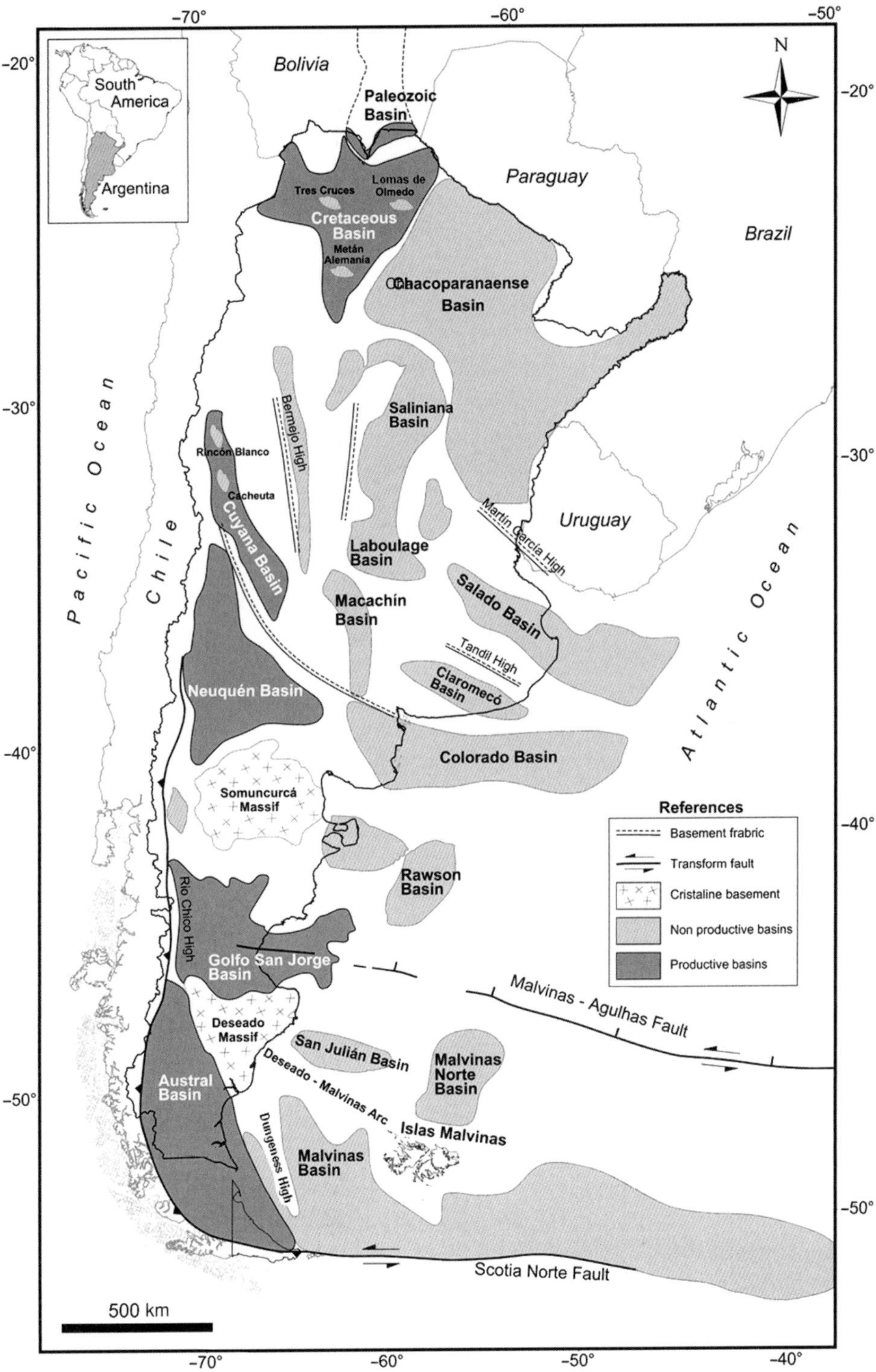

Figure 8.1 Map of the Argentinian sedimentary basins.

evolution, climate and local and/or global sea-level variation seem to have conditioned the accumulation and preservation of the organic matter.

The Cuyana Basin is 43 000 km^2 in area while Neuquén Basin covers a surface of approximately 115 000 km^2. They consist of rapidly subsiding, fault-bounded troughs (Barredo *et al.* 2008; Ramos and Kay, 1991; Rincón *et al.* 2011). The Cuyana Basin is a northwest trending elongate passive rift filled with continental deposits while the Neuquén Basin is a wide trough with embayment morphology, roughly triangular in plan view and filled with marine and continental sediments. The geometry of these two basins consists of border faults, derived for the most part from reactivated structures whose attitude to the extensional regime controlled the style of the rifts (mainly dip-slip or oblique to strike-slip dominated) (Barredo, 2012). The resulting depocentres consisted of scoop-shaped sub-basins separated by intrabasinal highs. The Cuyana Basin developed a hydrologically closed lacustrine environment under a semi-arid, seasonally humid, climate, leading to the accumulation of the kerogen type-I rich shale rocks of the Cerro de Las Cabras Formation during Middle Triassic times. This basin may reach a total organic carbon content (TOC) up to 10% but up to now the explored levels display low thermal maturation, which limits its potential as source rock. A reactivation in the fault system and a subtle change in the climatic conditions from sub-tropical to temperate and seasonally dry during Middle to Upper Triassic times gave rise to the development of balanced fill lakes subject to climatic base-level fall-to-rise turn-arounds (Barredo, 2012). This environment was ideal for the accumulation of the organic matter of the Cacheuta Formation – presently the most important source rock of the basin. This formation contains an average TOC that varies from 3% to 10%, a Romax of 0.6% to 1%, a hydrogen index of 600–900 mg HC/gTOC and an SPI of 3 to 10 t HC/m^2; its kerogen type is I and its visual kerogen analysis (VKA) is algal amorphous with limited terrestrial contribution (Legarreta and Villar, 2011). The thickness of the Cacheuta Formation ranges from 50 to 400 metres and extends all over the basin, although maturation took place in a very restricted depocentre.

During Upper Triassic to Lower Jurassic times a series of continental clastic and pyroclastic sediments, informally identified as Precuyo Cycle (Gulisano *et al.*, 1984) were deposited in a new rift, the Neuquén Basin, located southward of the Cuyana Basin. The climate was semi-arid seasonally humid; the lake was closed and underfilled and could accumulate the organic matter which is presently the source rock of Puesto Kauffman Formation or Precuyano. It has an average TOC that varies from 2% to 11%, a Romax of 0.4% to 0.8%, a hydrogen index of 300–900 mg HC/gTOC and an SPI of 10 t HC/m^2; its kerogen type is I to mixed I/III and its VKA suggests significant terrestrial woody input over the algal lacustrine production (Legarreta and Villar, 2011). Thickness ranges from 50 to over 1100 metres as a consequence of the marked rifting topography. The majority of wells that reached the Precuyano are located to the east and centre of the basin – many of them with oil and gas shows. Volume estimation assigns 2 to 5 Tcf to this formation. During the Middle Jurassic up to the Lower Cretaceous period, the basin experienced repeated drowning by the sea with the development of important source rock in the Los Molles, Vaca Muerta and Agrio formations, all with alternating periods of continentalization. These formations are the important unconventional reservoirs of our country, up to the time of writing. Los Molles Formation has an average TOC that varies from 1% to 5%, a Romax of 0.8% to 2%, a hydrogen index of 300–500 mg HC/gTOC and an SPI of 6 t HC/m^2; its kerogen type is II-III and its VKA is algal amorphous with variable terrestrial contribution (Legarreta and Villar, 2011). The thickness of the Los Molles Formation ranges from 100 to 800 metres. Tight reservoirs are distributed widely within the formation, being controlled by the paleotopography and the channelized submarine fans. The current development of unconventional reservoirs is focused on tight gas and oil reservoirs, mainly encountered within the vicinity of the depocentre of the basin, and is mostly related to gas-prone basin-centred gas systems. Volume estimates range from 130 to 190 Tcf.

The Vaca Muerta Formation is the most important source rock in Argentina and has an average TOC that varies from 3% to 8%, a Romax of 0.8% to 2%, a hydrogen index of 400–800 mg HC/g TOC and, an SPI of 5 to 20 t HC/m^2; its kerogen type is I/II and IIS in marginal areas and its VKA is high quality amorphous (Legarreta and Villar, 2011). The thickness of the Vaca Muerta Formation ranges from 25 to 450 metres, covering an area of at least 25 000 km^2. Thousands of wells have penetrated the formation partially or totally as part of the exploration and development of the basin. The majority of them have oil and gas shows and in many cases hydrocarbons are being produced. Beginning in 2010, different companies drilled exploration wells targeting the unconventional oil and gas shales of Vaca Muerta Formation. Its world class source rock characteristics have driven the economic activity within the basin. Volume estimates range from 170 to 220 Tcf.

The last Pacific ingression corresponds to the Agrio Formation (Middle Early Cretaceous) and took place during an important change in the subsidence regime: the gradual and pulsatile transition to the foreland phase under a dry climate (Barredo and Stinco, 2010). The Agrio Formation has an average TOC that varies from 2% to 5%, a Romax of 0.6% to 1%, a hydrogen index of 300–700 mg HC/Gtoc and an SPI of 4 to 12 t HC/m^2; its kerogen type is II to II/III and its VKA is algal amorphous with variable terrestrial contribution (Legarreta and Villar, 2011). The thickness of the Agrio Formation ranges from 50 to 400 metres. Oil and gas shows have been documented in many cases as well as the production of hydrocarbons. The current development of unconventional reservoirs is focused to the west and north of the basin. Volume estimates range from 20 to 40 Tcf.

Further south, during the Cretaceous first the Rocas Verdes Basin back-arc basin and then the Austral Basin (Figure 8.1), covering 146 000 km^2, developed synrift bituminous shales in underfilled to balanced-fill lakes, constituting the second most important source rock of the basin, the Serie Tobífera. It has an average TOC that varies from 1% to 3%, a Romax of 0.6% to 1.2%, a hydrogen index of 300–500 mg HC/gTOC and an SPI of 1 t HC/m^2; its kerogen type is II to III and its VKA is algal amorphous. The thickness of Serie Tobífera ranges from 5 to 25 metres. Due to its limited thickness and areal distribution, it is not considered as important as the Palermo Aike Formation – the main source rock developed during the thermal relaxation of the rift – yet it has decreasing fault activity and a high stand of the sea level (Barredo and Stinco, 2010). This formation has an average

TOC that varies from 0.5% to 2%, a Romax of 0.8% to 1.8%, a hydrogen index of 300–750 mg HC/GTOC and an SPI of 1 t HC/m^2; its kerogen type is II-III and its VKA is algal amorphous (Legarreta and Villar, 2011). The thickness of the Palermo Aike Formation ranges from 50 to 400 metres. Since 2011, oil and gas companies have been evaluating the feasibility of this unit as a shale gas/oil unconventional reservoir, with estimated technically recoverable resources of 160 Tcf.

The Golfo San Jorge Basin (Figure 8.1) is a complex continental and aulacogenic basin developed in a cratonic position (Barredo and Stinco, 2010) during the gradual breakup of Gondwana. The Golfo de San Jorge Basin covers an area of 170 000 km^2 (Sylwan *et al.*, 2008), one third of which is located offshore. The Pozo Anticlinal Aguada Bandera and Cerro Guadal formations were developed in balanced-fill strongly cyclic lakes. These rocks together are informally known as the Neocomian sequence and are one of the source rocks of the basin developed in a deep euxinic environment controlled by high angle limiting faults. The Neocomian sequence has an average TOC that varies from 0.5% to 3%, a Romax of 0.8% to 2.6%, a hydrogen index of 300–600 mg HC/gTOC and an SPI of 1 t HC/m^2; its kerogen type is II/III and its VKA is algal amorphous and structural terrestrial (Legarreta and Villar, 2011). The thickness of the sequence ranges from 500 to 1800 metres. The technically recoverable resources are of the order of 50 TCF for this sequence. The resulting deep basin geometry and its starved condition favoured the deposition of thick shales of the Pozo D-129 Formation, which is the main source rock and potentially the unconventional reservoir of the basin. In our present context this formation is probably the most relevant after the Neuquén Basin shales. It has an average TOC that varies from 0.5% to 3%, a Romax of 0.6% to 2.4%, a hydrogen index of 300–650 mg HC/gTOC and an SPI of 10 t HC/m^2; its kerogen type is I/II to II/III and its VKA is algal amorphous with minor terrestrial contribution (Legarreta and Villar, 2011). The thickness of the Pozo D-129 Formation reaches up to 2000 metres, which, combined with its areal distribution, is a significant plus. Current development is focused on the unconventional tight gas and oil reservoirs distributed all over the basin. Volume estimates are around 100 Tcf.

Finally, the Palaeozoic (or Tarija) Basin is around 25 000 km^2 in Argentina, although it has been developed mostly in Bolivia. When considered together with the Cretaceous basin they are referred to as the Northwest Basin (Figure 8.1). The geometry of the basin is a combination of lithospheric flexure and local deepening by reactivation of the old Cambrian–Ordovician rift faults. Several depocentres with local sedimentation were developed and infilled with marine and transitional deposits controlled mainly by eustatic variations. The high accommodation space reached during Middle to Upper Devonian times favoured the accumulation of a thick pile of shales, the Los Monos Formation, the source rock of the active petroleum system and potentially an unconventional reservoir. The formation has an average TOC that varies from 0.5% to 1.5%, a Romax of 1.8% to 2.2%, a hydrogen index of 300–400 mg HC/gTOC and an SPI of 1 to 3 t HC/m^2; its kerogen type is II/III to III/IV and its VKA suggests important terrestrial input during low stands

(Legarreta and Villar, 2011). The thickness of the Los Monos Formation ranges from 500 to 1000 metres and it extends all over the region, providing a high rock volume which compensates for the relatively low organic content. Volume estimates are around 40 Tcf.

The Cretaceous Basin covers an area of 53 000 km^2 and is composed of three significant depocentres: Metán-Alemanía, Lomas de Olmedo and Tres Cruces, all with an asymmetric shape. Sedimentation, represented by the Salta Group, comprises shallow-marine to non-marine and volcanic rocks controlled by fault propagation and sea-level variation, with minimal influence from the incipient lithospheric flexure (Barredo and Stinco, 2010). Thick black shales, carbonates and evaporates were developed in a complex geological setting composed of a hypersaline (hydrologically closed) lacustrine environment combined with a marked high frequency wet–dry cyclicity lacustrine environment, highly influenced by marine transgressions. The continental lacustrine and alternating marine clastic and carbonatic facies comprises the Yacoraite Formation, the source rock of the basin. It has an average TOC that varies from 0.6% to 6%, a Romax of 0.6% to 1%, a hydrogen index of 300–750 mg HC/Gtoc and an SPI of 1 t HC/m^2; its kerogen type is II to III and its VKA is algal amorphous (Legarreta and Villar, 2011). The thickness of the Yacoraite Formation ranges from 5 to 50 metres as it was developed mostly within the Lomas de Olmedo depocentre, hence limiting the total volume of rock.

In the Cretaceous and the Tertiary the opening of the Atlantic markedly accelerated South America's westward motion and, together with the convergent activity, created a new tectonic configuration in the productive basins as a consequence of flexural subsidence induced by the Andean orogenic overloading and by sediment charge through the Tertiary period. The basins were subjected to compression and foredeep deposition as they were forced to close (Rincón *et al.*, 2011). No important source rocks have yet been found. However, sediments coming from the shields to these profound foredeeps became texturally mature and highly quartzose along their long westward transport and today represent significant conventional reservoirs (Barredo and Stinco, 2013). Probably the central contribution of these tectonic events was the flexural subsidence induced by the orogeny, which triggered maturation of the overburden Palaeozoic and Mesozoic shales and their consequent partial expulsion – and the subsequent orogenic cycles that gave place to trap formation.

Tables 8.1 and 8.2 summarize the characteristics of the most important source rocks that are currently being considered as unconventional shale reservoirs of hydrocarbons.

Exploration and Production Activities

A significant number of operating companies are currently exploring various unconventional targets: the Vaca Muerta Formation (Figure 8.2), the most active, followed by the Los Molles (Figure 8.3) and the Agrio (Figure 8.4) formations, all located in the Neuquén Basin, and the Pozo D-129 Formation (Figure 8.5) in the Golfo San Jorge Basin.

Table 8.1 *Main characteristics of the source rocks for the Cuyana and Neuquén basins*

Unit	Cacheuta	Precuyano	Los Molles	Vaca Muerta	Agrio
Thickness (m)	50–400	50–1100	100–800	25–450	50–400
Average TOC (%)	3–10	2–11	1–5	3–8	2–5
K type	I	I to I/III	II-III	I/II	II to II/III
Romax (%)	0.6–1	0.4–0.8	0.8–2	0.8–2	0.6–1
HI (mgHC/gTOC)	600–900	300–900	300–500	400–800	300–700
SPI (t HC/m^2)	3–10	10	6	5–20	4–12
Age	Triassic	Upper Triassic Lower Jurassic	Early Jurassic	Upper Jurassic	Early Cretaceous
Environment	Lacustrine	Lacustrine	Marine	Marine	Marine

One point to highlight is that hundreds of wells have already penetrated these unconventional reservoirs and therefore, from a classic point of view, the current activities might not be considered as new frontier exploration.

Moreover, as unconventional reservoirs are the source rocks of the active petroleum systems, studies are focused on characterizing the rocks and facies and on understanding the heterogeneities and geomechanics in order to define the sweet spots and hence to help in designing the optimum completion.

Many studies are currently being undertaken by companies (Americas Petrogas, Capex, Chevron, Exxon, Medanito, O & G Developments, Pan American Energy, Pluspetrol, Petrobras, Roch, Shell, Total, Wintershall and YPF) in conjunction with universities (Universidad de Buenos Aires, Universidad de La Plata, Universidad Nacional de la Patagonia San Juan Bosco, Universidad Nacional del Comahue, Universidad Nacional del Sur, Universidad Nacional de Córdoba and Instituto Tecnológico de Buenos Aires) in the form of joint ventures or consortiums or as separate efforts.

The results of these coordinated studies can be in the form of unpublished reports not usually available to the public, or of published papers (Barredo, 2012; Barredo and Stinco, 2010; Barredo and Stinco, 2013; Barredo *et al.*, 2008; Belloti *et al.*, 2014; EIA, 2013; Gulisano and Gutierrez Pleimling, 1994; Gulisano *et al.*, 1984; Howell *et al.*, 2005; Kietzmann *et al.*, 2014; Kozlowski, 2011; Leanza, 2012; Legarreta and Villar, 2011; Rincón *et al.*, 2011; Santiago *et al.*, 2014; Sassali *et al.*, 2013; SEN, 2011; Spalletti *et al.*, 2000; Stinco and Mosquera, 2003; Stinco, 2013; Stinco, 2015; Stinco and Barredo, 2014a, 2014b; Stinco *et al.*, 2013; Utgé *et al.*, 2014; Veizaga-Saavedra *et al.*, 2014; Zencich *et al.*, 2008; among others).

Outcrop (Figure 8.6) and subsurface data integration are crucial in order to understand unconventional shale reservoirs.

Some authors (SEN, 2011; EIA, 2013; Barredo and Stinco, 2013, Stinco and Barredo, 2014), applying different methodologies, have estimated important amounts of resources in the unconventional shale reservoirs in Argentina.

Table 8.2 *Main characteristics of the source rocks for the Paleozoic, Cretaceous, Golfo San Jorge and Austral basins*

Unit	Los Monos	Yacoraite	Neocomiano	Pozo D-129	Serie Tobífera	Palermo Aike
Thickness (m)	500–1000	5–50	500–1800	1000–2000	5–25	50–400
Average TOC (%)	0.5–1.5	0.5–6	0.5–3	0.5–3	1–3	0.5–2
K type	II/III to III/IV	II-III	II/III	I/II to II/III	II to III	II-III
Romax (%)	0.5–1.5	0.6–1	0.8–2.6	0.6–2.4	0.6–1.2	0.8–1.8
HI (mgHC/gTOC)	300–400	300–750	300–600	300–650	300–500	300–750
SPI (t HC/m^2)	1–3	1	1	10	1	1
Age	Silurian Late Devonian	Upper Cretaceous	Early Cretaceous	Early Cretaceous	Middle/Upper Jurassic	Early Cretaceous
Environment	Marine	Lacustrine	Lacustrine	Lacustrine	Lacustrine	Marine

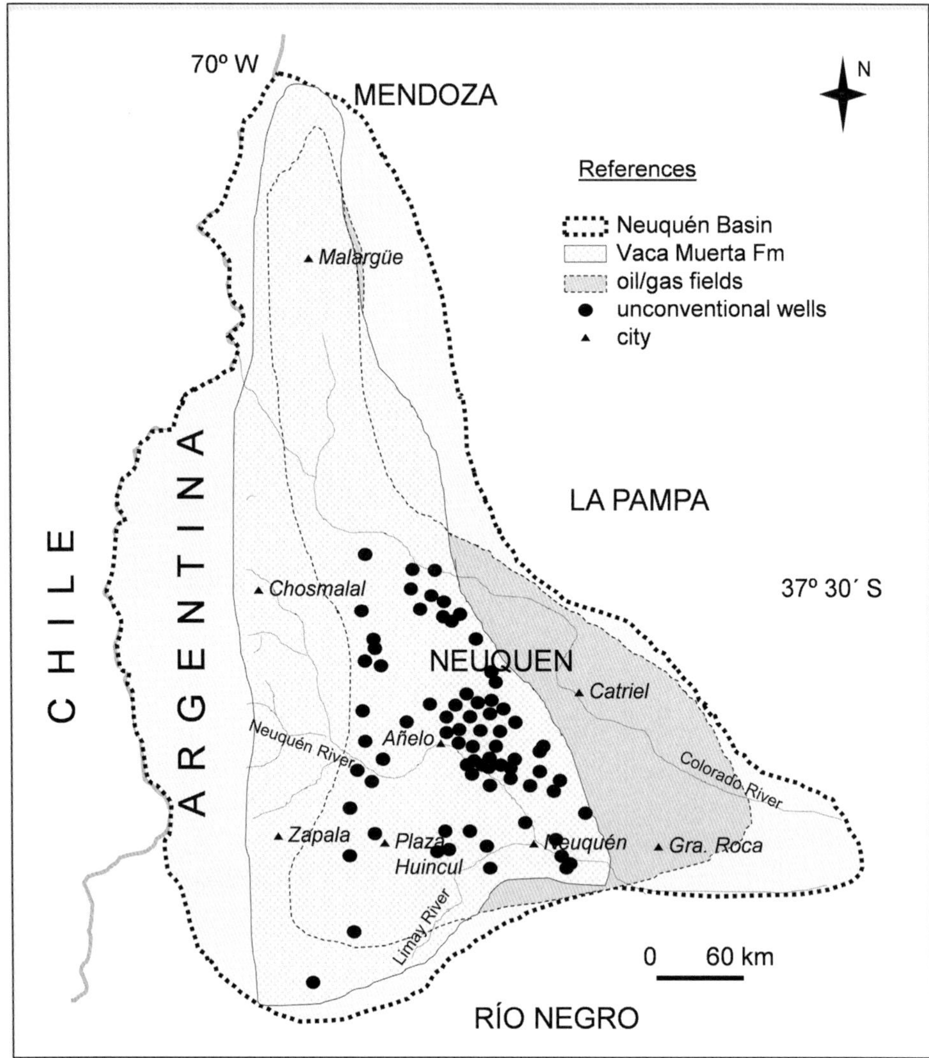

Figure 8.2 Vaca Muerta Formation location and activity map.

The paper SEN (2011) estimated 741 Tcf applying the methodology of Schmoker (1994), for the Los Monos, Los Molles, Vaca Muerta, Pozo D-129 and Palermo Aike formations (Table 8.3).

In EIA (2013) technically recoverable resources of 801.3 Tcf of gas and 27 bbo of oil were calculated (Table 8.3). The methodology included geological and reservoir characterization of the evaluated units within the basin, the areal extent of the shales, a definition of the prospective area, estimation of the risked shale gas and oil and an estimation of the technically recoverable shale gas and shale oil resources. Unfortunately, this report did not

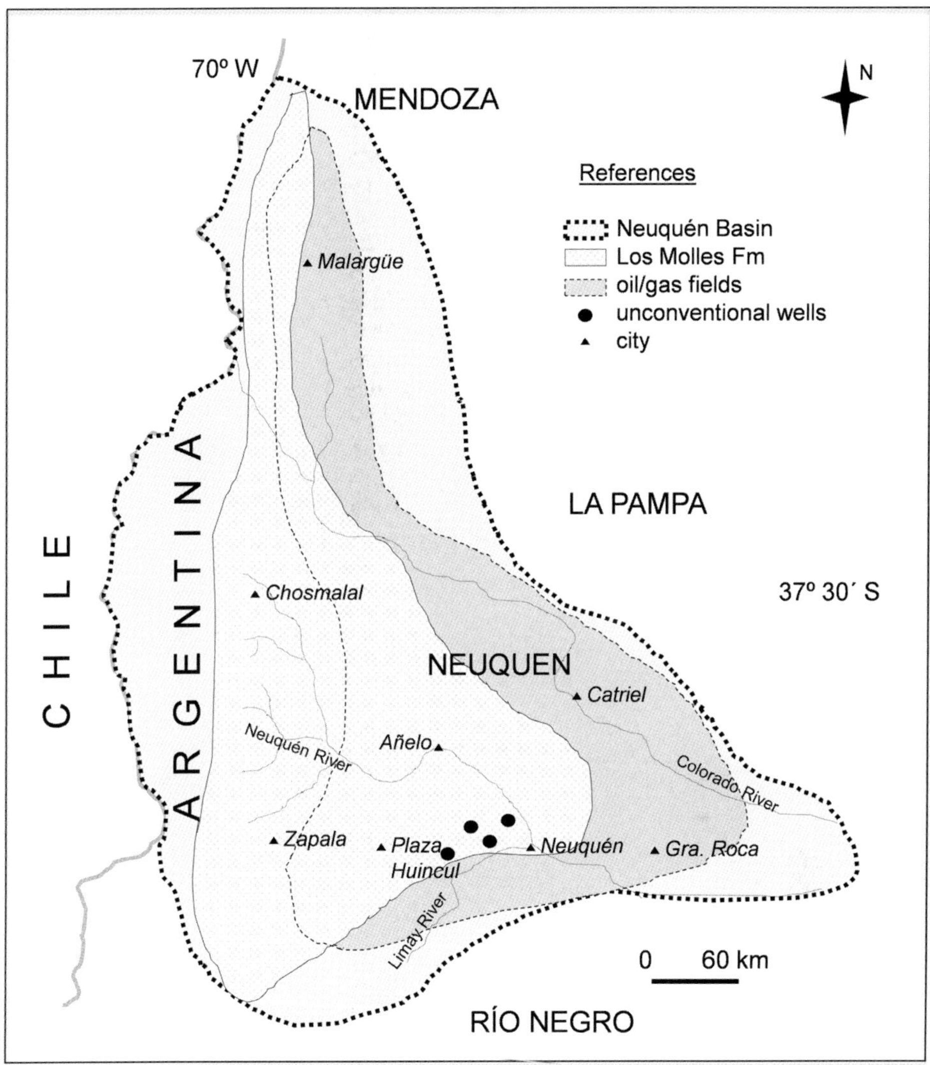

Figure 8.3 Los Molles Formation location and activity map.

take into consideration the resources associated with the Palaeozoic, Cretaceous and Cuyana basins, effective source rocks of proven petroleum systems that have produced (and are currently producing) a significant amount of oil and gas for more than a century.

Barredo and Stinco (2013) and Stinco and Barredo (2014) estimated gas resources of 800 Tcf (Table 8.3) applying a stochastic modelling on variables defined by Schmoker (1994) (area, thickness, source rock density, TOC, HI, retention factor, recovery factor, seal presence and efficiency, burial history and existing fields within the basin) and adjusted according to Kurchinskiy *et al.* (2012).

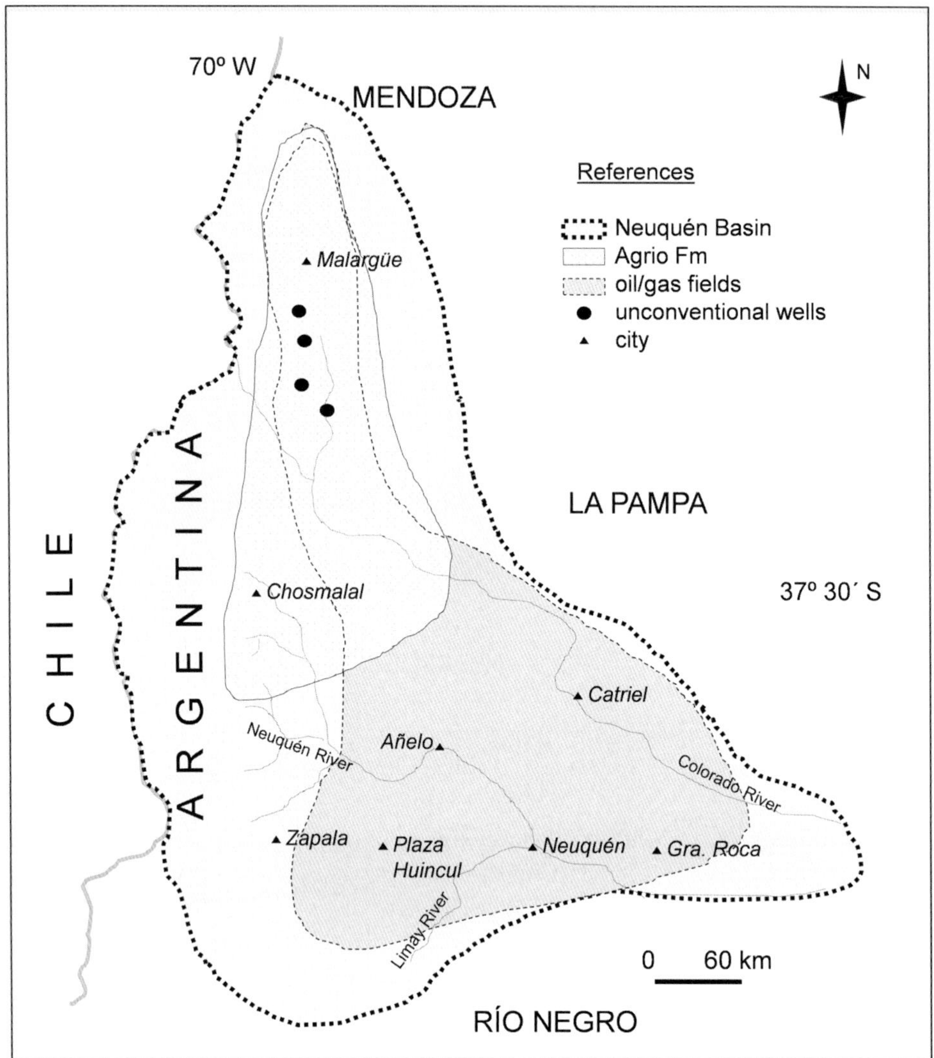

Figure 8.4 Agrio Formation location and activity map.

Annual national consumption is close to 1.5 Tcf and 0.204 bbo; therefore the implications of such large estimated resources are vast and could have a profound impact on future projects and in developing the country.

Two- and three-dimensional seismic data are available in the majority of the relevant areas. These data are being used to delineate the currently producing fields and, in this particular case, provide vital information regarding occurrence, depth, thickness, areal distribution, continuity, lateral variations, compartmentalization and faulting. However, microseismic data are also being acquired in order to understand the geomechanical behaviour of the rocks when hydraulic fracturing is deployed (Stinco, 2015).

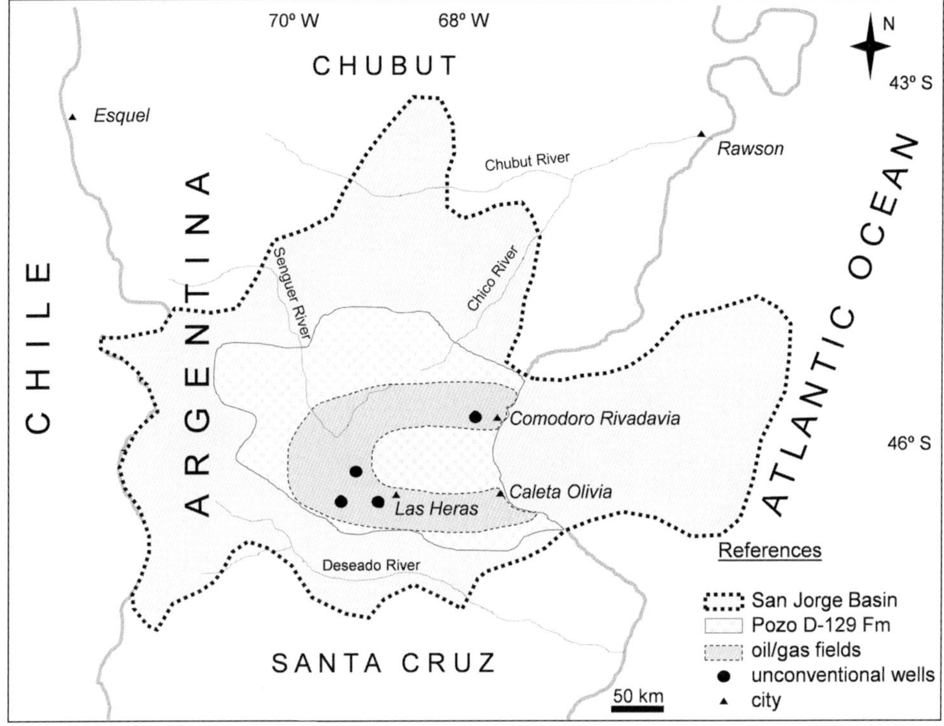

Figure 8.5 Pozo D-129 Formation location and activity map.

Furthermore, from mud and open-hole logging, side-wall core samples, core and test and production data, the following information may be acquired: TOC, kerogen and bitumen type, vitrinite reflectance, level of maturation of the organic matter, mineralogy, diagenesis, porosity, permeability, presence of natural fractures, type of fluids, the free and adsorbed proportions and the temperature, pressure and volumes of the hydrocarbons produced (Stinco, 2015). By processing these data the brittleness can be determined and hence a model for correct hydraulic fracking can be designed that incorporates water and chemical products usage and disposal in order to minimize any adverse effects.

The first wells targeting the Vaca Muerta Formation were drilled in 1971 but not with the modern unconventional concept, which started in 2010. At the time of writing there are more than 480 vertical and 40 horizontal wells, producing around 20 000 barrels per day of oil and more than 60 Mcf/day of natural gas, YPF being the most active company (representing 80% of the total activity). Average initial production for vertical oil wells is around 180 barrels per day of oil and 0.4 Mcf/day of gas, with an estimated ultimate recovery after 30 years of 260 000 barrels of oil equivalent. For horizontal oil wells the average initial production is close to 320 barrels per day of oil and 0.8 Mcf/day of gas, with an estimated ultimate recovery after 30 years of 517 000 barrels of oil equivalent. These numbers could vary significantly as current knowledge of the likely production history is very limited. The current vertical well spacing is on average close to 350 metres and

Figure 8.6 Rhythmic succession of marls and limestones of the Vaca Muerta Formation outcropping in the Los Catutos area, Neuquén Province. Scale in metres.

for horizontal parallel wells it is between 250 and 300 metres (Caligari and Hirschfeldt, 2015).

Less than 20 vertical and horizontal gas wells have been drilled in Vaca Muerta Formation, with an initial production of 2.9 Mcf/day and 4.7 Mcf/day respectively. Estimated ultimate recovery after 30 years for a vertical gas well is 2.2 bcf and for a horizontal gas well is around 3.8 bcf (Caligari and Hirschfeldt, 2015).

The main concern relating to drilling vertical or horizontal wells is still the lack of reliable data needed to diminish uncertainty when defining the sweet spots within Vaca Muerta Formation. Heterogeneity undoubtedly drives the development of the formation (Stinco and Barredo, 2014a).

Independently of the well type, hydraulic fracturing is necessary for the production of hydrocarbons. Considering intervals of formation between 50 and 120 metres, clusters of 50 centimetres separated by 10 to 25 metres are common for these operations.

Up to now, the Vaca Muerta Formation production data demonstrate that the best section is constrained to the lowest portion of the unit that coincides with the intervals with higher TOC content. Additionally, and in order to reduce costs, some operators are developing areas by placing up to four wells at the same location and with casing drilling.

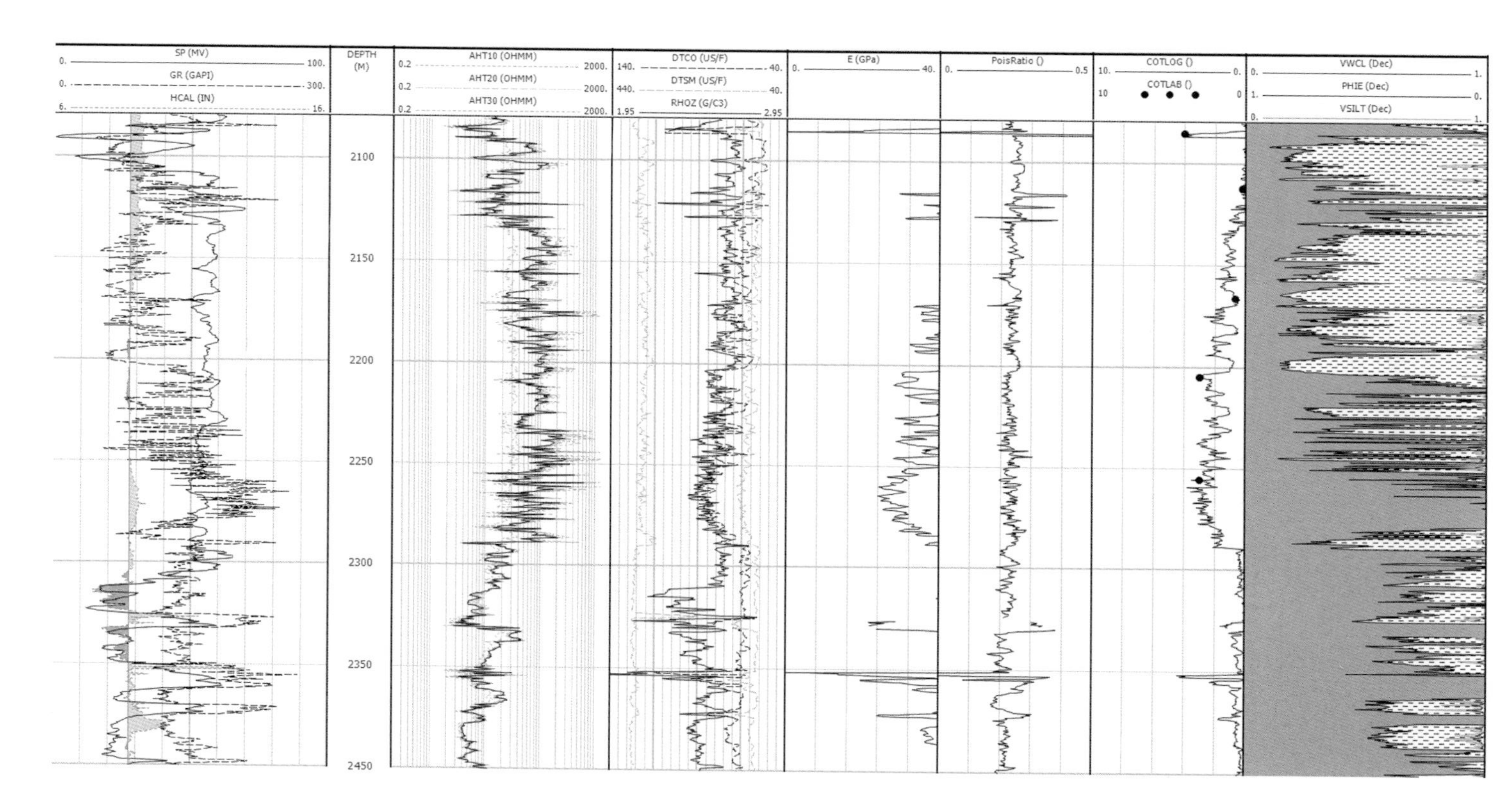

Figure 8.7 Electric and geomechanical response of the Vaca Muerta Formation. Left to right: track 1, SP (Spontaneous Potential, mV), GR (Gamma Ray, API), HCAL (calliper, inches), Bit (bit size, inches); track 2, metres below ground level; track 3, AHT10 – AHT20 – AHT30 – AHT60 – AHT90 (induction resistivities, ohms); track 4, DTCO (compressional transit time, microsec/foot), DTSM (shear transit time, microsec/foot), RHOZ (density, gr/cm3), PEFZ (photoelectric factor, barns/electron); track 5, E (Young´s modulus, gigapascals); track 6, PoisRatio (Poisson´s ratio); track 7, COTLOG (TOC from logs), COTLAB (TOC from laboratory); track 8, VWCL (clay volume, decimal), PHIE (porosity, decimal), VSILT (silt volume, decimal).

Table 8.3 *Argentina's hydrocarbon resources in Tcf estimated by SEN (2011), EIA (2013), Barredo and Stinco (2013) and Stinco and Barredo (2014a, b).*

SEN (2011)		EIA (2013)			Barredo & Stinco (2013)*;Stinco & Barredo (2014a, b)#		
Formation	Tcf	Formation	Tcf	bbo	Basin	Formation	Tcf
Los Monos	34	Ponta Grossa	3.2	0	Paleozoica	Los Monos #	40
Los Molles	259	Los Molles	275.3	3,7	Cretácica	Yacoraite #	5
Los Molles-Lajas	2	Vaca Muerta	307.7	16,2	Cuyana	Cacheuta #	15
Vaca Muerta	109	Neocomiano	50.8	0	Neuquén	Precuyano*	5
Pozo D-129	246	Pozo D-129	34.8	0,5	Neuquén	Los Molles*	190
Palermo Aike	91	Palermo Aike	129.5	6,6	Neuquén	Vaca Muerta*	220
Total	**741**	**Total**	**801.3**	**27**	Neuquén	Agrio*	40
					Golfo San Jorge	Neocomiano #	20
					Golfo San Jorge	Pozo D-129*	100
					Austral	Serie Tobífera #	5
					Austral	Palermo Aike*	160
					Total		**800**

SEN (2011), recoverable gas resources, EIA (2013), technically recoverable gas and oil resources; Barredo and Stinco (2013) and Stinco and Barredo (2014a, b), recoverable gas resources. TCF, trillions of cubic feet; BBO, billion barrels of oil.

Expectations are based on the initial production, and once a base volume is achieved then horizontal wells are drilled in the best areas confirmed by the vertical wells (Stinco and Barredo, 2014a).

The current development of fields involves more than 20 drilling rigs. However, future scenarios consider an increase in the number to 40 or even 60 units. Vertical development well costs are in the range of 7.5 to 8.5 MMUSD, while exploration well costs are close to 15.0 MMUSD. Costs for horizontal wells are generally double those for vertical wells.

In contrast, unconventional production from Los Molles, Agrio and Pozo D-129 is still at a very early stage as only a limited amount of wells have been drilled. In all cases, these wells are providing vital information regarding their viability as unconventional reservoirs.

Hydrogeology and Water Usage

Surface and subsurface water sources are available in the various basins and, although there can be geographically local restrictions due to arid to semi-arid climate conditions, the overall balance is positive. Different studies confirm that rivers (Figures 8.2, 8.3 and 8.4), dams and groundwater can provide enough water for developing the unconventional reservoirs in Neuquén Basin.

Hernández (2015), alluding to studies performed by the sub-secretary of water resources, states that the water drainage volumes for three rivers in Neuquén Basin (i.e., the Neuquén, Colorado and Limay) are 2192, 1903 and 3627 cubic hectometres per year respectively.

Present-day projects include the construction of an aqueduct with a diameter of 254 mm and 13 km long from the Neuquén River (near Añelo city) to a tank of 10 000 cubic metres capacity in one of the main areas, from which water will be distributed to the different drilling locations. This duct should guarantee water provision and (at the same time) reduce environmental impact. All such projects should be in line with provincial law 899 and decrees 790/99 and 1483/12.

In contrast, for developing the Golfo San Jorge Basin unconventional shales, the surface water is not sufficient and groundwater is a suitable option (Alvarez *et al.*, 2006; González *et al.*, 2009; Hernández, 2015). This alternative for the basin is not new: actually, for secondary recovery projects, water is taken from the Paleocene and Oligocene units. Incidentally, there are some advantages related to the use of groundwater; for example it provides thermal and compositional stability, colloidal material is almost absent, it is free of dissolved oxygen and organic matter and surface spills are reduced to a minimum. Furthermore, subsurface water is available close to the field operations.

According to Hernández *et al.* (2002), the hydrological balance model under a non-permanent regime should be adapted according to the particular conditions that exist in Patagonia (rapid concentration and infiltration, the reduction of consumptive losses, deferred recharges and the contribution of influent rivers).

The water needs for developing unconventional shale reservoirs are considerably higher than those for conventional reservoirs. It is one of the most important fluids involved during drilling and hydraulic fracture operations, particularly after flowback.

Aquifer protection takes into account two different aspects: (a) vulnerability according to GOD methodology (groundwater occurrence, overall aquifer class and depth) (Foster and Hirata, 1988), which is accomplished by acquiring mud and open-hole logging data for the first 500 metres, and (b) the integrity of the well-isolating design, which should be built according to international standards. In the Neuquén Basin, this type of analysis has shown that the possibility of affecting the aquifer is negligible (Hernández, 2015).

The amount of water usage during drilling varies within a range of 3% to 27% and for hydraulic fracture operations it fluctuates between 97% and 63%, depending on the different well situations (Hernández, 2015). Current studies show that the water volume for fracking can vary between 7500 and 30 300 cubic metres total (in the case of six vertical wells of 2000 metres below ground level); a horizontal section of 1200 metres requires volumes of water between 54 000 and 174 000 cubic metres (Hernández, 2015). Statistical data also show that flowback can vary from 15% up to 90% of the total amount of injected water, with a 40% reusage of these liquids, thus reducing the amount of surface or underground original water.

Hydraulic fracture design involves four to five fracture steps for vertical wells, requiring 4500 cubic metres of water. These fractures can reach a variable length between 80 and

150 metres. For horizontal wells with 10 to 12 steps, water usage can reach up to 11 000 cubic metres, with a fracture length between 20 and 50 metres. Projections of future requirements indicate that 1.20 cubic hectometres per year of water volume represents the minimum required for the above mentioned activities (Hernández, 2015).

The characteristics of hydraulic-fracking water currently used can be summarized as follows: pH within the range six to eight, total dissolved solids less than 100 000 parts per million, bicarbonates below 300 parts per million, sulphates under 500 parts per million, calcium and magnesium below 2000 parts per million, iron under 5 parts per million and bacteria below 105 milligrams per litre (Hernández, 2015).

Local regulations require that flowback water must be treated before final disposal or even if it is going to be reused for other oil field activities (hydraulic fracturing or injection in waterflooding projects).

These norms clearly state that it is completely forbidden to dispose of or store flowback water directly on the surface water or in open pits. However, jurisprudence permits the drilling of disposal wells over deep targets and the reusage of the flowback water in secondary recovery projects.

That there should be a water handling model for each unconventional project is proposed by Hernández (2015). The model would comprise different phases to take into consideration before hydrocarbon production starts: water production, aquifer protection, handling of the flowback water, control and monitoring, mathematical modelling and sustainability of the project.

There is room for improvement when considering the limited knowledge of the subsurface hydrogeology in some areas, especially for developing the proper management of fluids and wastes.

The control and monitoring of water is crucial to diminish any potential risk of incidents or accidents and in order to provide a proper tool for checking the effectiveness of good practice.

Nevertheless, the increasing field activities demonstrate the necessity of developing a systematic approach in order to evaluate environmental impact. Field-operating and service companies and local authorities are working together in order to minimize the effects of the massive use of water.

Environmental Issues

An important number of activities related to exploration for and the production of hydrocarbons should be analysed in order to determine their environmental impact. The majority of them correspond to routine activities not only for unconventional but also for conventional reservoirs, and hence are already included in local and national regulations.

The strict application of environmental management concepts, from the conceptual model of a project through to the production of hydrocarbons, could reduce any potential negative consequence on natural resources or the welfare and quality of life.

Project planning, engineering design and feasibility analyses (including technical, economical, environmental and social aspects) are crucial before starting any activity. Environmental impact studies are to be performed for each of the different stages involved in exploration, drilling, completion, production, infrastructure construction, waste disposal, well abandonment, transport and storage of hydrocarbons, as a requirement of the local and national authorities (Sarandón, 2015).

All the above-mentioned activities involve soil, human and equipment movements, as well as water and chemical product storage and transport. Figure 8.8 summarizes the key parameters (water, soil, effluents, flora and fauna, air quality and social) considered when evaluating the environmental impacts of the upstream and downstream activities.

According to Sarandón (2015), environmental issues to be assessed in advance, considering typology, areal extension and importance, are: water, air and drilling cutting contamination; the induction of geohazards; conflicts related to water usage and availability; the alteration of local ecosystems; modification of land usage; changes in the social context; traffic congestion; route deterioration; increase in solid waste; alteration of economic relations in the productive systems; destruction and/or lost of cultural, historical and paleontological heritage; a demand for environmental controls; a demand for technological development; conflicts of accessibility in disputed areas; and the consolidation of fossil fuels in the energy matrix.

The most pertinent document is the environmental impact study (EIS) prepared by the operating company. This study should have the authorities' approval to be valid and acts as a kick-off document. Generally, the EIS comprises an environmental management plan (EMP), an environmental management and monitoring plan (EMMP), environmental monitoring and control (EMC) and an environmental contingency program (ECP). After a period that varies between two and four months, the study is formally approved by the authorities and the company can start on any activity included in it.

In ANI (2013) it was proclaimed that the environmental issues can be solved using tools and mechanisms already available in the market. These include procedures, auditing and regulation of the activities performed during the different stages involved in the exploration and the production of unconventional reservoirs.

Sarandón (2015) grouped these tools and mechanisms into: (a) regional studies, in order to establish the environmental framework, ecosystem, natural elements, anthropic subsystem and vulnerable human groups; (b) the development of specific environmental management tools to complement the environmental impact study, such as: best practice procedures, independent certification of the drilling and completion activities, the recording of the chemical products used for fracking and the recording of incidents and accidents; (c) strengthening of the government control mechanisms according to federal and provincial laws; (d) the developing of interagency agreements in order to achieve a proper coordination of scientific transfers and counselling and the definition of standards, criteria and procedures; (e) effective communication strategies with the public; (f) the definition of compensation strategies for nonpredefined environmental impacts such as permanent land usage by infrastructure or landscape fragmentation by ducts or roads.

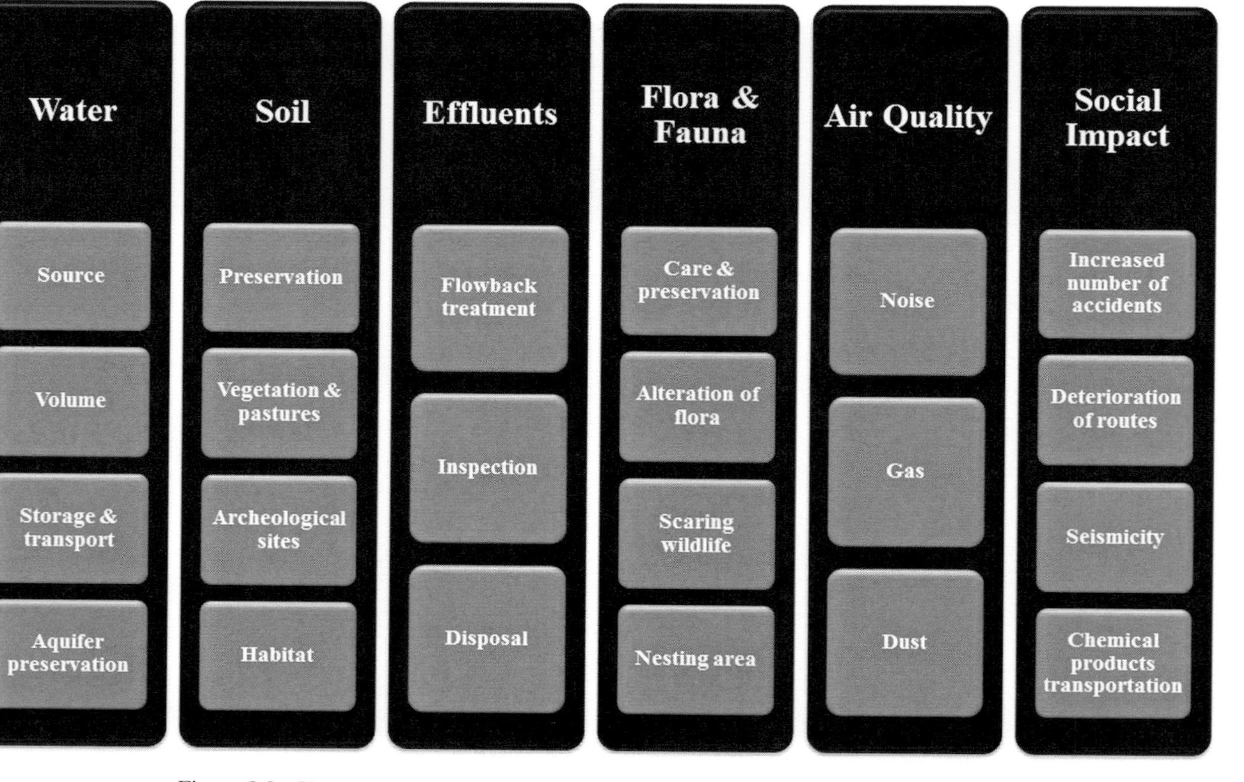

Figure 8.8 Key parameters to be considered when evaluating environmental impacts.

Laws, Decrees and Norms

According to Argentine law 17 319 (1967), hydrocarbons belong to the state therefore upstream and downstream activities should be ruled by the executive branch of the government. National and private oil companies can explore and produce hydrocarbons as long as they guarantee country energy supply and the replacement of proper reserves.

As earlier legislation did not take into account unconventional reservoirs, important modifications were introduced by law 27 007 (2014), favouring the concept of development activities over longer periods of time, such as: (a) a first exploration period for conventional reservoirs of up to three years, a second exploration period for conventional reservoirs of up to three years with extension time up to five years; (b) a first exploration period for unconventional reservoirs of up to four years, a second exploration period for conventional reservoirs of up to four years, with extension time up to five years; (c) an exploitation concession of 25 years for conventional reservoirs, 30 years for offshore activities and 35 years for unconventionals; (d) the introduction of a "pilot plan" figure of up to five years for defining the unconventional area to be developed; (e) an unlimited number of extensions for the operator over the same concession; (f) limiting the size of royalties to an average of 12%; (g) limiting carry-on to the exploration period; (h) establishing an investment promotion regime for projects of more than 250 MMUSD; (i) promoting contracts in line with the Guidance Notes of the AIPN (2014).

The National Constitution of 1994, in three different articles (41, 43 and 124) refers to environmental aspects and actions, providing a legal framework that rules on many activities including industrial and energy-related ones.

In addition, law 25 675 'General law regarding the environment' legislates on: biological diversity, sustainable development, principles of environmental policies, national environmental policy, congruence, prevention, intergenerational equity, responsibilities, subsidiarity, sustainability, solidarity, cooperation, environmental impact study, the control of human activities, environmental education, diagnostic and information systems, citizen participation, an environmental security and restoration fund, the federal environmental system, environmental damage and strict liability.

As this law was promulgated in 1994, there is a need to redefine some of its articles and to include new technologies and activities as a result of unconventional shale reservoir activities (Pigretti, 2015).

Economic Aspects

Argentina has a high dependence on hydrocarbons (89%) to fulfil current and future energy demands, with gas dominant (Figure 8.9). Incidentally, conventional hydrocarbon production has seen a very significant decline over recent years, and thus energy imports have grown since 2010.

Oil production peaked in 1998 followed by gas in 2001. Since then, in both cases production has declined with a clear tendency of not being able to recover previous levels unless

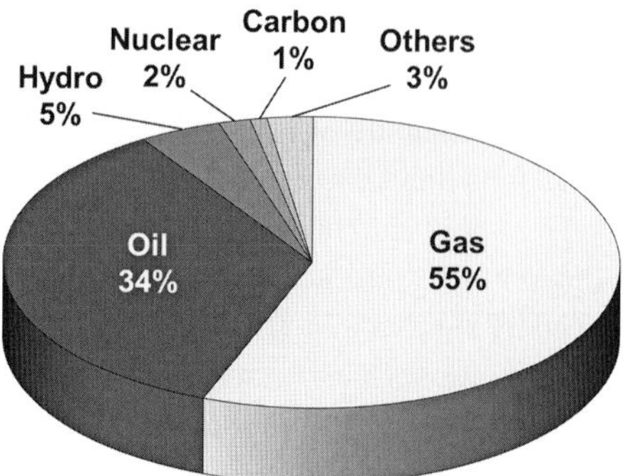

Figure 8.9 Energy matrix of Argentina.

new projects are successful, such as: new discoveries from exploration, implementation of improved and/or enhanced oil recovery plans and the exploitation of unconventional reservoirs.

For unconventional projects it is necessary to remember that investment flows are quite high and sustained in time, with long repayment periods (over 10 years) and low yields.

Riavitz (2015) demonstrated that Argentina needs to avoid importing energy; the current decline impacts negatively on the total energy balance.

Although there are some associated risks when developing unconventional shale reservoirs, mainly related to geological, engineering and environmental issues, the optimization of current operations can help in reducing the growth in energy imports.

Nowadays, the most important challenge is a combination of developing and adopting an appropriate learning curve, together with substantial cost reductions. Success with this latter aspect could help in the case of fluctuating oil and gas prices in the near future, which could otherwise have a negative impact in the early stages of development of Argentina's unconventional shale reservoirs. Current scenarios consider 100 USD/barrel of oil and for gas 7.5 USD/MMbtu for an average repayment of 13 years with an internal rate of return of 10%. However, as production history is quite recent and short, these projections could vary substantially with time.

Conclusions

There are a number of well-known risks associated with the exploration and development of shale reservoirs in Argentina. These include geological, engineering, environmental and social threats. Having identified the majority of them, applying best practice should minimize risks and correspondingly diminish the impact of potential problems. There is a clear and shared responsibility between the companies, together with the community and authorities, to be taken into account.

Proper regulation can provide the legal framework for investors to participate in the different projects with benefits starting from the very early stages of development.

The country has the resources, and needs to bring investors to the market to help it towards a sustainable industrial future. The current stage of the projects suggests a promising future, considering the growing energy demands, and consequently opens up the possibility of modifying the current negative hydrocarbon trade balance into a positive balance.

Finally, it is important and encouraging to observe that everyone – communities, organizations, companies and government – is promoting the development of these resources. There are still a number of steps to better identify and exploit our reserves and, up to now, all stakeholders are working together to accomplish that.

References

Academia Nacional de Ingeniería (ANI) (2013). Aspectos ambientales en la producción de hidrocarburos de yacimientos no convencionales. El caso particular de "Vaca Muerta" en la provincia de Neuquén. Academia Nacional de Ingeniería (Instituto de Energía), Documento N° 4, p. 37.

Alvarez, M., Hernández, L., Hernández, M. and González, N. (2006). Relación aguas subterráneas-aguas superficiales en Patagonia Extrandina. República Argentina. *Revista Latino-Americana de Hidrogeología*, 6, 43–48.

Association of International Petroleum Negotiators (AIPN) (2014). Unconventional Resources Operating Agreement "UROA" (2014).

Barredo, S. (2012). Geodynamic and tectonostratigrafic study of a continental rift: the Triassic Cuyana Basin, Argentina. In *Tectonics – Recent Advances*, ISBN 978–953–51–0675–3, pp. 99–130.

Barredo, S., Cristallini, E., Zambrano, O., Pando, G. and García, R. (2008). Análisis tectosedimentario del relleno de edad precuyana y cuyana inferior de la región septentrional del alto de Kauffman, Cuenca Neuquina. In *Proc. VII Congreso de Exploración y Desarrollo de Hidrocarburos*, Instituto Argentino del Petróleo y el Gas, pp. 443–446.

Barredo, S. and Stinco, L. (2010). *Geodinámica de las Cuencas Sedimentarias: Su Importancia en la Localización de Sistemas Petroleros en la Argentina*. Petrotecnia, Instituto Argentino del Petróleo y el Gas, pp. 48–68.

Barredo, S. and Stinco, L. (2013). A geodynamic view of oil and gas resources associated to the unconventional shale reservoirs of Argentina. In *Proc. Unconventional Resources Technology Conference (URTeC)*, CID: 1593090. American Association of Petroleum Geologists, Denver, Colorado.

Belloti, H., Rodríguez, J., Conforto, G., Pagan, F., Perez Mazas, A., Agüera, M. *et al.* (2014). La Formación Palermo Aike como reervorio no convencional en la Cuenca Austral, Provincia de Santa Cruz, Argentina. In *Proc. IX Congreso de Exploración y Desarrollo de Hidrocarburos, Simposio de Recursos No Convencionales,* Instituto Argentino del Petróleo y el Gas, pp. 137–159.

Caligari, R. and Hirschfeldt, M. (2015). Condiciones para la explotación de recursos hidrocarburíferos no convencionales en Argentina. In *Proc. Recursos Hidrocarburíferos no Convencionales Shale y el Desarrollo Energético de la Argentina. Caracterización, Oportunidades, Desafíos*. Eudeba, ISBN: 978–987–26841–4–3, pp. 209–301.

Energy Information Administration (EIA) (2013). Technically recoverable shale oil and shale gas resources: an assessment of 137 shale formations in 41 countries outside the United States. US Department of Energy, p. 730.

Foster, S. and Hirata, R. (1988). Determinación del riesgo de contaminación de aguas subterráneas. CEPIS/PAHO-WHO Technical Report, p. 81.

González, N., Hernández, L. and Hernández, M. (2009). Condicionantes climáticos y físicos de las regiones áridas en Argentina. Implicancias en el régimen hídrico superficial y subterráneo. In *Recarga de Acuíferos. Aspectos Generales y Particularidades en Regiones Áridas*, ISBN 978–987–1082–39–1, pp. 53–62.

Gulisano, C. and Gutiérrez Pleimling, A. (1994). Field trip guidebook, Neuquina Basin, Neuquén Province. In *Proc. Fourth International Congress on Jurassic Stratigraphy and Geology*.

Gulisano, C., Gutiérrez Pleimling, A. and Digregorio, R. (1984). Análisis estratigráfico del intervalo Tithoniano-Valanginiano (Formaciones Vaca Muerta, Quintuco y Mulichinco) en el suroeste de la provincia de Neuquén. In *Proc. 9 Congreso Geológico Argentino*, pp. 221–235.

Hernández, M., González, N. and Sánchez, R. (2002). Mecanismos de recarga de acuíferos en regiones áridas. Cuenca del Río Seco, Provincia de Santa Cruz. Argentina. In *Proc. XXXII IAH Congress – VIth Congreso ALHSUD*. Mar del Plata.

Hernández, M. (2015). Recursos hídricos. In *Recursos Hidrocarburíferos no Convencionales Shale y el Desarrollo Energético de la Argentina. Caracterización, Oportunidades, Desafíos*. Eudeba, ISBN: 978–987–26841–4–3, pp. 303–344.

Howell, J., Schwarz, E., Spalletti, L. and Veiga, G. (2005). *The Neuquén basin: and overview. In The Neuquén Basin, Argentina: A Case Study in Sequence Stratigraphy and Basin Dynamics*. Geological Society, London, Special Publications, vol. 252, pp. 1–14.

Keidel, J. (1916). La geología de las Sierras de la Provincia de Buenos Aires y sus relaciones con las montañas de Sudáfrica y Los Andes. *Ministerio de Agricultura de la Nación, Sección Geología, Mineralogía y Minería, anales*, XI (3), pp. 1–78.

Kietzmann, D., Palma, R., Riccardi, A., Martín-Chivelet, J. and López-Gómez, J. (2014). Sedimentology and sequence stratigraphy of a Tithonian–Valanginian carbonate ramp (Vaca Muerta Formation): a misunderstood exceptional source rock in the Southern Mendoza area of the Neuquén Basin, Argentina. *Sedimentary Geology*, 302, 64–86.

Kozlowski, E. (2011). ¿Qué se conoce, qué se puede inferir, qué perspectivas se visualizan en los recursos convencionales y no convencionales en las 5 cuencas más exploradas? In *Proc. SPE workshop Estimación de Recursos de Petróleo y Gas en la República Argentina*. SPE Argentina, Buenos Aires.

Kurchinskiy, V., Gentry, K. and Hill, R. (2012). Source rock evaluation technique: a probabilistic approach for determining hydrocarbon generation potential and in-place volume for shale plays. *AAPG Search and Discovery* article #41045.

Leanza, H. (2012). The Vaca Muerta Formation (Late Jurassic – Early Cretaceous): history, stratigraphic context and events of this emblematic unit of the Neuquén Basin, Argentina. *AAPG Search and Discovery* article #90165.

Legarreta, L. and Villar, H. (2011). Geological and geochemical keys of the potential shale resources, Argentina Basins. *AAPG Search and Discovery* article #80196.

Pigretti, E. (2015). Aspectos normativos ambientales sobre los recursos no convencionales. In *Recursos Hidrocarburíferos no Convencionales Shale y el Desarrollo Energético de la Argentina. Caracterización, Oportunidades, Desafíos*. ISBN: 978–987–26841–4–3, Eudeba, pp. 417–426.

Ramos, V. and Kay, S. (1991). Triassic rifting and associated basalts in the Cuyo basin, Central Argentina. In *Andean Magmatism and its Tectonics Setting, Geological Society of America*, Special Paper V, vol. 265, pp. 79–91.

Riavitz, L. (2014). El futuro de la energía en Argentina: alternativas. In *Recursos Hidrocarburíferos no Convencionales Shale y el Desarrollo Energético de la Argentina. Caracterización, Oportunidades, Desafíos.* ISBN: 978–987–26841–4–3, Eudeba, pp. 29–134.

Rincón M., Barredo, S., Zunino, J., Salinas, A., Reinante, S. and Manoni, R. (2011). Síntesis general de los bolsones intermontanos de San Juan y La Rioja. In *Proc. VIII Congreso de Exploración y Desarrollo de Hidrocarburos, Simposio de Cuencas Sedimentarias Argentinas*. IAPG, Instituto Argentino del Petróleo y el Gas, pp. 321–406.

Santiago, M., Rauzi, R., Lafitte, G. and Alvarado, O. (2014). La Formación Vaca Muerta como objetivo exploratorio no convencional en la Subcuenca de Picún Leufú, Neuquén, Argentina. In *Proc. IX Congreso de Exploración y Desarrollo de Hidrocarburos, Simposio de Recursos No Convencionales*, Instituto Argentino del Petróleo y el Gas, pp. 275–314.

Sarandón, R. (2015). Impacto ambiental de la explotación de los recursos no convencionales. In – *Caracterización, Oportunidades, Desafíos.* ISBN: 978–987–26841–4–3. Eudeba, pp. 293–358.

Sassali, L., Barredo, S. and Stinco, L. (2013). Análisis tectono-estratigráfico y del Sistema Petrolero del depocentro comprendido en el Yacimiento Diadema, Cuenca del Golfo San Jorge. In *Proc. I Jornadas Geológicas de la Cuenca del Golfo San Jorge*, p. 10.

SEN (Secretaría de Energía de la Nación) – (2011). Análisis de las potencialidades del desarrollo del gas no convencional en Argentina. D. Bogetti and J. Ubeda. *In Proc. VIII Congreso de Exploración y Desarrollo de Hidrocarburos*, Instituto Argentino del Petróleo y el Gas.

Schmoker, J. (1994). Volumetric calculation of hydrocarbons generated. In *The Petroleum System – from Source to Trap. American Association of Petroleum Geologists Memoir* 60, pp. 323–326.

Spalletti, L., Franzese, J., Matheos, S. and Schwarz, E. (2000). Sequence stratigraphy of a tidally dominated carbonate siliciclastic ramp; the Tithonian–Early Berriasian of the Southern Neuquén Basin, Argentina. *Journal of the Geological Society,* 157, 433–446.

Stinco, L. and Mosquera, A. (2003). Estimación del contenido total de carbono orgánico a partir de registros de pozo para las formaciones Vaca Muerta y Los Molles, Cuenca Neuquina, Argentina. In *Proc. II Congreso de Hidrocarburos*, Instituto Argentino del Petróleo y el Gas, p. 17.

Stinco, L. (2013). Cómo son los reservorios no convencionales en la Argentina. Petrotecnia, LIV No. 3, Instituto Argentino del Petróleo y el Gas, pp. 63–71.

Stinco, L., Schiuma, M., Cabanillas, L., Casalis, D., Rabanaque., L., Velasco, G. *et al.* (2013). Aspectos técnicos, estratégicos y económicos de la exploración y producción de hidrocarburos. Instituto Argentino del Petróleo y el Gas. ISBN 978–987–9139–63–9, p. 262.

Stinco, L. and Barredo, S. (2014a). Vaca Muerta Formation: an example of shale heterogeneities controlling hydrocarbon´s accumulations. CID 1922563. In *Proc. Unconventional Resources Technology Conference (URTeC)*, American Association of Petroleum Geologists, Denver.

Stinco, L. and Barredo, S. (2014b). Características geológicas y recursos asociados con los reservorios no convencionales del tipo "shale" de las cuencas productivas de la Argentina. Petrotecnia. LV No. 5, Instituto Argentino del Petróleo y el Gas, pp. 44–66.

Stinco, L. (2015). Los recursos hidrocarburíferos en la Argentina y las características de los reservorios no convencionales del tipo shale. In *Recursos Hidrocarburíferos No Convencionales Shale y el Desarrollo Energético de la Argentina. Caracterización, Oportunidades, Desafíos*. ISBN: 978–987–26841–4–3, Eudeba, pp. 135–207.

Sylwan, C., Rodríguez, J. and Strelkov, E. (2008). Petroleum system of the Golfo San Jorge basin, Argentina. In *Proc. VII Congreso de Exploración y Desarrollo de Hidrocarburos. Simposio Sistemas etroleros de las Cuencas Andinas*. Instituto Argentino del Petróleo y el Gas, pp. 53–78.

Utgé, S., Martinez Cal, S., Basile, Y. and Ponce, C. (2014). Evaluación del potencial de la Formación Pozo D-129 como reservorio no convencional shale gas/oil. In *Proc. IX Congreso de Exploración y Desarrollo de Hidrocarburos, Simposio de Recursos No Convencionales*, Instituto Argentino del Petróleo y el Gas, pp. 117–135.

Veizaga-Saavedra, J., Poiré, D., Vergani, G. and Salfity, J. (2014). Formación Los Monos (Devónico), Cuenca de Tarija: aproximación geoquímica y mineralógica del potencial como shale gas y shale oil. In *Proc. IX Congreso de Exploración y Desarrollo de Hidrocarburos, Simposio de Recursos No Convencionales*, Instituto Argentino del Petróleo y el Gas, pp. 93–115.

Zencich, S., Villar, H. and Bogetti, D. (2008). Sistema petrolero Cacheuta-Barrancas de la Cuenca Cuyana, Provincia de Mendoza, Argentina. In *Proc. VII Congreso de Exploración y Desarrollo de Hidrocarburos, Sistemas Petroleros de las Cuencas Andinas*, Instituto Argentino del Petróleo y el Gas, pp. 109–134.

9

Unconventional Gas in the United Kingdom

MICHAEL BRADSHAW[1]

Introduction

Despite the fact that the unconventional gas industry in the United Kingdom (UK) is at the very earliest stages of development it has attracted a huge amount of public attention and is now the subject of a highly polarised debate. One the one side, the national government, industry organisations and companies have talked up the economic opportunities that would result from commercial shale gas development as well as the benefits it would bring in terms of energy security and climate change policy. The then Prime Minister David Cameron famously said in January 2014 that his government was 'going all out for shale'. The UK government also maintained that the risks associated with shale gas development were manageable within the existing regulatory regime and that the early stages of exploration would be very carefully monitored. On the other side, opposition parties, environmental groups and the public in areas of potential drilling have raised concerns about the negative environmental, social and economic impacts and see such development as incompatible with the UK's climate change legislation. They also question the efficacy and capacity of the regulatory regime.

Some of those opposed to unconventional oil and gas would like to see a ban on shale gas activity, while others call for a more precautionary approach and raise concerns about the current regulatory regime. Already, parliamentary enquiries, government departments, environmental groups, industry lobby groups, think tanks and academia have produced a huge volume of material; the media has also been attracted to the issue with varying degrees of impartiality (Jaspal and Nerlich 2014). As of September 2015, however, there had only been one well drilled and hydraulically fractured, by Cuadrilla at Preese-Hall in Lancashire in early 2011, but operations had to be stopped when they triggered two minor seismic events. This subsequently resulted in the government's imposing a moratorium between June 2011 and December 2012. Although various vertical wells have been drilled onshore, often for conventional resources, that have sought to appraise the potential of shale gas, it remains the case that the UK has yet to see a programme of horizontal drilling

[1] I wish to acknowledge the support of a grant from the UK Energy Research Centre (UKERC), which is funded by RCUK's Energy theme, and also of a grant from the M4ShaleGas project, funded by the European Union's Horizon 2020 Research and Innovation Programme under grant agreement number 640715.

and high volume hydraulic fracturing to test shale gas flow rates. Thus, in the UK in 2015, the issue was the potential of shale gas and the challenges that would have to be overcome just to enable a programme of exploration drilling and appraisal, rather than commercial development.

This chapter follows the common structure of country case studies in this volume. The first, substantive, section considers the UK's unconventional gas potential; the second section provides some historical background and explores the contours of the shale gas debate in the UK; the third section describes the current regulatory regime and its recent evolution; the fourth and final section concludes the chapter by considering the challenges that must be overcome if the UK shale gas industry is to gain social licence to frack.

The UK's Unconventional Gas Potential

The UK has global status as a producer of offshore oil and gas on the continental shelf of the North Sea, and there is a long history of onshore conventional oil and gas drilling, with some 2100 wells drilled in the last 100 years. Nevertheless, the UK shale gas industry is still in its infancy. Selley (2012) reported that research in the United States (US) in the 1970s and 1980s stimulated interest in the UK and that work at Imperial College did identify potential resources in shale oil and gas, but at the time it was not considered economically viable. The surge in unconventional oil and gas production in the US a decade or so ago eventually triggered renewed interest in the UK and in 2008 the British Geological Survey (BGS) began a review of the unconventional hydrocarbon resources of Britain's onshore basins. This resulted in two reports produced by the BGS for the Department of Energy and Climate Change (DECC)[2] one on shale gas and a second on coal bed methane (CBM) (Harvey and Gray 2013a, 2013b).

The concerns in the UK are directed primarily to shale gas as there is already some CBM production, which has, so far, proved less controversial. Harvey and Gray (2013a, p. 1) noted that: 'ahead of more drilling, fracture stimulation and testing there are no reliable indicators of potential productivity'; however, they conclude that: 'by analogy with similar producing plays in America, the UK shale gas reserve could be as large as 150 bcm (5.3 Tcf) – very large compared with the 2–3 bcm estimate of undiscovered gas resources for onshore conventional petroleum'. A study for the US Energy Administrations (USEIA 2013, XI-1–XI-25) estimated the UK's technically recoverable shale gas reserves to be 740 bcm (26 Tcf). Given the current status of exploratory drilling in the UK there are no reliable reserve estimates, but to inform further development the DECC commissioned the BGS to produce a number of studies aimed at estimating the size of the resource base (the total gas in place) in key regions.

The BGS has now completed four Gas-in-Place Resource Assessments, two in England, one in Scotland and one in Wales. The first study examined the Bowland–Hodder shale

[2] In mid-July 2016, in the aftermath of the EU Referendum and the goverment reshuffle following Theresa May's appointment as Prime Minister, DECC was merged with another ministry to create the Department of Business, Energy & Industrial Strategy (BEIS).

Table 9.1 *The BGS Bowland–Hodder gas-in-place assessment*

	Low		Central		High	
	Tcf	Tcm	Tcf	Tcm	Tcf	Tcm
Upper limit	164	4.6	264	7.5	447	12.7
Lower limit	658	18.7	1065	30.3	1834	51.9
TOTAL	822	23.3	1329	37.6	2281	64.6

Note: There is 50% chance that the true value will lie above the central estimate and a 50% chance that it will lie below. There is a 10% chance that the true value will lie below the low estimate and a 10% chance that it will be above the high estimate.
Source: Andrews (2013, p. 3).

formation in northern England. According to the BGS (Andrews 2013): 'over 15 000 miles of seismic data were integrated with BGS outcrop and fault mapping, well data, historical and newly-commissioned laboratory studies to identify the potential volumes of shale gas in the Bowland–Hodder formation'. The BGS then used a three-dimensional geological model and a statistical stimulation to estimate the amount of gas that is physically in the rock. This is not the amount that is technically recoverable, which would be a smaller number, or the proven reserve that is technically recoverable and economically viable to produce, which is an even smaller number. As noted above, more precise estimates require a substantial programme of exploratory and appraisal drilling. Two sets of estimates were produced, one for the upper unit, which is typically 150 m thick and is distributed over a large area across the north of England, and one for the lower unit, which is deeper and about which a lot less is known. However, the latter could be up to 3000 m thick, and this would enable multiple fractures on the vertical axis, which could enhance its commercial viability. The results of the Bowland–Hodder assessment are shown in Table 9.1.

These are substantial potential resources when one considers that the UK's current level of gas consumption was 66.7 bcm in 2014 and that some 45 per cent was imported (BP 2015, pp. 22–23). However, the BGS is very careful to point out that it currently does not have a sufficient understanding of the geological or commercial factors to generate an acceptable reserve estimate. Guessing at a recovery rate of 10 per cent, as some have done, is fanciful because it is not known how many wells could feasibly be drilled in a year and what their individual flow and recovery rates would be. Some indication of the level of uncertainty is provided by the National Grid's (2015, p. 152) latest *Future Energy Scenarios*, which show a range of shale gas production in the UK in the 2020s between zero and 32 bcm a year, the latter figure is almost half current consumption.

Following their Bowland–Hodder work, the BGS also produced assessments for DECC of the Central Valley of Scotland (Monaghan 2014, p. 2) and the Weald Basin in the South of England (Andrews 2014, p. 2). Both regions have a long history of oil and gas exploration: the West Loathian oil-shale formation was the birthplace of the oil-shale industry in the 1850s, and in the Weald Basin there are also historic wells and currently 13 producing sites. The results of the Central Valley assessment showed both shale gas and shale oil in place. The range of assessments for shale oil is 3.2–6.0–11.2 billion barrels in place (421–793–1497 million tonnes) and for shale gas is 49.4–80.3–134.6 tcf (1.40–2.27–3.81 tcm). The gas-in-place number is much smaller than for the Bowland–Hodder, but it is of local significance as there are major gas-consuming industries in the region. The assessment of the Weald Basin concluded that there is unlikely to be any shale gas potential although there could be significant shale oil in place, in the range of 2.2–8.5 billion barrels (290–1110 million tonnes). The Welsh Government commissioned the final study by the BGS (2014), but it did not make any new reserves or resources estimates because there are insufficient publicly available data to reach reliable conclusions. In addition, most of the unconventional gas in Wales is likely to come from CBM associated with the coalfields in south Wales. There is also some commercial interest in the shale gas potential in north Wales, around Wrexham in Cheshire, which is an extension of the Bowland–Hodder shales.

As a result of the efforts of the BGS, it is evident that there are significant amounts of shale gas in place across the north of England, that there is modest – but locally significant – potential in the Central Valley of Scotland and that the Weald basin is unlikely to produce shale gas, but not enough is known about the resource base in Wales to reach any conclusions. The geological potential and political acceptability of unconventional gas in the UK are aligned, as both the Scottish and Welsh governments have placed moratoria on drilling while further evidence is gathered (a similar situation prevails in Northern Ireland). This means that England is the only region in the UK where it is currently possible to carry out shale gas exploration.

The Shale Gas Debate in the UK

The shale gas industry in the UK dates back to the 13th Onshore Licensing round, which was held in 2008 when a number of companies were granted licences with a view to exploring the UK's shale gas potential. The front-runners were companies such as Dart Energy, Celtique Energie, Cuadrilla, IGas, Egdon Resources, Third Energy and Rathlin Energy; these were then small companies with little or no public profile. They hoped that successful exploration would attract the attention of bigger investors, who could finance further activity. To some degree the strategy has succeeded, but more on the basis of potential than actual drilling results. Companies including Centrica, Total, EDF and GDF Suez have all bought into 13th-Round licence activities in the UK. There has also been a degree of consolidation, with IGas Energy acquiring Dart Energy in 2014 and, most recently, Ineos purchasing some of the IGas Energy licences and establishing itself as a serious player in the shale industry in the UK. As the industry association UK Onshore Oil and Gas (UKOOG 2015, p. 2) makes clear, there is already an established onshore oil and gas industry in the UK, which in 2014 at existing conventional sites produced an average of 22 000 barrels of oil equivalent a day from 230 operational wells. The general public are unaware of this onshore industry, but opinion poll research shows that there is now a growing awareness of shale gas and hydraulic fracturing. In 2012, when DECC first asked a question about shale in its *Waves Public Attitudes Tracker*,[3] only 42 per cent of the population said that they knew anything about it; in their August 2015 survey 75 per cent of those polled said that knew about shale gas. What has triggered this increasing awareness and what are the main areas of concern?

On 1 April and 27 May 2011 two small seismic events were detected in the Blackpool area. It was later determined that they were related to the hydraulic fracturing activities of Cuadrilla at its Preese Hall well. It was also revealed that the events had damaged the casing of the well, which was then abandoned in December 2012, securely plugged in November 2014 and fully restored in April 2015. A subsequent review recommended a traffic light system to monitor any future induced seismic activity triggered by shale gas activities (Green *et al.* 2012); this system has now become part of the regulatory regime. The two seismic events were very minor and many similar tremors occur across the UK on a daily basis (for a review of anthropogenic earthquakes in the UK see Wilson *et al.* (2015)). However, they aroused public opinion and in retrospect can be seen as the starting point of the shale gas debate in the UK, not least because the Government imposed a moratorium on further drilling that was not lifted until late 2012. Annex 9.1 presents a timeline of key shale gas developments over the 4–5 years from 2010, when Cuadrilla commenced its drilling operation in Lancashire, to the time of writing in September 2015 when the Government released the initial results of the 14th Onshore Oil and Gas Licensing Round (this is explained below in the section on regulation). The table (annex 9.1) also references the key reports and publications associated with the evolution of the shale gas debate in the UK.

The Case in Favour of Shale Gas Development in the UK

Given the current status of the industry in the UK, it is not possible to access present progress; rather, this section examines the arguments for and against shale gas exploration and potential development. Table 9.2 provides a summary of the key issues that are being marshalled in support of, or in opposition to, shale gas development. This section reviews each position and the key sources of evidence that are being used. Given the wide-ranging scope of the debate, we do not examine each issue in detail, the purpose here being to map out the contours of the debate and consider the claims and counter claims.

The key proponents of shale gas development are the UK Government – initially the Conservative Party elements of the previous coalition government and at the time of writing the majority Conservative Government. The 2015 Conservative Party Manifesto (2015, p. 57) had the following to say: ‘We will continue to support the safe development of shale gas, and ensure that local communities share the proceeds through generous community benefit packages. We will create a Sovereign Wealth Fund for the North of England,

[3] Details of this can be found at: https://www.gov.uk/government/collections/public-attitudes-tracking-survey. Accessed 19 August 2015.

so that the shale resources of the North are used to support development of the North.' Shale gas development is also supported by industry organisations, most notably the Institute of Directors and the CBI. Obviously, the onshore industry itself is making the case for development, both individually and through the UKOOG. A review of Government statements, industry reports and attendance at industry conferences suggests that the arguments in favour of shale gas development focus on energy security, economic benefits, investment, employment and positive local economic impacts. At the same time, shale gas development is seen as compatible with the UK's climate change strategy as it is cleaner than coal and has lower life-cycle emissions than imported LNG or long-distance pipeline supplies.

During the 1990s the UK experienced a 'dash for gas' as gas-power generation replaced a significant amount of coal-fired capacity and gas became the dominant fuel for heating in households. This was all enabled by the development of offshore oil and gas on the UK continental shelf. Natural gas production in the UK peaked in 2000 and in 2004 the UK became a net importer. In 2014 gas accounted for 32 per cent of the UK primary energy consumption and 29 per cent of electricity production and is used for heating in over 80 per cent of households (DECC 2015a). In 2014 net imports accounted for 45 per cent of gas consumption and this has been enabled by investment in pipelines from the Norwegian continental shelf, interconnectors to Belgium and the Netherlands and the construction of three liquefied natural gas import terminals (see Bradshaw *et al.* (2015) for an assessment of the UK's gas security challenges).

In the view of the politicians, this growing import dependence is a cause for concern as it exposes UK consumers to price volatility in international gas markets and makes them vulnerable to the geopolitical manipulation of energy supplies (by Russia). An added problem is that oil and gas production in the North Sea is continuing to decline and projections suggest that import dependence could be as high as 70 per cent by the mid to late 2020s. Thus, the possibility of a new source of indigenous gas supply is seen as making a significant contribution to improving the UK's energy security, understood as constraining the growth in gas import dependence. A reduction in the volume of imported gas and an increase in domestic production would also improve the country's balance of payments and generate tax revenue for national and local governments.

Early on in the shale oil and gas debate the Prime Minister David Cameron and the Chancellor of the Exchequer George Osborne claimed that a 'UK shale gas revolution' might lower energy bills. This was a time of growing public dissatisfaction with increasing energy costs. This view was widely criticised by both supporters and opposers alike; for example, Lord John Brown, then Chairman of Cuadrilla, suggested that: 'unless it is a gigantic amount of gas, it is not going to have material impact on price' (Carrington 2013). Lord Nicholas Stern was more strident in his criticism, describing the Chancellor's view as 'baseless economics' (Bawden 2013). The Government did commission a report by Navigant (Rathbone and Bass 2012) on the potential impact of shale gas on the UK gas market, which suggested the possibility of a modest impact, but this judgement was made on the basis of questionable assumptions about future developments in global gas markets.

A study by Pöyry (2012) suggested that prices might be between 2 and 4 per cent lower from 2021 owing to production from Lancashire, which was expected to reach 12 bcm/year (this now seems unlikely).

The consensus view amongst industry and experts is that the open nature of the UK gas market – as opposed to the closed nature of the North American gas market – means that UK shale gas production would not isolate UK consumers from market conditions in the European market or the global LNG market. More recently the new Conservative Government has been more restrained in its comments, focusing instead on energy security and economic benefits. At the announcement of the initial outcomes of the 14th Onshore Oil and Gas Licensing Round, the new UK Energy Minister, Lord Bourne, said: 'As part of our long-term plan to build a more resilient economy, create jobs and deliver secure energy supplies, we continue to back our onshore oil and gas industry and the safe development of shale gas in the UK.'[4]

There have been a number of reports that have been used by both the national government and industry to highlight the economic benefits of shale gas development in the UK. Three sources are particularly important: the Institute of Directors' (2013) report 'Getting shale gas working', the Strategic Impact Assessment for the 14th Round conducted by Amec on behalf of DECC (Amec 2013) and the report by EY (2014) 'Getting ready for UK shale gas', which was commissioned by UKOOG and part funded by the Department of Business Innovation and Skills. The Institute of Directors' (2013, p. 17) report presents a number of scenarios; the central scenario – which is the one most often quoted by the UK Government – would see a multi-year development of 100 shale gas pads of 40 laterals each, with peak production 1121 bcf (31.4 bcm). Capital expenditure and operating expenditure would peak at around £3.7 billion a year, supporting 74 000 jobs . . . Gas import dependence could be reduced from 76 per cent to 37 per cent in 2030, and the cost of net gas imports in 2030 could fall from £15.6 billion (2012 prices).

The Institute of Directors' (2013) report thus made a strong economic case for shale gas development while the Amec (2013) report described and evaluated the likely significant environmental effects associated with the 14th Licensing Round. Again, a series of scenarios are considered and, under the 'high activity' scenario: 'It is estimated that at its peak, some 16 000–32 000 full time equivalent positions could be created, which would represent an increase between 3.5 per cent and 7 per cent in the level of employment supported in the UK oil and gas industry' (Amec 2013, pp. xii–xiv). The lower job creation numbers in comparison with those in the Institute of Directors' analysis are a result of different estimates of the number of wells in operation. It is noteworthy that the UK Government continues to quote the higher figures, from the Institute of Directors, report.

The EY (2014, p. 1) report: 'focuses on what is required to ramp-up for production; and build on existing industrial and regulatory resources already in place to support the exploration phase' and its headline findings support the earlier work by the Institute of

[4] New onshore oil and gas blocks to be offered, https://www.gov.uk/government/news/new-onshore-oil-and-gas-blocks-to-be-offered. Accessed 27 August 2015.

Directors. It suggests that: 'over 2016–32 c. £33 billion of spend would be required to bring up to 4000 wells into production. At peak this equates to a spending of around £3.3 billion and some 64 500 jobs (around 6100 of which are direct roles). Much is also made of the fact that the economic impact will benefit the north of England, which is perceived by the Government in Westminster as economically depressed and is currently the subject of efforts to create a 'Northern Powerhouse'.

One of the most significant differences between the situation in the UK and in the US is that in the latter the subsoil rights belong to the landowner (though the surface landowner may have sold the subsoil rights), while in the former the subsoil rights reside with the Crown (the state). While it is the case in the UK that landowners will be compensated for allowing drilling activities on their land, the revenues generated by any subsequent production are shared between the operating company and the state. In June 2013 the industry launched a Community Engagement Charter to provide an incentive to local communities that would be affected by shale gas drilling operations. The Charter included a commitment to pay £100 000 per well site where there is hydraulic fracturing and, should the well enter into production, the local community will receive one per cent of revenues (allocated as approximately two thirds to the local community and one third at county level). It is estimated that this could be worth £5–10 million over the lifetime of a site (UKOOG 2013). This industry offer has now become part of government policy and, in January 2014, the Prime Minister announced that councils would be able to keep 100 per cent of the business rates that they collect from shale gas activities (which is double the current figure, 50 per cent); this could be worth £1.7 million a year for a typical site.[5] Thus, both the industry and government have sought to provide cash incentives so that local communities and councils benefit from shale gas development in their locality.

The environmental benefits of shale gas development rest on the fact that natural gas (methane) is a 'cleaner' fossil fuel than coal when used in power generation or as a transportation fuel. Nevertheless, methane is also an aggressive greenhouse gas and any leakage (fugitive emission) during production, processing or transportation is likely to reduce the positive environmental benefits associated with shale gas development. Shale gas development may also bring with it a range of other environmental impacts, which are considered later; here we are just concerned with the two issues of fugitive emissions and cumulative life-cycle emissions. Both are highly controversial, and there is no definitive position on either, in large part because the current evidence base is based on US experience, which suffers from a lack of baseline monitoring, limited data and a wide range of different methodologies producing contradictory results (Balcombe *et al.* 2015).

In the context of the UK debate, the key environmental report is that of MacKay and Stone (2013), 'Potential greenhouse gas emissions associated with shale gas extraction and use'. The Secretary of State for the Department of Energy and Climate Change requested the study. It is now used widely by the UK Government and shale gas supporters to demonstrate

[5] 'Local councils receive millions in business rates from shale gas developments', https://www.gov.uk/government/news/local-councils-to-receive-millions-in-business-rates-from-shale-gas-developments. Accessed 27 August 2015.

the climate change benefits of domestic shale gas development. The study considered both the emissions associated with shale gas exploration and production and also the effect of shale gas use on overall GHG emission rates and cumulative emissions. In the absence of actual drilling activity in the UK, it was based on the US experience. Furthermore, because it maintains that the: 'principal effect of UK shale gas production and use will be that it displaces imported LNG, or possibly piped gas from outside Europe' (MacKay and Stone 2013, p. 5); its concern is to compare the life-cycle emissions of domestic shale gas production against these sources of imported gas. It concludes that: 'local GHG emissions from shale gas operations should represent only a small proportion of the total carbon footprint of shale gas, which is likely to be dominated by CO_2 emissions associated with its combustion'. The conclusions of this analysis were that shale gas's overall carbon footprint is comparable with that of gas extracted from conventional sources and lower than the carbon footprint from LNG and that, when shale gas is used for electricity generation, its carbon footprint is significantly lower than that of coal (MacKay and Stone 2013, p. 3). This is in agreement with an early study conducted for the European Commission (Forster and Perks 2012, p. iv), which estimated the GHG emissions per unit of electricity generated from shale gas to be around four to eight per cent higher than for electricity generated by conventional pipeline gas from within Europe. Furthermore, emissions from shale gas generation in Europe are two to ten per cent lower than emissions from generation using imported conventional pipeline gas from Russia or Algeria and 10 per cent lower than emissions from imported LNG.

When the MacKay and Stone paper was published, the Climate Change Committee, an independent body that monitor's the UK's compliance with its legally binding carbon budgets, made clear its own views on the potential role of shale gas in a low carbon economy. They noted that into the 2030s the UK will continue to use gas and that: 'if anything, using well-regulated UK shale to fill this gap [caused by falling North Sea production] could lead to lower lifecycle greenhouse gas emissions than continuing to import LNG' (Joffe 2013). Most recently, the third report from the Task Force on Shale Gas (2015c, p. 26) 'Assessing the impact of shale gas on climate change', concluded that: 'The evidence suggests that the impact of shale gas on the climate is similar to that of conventional gas and less than that of LNG.' Overall the message is clear: first, shale gas emissions, if properly managed, are not significantly higher than those from conventional gas production; second, using shale gas to generate electricity still results in significantly lower emissions than using coal; and third, using domestic shale gas results in lower life-cycle greenhouse emissions than using distant pipeline imports or LNG. However, from a global perspective there is only a net benefit if the gas imports displaced by domestic shale gas are not burnt elsewhere. The Government's response to this conundrum is to support a global climate change agreement.

The final element in support of shale gas development is the proposition that the UK's regulatory regime is fit for purpose and will minimise the impact of shale gas development on people's health and on the natural environment. Natural Health England published a review of the potential public health impacts of exposures to chemical and radioactive pollutants as result of shale gas extraction. The report concluded that: 'The currently available evidence indicates that the potential risks to public health from exposure to the emissions with shale gas extraction are low if the operations are properly run and regulated' (Kibble *et al.* 2014, p. iii). The second report by the Task Force on Shale Gas (2015b, p. 30) on local environment and health impact said that the Task Force were: 'satisfied that the risk levels associated with the public health hazards outlined above are acceptable provided that the well is properly drilled, monitored and regulated'.

The Environment Agency (2013) conducted an environmental risk assessment for shale gas exploratory operations in England and, anticipating a production phase, commissioned a study entitled: 'Unconventional gas in England: description of infrastructure and future scenarios' (Broomfield *et al.* 2014). Earlier a report by the Royal Society and the Royal Academy of Engineers (2012, p. 4) entitled 'Shale gas extraction in the UK: a review of hydraulic fracturing' concluded that: 'The health, safety and environmental risks associated with hydraulic fracturing (often termed "fracking") as a means to extract shale gas can be managed effectively in the UK as long as operational best practices are implemented and enforced through regulation.' This view appears to be shared by the UK Government, which, in late 2013, as the previous coalition government, set up the Office of Unconventional Gas and Oil (OUGO) within DECC with the aim of joining up responsibilities across government (these are explained in the section on regulation). The OUGO provides a single point of contact for investors and ensuring a streamlined regulatory process. In addition, DECC has produced a series of guides and videos for the interested public that address the various concerns raised by unconventional oil and gas exploration and development. However, any judgement on the efficacy of the regulatory regime and attempts to address public concerns must follow a discussion of the views of those that oppose shale gas development and the detailed discussion of the regulatory regime, which is presented in the third substantive section of this chapter.

The Case against Shale Gas Development in the UK

Those that campaign against shale gas development in the UK are a far less coherent and organised group than the official proponents of shale gas development. Much of the opposition comprises non-governmental organisations (NGOs) and campaign groups that include celebrities, certain media outlets, national and local politicians, professionals, academics and the concerned public. For the most part the opposition is expressed by protests, media engagement and the creation of websites and blogs. In its first report on planning, regulation and local engagement, the Task Force on Shale Gas (2015a, p. 5) noted that: 'At a local level more than 180 groups have been identified that oppose shale gas extraction.' The following organisations provided a written response to the strategic environmental assessment conducted by Amec (2013): Campaign for National Parks, Campaign to Protect Rural England (CPRE), CPRE-Kent, Concerned Communities of Falkirk, Frack Free Balcombe Residents Association, Frack Free Lincolnshire, Frack Free Wales, Friends of the Earth (FoE), FoE Scotland, the Geological Society, Gower Society, Greenpeace, Keep Kirdford and Wisborough Green, National Association of Areas of Outstanding National Beauty, National Trust, the Planning Officers' Society, Royal Society for the Protection of Birds, Safety in Fossil Fuel Exploitation Alliance, Scottish Environment LINK, Stretton Climate Care, Sussex Wildlife Trust, Swansea Environmental Forum, Transition Mayfield, Transition Town Louth, Woodland Trust. This list conveys a sense of the diversity of the opposition and that it involves national, regional and local organisations. National environmental NGOs such as Friends of the Earth (FoE), WWF and Greenpeace count shale gas amongst their campaign targets. The above list includes organisations with an interest in the rural environment and wildlife and ecology, such as the Royal Society for the Protection of Birds (RSPB), the Countryside Alliance and the National Trust, to name just three. There are also many single-issue campaign groups opposed to shale gas development in their specific locality. There are also a number of key websites which provide information to counter the pro-shale-gas lobby and which coordinate protests and campaign.

'Frack-Off: the extreme energy action network' (http://frack-off.org.uk) is one of the higher profile and longer-standing websites and includes an interactive map of fracking-related activity and local groups. Their website serves as a clearing house for information, including reactions to the various reports produced by the government that are discussed above. 'Talk Fracking' (http://www.talkfracking.org/about/) is another example. According to its website it is committed to: 'highlighting the issues surrounding fracking in the UK, holding the policy makers and industry to account and providing the forum for the debate that the British public deserves before it's too late'. The Talk Fracking website also hosts 'Frackademics', which it commissioned and which is a study of the relationships between academia, the fossil fuels industry and public agencies. It focuses, in part, on four reports: the Royal Society/Royal Academy of Engineering (2012) review, the MacKay–Stone (2013) review, the Public Health England (Kibble *et al.* 2014) review and a report commissioned by the Scottish Government on unconventional gas (this is not considered here due to their moratorium on unconventional oil and gas). The author of Frackademics, Paul Mobbs (2015, p. 24) argued that the academic community is being co-opted into producing and supporting the body of evidence that is being used by the Government and industry to make the case for shale gas development. He concludes that:

> 'There is no objective case to support the development of unconventional gas and oil in Britain. At best, reviewing evidence from around the world, what we can say is that there is a great deal of uncertainty as to the scope and severity of impacts from these processes. Therefore the use of scientists by Government and industry to promote a positive view of these technologies is misleading, since, in nearly all cases, that uncertainty is not being represented to the public. This risks further diminishing the public's trust in science, as it is increasingly being used to support developments which arguably have an uncertain – but likely negative – impact on the public's interests.'

Talk Fracking, together with another 172 organisations – 19 national campaign organisations and 153 residents' community groups – is also deeply suspicious of the Shale Gas Task

Force, which it characterises as 'a PR stunt to push through Fracking undemocratically'. So what is the case against shale gas development in the UK?

If you follow the logic of the argument above, given the absence of any drilling activity in the UK, just as there is no objective case to support shale gas development, so an equally provocative view might be that, if you believe that the regulatory regime is fit for purpose, at present there can be no objective case to ban such development. Equally, just as the argument in favour relies on evidence from elsewhere, so the argument against relies on evidence of negative economic, social and environmental impacts elsewhere. It is also the case that those opposing shale gas development are seldom subject to the same degree of scrutiny as they apply to representatives from government and industry. The defence of the opposition groups might be that the precautionary principle requires government and industry to demonstrate that the risks are being adequately managed and that their role is to hold them to account of on behalf of wider society. Ongoing personal involvement of the author in the debate and engagement with the NGOs, campaign groups and the concerned public suggests that there is a range of issues that summarise the opposition to shale gas in the UK; these are presented in Table 9.2.

There are two scales of issues and concern here: first, at the national scale there is the issue of the compatibility of shale gas development with the UK's climate change policy and binding carbon targets, which require the decarbonisation of the UK energy system and involves the claim that shale gas development would contribute to improving energy security; second, there is a range of issues that are regional and local in scale and relate to the negative economic, social and environmental impacts of shale gas activities in particular localities. There is one issue that transcends these scalar concerns: the question of the effectiveness of the regulatory regime in managing and mitigating the risks associated with shale gas development. The opposition view is discussed below. The discussion draws on a submission made to the House of Commons, i.e. the Environmental Audit Committee's (2015) enquiry 'The environmental risks and fracking', published in early 2015. The discussion also draws on published papers and briefing documents produced by various environmental non-governmental organisation (ENGOs) and others who have spoken out against shale gas development.

The first four issues listed in Table 9.2 relate to energy security and climate change. Those that oppose shale gas development tend to favour the evidence base from the US that suggests that the level of fugitive emissions is such that it negates any positive decarbonising benefits (for example, Howarth *et al.* 2011), although some of the more benevolent commentators are prepared to accept that the 'jury is still out' on the issue (Friends of the Earth 2015). At the same time, opponents point out that shale gas is a fossil fuel and therefore it is part of the problem of climate change rather than a solution to it. A quote from John Ashton (2014) – who served as Special Representative for Climate Change to three successive UK Foreign Secretaries – aptly summarised the opposition view when briefing Lancashire County Council as an independent commentator and adviser on the politics of climate change: 'You can be in favour of fixing the climate. Or you can be in favour of exploiting shale gas. But you can't be in favour of both at the same time.' The

Table 9.2 *The contours of the shale gas debate in the UK*

Arguments in favour of shale gas development	Arguments against shale gas development
1. It will improve the UK's energy security by reducing the level of gas import dependence. 2. It will generate tax revenue and improve the UK's balance of payments. 3. It will result in lower energy bills. 4. It will attract new investment and create new jobs. 5. It will promote local (community payments) and regional economic development (the 'Northern Powerhouse'). 6. It will make a positive contribution to decarbonisation and the UK's climate change policy. 7. The regulatory regime will minimise the risks to the environment and health and safety.	1. Shale gas (methane) as a hydrocarbon contributes to climate change. 2. The problem of fugitive emissions reduces its decarbonising benefits. 3. Investment in shale gas may reduce investment in renewable energy and should not receive tax breaks. 4. Shale gas will not reduce energy bills or contribute to energy security. 5. There is a risk of induced seismicity. 6. There is a risk of groundwater pollution and it places additional stress on water resources and waste treatment facilities. 7. There are significant negative local impacts in relation to air, noise and light pollution and traffic congestion and loss of amenity. 8. It will have a negative impact on the local economy and depress property prices and agricultural land values. 9. Community payments amount to bribes and create a conflict of interest for County Councils. 10. The regulatory regime in is inadequate and there is insufficient regulatory capacity.

argument here is as much about the future role of natural gas in the UK's energy mix as it is specifically about shale gas.

When countering the Government's argument that shale gas can be a bridge to a low carbon future, opponents point out that the UK is already a considerable way across that bridge having already experienced a 'dash for gas' in the 1990s and that by the mid 2020s – when commercial scale shale gas development might be expected – there will be very little coal left in the UK's power generation mix (FoE 2013, Ottery 2016). Furthermore, towards the end of that decade – with or without CCS – the level of natural gas consumption will have to be significantly lower to be compliant with the UK's carbon budget. Thus, they argue that shale gas development is simply not needed (FoE 2015). The argument that shale gas development is essential to address growing energy security concerns as a result of falling UK CS production is also refuted, on the basis that the best solution is to promote energy demand reduction and efficiency and encourage low carbon alternatives – both in domestic heat and power generation – in order to significantly reduce the overall level of gas demand in the UK in the 2020s. This is an argument in favour of using less gas, rather than securing new gas. There is also a concern that investing in shale gas development will divert funds away from renewable energy sources and lock the energy system into a higher carbon trajectory that would threaten climate change targets. At present, neither claim is supported by a rigorous evidence base but the fact that the new Conservative Government has retained tax incentives for shale gas exploration while cutting subsidies for onshore wind and solar has been widely criticised by environmental groups. In their third report, the Task Force on Shale Gas (2015c, p. 26) recommended that: '... government should commit to deploying the government specific revenue derived from a developed shale gas industry to investment in R&D in innovation in CCS and low carbon energy generation, storage and distribution.'

A final line of argument in relation to shale gas and climate change relates to the UK's international leadership. It is suggested that promoting an international agreement on climate change, as the UK Government does, is not compatible with encouraging a new round of domestic fossil fuel development, especially when it is clear that at a global level there are already more than sufficient conventional gas reserves to meet the level of future demand that is compatible with constraining global warming to 2 °C. Thus, the debate around stranded assets is now being deployed in the shale gas debate to argue that the gas should be left in the ground.

A second set of concerns relates to the local impacts of shale gas development. These are largely focused on the environmental, economic and social (including health) impacts in the localities where shale gas development might take place (for a recent review see Hays *et al.* (2015)). There is also a related concern that the cumulative impacts of commercial shale gas development, with potentially thousands of wells being drilled, have not been adequately considered. Table 9.2 identifies the key local concerns: induced seismicity, the risk of groundwater pollution from poor well completion, the impact of increased demand for water and increased pressure upon local waste treatment facilities, the impacts of increased noise and traffic movements and the potential health risks associated with some chemicals used by the industry. Again, widespread use is made of evidence from the US, often – the likes of UKOOG would argue – without taking into consideration the fact that the regulatory environment in the UK is very different. The issue of induced seismicity seems to be less of a concern, maybe because a mitigation strategy is in place but more likely because there has been no hydraulic fracturing in the UK since the Preese Hall event. Furthermore, much of the seismicity reported in the US is related to waste water injection and that is unlikely to be permitted in the UK.

Presentations to the Lancashire County Council in relation to Cuadrilla's planning applications suggest that ground water pollution, noise and light pollution and traffic movements and potential health problems remain the most significant concerns raised by opposition groups and local communities. On the issue of well completion, which is the key problem relating to ground water pollution, opponents have demanded independent inspection of well integrity. The issues of air, noise and light pollution are central to the planning process at the local level. Understandably, the health issue has proved particularly emotive and opponents are making much of the decision by New York State to ban shale gas development (New York State Department of Health 2014). Health care professionals in the UK (Kovats *et al.* 2014, Stott *et al.* 2014) have been critical of Public Health England's assessment, and a review by the NGO Medact (McCoy and Saunders 2015) recommended a ban, on the basis that: 'The risks and serious nature of the hazards associated with fracking, coupled with the concerns and uncertainties about the regulatory system, indicate that shale gas development should be halted until a more detailed health and environmental impact assessment is undertaken.' A recent briefing by Breast Cancer UK (2015) summarises the situation: '. . . most of the evidence concerning fracking comes from the USA where regulatory procedures are different to those in the EU. Currently, there is no direct evidence yet available that fracking leads to an increased incidence of breast cancer. However, the process does introduce chemicals into the environment which may lead to an increased risk. Although Breast Cancer UK were considering the risk of breast cancer, the general concern is that the shale gas industry uses chemicals that are known to be harmful to human health and, therefore, that there is a potential health risk to those who come in contact with them. A report by Chemtrust (Warhurst and Buck 2015) evaluated how toxic chemicals from fracking could affect wildlife and people in the UK and the EU. They made a number of recommendations and suggested that there should be a EU-wide moratorium until their recommendations are in place. A recent review of the environmental health impacts of unconventional natural gas development in the US (Werner *et al.* 2015, p. 1127) was critical of the methodological rigour of studies that suggested a link between adverse health outcomes and such activities, but it noted that: '. . . there is also no evidence to rule out such health impacts'. This remains a key issue for local communities, and the burden of proof that shale gas development is not detrimental to human health falls on the regulators and the industry.

Both government and the industry have stressed the positive impacts of shale gas development in terms of investment, employment and community benefits. Opponents, while sceptical of such claims, tend to stress the potential negative economic impacts that shale gas development might have on the existing economy, for which there is ample evidence from the US experience (Jacquet 2014). The concerns about traffic and congestion relate to the possible loss of local amenity and the impact on local industries, such as tourism and farming, that might be detrimentally affected by shale gas activities. The National Farmers' Union (2015) suggested that there are significant potential direct impacts that include: the possible risk of contaminating water resources or land, competition for water, and management and disposal of contaminated flow-back waste water. They also identified more general direct impacts on the rural economy that include: reduction in the open market value of land because of the current (negative) perception of fracking, economic damage to supply chain relationships or farm diversification income (such as tourism) and short-term damages and costs incurred by anti-fracking protestors on farmers' land. It should be noted that, while a farmer would be paid to allow shale gas activities in their land, they would not profit from any future gas production. A study for the Countryside Alliance (Petrenel 2014, p. 8) suggested that particular attention should be paid to well site selection and recommended that: 'land that is least susceptible to incremental damage be utilised in the early stages'. The experience of iGas, drilling on a brownfield site at Elsmere Port, compared to that of Cuadrilla, seeking to drill in open farmland, suggests that this is an important consideration.

The impact of shale gas activities on property prices has also proved an emotive issue and is linked to a heavily redacted (censored) study entitled 'Shale gas rural economy impacts', produced for internal use by the Department of the Environment, Food and Rural Affairs (Defra) in March 2014 (Defra 2014). The UK's transparency watchdog, the Information Commissioner Officer, ruled on 8 June 2015 that the report should be published in full (Defra 2015). This was a week before the Lancashire County Council decided on Cuadrilla's planning application, but Defra did not release it until after the hearings. The report itself is underwhelming. It came with a covering note that explained that it was a draft paper intended as a review of existing literature, which contained no new evidence and referred to data from overseas studies that, Defra maintains, 'cannot be used to predict impacts in the UK with any degree of reliability'. Nonetheless, despite noting on p. 14 that the 'overall evidence on the impact on property prices in the literature is quite thin and the results are not conclusive', the report goes on to state that: 'House prices in close proximity to the drilling operations are likely to fall. However, rents may increase due to additional demand from site workers and supply chain' (Defra 2015, p. 15). All this discussion has served to highlight that the proponents of shale gas have tended only to consider the potential positive economic benefits in terms of new jobs and investment and have largely failed to consider the potential negative impacts on the existing economy and community in potential shale gas areas.

The final set of concerns relates to the efficacy of the regulatory regime itself, which is discussed in detail in the next section. Opponents point out that support for shale gas development is premised on the assumption that the regulatory regime will be effective. Concerns about regulation revolve around four issues: the approach adopted, risk management versus precautionary; whether all the risks are covered by the regulations or whether a special regime is needed; where shale gas exploration will be allowed, which involves the issues of habitat protection and of watershed; and, finally, regulatory capacity. There is also a concern that the proposed payments to the County Councils and local communities in shale gas licence areas will compromise the planning process, as those same Councils have to grant planning permission to the operators.

In sum, it is sufficient to note that the opposition to shale gas does not agree with the Government that the UK has a 'gold standard' when it comes to regulating shale gas development (FoE 2014). In their view a ban on shale gas development is the only way forward, either because a more thorough assessment of the risks and uncertainties is required or because there is no benefit from developing shale gas, only costs, so why take the risk? This does not leave any room for compromise, which suggests that future shale gas planning applications are likely to face a substantial and very vocal opposition.

The Shale Gas Regulatory Regime

This final substantive section examines the current shale gas regulatory regime in England (and Wales). The emphasis here will be on identifying the different actors and their responsibilities, rather than the detail of individual pieces of legislation. The discussion links the issues raised previously with the different elements of the decision-making process. The shale gas 'regulatory roadmap' (DECC 2013), as the UK Government likes to call it, is largely based on pre-existing legislation rather than a specific set of regulations for shale gas exploration and development. The position of the UK Government and its ministries and agencies is that all the risks posed by shale gas operations are familiar and, therefore, that they are covered by existing legislation.

The UK Government is hostile to the idea of a new EU Directive on shale gas, and for the moment the EU has only made a recommendation: 'On minimum principles for the exploration and production of hydrocarbons (such as shale gas) using high-volume hydraulic fracturing' and is maintaining a watching brief on development in member states (European Commission 2014). In a recent guide to shale gas, the Energy Institute (2015, p. 15) pointed out that there are currently 16 Acts and regulations relevant to the UK oil gas industry as well as 14 separate pieces of European legislation. These are not specific to shale gas but are applicable to any form of oil and gas activity onshore. The key elements of the legal framework include the Petroleum Act 1998, the Environmental Planning Regulations, the Town and Country Planning Act and, most recently, the Infrastructure Act 2015. Opponents argue that the regulations are fragmented and many pre-date the development of high volume hydraulic fracturing. The situation is further complicated by the fact the regulatory responsibility is split between various government ministries and agencies and between central and local government.

Figure 9.1 presents a simplified version of the current regulatory regime relating to shale gas exploration in England. The starting point at the top of the diagram relates to the initial award of a licence to explore for oil and gas, known as a Petroleum Exploration and Drilling Licence (PEDL). In the UK these licences are awarded through a competitive bidding process known as a 'licensing round'. The current onshore oil and gas activity dates back to the 13th Landward Licensing Round in 2008. Most recently, in 2014, the 14th Onshore Oil and Gas Licensing Round took place (Figure 9.2 shows the existing 13th Round block licences and the blocks auctioned in the 14th Round). According to DECC (2015b), a total of 95 applications were received for 295 blocks (each block is 10 km × 10 km). The outcome was delayed by the UK 2015 General Election, but the initial results were announced in August 2015: 27 blocks were awarded, with a further 132 further blocks subject to detailed assessment under the Conservation of Habitats and Species Regulations 2010.[6] In December 2015, the Oil & Gas Authority (OGA) – the UK's oil and gas regulator – announced that a total of 159 onshore blocks were being offered under the 14th Onshore Oil and Gas Licensing Round and that these blocks would be incorporated into 93 onshore licences (about 75 per cent of the blocks related to unconventional oil and gas).

[6] Details of the initial outcome can be found at: https://www.gov.uk/government/news/new-onshore-oil-and-gas-blocks-to-be-offered. Accessed 15 September 2015.

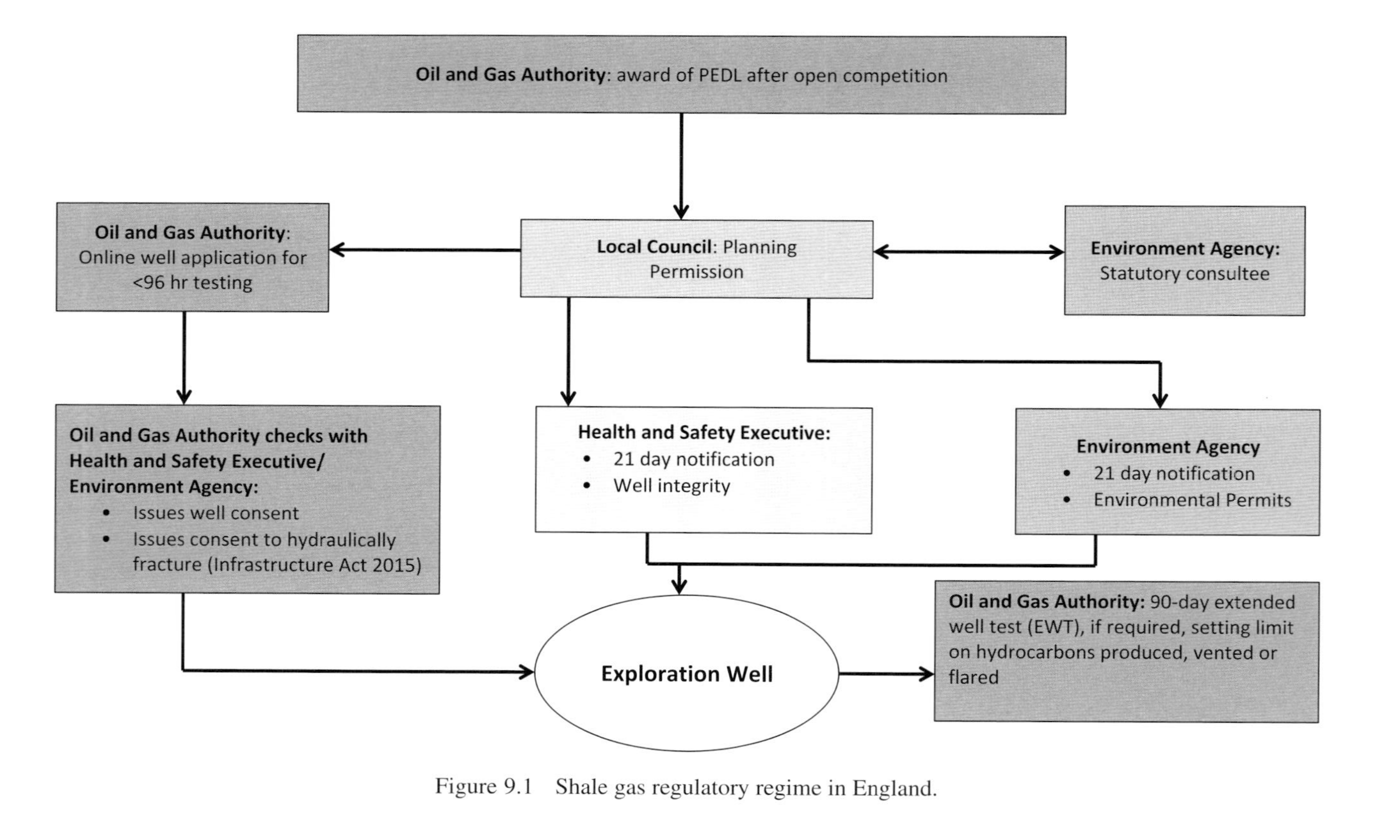

Figure 9.1 Shale gas regulatory regime in England.

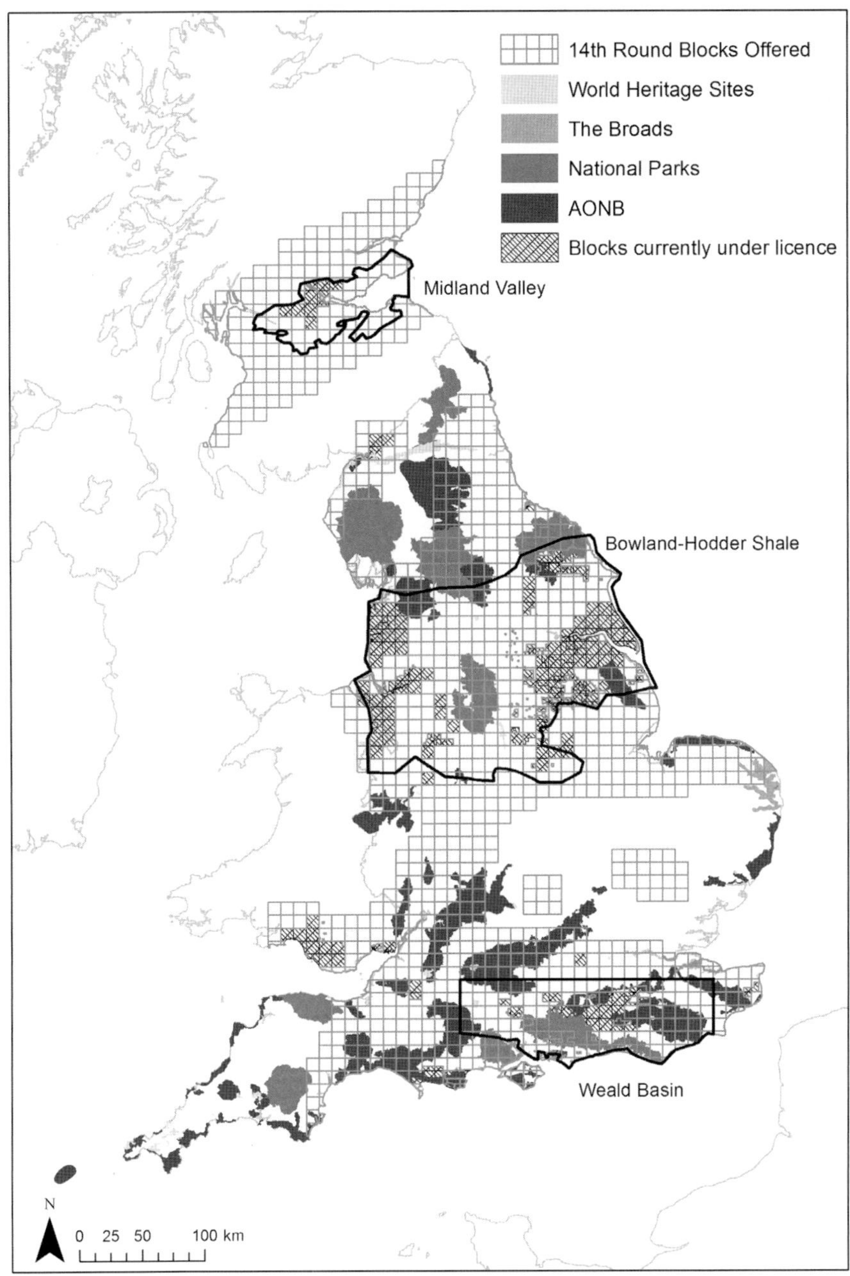

Figure 9.2 Licence blocks and protected areas.

After a company (operator) has secured a PEDL it needs to determine the optimal well site. This is a critical decision and requires negotiated access with landowners to purchase or secure rights (lease or licence) to access. With access agreed, the operator starts the process of gaining permission to carry out exploratory drilling, which may or may not initially involve high volume hydraulic fracturing or horizontal drilling. At this stage four actors are critical, the Oil and Gas Authority (created in April 2015), which is currently an Executive Agency of DECC, the Environmental Agency and the Health and Safety Executive, both of which are part of Defra, and the Mineral Planning Authority (MPA). According to the Government's planning practice guidance the responsibilities of these different actors are as follows.

The Oil and Gas Authority (DECC) issues the PEDL and gives consent to drill under the licence, once other permissions and approvals are in place, and also has responsibility for assessing the risk of, and the monitoring of, seismic activity as well as granting consent to flaring or venting. The Environment Agency is responsible for protecting water resources, including ground water (aquifers), as well as for assessing and approving the use of the chemicals used in drilling operations, including hydraulic fracturing. It also ensures that appropriate measures are taken in relation to the treatment and disposal of mining waste (which would include the produced water), emissions to air (including methane) and the suitable treatment and management of any naturally occurring radioactive materials (NORMs) that are returned with the produced water.

The Environment Agency will provide environmental permits that indicate that they are satisfied with the proposed measures in relation to: ground water activity, mining waste activity, industrial emissions activity, radioactive substances activity, water discharge activity, ground water investigation consent, water abstraction licences and flood risk consent.

The Health and Safety Executive regulates the safety aspects of all phases of extraction and is responsible for ensuring that the design and construction of a well casing for any borehole is appropriate. This is a critical factor in relation to potential ground water contamination, and the HSE will continue to monitor compliance during the drilling phase. As part of this process, the operator has to carry out an environmental risk assessment: 'to provide at an early stage a comprehensive review of all potential safety and environmental (including health) risks relevant to the proposed shale gas activities, and to show how these will be mitigated and managed' (UKOOG, 2014, p. 1). The industry has agreed that a full Environmental Impact Assessment (EIA) is required for all exploration wells that involve high volume hydraulic fracturing along with an Environmental Statement that details the findings of the EIA for the local MPA.

The MPA has general responsibility for: formulating policies and plans to guide future development ('forward planning'); regulating individual developments that are proposed through deciding planning applications ('development management'); and the policing of existing developments to ensure that they are working within any legal constraints outlined in the planning permission ('monitoring and enforcement'). In the case of a shale gas application the MPA must grant permission for the location of any wells and well pads and must impose conditions to ensure that the impact on the use of land is acceptable. The MPA must make its decisions with reference to planning law, and key issues may include: noise, dust, air quality, lighting, visual intrusion into the local setting and wider landscape, landscape character, archaeological and heritage features, traffic, land contamination, impact on best agricultural land, flood risk, land stability and subsidence (seismic risk), impact on wildlife and habitats, nationally protected geological and geomorphological sites and features and site restoration and aftercare. This is a very long list of issues and there is overlap and potential for disagreement amongst the various actors. The fall-out from the decisions against Cuadrilla in Lancashire in the summer of 2015, when the EA had granted its permits, suggests that the current Conservative government is concerned that the MPA should confine itself to making planning decisions rather than passing general judgment on the wisdom of shale gas development. If the MPA refuses to grant planning permission, the operator has the right to appeal to the Secretary of State for Communities and Local Government. This is what Cuadrilla has done in relation to its unsuccessful applications. There are also other statutory authorities and organisations that are asked for consent, including: the Coal Authority, should there be any drilling through a coal seam; Natural England; the British Geological Survey; and the Hazardous Substances Authorities. In the case of the Cuadrilla planning process, the Lancashire County Council's Planning Officers consulted no less than 23 separate bodies in reaching its decisions (Development Control Office, Lancashire Country Council 2015).

The previous coalition Government, and then the Conservative Government, consistently maintained that the regulatory regime is fit for purpose and that shale gas development in the UK will be subject to very strict controls, to minimise the risk to people and the environment. At the beginning of 2015 the House of Commons' Environmental Audit (2015, p. 3) published a report entitled 'Environmental risks of fracking', which was of a contradictory opinion and identified: '. . . a need for a more coherent and more joined-up regulatory system . . . that needs to be put in place before further fracking activity can be contemplated'. A wide range of recommendations was made that informed the subsequent debate in the House of Commons in relation to the Infrastructure Bill. The Labour Party tabled 13 new requirements in relation to shale gas development but stopped short of voting in favour of a ban (although the Scottish Government did impose a moratorium).

Shale gas opponents maintain that the British Labour Party requirements were 'watered down' when the Bill returned to the House of Lords, but the subsequent Infrastructure Act 2015 does contain a set of additional requirements in relation to shale gas development.[7] The Act requires the Secretary of State (for Energy and Climate Change) to consult with the Climate Change Committee regarding the impact of onshore petroleum (oil and gas) on the UK carbon budget target for 2050. The Act also revised the law on trespass, granting an automatic right of access to 'deep level' land (300 m or lower) 'for the purposes of exploiting petroleum or deep geothermal energy'. A number of additional safeguards were introduced in relation to onshore hydraulic fracturing. Hydraulic fracturing is not permitted

[7] The full *Infrastructure Act 2015* can be found at: http://www.legislation.gov.uk/ukpga/2015/7/enacted. Accessed 15 September 2015.

Table 9.3 *Onshore hydraulic fracturing safeguards in the Infrastructure Act 2015*

1. The environmental impact of the development which includes the relevant well has been taken into account by the local planning authority.
2. Appropriate arrangements have been made for independent inspection of the integrity of the relevant well.
3. The level of methane in the ground water has, or will have been, monitored in the period of 12 months before the associated hydraulic fracturing begins.
4. Appropriate arrangements have been made for monitoring of emissions of methane into the air.
5. The associated hydraulic fracturing will not take place within other protected ground water source areas.
6. The associated hydraulic fracturing will not take place within other protected areas.
7. In considering an application for the relevant planning permission, the local planning authority has (where material) taken into account the cumulative effects of (a) that application and (b) other applications relating to the exploitation of onshore obtainable by hydraulic fracturing.
8. The substances used, or expected to be used, in associated hydraulic fracturing (a) are approved or (b) are subject to approval by the relevant environmental regulator.
9. In considering an application for the relevant planning permission, the local planning authority has considered whether to impose a restoration condition in relation to that development.
10. The relevant undertaker has been consulted before granting of the relevant planning permission.
11. The public was given notice of the application for the relevant planning permission.

Source: Infrastructure Act 2015, Chapter 7, Part 6, Energy. Available at: http://www.legislation.gov.uk/ukpga/2015/7/enacted.

at depths of less than 1000 metres unless there is specific consent from the Secretary of State. A list of the specific additional measures that must be met is listed in Table 9.3. Subsequent regulations to clarify the requirements have angered opposition groups, as the protected groundwater source areas have been confined to 'Source Protection Zone 1', and the other protected areas have been defined to include National Parks, the Broads (Norfolk and Suffolk Broads), areas of outstanding natural beauty and World Heritage sites (these are shown in Figure 9.2). It does not automatically include Sites of Special Scientific Interest (SSSI) and it is possible to drill under protected areas using rigs located outside. In all cases, hydraulic fracturing is not permitted above a depth of 1200 metres.

Soon after the initial results of the 14th Round were announced, the Government released advice to planning authorities reminding them of the need to make planning decisions in relation to shale gas applications within the statutory timeframe of 16 weeks. This was widely perceived by the media and the opposition as an attempt to 'fast track' fracking; however, it did not involve the introduction of any new requirements nor did it take away any authority from the MPA. Nevertheless, the implication was that a failure to meet the time lines would result in the decision being reviewed by central government. In her blog, the new Secretary of State for Energy and Climate Change, Amber Rudd (2015a), stated:

'Getting shale exploration up and running is a key part of our long-term plan to build a stronger, more competitive economy, create jobs by backing the industries of the future and take our country forward.' In a subsequent written statement to parliament, Amber Rudd (2015b, p. 2) stated that: 'The Government therefore considers that there is a clear need to seize the opportunity now and to explore and test our shale gas potential.'

Conclusions: A Social Licence to Frack?

Commentators at the numerous shale gas conferences in the UK often joke that the only people making money at the moment are the security companies that are employed to protect drilling sites from protestors and those companies that organise such conferences, which bring people together to talk about why nothing is happening! However, the tenor of the discussions at such conferences has certainly changed over the last couple of years and the first thing that an operator is likely to talk about today is the need to obtain 'a social licence to operate'. It is not always clear what they mean by this statement, but it usually indicates recognition that without public acceptance there is not going to be a shale gas industry in the UK. A recent report by the European Academies' Science Advisory Council (2014, pp. 10–11) notes that central to the 'social licence' is the proposition that: 'even if fully compliant with laws and regulation, activities that are particularly intrusive or perceived to carry significant risks can be vetoed by a hostile public through campaigns, legal actions, demonstrations or other democratic pressures'. This is an apt description of the current situation in the UK and it makes clear that the social licence goes beyond the existence of a regulatory regime and relates to wider social acceptance of the need to explore – and potentially exploit – the UK's shale gas resources, as well as an acceptance of the level of risk to the environment and people associated with shale gas activities. The shale gas industry clearly does not currently have a social licence to operate in the UK. A report by the Council of Canadian Academies (2014, p. xvi) notes that: 'Public acceptance of large-scale shale gas development will not be gained through industry claims of technological prowess or through government assurances that environmental effects are acceptable. It will be gained by transparent and credible monitoring of the environmental impacts.'

This chapter has not dealt in detail with the issue of public attitudes to shale gas but even DECC's own opinion poll research shows a growing rejection of shale gas by the general public. Part of the problem appears to be the fact the potential benefits of energy security, improved balance of payments and decarbonisation are national goods while the potential environmental and socio-economic costs associated with shale gas fall on particular regions and communities in the UK. This raises a fundamental question about environmental justice. It also plays to other geographical divisions within British society, such as the 'north–south divide' and threats to rural communities. The Government has tried to counter this by stressing the potential economic benefits that would come from shale gas investment, but, as discussed above, local communities are far more concerned about the potential negative impacts on their current way of life. The representatives of local government now find themselves caught between on the one hand a concerned and vocal local electorate urging caution and a more precautionary approach and on the other hand a national government that favours shale gas exploration and that wishes to speed up decision-making. What happens next is far from certain.

The government and the industry hope that they can make sufficient progress with a programme of exploratory drilling to demonstrate that shale gas development in the UK is safe and effectively regulated. However, this will require local governments to grant the planning permissions needed for drilling to take place and these are bound to be contested. A programme of exploration drilling will afford an opportunity to monitor the environmental impacts of high-volume hydraulic fracturing and horizontal drilling for shale gas in the UK. Thus, it will provide the much needed evidence base for a better-informed debate about the costs and benefits of shale development in the UK. It may turn out to be the case that the costs of development are too high to support the commercial development of the UK's shale gas resources; equally, it may be that the environmental impacts do prove to be too great to secure a social licence to operate. But it might turn out to be the case that shale gas development can be both profitable and socially acceptable. The key point is that we will never know unless there is a programme of exploratory drilling and appraisal that enables the collection of evidence to inform government policy and public opinion.

Annex 9.1 *UK Shale Gas Debate Time Line and Key Publications*

Event	Associated Publication/Website
August 2010: Cuadrilla starts drilling at Preese Hall in Lancashire with the intention of conducting hydraulic fracturing.	
April/May 2011: Cuadrilla's activities trigger two minor seismic events that result in June 2011 in the Government's imposing a moratorium on further drilling.	
May 2011: The House of Commons, Energy and Climate Select Committee report 'Shale gas' concludes that in the UK is it is likely to be a 'game changer'.	House of Commons (2011)
April 2012: Report on Preese Hall published along with recommendations to mitigate against further events.	Green *et al.*(2012)
June 2012: The Royal Society and Royal Academy of Engineering report: 'Shale gas extraction in the UK: a review of hydraulic fracturing' is published.	Royal Society and Royal Academy of Engineering (2012)
September 2012: The Institute of Directors publish their report 'Britain's shale gas potential' that suggests significant economic benefits from shale gas development.	Institute of Directors (2012)
December 2012: The UK Government lifts the moratorium on shale gas drilling with a new traffic light system in place to monitor seismicity.	

Annex 9.1 *(cont.)*

March 2013: The Office of Unconventional Oil and Gas is created in the Department of Energy and Climate Change.	
April 2013: The House of Commons, Energy and Climate Change Committee report 'The impact of shale gas on energy markets' provides cautious and qualified support for shale gas development in the UK.	House of Commons (2013)
May 2013: The Institute of Directors publish a second report 'Getting shale gas working' that suggests that the industry could attract £3.7 billion in investment a year and create 74 000 jobs.	Institute of Directors (2013)
June 2013: Publication of the BGS Gas-in-Place Assessment for the Bowland Shale.	Andrews (2013)
June 2013: UKOOG announces the industry's Community Engagement Charter, which will provide £100,000 per well site and 1% of revenues at the production stage.	www.ukoog.org.uk
August 2013: Public protests at Balcombe in West Sussex in relation to the activities of Cuadrilla, who were drilling a conventional oil well.	
September 2013: DECC publishes the Mackay and Stone report 'Potential greenhouse gas emissions associated with shale gas extraction and use'.	MacKay and Stone (2013); Joffe (2013)
October 2013: Public Health England publishes their report 'Review of the potential public health impacts of exposures to chemical and radioactive pollutants as a result of shale gas', which concludes that good regulation can minimise the impacts on the environment and public health.	Public Health England (2013)
November 2013: Chancellor's Autumn Statement introduces tax breaks for shale gas exploration.	HM Treasury (2013)
December 2013: The DECC published AMEC's 'Strategic environmental assessment for further onshore oil and gas licensing' in preparation for the 14th Licensing Round.	Amec (2013)
January 2014: Prime Minister David Cameron declares that his Government is 'going all out for shale' and the Government announces that local governments can retain 100% of the business rates (instead of 50%) associated with shale gas activity.	
March 2014: The European Commission publishes a Communication on hydraulic fracturing in the EU that lays out principles of best practice with a commitment to review the polices of member states.	European Commission (2014)
March 2014: A coalition of environmental and other interest groups publish a report 'Are we fit to frack?' that calls for a much more cautious approach to shale gas development.	RSPB (2014)

(cont.)

Annex 9.1 *(cont.)*

April 2014: EY publish their report 'Getting ready for UK shale: supply chain and skills requirements and opportunities', which forecasts £33 bn needed to bring 4000 wells into production and 64 500 jobs.	EY (2014)
May 2014: House of Lords, Economic Affairs Committee, report 'The economic impact on UK energy policy of shale gas and oil' supports the development of shale gas and asked the Government to do more.	House of Lords (2014)
June 2014: Cuadrilla submits planning applications to carry out drilling and hydraulic fracturing at two sites in Lancashire.	
July 2014: 14th Onshore Oil and Gas Licensing Round announced.	
September 2014: The Task Force on Shale Gas launched, chaired by Lord Chris Smith, to provide an impartial, transparent and evidence-based assessment of the potential benefits and risks of shale gas extraction in the UK.	https://www.taskforceonshalegas.uk
October 2014: Competition for 14th Licensing Round closed with 95 applications received for 295 blocks.	
November 2014: Chancellor's Autumn Statement provides additional support for shale gas monitoring, regulation and engagement.	
January 2015: House of Commons Environmental Audit Committee report 'Environmental risks of fracking' recommends a moratorium on shale gas drilling.	House of Commons (2015)
January 2015: Decision on Cuadrilla's planning application is delayed at request of applicant.	
January 2015: The Scottish Government places a moratorium on unconventional oil and gas developments.	
February 2015: Passage of the Infrastructure Act with new conditions for unconventional oil and gas.	
March 2015: Medact publishes their report 'Health & fracking: impacts and opportunity costs', which recommends a halt to shale gas development until more detailed health and environmental impact assessment is undertaken.	Medact (2014)
March 2015: The Task Force on Shale Gas publishes its first report 'Assessing planning, regulation and local engagement', which recommends a new bespoke regulator for onshore underground energy.	Task Force on Shale Gas (2015a)
May 2015: The Conservative Party wins a majority in the General Election with a manifesto commitment to develop a shale gas industry in the UK.	

Annex 9.1 *(cont.)*

July 2015: The Task Force on Shale Gas publishes its second interim report 'Assessing the impact of shale gas on the local environment and health', recommending greater transparency and monitoring.	Task Force on Shale Gas (2015b)
July 2015: The Department of Environment, Food and Rural Affairs publishes full version of the previous redacted draft report 'Shale gas rural economy impacts'.	Department of Environment, Food and Rural Affairs (2015)
July 2015: Lancashire County Council rejects all Cuadrilla's planning applications (Cuadrilla subsequently announces that it will appeal the decision).	
August 2015: Government issues planning advice to speed up shale gas related planning decisions.	
August 2015: Government announces initial results of the 14th Licensing Round.	

References

Amec (2013). Strategic environment assessment for further onshore oil and gas licensing: environmental report. London: Department of Energy and Climate Change. Available at: https://www.gov.uk/government/consultations/environmental-report-for-further-onshore-oil-and-gas-licensing. Accessed: 27 August 2015.

Andrews, I.J. (2014). *The Jurassic Shales of the Weald Basin: Geology and Shale Oil and Gas Estimation*. London: British Geological Survey for the Department of Energy and Climate Change.

Andrews, I.J. (2013). *The Carboniferous Bowland Shale Gas Study: Geology and Resource Estimation*. London: British Geological Survey for the Department of Energy and Climate Change.

Ashton J. (2014). *Briefing for Lancashire County Council*. London: E3G.

Balcombe, P., Anderson, K., Speirs, J., Brandon, N. and Hawkes, A. (2015). Methane and CO_2 emissions for the natural gas supply chain: an evidence assessment. London: Sustainable Gas Institute, Imperial College.

Bawden, T. (2013). '"Baseless economics": Lord Stern on David Cameron's claims that a fracking boom can bring down price of gas.' *The Independent*. Available at: http://www.independent.co.uk/news/uk/politics/baseless-economics-lord-stern-on-david-camerons-claims-that-a-uk-fracking-boom-can-bring-down-price-of-gas-8796758.html. Accessed: 27 August 2015.

Bradshaw, M., Bridge, G., Bouzarvoski, S., Watson, J. and Dutton, J. (2014). The UK's global gas challenge – Research Report. London: UKERC.

Breast Cancer UK (2015). BCUK Position Paper: Fracking. Available at: http://www.breastcanceruk.org.uk/science/bcukfs-fracking/. Accessed: 27 August 2015.

BP (2015). BP statistical review of world energy June 2015. London: BP.

BGS (British Geological Survey) (2014). A study of potential unconventional gas resource in Wales. Cardiff, Geology and Regional Geophysics, Commissioned Report CR/13/142 for the Welsh Government.

Broomfield, M., Hamilton, S. and Kirsch, F. (2014). *Unconventional Gas in England: Description of Infrastructure and Future Scenarios*. Didcot: Ricardo-AEA Ltd.

Carrington, D. (2013). 'Lord Browne: fracking will not reduce UK gas prices.' In *The Guardian*, 29 November. Available at: http://www.theguardian.com/environment/2013/nov/29/browne-fracking-not-reduce-uk-gas-prices-shale-energy-bills. Accessed: 27 August 2015.

Conservative Party (2015). *The Conservative Party Manifesto 2015*. London: The Conservative Party.

Council of Canadian Academies (2014). *Environmental Impacts of Shale Gas Extraction in Canada*. Ottawa: Council of Canadian Academies.

DECC (2013). *Onshore oil and gas exploration in the UK: regulation and best practice*. London: DECC. Available at: https://www.gov.uk/government/publications/regulatory-roadmap-onshore-oil-and-gas-exploration-in-the-uk-regulation-and-best-practice. Accessed: 5 August 2016.

DECC (2015a). *UK energy in brief 2015*. London: DECC.

DECC (2015b). New onshore oil and gas blocks to be offered. London: DECC. Available at: https://www.gov.uk/government/news/new-onshore-oil-and-gas-blocks-to-be-offered. Accessed: 25 September 2015.

Defra (2014). Shale gas rural economy impacts – redacted version. London: Defra.

Defra (2015). Draft shale gas rural economy impact paper – full version. London: Defra.

Development Control Committee, Lancashire County Council (2015.) Flyde Borough: application number. LCC/2014/0096. Preston: Lancashire County Council.

Energy Institute (2015). *A guide to shale gas*. London: Energy Institute.

Environment Agency (2013). *An environmental risk assessment for shale gas exploratory operations in England*. Bristol: Environment Agency.

European Academies' Science Advisory Council (2014). *Shale gas extraction: issues of particular relevance to the European Union*. Brussels: EASAC.

European Commission (2014). *Communication from the Commission to the Council and the European Parliament on the exploration and production of hydrocarbons (such as shale gas) using high volume hydraulic fracturing in the EU*. Brussels: European Commission.

EY (2014). *Getting ready for UK shale gas: Supply chain and skills requirements and opportunities*. London: EY.

Forster, D. and Perks, J. (2012). *Climate impact of potential shale gas production in the EU*. Didcot: AEA Technology plc.

Friends of the Earth (FoE) (2015). *No need to step on the gas*. London: Friends of the Earth. Available at: https://www.foe.co.uk/sites/default/files/downloads/no-need-step-gas-76983.pdf. Accessed: 11 September 2015.

FoE (2014). All that glitters... Is the regulation of unconventional gas and oil exploration in England really 'gold standard'? London: Friends of the Earth. Available at: https://www.foe.co.uk/sites/default/files/downloads/all-glitters-critique-fracking-regulation-46660.pdf. Accessed: 11 September 2015.

FoE (2013). Unconventional, unnecessary and unwanted: why fracking for shale gas is a gamble the UK does not need to take. London: Friends of the Earth. Available at: http://www.foe.co.uk/sites/default/files/downloads/fracking_summary_2013.pdf. Accessed: 11 September 2015.

Green, C.A., Styles, P. and Baptie, B.J. (2012). Preese Hall shale gas fracturing: review and recommendations for induced seismic mitigation. London: Department of Energy and Climate Change.

Harvey, T. and Gray, J. (2013a). The unconventional hydrocarbon resources of Britain's onshore basins – shale gas. London: Department of Energy and Climate Change.

Harvey, T. and Gray, J. (2013b). The unconventional hydrocarbon resources of Britain's onshore basins – coalbed methane (CBM). London: Department of Energy and Climate Change.

Hays, J. Finkel, M.J., Depledge, M., Law, A. and Shonhoff, S.B.C. (2015). Considerations for the development of shale gas in the United Kingdom. *Science of the Total Environment*, 512–5134, 36–42.

House of Commons, Environment Audit Committee (2015). *Environmental Risks of Fracking: Eighth Report of Session 2014–15*. London: The Stationery Office Limited.

Howarth, R.W., Santoro, R. and Ingraffea, A. (2011). Methane and the greenhouse-gas footprint of natural gas from shale formation. *Climate Change* 106, 679–690.

Institute of Directors (2013). *Getting shale gas working*. London: Institute of Directors.

Institute of Directors (2012). *Britain's shale gas potential*. London: Institute of Directors.

Jacquet, J.B. (2014). Review of risks to communities from shale energy development. *Environmental Science & Technology* 48, 8321–8333.

Jaspal, R. and Nerlich, B. (2014). Fracking in the UK press: threat dynamics in an unfolding debate. *Public Understanding of Science* 23 (3), 348–363.

Joffe, D. (2013). *A role for shale gas in a low-carbon economy?* London: Climate Change Committee. Available at: https://www.theccc.org.uk/2013/09/13/a-role-for-shale-gas-in-a-low-carbon-economy/. Accessed 27 August 2015.

Kovats, S., Depledge, M., Haines, A., Fleming, L.E., Wilkinson, P., Shonkoff, S.B. *et al.* (2014). The health implications of fracking. *The Lancet* 383 (1), 757–758.

Kibble, A., Cabianca, T., Daraktchieva, Z, Gooding, T., Smithard, J., Kowalczyk, G. *et al.* (2014). *Review of the potential public health impacts of exposures to chemical and radioactive pollutants as a result of shale gas extraction*. Didcot: Centre for Radiation, Chemical and Environmental Hazards, Public Health England.

MacKay, D. J.C. and Stone, T.J. (2013). *Potential Greenhouse Gas Emissions Associated with Shale Gas Extraction and Use*. London: Department of Energy & Climate Change.

McCoy, D. and Saunders, P. (2015). *Health & Fracking: The Impacts & Opportunity Costs*. London: Medact.

Mobbs, P. (2015). *'Frackademics' – a study of the relationship between academia, the fossil fuels industry and public agencies*. Banbury: Mobb's Environmental Investigations.

Monaghan, A.A. (2014). *The carboniferous shales of the Midland Valley of Scotland: geology and resource estimation*. British Geological Survey for the Department of Energy and Climate Change. London.

National Grid (2015). *Future energy scenarios 2015*. Warwick: National Grid.

National Farmers' Union (NFU) (2014). Written evidence submitted by the National Farmers' Union, Environmental Audit Committee Enquiry: Environment Risks of Fracking Enquiry. Available at: http://www.parliament.uk/environmental-risks-of-fracking-inquiry. Accessed: 11 September 2015.

New York State Department of Health (2014). *A Public Health Review of High Volume Hydraulic Fracturing for Shale Gas Development*. Albany: New York State Department of Health.

Ottery, C. (2016). Factcheck: Is fracking compatible with action on climate change? Greenpeace. Available at: http://energydesk.greenpeace.org/2015/09/30/factcheck-is-fracking-compatible-with-action-on-climate-change/. Accessed: 6 October 2015.

Petrenel (2014). *A Review of the Potential Impact of Shale Gas and Oil Development on the UK's Countryside*. Ascot: Petrenel.

Pöyry (2012). How will Lancashire shale gas impact the GB energy market? London: Pöyry. Available at: http://www.poyry.com/sites/default/files/imce/files/shale_gas_point_of_view_small.pdf. Accessed: 27 August 2015.

Rathbone, P. and Bass, R. (2012). *Unconventional Gas: The Potential Impact on UK Gas Prices*. London: Navigant Consulting (Europe) Ltd.

Royal Society and Royal Academy of Engineering (2012). *Shale Gas Extraction in the UK: A Review of Hydraulic Fracturing*. London: The Royal Society.

RSPB (Royal Society for the Protection of Birds) (2014). Are we fit to frack? Policy recommendations for a robust regulatory framework for the shale gas industry in the UK. Sandy, Bedfordshire: RSPB.

Rudd, A. (2015a). Amber Rudd. Secretary of State for Energy and Climate Change on Shale Gas. London: DECC. Available at: https://decc.blog.gov.uk/2015/08/10/amber-rudd-secretary-of-state-for-energy-and-climate-change-on-shale-gas/. Accessed: 25 September 2015.

Rudd, A. (2015b). Written Statement to the House of Commons: Shale Gas and Oil Policy – HCES202. London: House of Commons. Available at: http://www.parliament.uk/business/publications/written-questions-answers-statements/written-statement/Commons/2015–09–16/HCWS202/. Accessed: 17 Spetember 2015.

Selley, R.C. (2012). UK shale gas: the story so far. *Marine and Petroleum Geology* 31 (1), 100–109.

Stott, R. (and 17 others) (2014). Public Health England's draft report on shale gas extraction. *British Medical Journal* 348, 27–28.

Task Force on Shale Gas (2015a). *Planning, regulation and local engagement: first interim report*. London: Task Force on Shale Gas. Available at: https://www.taskforceonshalegas.uk/reports Accessed: 11 August 2015.

Task Force on Shale Gas (2015b). *Assessing the impact of shale gas on the local environment and health*. London: Task Force on Shale Gas. Available at: https://www.taskforceonshalegas.uk/reports Accessed: 11 August 2015.

Task Force on Shale Gas (2015c). *Assessing the impact of shale gas on climate change*. London: Task Force on Shale Gas. Available at: https://www.taskforceonshalegas.uk/reports Accessed: 25 September 2015.

UK Onshore Oil and Gas (UKOOG) (2015). UKOOG Annual Report 2014. London: UKOOG.

UKOOG (2014). Press release: How to engage with shale gas/hydraulic fracturing planning and permitting. Available at: http://www.ukoog.org.uk/about-ukoog/pressreleases/66-how-to-engage-with-shale-gas-hydraulic-fracturing-planningand-permitting. Accessed: 25 September 2015.

UKOOG (2013a). A community engagement charter. London: UKOOG.

US Energy Information Administration (USEIA) (2013). *Technically Recoverable Shale Oil and Shale Gas Resources: An Assessment of 137 Shale Formations in 41 Countries Outside the United States*. Washington DC: USEIA.

Warhurst, M. and Buck, G. (2015). *Fracking Pollution: How Toxic Chemicals from Fracking Could Affect Wildlife and People in The UK and EU*. London: Chemtrust.

Werner, A.K., Vink, S., Watt, K. and Jagals, P. (2015). Environmental health impacts of unconventional natural gas development: A review of the current strength of the evidence. *Science of the Total Environment* 505, 1127–1141.

Wilson, M.P. *et al.* (2015). Anthropogenic earthquakes in the UK: a national baseline prior to shale exploitation. *Marine and Petroleum Geology A* 86, 1–17.

10

Alberta Natural Gas – Landlocked Largesse

MICHAL C. MOORE

Province and Description

Alberta has an area of 661 848 km^2 and is the fourth largest Canadian Province after Quebec, Ontario and British Columbia. The Province borders British Columbia on the west, Saskatchewan on the east and the Northwest Territories to the north. Its southern border separates it from the US State Montana.

Population growth rates in recent years have been impressive: the population was 3 645 257 in 2011 and had grown to 4 145 900 by 2014 (making it Canada's fourth largest province), driven in large measure by oil- and gas-derived revenues. Alberta is one of three Canadian provinces and territories that border only a single US State and one of only two provinces that are landlocked with no port or tidewater access.

Alberta hosts a rich reserve range of accessible hydrocarbons, from traditional to unconventional oil and gas resources as well as widespread high quality coal deposits.[1] In recent years, the province has taken advantage of international investment interest in developing reserves through a combination of new technology deployment, aggressive new exploration and development and a governmental and regulatory structure designed to deal with specialized oil and gas operations. The result has been a historically unprecedented growth in reserve identification, exports, and associated economic expansion within the province. In terms of energy investments and export sales, Alberta is easily the most energy intensive province in Canada and a significant economic engine for the entire country.

Canada has an impressive energy-production ranking, now fifth largest in the world, with nearly a third of all hydrocarbon output destined to US refining and direct use. At the same time the country is the seventh largest energy consumer in the world, based on per capita demand (International Monetary Fund, 2014).

The highest concentration of hydrocarbon resources in Alberta and British Columbia is located in a deep northwest trending sedimentary basin, bounded on the west by the Rocky Mountains and extending at shallow depths over the prairie regions east of Alberta. This

[1] High quality coal is a broad reference to deposits that are relatively low in sulphur.

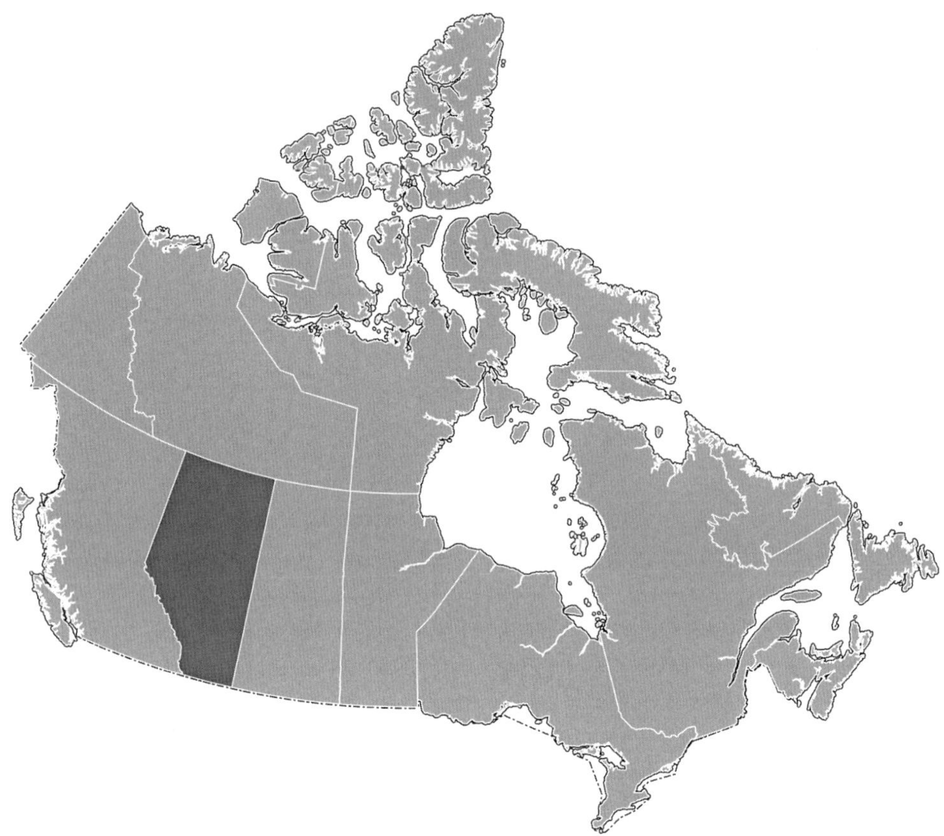

Figure 10.1 The province of Alberta, Canada. *Source:* canadacitiesmap.blogspot.com.

basin also extends well into the US and is the basis for vast conventional and unconventional oil and gas operations in Montana and North Dakota (Hamblin, 2006; Heffernan and Dawson, 2010; Langenberg *et al.*, 2002; National Energy Board, 2013a) as well.

Alberta's hydrocarbons are located in a long linear band in the western regions of North America. Figure 10.2a shows the general trend on the continent, while Figures 10.2b and 10.2c illustrate the sub-basins in more depth and their distribution throughout Alberta, British Columbia, Saskatchewan and North Dakota.

As late as the 1990s, Canada was the third largest global conventional natural gas producer and was supplying up to 28% of overall US demand, but this has declined owing to the massive growth in unconventional gas production in the US. The largest fraction of the natural gas produced came from the Western Canadian Sedimentary Basin (WCSB), located largely within the boundaries of Alberta. In terms of all North American capacity, especially when unconventional resources are included, the resources found in Canada represent a significant fraction of *all* reserves.

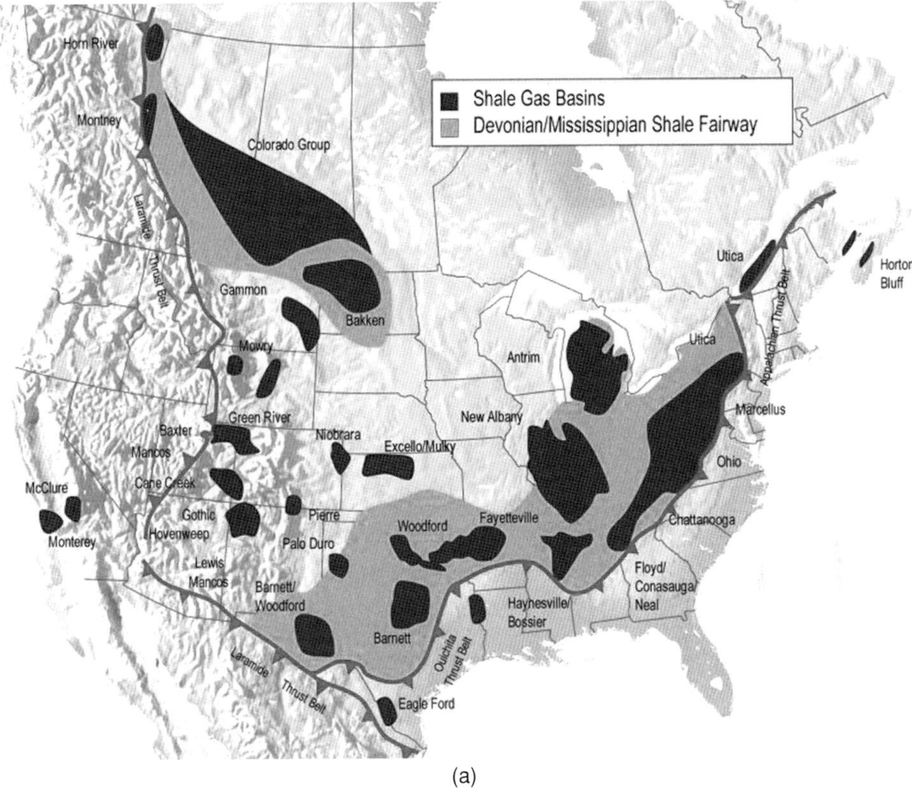

Figure 10.2a Major North American basins. *Source:* AGS and Geological Survey of Canada.

The most recent estimates of proved reserves are shown in Table 10.1. The Canadian total shows a wide variance, reflecting some uncertainty about formations outside Alberta, whereas within Alberta more historical data on drilling and extraction are available. The US estimates follow a renewed interest in and discovery of unconventional resources and the better recovery that is possible with newer technology.

Alberta and Unconventional Natural Gas Markets

Canadian natural gas producers face significant challenges in coming years. Many of their traditional markets in eastern Canada and the US midwest have political obstacles that compound the future expansion of pipeline capacity and storage. Environmental concerns and delays in developing west coast ports are slowing plans to develop LNG ports in British Columbia. This is exacerbated by competition from US LNG producers that threatens to capture a significant fraction of the Asian and European demand.

Most of the gas produced in Alberta has traditionally flowed east and south, much of it to Ontario and Quebec and to gas consumers south of the border. Some gas also has flowed

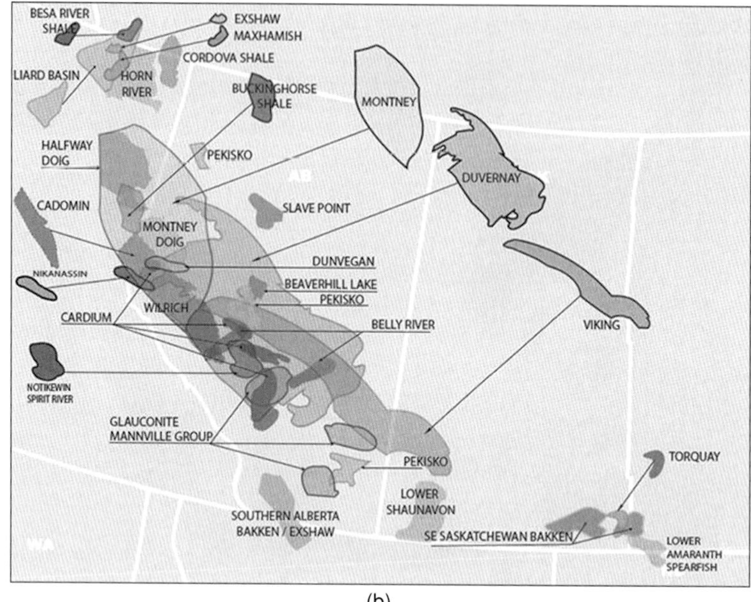

(b)

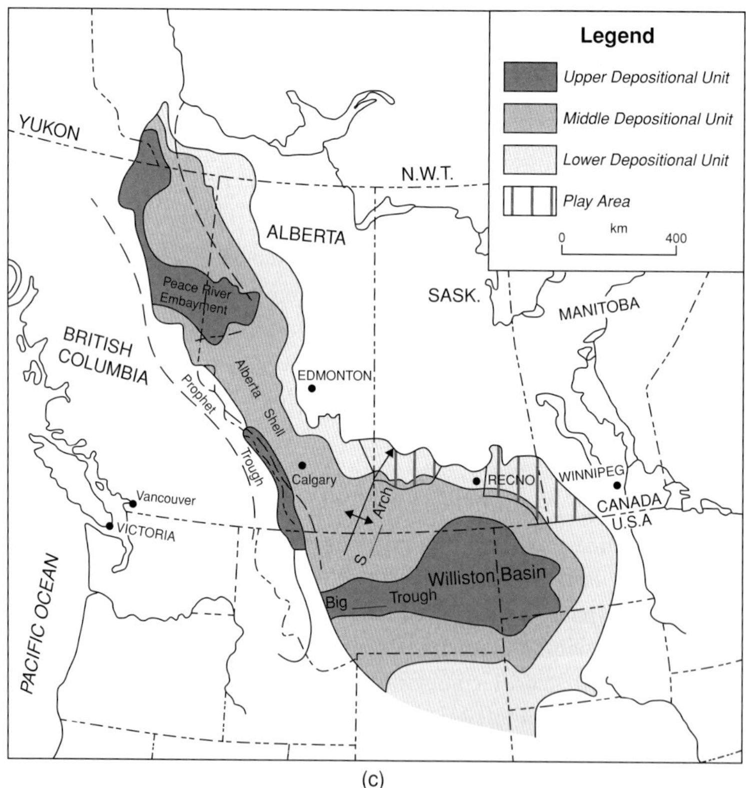

(c)

Figures 10.2b and 10.2c Oil and gas basins in western provinces and the US. *Source:* NEB and Geological Survey of Canada.

Table 10.1 *Summary of natural gas reserves in Tcf for Canada and the US*

	Proved reserves (at the end of 2012)*
Canada	71
US	308
Total	379
	Technically recoverable resources†
Canada total	885–1,566
Canadian conventional	357–436
Canadian unconventional	528–1,130 (coal bed methane, shale and tight gas)
US total	2,431
Portion that is shale and tight gas	664
Portion that is "other"	1766
World Total	28 605
Portion that is conventional	16 527
Portion that is unconventional	12113

* Reserves known to exist and that are recoverable under current technological and economic conditions.

† Gas estimated to be recoverable as drilling and infrastructure expands.

Source: Energy Markets Fact Book, 2014–2015, Natural Resources Canada, Cat. No. M136–1/2014E-PDF.

southwest, to California, and some has remained in-province for a mix of uses: primarily power generation and industrial consumption, typically in terms of steam for *in situ* oil sands production as well as residential and commercial space heating.

Currently, gas demand from Alberta's oil fields is unstable owing to potential cutbacks in production following the softening of world oil prices. This is likely to be reversed, given the capability of the established industry to rebound from periods of diminished demand. Future expansion and growth may be affected, however, since the capital costs necessary to develop projects and distribution capacity to distant markets are high.

The decline of Canadian proven and forecast reserves of natural gas began in the middle 1990s, with the depletion of conventional reserves in the WCSB. This situation began to reverse with the advent of unconventional drilling, with a gradual reassessment and estimate of overall reserves. Nevertheless, increased competition and costs together with declines in demand from the principal importer of Canadian natural gas (the United States) have created a relative surplus of pipeline capacity with resulting contraction in drilling activity and actual volumes extracted. This phenomenon is shown in Figure 10.3 below; these trends underpin the desire for Canadian producers to find tidewater access points where LNG will become economic for export to both Europe and the Pacific Rim.

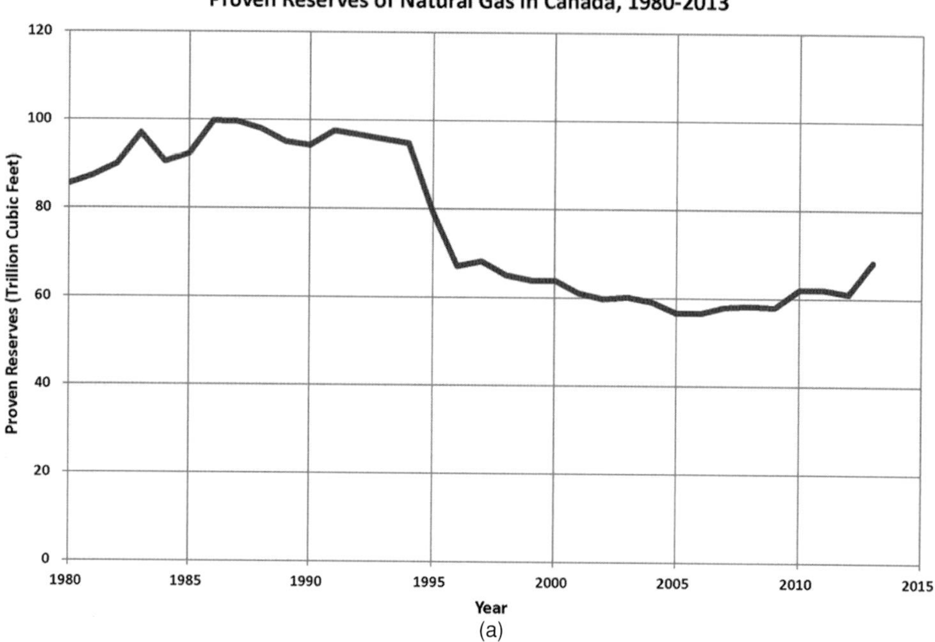

(a)

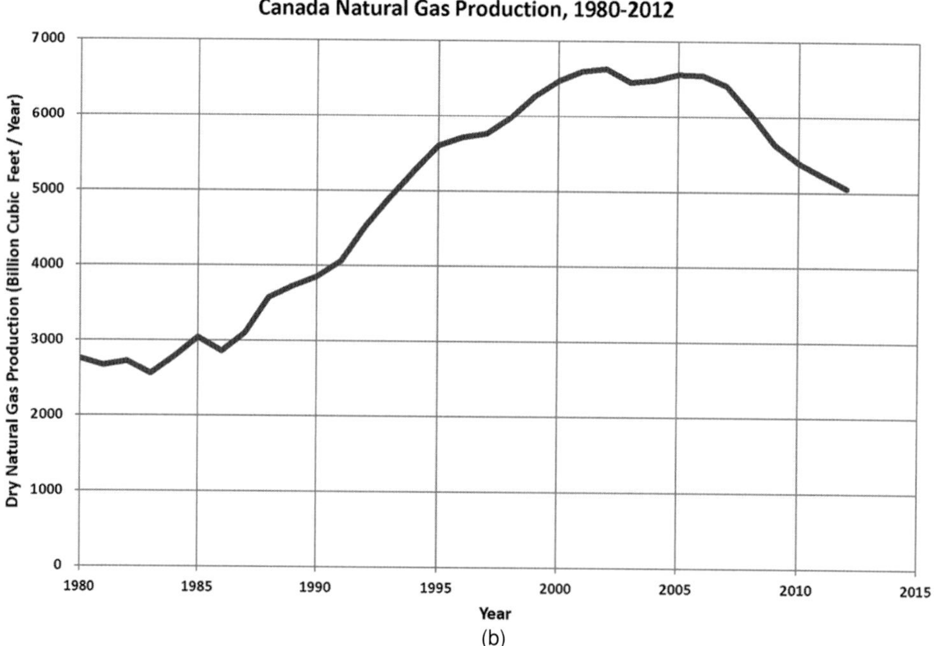

(b)

Figure 10.3 (a) Reserves and (b) production. *Source:* NEB, US EIA (2012).

Alberta dominates the estimates of unconventional natural gas reserves in North America and Canada. The availability of competitively priced gas reserves along with the declining use of coal[2] supports the operation of a vast and competitive electricity generating system.

In terms of areal extent, Alberta's natural gas resources are shared by a geographic and lithographic connection with British Columbia, a fact that complicates the ongoing political dialogue between the two provinces concerning future transport and tariff charges for interprovincial pipelines (Chow, 2006; ERCB, 2012; National Energy Board, 2013a).

According to the US EIA (2014), the focus on resources in the future will clearly be dominated by unconventional resources, where, "although production of natural gas is undergoing decline[s] as a result of reserve depletion, technological advances have spurred rapid investment in the region, and natural gas production from the WCSB will increasingly come from shale gas, tight gas, and CBM".

The major sub-basins (referred to as geologic "provinces") are shown in Figure 10.4 below, and reveal the major lithographic trend of oil and gas bearing sediments from southeast to northwest across the four western Canadian provinces and their continuation to the upper midwest of the US.

Within Alberta, unconventional natural gas resources are present in well-explored sedimentary layers, as illustrated in Figure 10.5 (ERCB, 2012; Mossop *et al.*, 2004). The Rocky Mountains represent a clear geographic as well as lithographic boundary for the WCSB. This reflects the presence of sedimentary rock, acting as a massive wedge extending as far as the Canadian Shield in the east. At the thickest point under the Rocky Mountains this wedge extends approximately six kilometers deep (3.7 mi), thinning to zero at its eastern margins.

The volume of gas represented by these formations is impressive, as shown in Table 10.2 below, giving the most recent estimates available. This table also highlights the interconnected nature of the resource, between provinces and, of course, extending well down into the United States.

Unconventional Hydrocarbon Resources

Shales are the primary source rock for accessing conventional hydrocarbons, and they provide a seal for conventional reservoirs. In recent years, with the addition of new drilling and extraction techniques, olefin-rich shales have become an economic resource for the production of gas, natural gas liquids, and oil. These fine-grained formations tend to have very low permeability; consequently, hydraulic fracturing and stimulation are required to access fluids within the rock and allow fluids to flow to wellbores for extraction. Shale gas or shale oil is not restricted to shale alone; clays, mudstones, silt and fine-grained sandstones and, in some formations, carbonates are associated with hydrocarbon-rich strata.

[2] Canadian coal supplies are extensive and relatively well distributed. Coal for domestic power generation as well as export to the US has declined in the face of increasingly stringent environmental standards and the capture of rail capacity for heavy oil shipments.

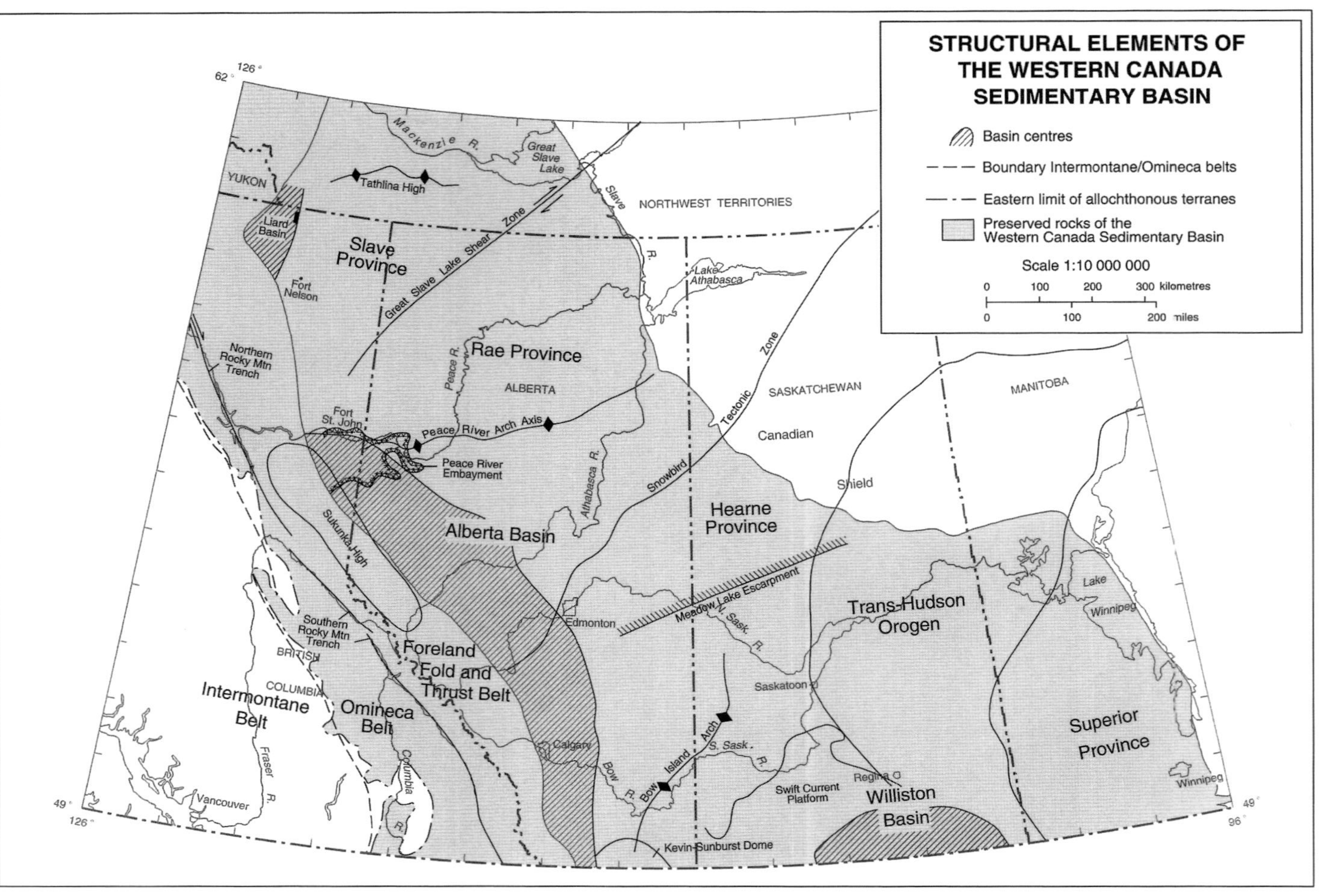

Figure 10.4 Structural sub-basins in the WCSB. *Source:* Alberta Geological Survey.

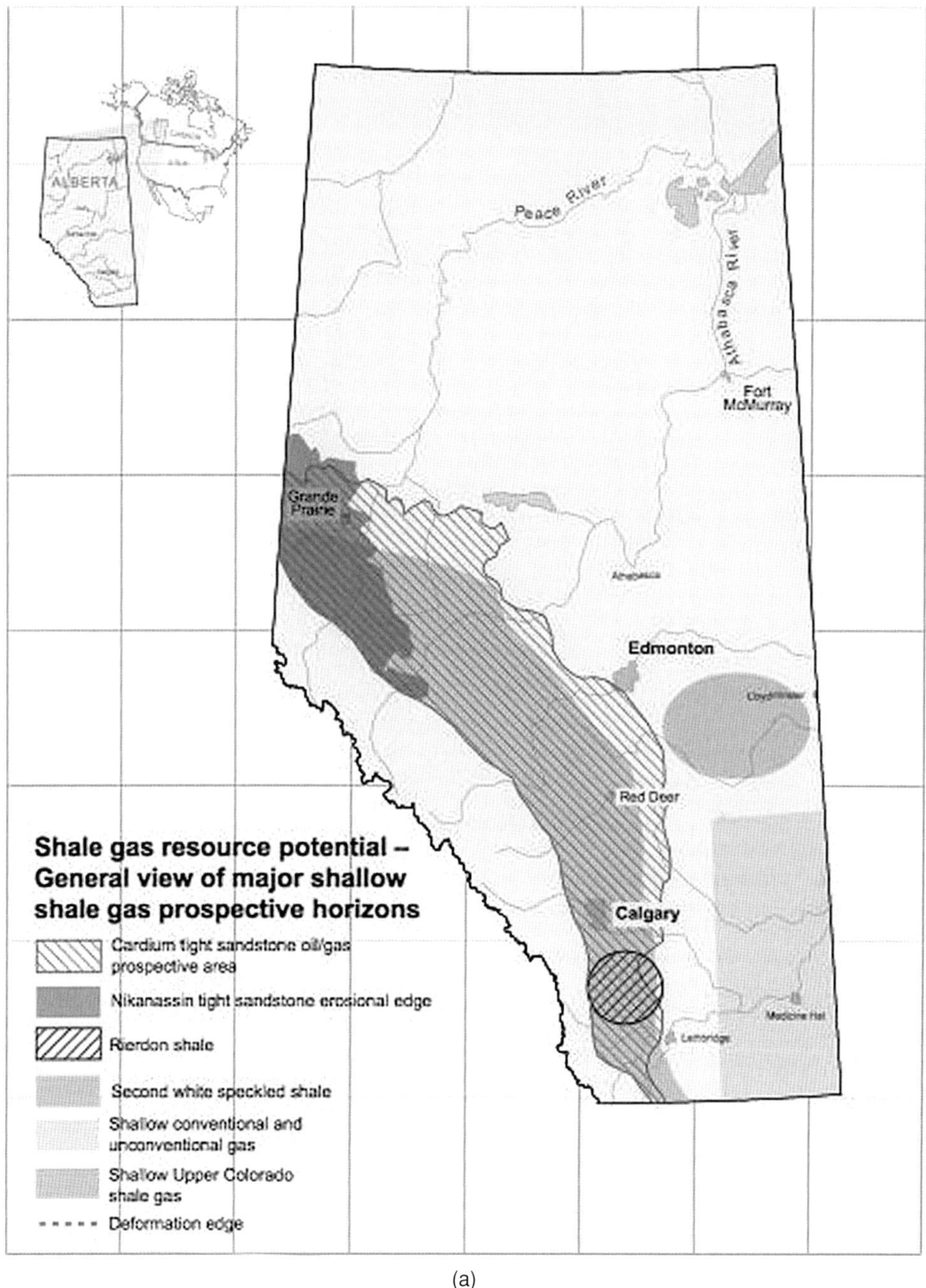

(a)

Figure 10.5 Unconventional shale gas resources. *Source:* ERCB (2012).

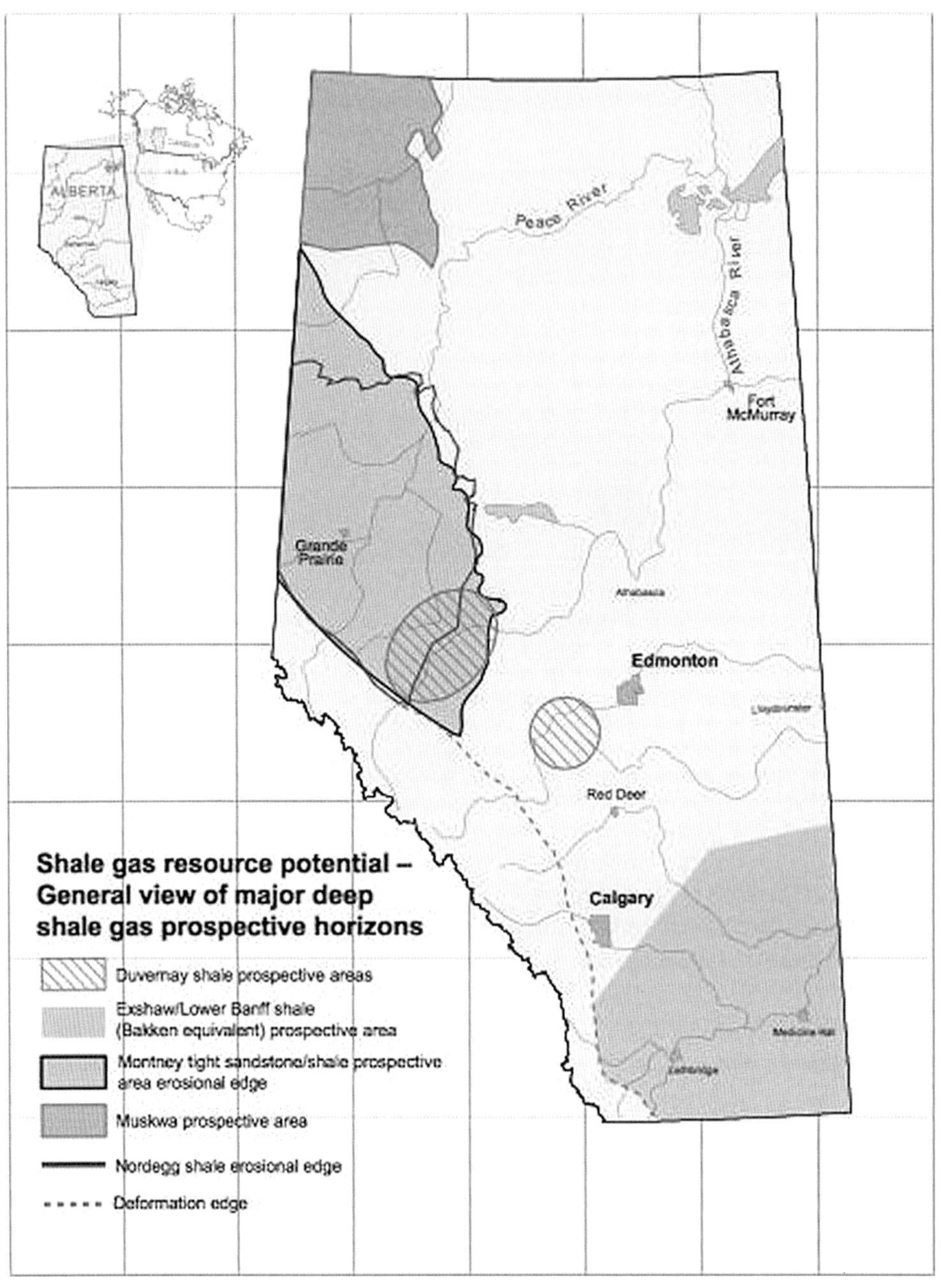

(b)

Figure 10.5 (*cont.*)

Table 10.2 *Estimate of ultimate potential for marketable natural gas in the WCSB in billions of cubic metres (bcm) and in trillions of cubic feet (Tcf)*

Area		bcm Ultimate Pot	bcm Cumulative Production	bcm Remaining	Tcf Ultimate Pot	Tcf Cum. Prod.	Tcf Remaining
Alberta	Conventional	6276	**4425**	**6994**	222	**156**	**247**
	Unconventional						
	CBM	*101*			*4*		
	Montney	*5042*			*178*		
	Unconventional total	5143			182		
	Total	**11419**			**403**		
British Columbia	Conventional	1462	**695**	**10 642**	52	**25**	**376**
	Unconventional						
	Horn River Basin	*2198*			*78*		
	Montney	*7677*			*271*		
	Unconventional total	9875			349		
	Total	**11 337**			**400**		
Sask	Conventional	**297**	**211**	**86**	**10**	**7**	**3**
So Terr	Conventional	**196**	**20**	**176**	**7**	**1**	**6**
WCSB tot		**23 249**	**5351**	**17 898**	**821**	**189**	**632**

Correct at year end 2012.
Source: National Energy Board, November 2013.

Canadian Petroleum Districts and Hydrocarbon Inventories

Natural gas is found throughout Canada and may exist in marketable quantities in the as-yet-to-be-explored North West Passage areas beyond the Territories. Natural gas is included in reserve calculations along with other hydrocarbons and is associated with sub-regions for reporting purposes. The regional distribution of initial established reserves, remaining established reserves, and yet-to-be-established reserves is shown, within the Petroleum Services Association of Canada (PSAC) geographic areas, in Figure 10.6.

Alberta PSAC Area 2 contains roughly 40 per cent of the remaining established reserves, and PSAC Area 7 contains 28 per cent of the yet-to-be-established (frontier) reserves. To date, most gas wells have been drilled in the southern plains (PSAC Areas 3, 4, and 5). However, there is wide speculation that in the Northern Territories and in the Northwest Passage future reconnaissance will discover significant new reserves of conventional as well as unconventional reserves. The implications for price impacts and shipment capacity are large and, if true, would redefine Canadian as well as North American natural gas markets.

The gas found in Alberta is produced from conventional as well as unconventional reserves. Recently the balance has changed to favour the development of primarily unconventional sources, including coal, coal bed methane, shale, and tight sand formations. In the province more than 15 shale formations exhibit a potential for shale gas, natural gas liquids, or oil. The generalized distribution of discovered "plays" shown in Figure 10.7 highlights the economic formations most likely to produce gas or oil in the future. According to the Alberta Energy Regulator (AER), not all these formations are source rocks (i.e., are organic rich); some contain small amounts of organic matter and may be more like low-permeability strata or aquitards than organic-rich shale (AER, 2014).

Development History

Natural gas has been the primary driver in Alberta's growth over the past 15 years. Today, the WCSB produces 97% of all Canadian gas; Alberta contributes nearly 80% of that supply. The majority of all Canadian gas resources, conventional and unconventional, are associated with sub-formations in the WCSB, the largest geologic basin in North America. In turn, approximately 80% of the WCSB is located in Alberta. The WCSB is considered a mature basin, with oil production for more than 100 years in the province. According to the ERCB, the production of conventional oil and of natural gas in Alberta peaked in 1973 and 2001 respectively.

The average conventional natural gas pool already discovered has declined in size over time. Since 1986, the mean gas pool size declined from 2.96 bcf to 0.88 bcf by 2005. From that same period, the median pool size has remained comparatively stable, at 0.45 bcf, suggesting larger average unconventional resources present in the new fields.

The decrease in aggregate average pool sizes is strongly correlated to production rates from conventional gas wells. Average initial productivity per well has declined from about 576 Tcf per day in 1996 to the current levels of approximately 166 Tcf per day.

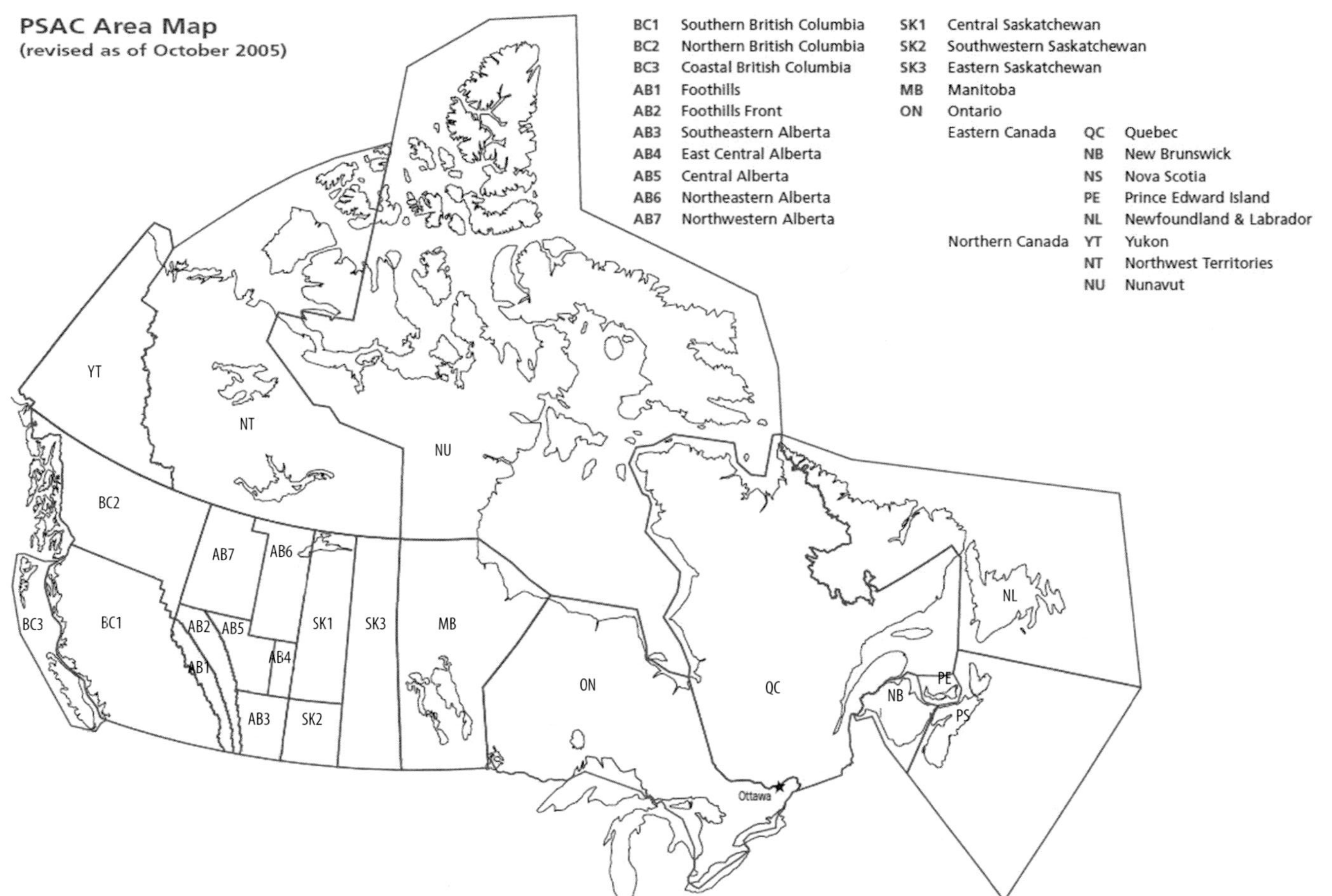

Figure 10.6 PSAC regions in Canada. *Source:* Petroleum Services Association of Canada.

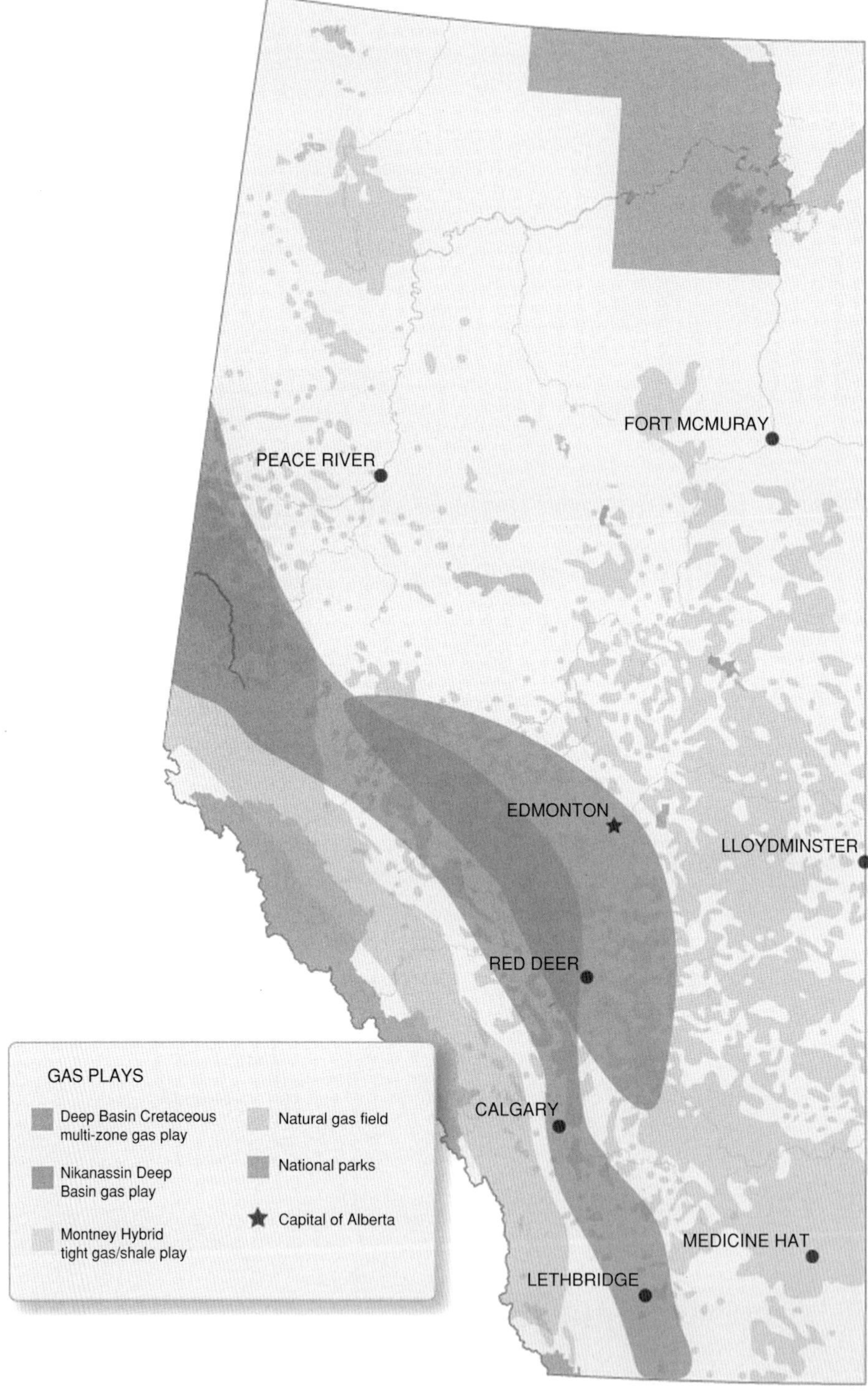

Figure 10.7 Major natural gas plays in Alberta. *Source:* June Warren-Nickles, Natural gas plays, *Alberta Oil and Gas Quarterly Update*, Winter 2016, 4.

Table 10.3 *Wells drilled by type*

Year	Oil	Gas	Dry	SRV*	Total	% Dry to total
2001	4689	11 177	1759	308	17 933	1.7%
2002	3832	9073	1289	265	14 459	1.8%
2003	4473	13 944	1233	201	19 851	1.0%
2004	4427	15 645	1266	255	21 593	1.2%
2005	4822	15 359	1414	330	21 925	1.5%
2006	5599	15 289	1072	167	22 127	0.8%
2007	5429	12 621	935	159	19 144	0.8%
2008	6214	12 326	1689	450	20 679	2.2%
2009	3190	5060	756	336	9 342	3.6%
2010	6522	5856	625	563	13 566	4.2%
2011	10 022	4449	374	1226	16 071	7.6%
2012	8802	1751	215	883	11 651	7.6%
2013	8267	1618	142	856	10 883	7.9%
2014	7769	2130	204	817	10 920	7.5%

* includes injectors, disposal, observation and data collection wells.

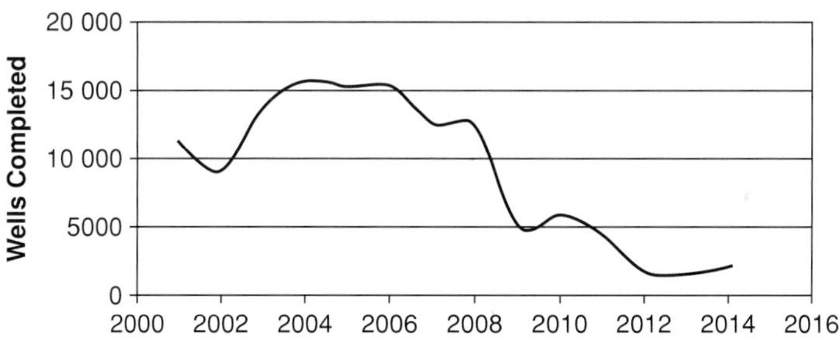

Figure 10.8 Alberta well activity. *Source:* AER and AGS.

Well Activity

The drilling and completion of wells provides a relatively reliable indicator of the trends and economic vitality of the industry as well as of distant demand centres. In the case of Alberta, some of the declines mirror the change to unconventional gas operations in the largest export market, situated in the US. This is clearly shown in Table 10.3 and in Figure 10.8, where the actual number of wells drilled has fallen, with a corresponding increase in dry gas wells, relative to the total.

Alberta Timelines and History

The majority of Canadian drilling activity, infrastructure and export activity is based in the province. The history of energy development in the province reflects the focus and

Table 10.4 *Alberta natural gas timeline*

Year	Significant event(s)
1883	First gas found in Alberta – at Langevin, near Medicine Hat – while drilling water well for the Canadian Pacific Railroad!
1901	First commercial gas field developed at Medicine Hat.
1909	"Old Glory" well drilled at Bow Island – largest gas well to that time in Canada. Drilled by the CPR – reportedly on the wrong location.
1912	Gas pipeline built from Bow Island to Calgary (275 km) by Canadian Western Natural Gas.
1914	First discovery of gas-condensate reservoir at Turner Valley (from Cretaceous) – known as Dingman #1.
1923	City of Edmonton converted to natural gas.
1924	Gas condensate reservoir discovered in Mississippian at Turner Valley.
1930	Bow Island field exhausted having produced 336 bcf into the Calgary market.
1938	Alberta Petroleum and Natural Gas Conservation Board formed by Social Credit Government – significantly reduced flaring of natural gas as a by-product of oil production.
1944	Jumping Pound Formation discovered by Shell – first efforts to recover sulphur from sour gas.
1957	First gas exported to eastern Canada by TransCanada Pipelines.
1982	Alberta's remaining conventional natural gas reserves peaked at 65 Tcf.
2000	(11 December) Spot price for Alberta gas closed at a record $16.95 (Cdn) per GJ.
2001	Alberta gas production peaked at 14 bcf/day marketable gas.
2003	(1 September) 337 gas wells (about 100 Mcf/day) in the Fort McMurray area shut-in by the AEUB to preserve reservoir pressure for future bitumen extraction.

Source: Jaremco (2013).

commitment of the industry to employ new technology and direct investment with the cooperation and support of both policy-makers and regulators in the province.

Some of the key dates in the development of Canada's natural gas industry[3] are listed in Table 10.4. They provide insight concerning the growth of the industry, as well as the development of an innovative and responsive regulatory institution to match and control drilling and shipping activities. The list is notable for the decline in the time needed to bring frontier gas discoveries to market as well as for highlighting the relatively recent interest in unconventional gas development. According to the AEUB, Alberta's remaining (conventional) reserves peaked in 1982 while production from them peaked in 2001.

Reserves

Estimating reserves is a dynamic process, changing as a function of new exploration, technology and relative estimates of quality. The majority of Canadian reserves "estimated"

[3] A more complete chronology and extensive bibliography can be found on Geo-Help's website: www.geohelp.ab.ca

by the National Energy Board (NEB) and reflected in Statistics Canada and US EIA publications continue to be within, or associated with, the Western Canadian Sedimentary Basin. Speculation about reserves in the North West Passage remains unverified at this date.

Calculating the value of natural gas reserves involves rules of Securities Exchange, both in Canada and the US, that depend in part on estimates of volume as well as of the money spent on developing them.

An excerpt from the rules adopted by the Society of Petroleum Engineers Oil and Gas Reserves Committee (OGRC) in 2005 illustrates this:

Commercial Producibility and Economic Efficiency – based on level of commercial producibility of a deposit and future net discounted cash flow (NPV) based on predicted performance indicators and fixed discount rates. Resources are grouped according to Expected Monetary Value (EMV).

Reserves are separated into three groups:

- Economic Normally Profitable are reserves that according to technical/economic calculations have been assessed, on a given date, to be commercially recoverable if brought to production under competitive market conditions, with use of equipment and technology of recovery and treatment ensuring that the requirements for rational use of the subsoil and environmental protection are observed.
- Economic Contingent Profitable are reserves not considered, on a given date, to ensure viability under competitive market conditions due to low performance characteristics but the development of which may be feasible through changing prices, new markets, or new technologies.
- Sub-economic are reserves the development of which, on a given date, is not considered feasible for economic, technical, or technological reasons. This includes not only non-commercial accumulations [but] also those shut-in within the limits of water protection zones, populated areas, national parks, historical/cultural monuments, and deposits located far from transportation lines and producing infrastructure.

In economic deposits, on the basis of technological and technical/economic assessments, recoverable reserves are calculated and booked. Recoverable reserve is that portion of geological reserves which, on a date of calculation, proves commercially efficient for recovery under competitive market conditions with up-to-date equipment and technologies rationally applied and subsoil and environmental protection requirements observed.

In sub-economic deposits, geologic reserves (in-place) are calculated and booked but no estimates of recoverable reserves are made.

Petroleum Resources are subdivided into Potentially Profitable (positive EMV) and Indefinitely Profitable (insufficient information to compute EMV). Recoverable resources are only calculated for Potentially Profitable.

Canada and North American Markets

Canada is part of an integrated North American natural gas market with relatively seamless flows across borders from supply basins to demand. According to Statistics Canada, increasing natural gas production from deeper, higher-productive, liquids-rich wells is outpacing

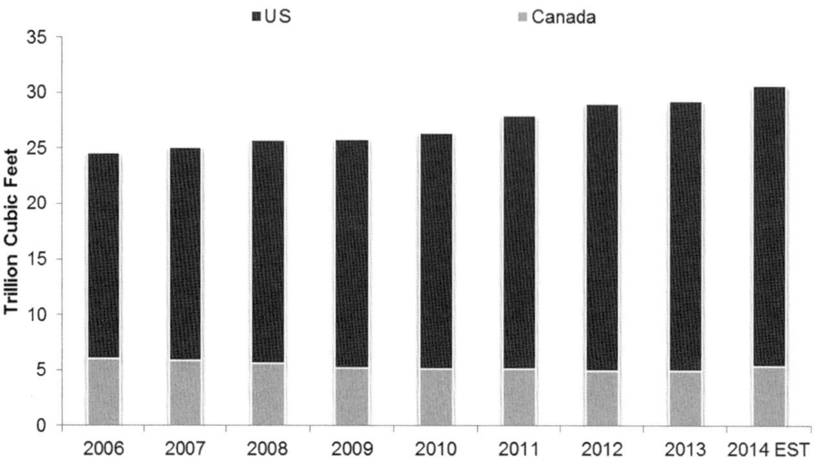

Figure 10.9 North American gas production.

the decline in conventional production. The result is an overall increase of net capacity estimates, as shown in Figure 10.9. In recent years, higher production per well has been achieved using a combination of horizontal drilling and hydraulic fracturing. The average length drilled per well in 2013 was a record 3444 metres (NRCan, 2014).

With estimates of 18.3 Bcf flowing per day in 2011, Canada was ranked as the fourth largest producer of natural gas in the world (US EIA, 2013). Proven reserves exceeded 58 Tcf over an area of 1650 km^3 at the close of 2006.

Alberta

The majority of drilling and extraction in Alberta occurs within the Western Canadian Sedimentary Basin (WCSB). In terms of unconventional exploration, more than 174 000 wells have been drilled and fractured in the province since the early 1950s, with horizontal multistage fracturing used in over 7700 of them (ERCB, 2014; Rivard *et al.*, 2014; Ross and Bustin, 2008).

The WCSB contains one of the world's largest reserves of petroleum and natural gas and supplies much of the North American market, producing more than 16 000 000 000 cubic feet (450 000 000 m^3) per day of gas in 2000. It also has huge reserves of coal. Of the provinces and territories within the WCSB, Alberta holds most of the oil and gas reserves and virtually all the oil sands formations within its borders.

The EUB of Alberta and the NEB have estimated reserves on a basin-by-basin and play-by-play basis, assessing both gas in place and marketable gas. The low/medium/high

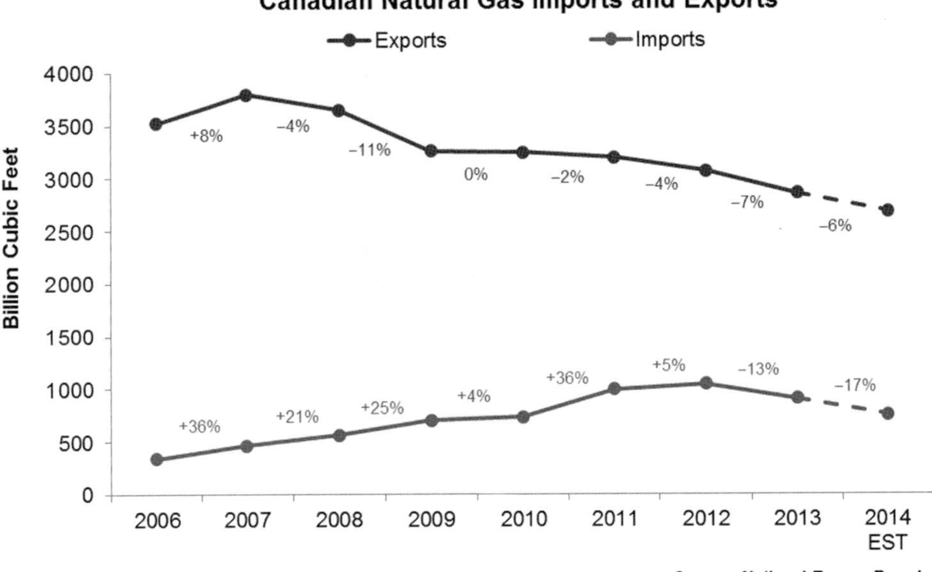

Figure 10.10 Canadian natural gas import and export trends.

estimates for gas in place were 345/376/426 Tcf respectively; the medium-case value was selected for regulatory purposes and is the figure most often cited as the most reliable reserve value.

Domestic Demand for Natural Gas

Domestic demand for natural gas is split amongst a variety of sectors. Industrial demand (including the gas used in oil sands extraction) is highest, at 41%, followed by residential at 20%, electricity generation at 18%, and commercial at 15%. Transportation and agriculture make up the remaining 5% of demand.

Over the 2015–2016 heating season, industrial demand is likely to be reduced as a result of relatively poor prevailing economic conditions. Residential and commercial use will primarily be affected by winter space heating requirements.

Export and Import Activity

Total Canadian natural gas exports are currently in decline. The vast majority of these exports are sent to the US, where increased access to unconventional natural gas supplies has reduced the overall demand for Canadian supplies. As of September 2014, year-to-date Canadian exports to the US were 6% lower than at the same time the previous year. Figure 10.10 shows the change in this trend in terms of overall exports and imports from the US.

Canadian consumers, primarily in the eastern provinces, have been importing less gas from the US than in 2013. Year-to-date imports (as of September) are 16.9% lower than the same time last year. This continues the trend reversal, first seen last year, of significant natural gas import growth. Overall, from 2009 to 2013, Canadian gas exports to the US have declined by 12% and imports from the US have increased by 28%, yielding a 24% decline in net exports.

Canadian gas prices regularly experience a greater discount to US prices in North America in competitive markets. For instance, Canadian gas can be up to 30 cents per Mcf (million cubic feet) cheaper than US gas in winter and as much 50 cents cheaper in summer. Currently this discount is higher and may reach 90 cents by the end of 2015, as expressed in the futures market.

Natural Gas Storage

In the summer, when natural gas demand is low, gas is injected into storage facilities so that it can be easily withdrawn during the peak demand winter months. Canada's underground storage facilities can store over 30% of Canada's annual natural gas demand. New storage capacity has been increasing at a rate of about 5% per year over the past decade (Figure 10.11).

The winter of 2013–2014 was unusually cold and long in North America, depleting natural gas storage inventories more than usual, but it was followed by a relatively mild summer (Figure 10.11). Both Canada and the US entered the winter of 2014–2015 with storage levels 14.2% and 8.3% below their previous five-year minimums respectively. As with oil supplies, 2014/2015 storage capacity is constrained due to the surplus available.

Infrastructure

Both the oil and gas industry are critically dependent on pipeline networks, both regional and local, to transport their products for export or domestic consumption. This distribution network is not confined solely to pipelines; it includes electricity generation for pumping and compression, roads and rail transport and upgrading facilities that process gas, removing water, hydrogen sulfide and carbon dioxide when present, or other impurities, and temporary storage facilities to match market demand and seasonal fluctuations.

Current Infrastructure Systems

Canada has a vast, sophisticated, gas pipeline delivery and storage system, as shown in Figure 10.12. This system is currently the second longest in the world (242 000 km), ranking after the US (2.08 million km). Within Canada, Alberta has the most advanced and extensive system, with 105 850 km. Within Canada, the majority of gathering and transfer systems originate in or pass through Alberta on their way to other markets, primarily in eastern Canada (via the Mainline) or to hubs in the US. A small fraction of natural gas originating in Alberta is actually exported through British Columbia via the Spectra Main Line running from north to south in central British Columbia.

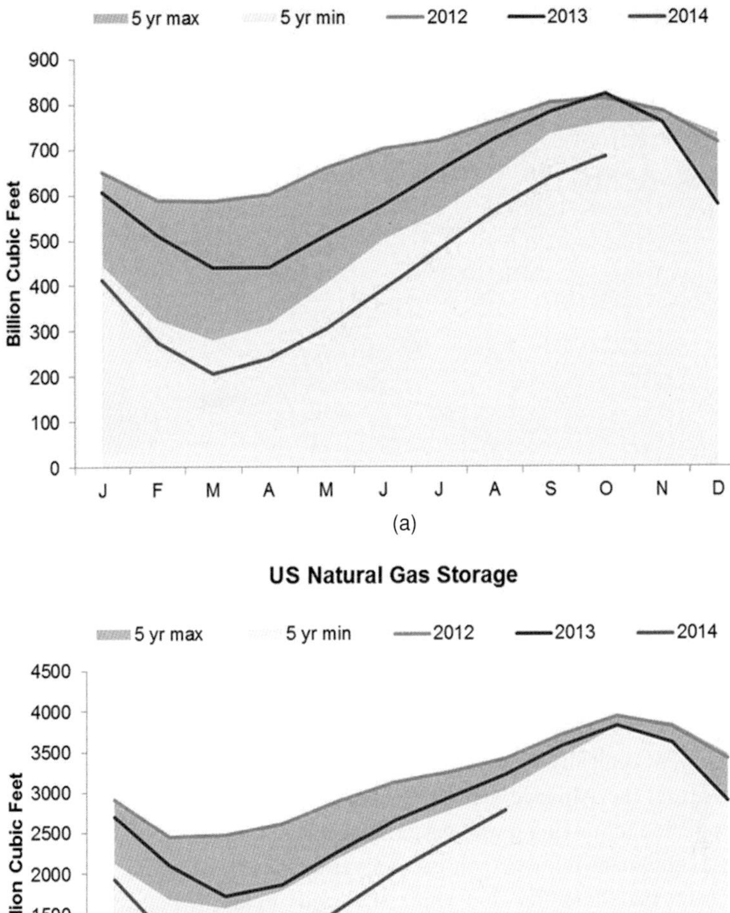

Figure 10.11 (a) Canadian and (b) US natural gas storage. *Sources:* CGA (2014) and First Energy Capital Corp. (2014).

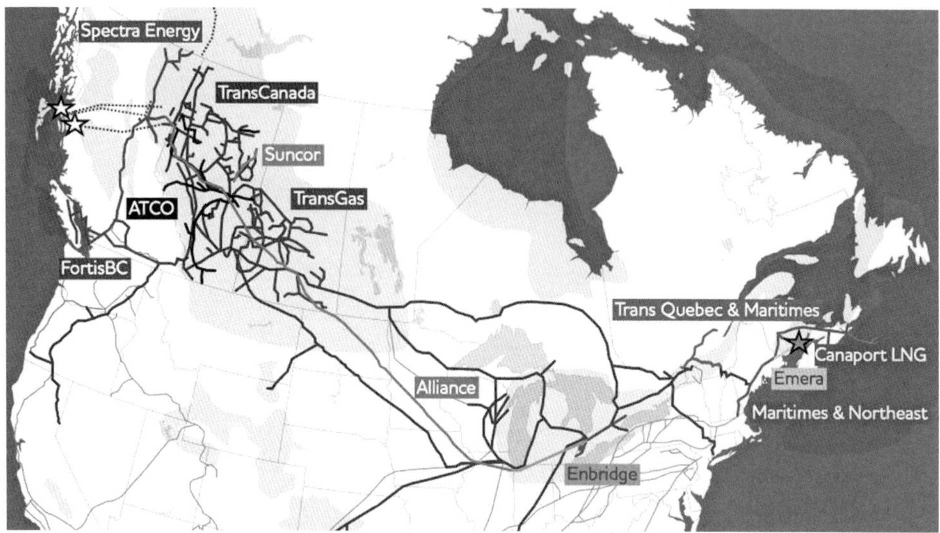

Figure 10.12 Major gas distribution networks in Canada. *Source:* Canadian Energy Pipeline Association (2014).

Trans-Canada Mainline

The Mainline is one of largest natural gas systems in the North American continent. In its first year of operation (1959), the peak volume in the line was 19 billion m^3 per day of natural gas. The line operated until 1998 as a regulated enterprise without competition. Load factors were typically high, sustained by long-term service contracts with shippers. Competition from the Alliance and Vector pipelines in 2000 challenged the competitive role of the pipeline in moving gas to eastern Canada and the northeast US.

The most significant influence on gas deliveries via the Mainline, however, originates in the US from the Marcellus and Utica shale gas fields in Pennsylvania, diminishing throughput and creating a need for increased tolling by the operators. In 2006, shale gas production in this region was minor, averaging just 849 million m^3/day. This is forecast to be at least 1.3 billion m^3/day by 2020, and TransCanada has proposed to take advantage of the eastern Canada demand by constructing new south-to-north pipeline capacity to handle it.

In terms of the Mainline, the current market has been unable to sustain the increased tolls, forcing the pipe owners (TransCanada Corporation) to submit a broad and historically unprecedented application to restructure the nature and extent of future tolls on the line. The TransCanada Mainline was designed originally to transport 198 million m^3/day of gas. By the time of the NEB hearing in 2014, the volume had declined to 42 million m^3 /day.

The fixed costs on this pipeline were high and had to be recouped in spite of lower firm transportation volumes. Kaiser (2013) pointed out that the "application proposed to shift $400 billion per year of costs to users of the Alberta system by extending the Alberta system to Saskatchewan and Manitoba. The application also proposed re-allocating $1.2 billion of accumulated depreciation from the prairies and Eastern triangle segments to a northern Ontario segment that was underutilized and had a large un-depreciated balance. This would reduce the book value of the northern Ontario line and shift costs to Western producers and Eastern consumers by increasing depreciation payments in the Prairie and Eastern segments of the Mainline" (Kaiser, 2013).

Infrastructure and the LNG Future

Natural gas markets are volatile by their very nature. They fluctuate seasonally, with changes in weather and demand for power generation and in the plastics and chemical markets. Alberta suppliers have experienced this roller coaster many times in the past, but have always been rescued by resurgent demand from the embedded base in the US. However, as demand from eastern Canada and the US has declined, the land-locked province of Alberta has begun to look further afield, specifically to Asia–Pacific countries and to Europe as well.

The advent of unconventional natural gas has expanded supply capacity, price differentials in the Asia–Pacific marketplace have widened (until recently) and the technology for liquefaction and regasification facilities have fallen in price. Assuming that a host of political and geographic hurdles can be overcome, serving these distant markets via LNG shipping routes presents an attractive, lucrative and strategic opportunity for Alberta.

The difficulty, beyond domestic jurisdictions, is to capture some fraction of this higher-priced Asian market before the near and mid-term contracts for delivery are completed. Considering the time involved (20–24 years) and the lack of a deep and robust spot market, failing to develop pipelines and ports in time presents an almost unsolvable dilemma for both Alberta and British Columbia developers (Moore *et al.*, 2014).

Summary and Future Issues

Alberta has weathered the decline in conventional reserves, has reallocated capital and exploratory efforts and has identified world-class reserves of unconventional natural gas. This is attributable to a combination of breakthroughs in horizontal drilling and multistage hydraulic fracturing technologies. As a result, the reserves of both shale and tight gas have almost doubled. This, combined with a well-established gathering and distribution system, has created a formidable store of both dry and wet gas (NGLs). The marginal costs of production are competitive with other resources throughout North America.

Outside market forces are making future profits and expansion of the industry uncertain and are threatening the maturation of the unconventional gas industry in the province. As a result there is now a relatively high-risk premium assigned to future development by investors.

Several conflicting forces are contributing to risks in the Alberta gas sector at present. Although the record of natural gas discovery and development in Alberta roughly parallels the worldwide market, recent advances in unconventional gas drilling in the US have curtailed growth in demand there; this is most strongly felt in eastern Canadian markets, where the incentive for exports from Canada has been reversed. The western gas supplies from Alberta are currently not competitive in the near term with supplies from the Marcellus and Utica formations in the US northeast. This is highlighted by the need to repurpose the role of the Canada Mainline and redesign the tolling system to recover fixed and variable costs. Further, any anticipated new or expanded demand in the southeast and southwest portions of the US are being met with new supplies from the Barnett and new unconventional plays from the Midwest.

Market demand in the Asia–Pacific basin has been increasing, with strong and attractive price differentials for natural gas received via LNG facilities. There is a strong interest in Canadian firms (along with a wide range of foreign investment as well) to serve this market place. However, pipeline approvals and port development are constrained by a lack of agreement with First Nation bands and by interprovincial fiscal and permit agreements. As well, a series of lawsuits from environmental groups protesting inferred land use impacts from the infrastructure construction, and unconfirmed public health threats from hydraulic fracturing, are development risks. This is compounded by general risks in terms of climate change policies in Canada and globally.

In every jurisdiction at or near hydraulic fracturing operations, public interest, scrutiny and information demands have combined to slow and/or stop many unconventional gas development operations. Hydraulic fracturing has been used in Canada for decades (Auditor General of Canada, 2012; Council of Canadian Academies, 2013; 2014) and has resulted in a reversal of estimated economic reserves (Rivard *et al.*, 2014; Maule *et al.*, 2013). Nevertheless, the land-locked nature of Alberta Province and difficulties developing strategic pipeline capacity to reach new markets could effectively leave existing and expected reserve capacity stranded. If development does not occur soon then competition for contract agreements may close these markets, for the foreseeable future, to Alberta producers.

Successful contract negotiations are synonymous with the removal of some equivalent amount of import demand for 20+ years, because each contract effectively binds both buyer and seller for that period of time. In addition, price differentials for delivered LNG shipments have collapsed and may not be attractive for investment over the time needed for construction of all the processing and transport facilities that would be required (Moore *et al.*, 2014).

In sum, the prodigious reserves of natural gas in Alberta rival that of Middle East suppliers, but formidable challenges to remain competitive and accessible over the short to middle term must be faced.

Bibliography and References

Alberta Energy, 2013. What is natural gas? Government of Alberta, Edmonton: http://www.energy.alberta.ca/naturalgas/723.asp, 2013.

Alberta Energy and Utilities Board (AEUB), 2005. Alberta's energy reserves 2005 and supply/demand outlook 2006–2015.

Alberta Energy and Utilities Board, National Energy Board, 2005. Alberta's ultimate potential for conventional natural gas. EUB/NEB Report 2005-A.

Alberta Geological Survey, 2014. Geological Atlas of the Western Canadian Sedimentary Basin: http://www.ags.gov.ab.ca/publications/wcsb_atlas/atlas.html.

Auditor General of Canada. 2012. Chapter 5 – Environmental petitions. Part II – Update on government responses to petitions on hydraulic fracturing. Exhibit 5.3 – The hydraulic fracturing process. Fall Report of the Commission of the Environment and Sustainable Development.

Beaton, A., 2003. Production potential of coalbed methane resources in Alberta. Alberta Energy and Utilities Board, EUB/AGS Earth Sciences Report 2003–03.

Boyd, L. 2008. Canada's CBM experience: from the success of Horseshoe Canyon Coals, to the challenges of Mannville Coals. AJM Petroleum Consultants (in *Proc. IndoCBM 2008, the 2nd International Conference*), 26 pp.

Canadian Association of Petroleum Producers (CAPP), 2009. Statistical handbook for Canada's upstream petroleum industry.

Canadian Centre for Energy Information, 2000. Canada's natural gas resources, November, 30 pp.

Canadian Natural Gas Association, 2014. Annual natural gas sector report, Issue 2, 2014: www.cga.ca/wp-content/....../CGA-Natural-Gas-Annual-EN-Final-2014.pdf.

Chow, K., 2006. Unconventional gas in Canada: past, present, future. *Canadian Society for Unconventional Gas* (presentation), March 16, 43 pp.

Council of Canadian Academies, 2013. Environmental Impacts of Shale Gas Extraction in Canada. Ottawa, Ontario.

Council of Canadian Academies, 2014. Report of Expert Panel on Harnessing Science and Technology to Understand the Environmental Impacts of Shale Gas Extraction. http://www.scienceadvice.ca/uploads/eng/assessments%20and%20publications%20and%20news%20releases/shale%20gas/shalegas_fullreporten.pdf.

Encana Corporation, 2013, 2013 Summer Alberta Oil & Gas Industry Quarterly Update, June 2013.

Energy Resources Conservation Board, 2009. Alberta's energy reserves 2008 and supply/demand outlook 2009–2018. Report ST98–2009.

ERCB/AGS, 2012. Summary of Alberta shale and siltstone-hosted hydrocarbon resource potential. Open File Report 2012–06.

ERCB, 2013. Alberta's energy reserves 2012 and supply demand outlook, 2013–2022. Report ST 98–2013.

First Energy Capital Corporation, 2014. Financial forecast, published monthly, http://www.firstenercastfinancial.com/energy/.

Government of Canada, 2012. Fall Report of the Commissioner of the Environment and Sustainable Development. Ottawa, ON: Government of Canada; http://www.oag-bvg.gc.ca/internet/English/parl_cesd_201212_05_e_37714.html#hd3e.

Hamblin, A.P., 2006. The "shale gas" concept in Canada: a preliminary inventory of possibilities. Geological Survey of Canada Open File 5384.

Heffernan, K. and Dawson, F.M. 2010, An overview of Canada's natural gas resources, *Canadian Society for Unconventional Gas*, May 2010.

International Monetary Fund, 2014. Data and Statistics, 2014, http://www.imf.org/external/data.htm.

Jaremko, G., 2013. *75 Years of Alberta Energy Regulation*, ERCB, Alberta.

Kaiser, G. 2013, *The TransCanada Mainline Decision: Toward Hybrid Regulation*, November 2013. Volume 1, Fall 2013.

Langenberg, C.W., Beaton, C. and Berhane, H. 2002, Regional evaluation of the coalbed methane potential of the foothills/mountains of Alberta (Second Edition). Alberta Energy and Utilities Board and Alberta Geological Survey (EUB/AGS Earth Sciences Report 2002–05), 90 pp.

Maule, A.L., Makey, C.M., Benson, E.B., Burrows, I.J. and Scammell, M.K. 2013. Disclosure of hydraulic fracturing fluid chemical additives: analysis of regulations. *New Solut*. 23 (1): 167–87.

Moore, M.C., Hackett, D., Noda, L., Pilcher, M., Winter, J. and Karski, R. 2014. Risky business: the issue of timing, entry and performance in the Asia-Pacific LNG market, *School of Public Policy* 7(18), July 2014.

Mossop, G.D., Wallace-Dudley, K.F., G.G., Harrison, J.C., (compilers), 2004. Sedimentary basins of Canada. Geological Survey of Canada Open file Map 4673.

National Energy Board, 2004. Canada's conventional natural gas resources: a status report, 38 pp.

National Energy Board, 2012. Estimate of ultimate potential for marketable natural gas in the WCSB, year end 2012, December 2012.

National Energy Board, BC Oil and Gas Commission, Alberta Energy Regulator, British Columbia Ministry of Natural Gas Development, 2013a. Energy briefing note, the ultimate potential for unconventional petroleum from the Montney formation of British Columbia and Alberta, November 2013.

National Energy Board, 2013b. Canada's energy future 2013.Energy supply and demand projections to 2035. Calgary, Alberta: http://www.neb-one.gc.ca/clf-nsi/rnrgynfmtn/nrgyrprt/nrgyftr/2013.

Natural Resources Canada, 2013. Additional statistics on energy. Ottawa, Government of Canada: http://www.nrcan.gc.ca/publications/statistics-facts/1239.

Natural Resources Canada, 2013. Shale gas development in Canada – an NRCan perspective, large Canadian natural gas resources (Presentation from May 2013).

Natural Resources Canada, 2014, Reserve summary, Canada, US and world, Energy Markets Fact Book, 2014–2015, Cat. No. M136–1/2014E-PDF.

PacWest Consulting Partners, 2014. Map of natural gas resources in Canadian basins, Government of Canada, November 2014: http://www.nrcan.gc.ca/energy/natural-gas/14186.

Petroleum Services Association of Canada, 2014, Canadian petroleum districts: www.psac.ca, 2014.

Rivard, C., Lavoie, D., Lefebvre, R., Séjourné, S., Lamontagne, C. and Duchesne, M. 2014. An overview of Canadian shale gas production and environmental concerns. *Int. J. Coal Geol*. 126 (0), 64–76.

Ross, J.K. and Bustin, M.R., 2008. Characterizing the shale gas resource potential of Devonian-Mississippian strata in the Western Canada sedimentary basin: application of an integrated formation evaluation. *American Association of Petroleum Geologists Bulletin* 92 (1) (January 2008): 87–125.

Schmoker, J. 2002. Resource-assessment perspectives for unconventional gas systems. *American Association of Petroleum Geologists Bulletin* 86, 1993–1999.

Society of Petroleum Engineers Oil and Gas Reserves Committee (OGRC), 2005. Comparison of selected reserves and resource classifications and associated definitions, December 2005.

US Energy Information Administration (EIA), 2012. International energy statistics., 2012.

US Energy Information Administration (EIA), 2013, North America leads the world in production of shale gas. Washington, DC: http://www.eia.gov/todayinenergy/detail.cfm?id=13491.

US Energy Information Administration (EIA), 2014. Country analysis:, http://www.eia.gov/countries/cab.cfm?fips=ca.

US Environmental Protection Agency (EPA), 2012. Study of the potential impacts of hydraulic fracturing on drinking water resources. Progress Report. Washington, DC: http://www2.epa.gov/sites/production/files/documents/hf-report20121214.pdf.

Ziff Energy, 2015. North American economic ranking study (Third Edition), www.ziffenergy.com.

Other Reading

Alberta Energy page on publications and maps of unconventional gas and oil: http://www.energy.alberta.ca/NaturalGas/940.asp.

Alberta regulator site – play based regulation: http://www.aer.ca/about-aer/spotlight-on/pbr-pilot-project.

Article on fracking: http://eaglefordtexas.com/news/id/145968/geologists-look-earthquake-fracking-correlation/.

CAPP page on hydraulic fracturing: http://www.capp.ca/aboutUs/mediaCentre/NewsReleases/Pages/operating-practices-for-hydraulic-fracturing.aspx.

Discussion paper on unconventional gas in Alberta: http://www.aer.ca/documents/projects/URF/URF_DiscussionPaper_20121217.pdf.

General statistics on Canadian energy production, Statistics Canada, NRCan, www.statcan.gc.ca.

Hydraulic fracturing page: http://www.aer.ca/rules-and-regulations/by-topic/hydraulic-fracturing.

NRCanpage on unconventional resources: http://www.nrcan.gc.ca/energy/natural-gas/14186.

Pipeline performance: http://www.aer.ca/documents/reports/R2013-B.pdf.

Pipeline safety review: http://www.energy.alberta.ca/Org/pdfs/PSRfinalReportNoApp.pdf.

Presentation on unconventional gas in Alberta: http://www.aer.ca/documents/projects/URF/URF_Powerpoint.pdf.

The evolution of Canadian oil and gas programs: http://www.energybc.ca/cache/oil/www.centreforenergy.com/shopping/uploads/122.pdf.

US plan on reducing methane emissions: http://climateobserver.org/us-steps-forward-plan-reduce-methane-emissions.

Wiki page on petroleum development in Canada: http://en.wikipedia.org/wiki/Petroleum_production_in_Canada.

11

Managing the Regulatory Risk of Unconventional Natural Gas

MICHAL C. MOORE

'So what are policy-makers to do? First and foremost, reduce uncertainty.'
Oliver Blanchard, chief economist, IMF, *The Economist*, January 31, 2009

Introduction

Unconventional natural gas development has emerged as a potentially cost effective technique for mobilizing and recovering gas molecules from formations that were inaccessible in the past. The gas resources in shale or coal beds can now be mined as a result of technical solutions involving directional drilling and hydraulic fracturing. This process in shale beds requires substantial volumes of water, plus ways to dispose of the used water. Consequently regulators have been forced to create new standards[1] for reviewing projects. The reason is that there is a great deal of uncertainty about the long-term impacts of intensive, but relatively short-term, drilling and hydraulic fracturing.

This chapter highlights the issues facing regulators as they manage an industrial process with the potential to create external costs that are difficult to forecast (imperfect causation) and are even more difficult to associate with outcomes (imperfect correlation). Another key challenge with unconventional gas extraction is a heightened perception or sense that the risks involved somehow lie outside the conventional definition of risk.[2] A negative perception of risk can override experience, potentially transforming decisions from a category of permissive regulation with mitigation to a defensive project with denials or delays. The ultimate result is that guarantees of safety or performance result in no development. This may be viewed as an application of the precautionary principle but this only holds true if there is inadequate evidence or information about risks.

[1] Not every regulatory agency has undertaken this review, and one consequence has been a series of judicial challenges regarding standards, new legislation to establish appropriate standards and a growing public sentiment that regards the entire process, permitted or not, as a dangerous, unnecessary, tool for developing energy resources.

[2] The basic convention adopted here is that "risk" implies a condition in which there is a *possibility* that persons or property could experience adverse consequences from a given action. Risk can be estimated with experience, experiment or extrapolation in order to assign probabilistic values. This is in contrast with "uncertainty", where, given the existing state of knowledge or information, any event and its consequences are unpredictable and realistic probabilities regarding outcomes cannot be assigned.

The dilemma of the regulation of unconventional gas is due, in part, to the nature of decision-making when data, information and expertise are typically represented in probabilistic calculations yet ultimately require some deterministic action from the regulator. Unjustified delays or project rejection ultimately have the potential to affect long-term power delivery, potentially increasing costs and the efficient operation of energy systems. Some potential delays associated with claims or testimony regarding assumed risk(s) can be minimized using focused *causal risk analysis* to identify the range of uncertainty and risk in any given project. Here, the use of objective, data-driven, methods of causal analysis can reduce bias and increase the credibility and realism of assessments in areas such as health effects or even localized seismic risk (Cox, 2013).

In sum, a combination of public concern and decision-maker uncertainty has created confusion and conflicting claims about the future impacts of unconventional gas extraction. In response to public pressure, whether informed or not, a predictable reaction of decision-makers has been to defer, delay or deny projects for extracting unconventional resources. This chapter considers this risk response and reviews alternatives.

Uncertainty and Risk

The world is characterized by uncertainty and risk. While uncertain events cannot be controlled, we can anticipate them and in turn foresee and estimate the nature of the risks that may follow. Each step requires data and information in order to be useful. Unfortunately, no information stream is perfect, and no time frame is short enough (or long enough) to avoid or anticipate the impact of every decision or choice.

In response to risk, the insurance industry indemnifies individuals or firms and spreads its probability of loss among all ratepayers and sometimes taxpayers, depending on the government involved. Many times people pay premiums that in hindsight appear unnecessary because nothing happened to trigger a claim. For most individuals or firms, these premiums and the insurance involved focus on relatively short time frames and are renewed or revised as circumstances change.[3]

Future events are uncertain but greater experience and learning can increase confidence. Nevertheless, even with experience, as we project further into the future the characteristics of, and confidence in, events become less definitive and more descriptive. Consequently, describing the future is a combination of unpredictable, unreliable, inconsistent, inconstant or variable events. Thus, in order to manage and cope with this uncertainty, we can and do assign risk assessments to a known, expected or imagined range of outcomes. In this way we anticipate the potential nature and magnitude of loss that may follow decisions or their implementation.

In the case of energy systems, the dilemma is how to develop very large, long-term and immobile capital facilities and simultaneously to adopt or integrate imperfect and evolving

[3] The opposite time commitment is implied by "life insurance", where the eventual outcome is absolutely certain, but the time frame is not.

technologies. Typically, decision-makers make heroic assumptions about the interaction and impact of dynamic environmental systems, taking into account past experience, forecast performance and, in the case of unconventional resources, a series of less understood externalities such as induced seismicity and water quality changes.

Policy makers and regulators face new challenges when responding to the regulation of unconventional gas. In the world of unconventional fuels there is mixed experience and field data regarding some risks such as water contamination and induced seismicity. This results in the perception of negative aspects, which influences public opinion in ways that public officials cannot ignore.

Risk and Risk Tolerance in Decision-Making

Energy systems, ranging from exploration and site development to transportation and ultimate energy delivery, are extremely capital intensive. Typically, they involve significant commitments of land and ongoing operational costs but also reflect the 'normal' risk profiles for all parties involved. Broadly these risks are in four categories: market and finance; political and regulatory; performance and system failure; and public support or what is commonly called social licence.

Responding to risks in any of these categories is a process that is dynamic, yet any approval(s) or permits must ultimately be considered temporal in nature. For instance, permits can be terminated, a failure due to a natural disaster may result in the termination of land or development rights or a change in public sentiment may influence a policy change. In many cases the drivers are predictable or at least foreseen events, and risk has a proxy value as represented by discount rates or option price.

Table 11.1 highlights the characteristics and metrics of the traditional risk categories involved in energy systems. In the case of unconventional fuels, changes in risk perception, where experience and data are limited and where forecasting is inconsistent or incomplete, pose challenges for the decision process and ultimately for investors and consumers.

The operation of energy systems involves a variety of agents, from local to distant and from political to market. All have interests that may change over time, depending on circumstances, perceived benefits and service levels and that may cause or be affected by changes in the external environment, including natural systems such as climate, or acts of God or even geopolitical conflict. These are briefly described in Table 11.2 below.

The discount rate preference can be treated as a proxy for the confidence that each agent has in the performance and future events, including externalities. These views about the future state of the world may impact returns to investment, trust and concerns on human health and safety.

The Emergence of Risk Perception and Communication

Policy-makers, legislators and appointed or elected regulators have statutory (as well as moral or ethical) and legal responsibilities to the communities or regions they serve. In general, individuals acting as decision-makers in each of these categories, unlike corporate

Table 11.1 *Influence of uncertainty on risk costs*

Risk Type	Influence on Risk Profile	Time Frame	Confidence Levels and Risk Rates
Market and finance	Demand and supply Interest rates and competitive opportunities	Term of loans or bonds compared with long term changes in demand	Investments in traditional have low risk rates Unconventional with high uncertainty have high, perhaps uncompetitive, rates
Political (tactical) and regulatory	Policy plans Commitments of support Consistency of rules, regulation and policy over time	Political – terms of office Regulatory – Investment or Service plans	Political – high Regulatory – medium
Performance and system resilience	Past performance Nature of technology (conventional vs. emerging or experimental)	20–50 years	Generally high, with exception of new policy mandates or disruptive technology Medium if subject to natural disasters
Social licence	Increased uncertainty and higher costs	Historically low rate of change	Reinforces use of existing technologies

directors, do not bear or face personal liability for decisions they make in the public interest.[4] Nevertheless, decision-makers must operate within the framework of rules of conduct and in the context of established goals or plans.

When considering standard projects, or even unconventional projects such as nuclear facilities, the testimony and submittals of area experts underpin decisions, mitigation conditions and the nature of enforcement standards. Given the experience of over nearly a century of fuel extraction, processing and transportation, the risks and risk factors can be accurately assessed for most projects and dealt with by regulatory agencies.

By contrast, the risk from projects that are imperfectly quantified and/or predicted (i.e. with low-reliability probability-of-occurrence estimates), combined with low public confidence in the regulatory process, increases the perceived risk. This is true of unconventional gas developments, where publicized events associated with unconventional fuels have increased the perceived threat and level of risk to the public.

To many, mitigation and measurement techniques appear to have lagged behind the expansion and development of the industry. As a result there are increasingly higher levels of scrutiny and questions about the capability of the regulatory institutions to manage

[4] I exclude the obvious issue of malfeasance or criminal activity from consideration in this discussion.

Table 11.2 *Risk tolerance*

Actor/Agent	Risk Tolerance	Time Period of Interest	Discount Rate Preference	Characteristic
Developers	High	Project approval and construction phase	Short time period high rates of return	Reflect needs assessment or utility contracts
Finance institutions or investors	Medium to high	Life of project	Long time period competitive rates of return	Stable interest in system performance
Policy-makers	Low	Less than election cycle	Short period of demonstrable public benefits	Interest in demonstrating public needs and affordable benefits are being met
System operators	Low	Life of project	Long time period competitive rates of return	Interest in dependable fuel supply and competitive price
Regulators	Low	Project permit process through end of project life	Long time period competitive rates of return	Projects must meet test of providing competitive power systems, must meet public health and safety standards
Public	Medium	Indefinite	Low time period, indifferent beyond affordable rates and implied low risk to consumers	Typically less informed than other groups but more responsive to alerts or inflammatory material

appropriately the public health and safety concerns that may be present. This scrutiny is compounded by the difficulty of communicating the nature of risk, which extends from acceptable and mitigable risk to unacceptable and prohibited risk and includes background risk and tolerated risk. Development of unconventional fuel extraction has also revealed that there is no standard convention about measuring risk categories.[5] Given that most regulators are risk averse and will err on the side of public safety,[6] it is worth reviewing tools to address unconventional risk[7] that provide some level of both insurance and assurance over the expected life of an unconventional gas project.

[5] Graham *et al.* (2014). [6] This is not always the perception held by the public.

[7] Unconventional risk is defined here as risk of public harm from the development of unconventional resources, where the risk characteristic, timing, magnitude or recurrence cannot be estimated with experience or confidence.

Decision-Makers and Risk

An open approach to risk-governance requires a transparent and informed decision and consultation process. This demands good information and reliable and reproducible predictive techniques that can make decision-makers confident in outcomes over the life of the project. Shrader-Frechette (1991) suggested that lack of transparency occurs "... when the safety issues are so complex that only a small number of specialized experts can understand the issues and participate meaningfully in risk management, and when the experts in a field are seen as 'too close' (e.g., through financial consulting arrangements) to companies that are working to commercialize a potentially hazardous technology".

Water Quantity and Quality

Unconventional hydrocarbon production from shale formations demands large volumes of water that, after use, becomes unpotable and must ultimately be reinjected or treated and disposed of in a way that avoids high external costs. In particular, waste water and fluid used up in the process can potentially contaminate aquifers that are penetrated when wells are drilled. Operational failures in this process include:

- fluid migration around the production casing of a well;
- improper closure and plugging of an abandoned well;
- direct communication (flow of contaminants) between the reservoir (during and/or after production) and the groundwater resources;
- surface spills and leaks near the wellsite or during transport to and from the wellsite; and
- improper handling and disposal of wastes.[8]

Making decisions about operations is confounded by the tendency of some regulatory agencies to fit requirements for water treatment to existing sewer facilities, most of which were never designed to handle large industrial-type loads. In addition, the oversight of water volume use, in addition to well completions, may reside in competing non-overlapping agencies. As a result, this can generate conflicts and difficulties in permission standards, review periods, penalties and enforcement and reporting requirements.

As Shrader-Frechette notes "... in general, the consensus in the technical and regulatory communities is that with the appropriate implementation of best practices and regulatory compliance, risks to water resources via loss of borehole integrity can be effectively managed and minimized". Such confidence, however, may not always be conveyed in a reliable or consistent manner to citizens and their representatives.

[8] For further treatment of this see Small *et al.* (2014), *Risk and Risk Governance in Unconventional Shale Gas Development.*

Induced Seismicity

A key feature of extracting unconventional hydrocarbons is the use of hydraulic fracturing. In this process the injection, or withdrawal, of fluids from a subsurface formation can result in some level of local or even sub-regional seismic events (Kim, 2013; Majer, 2007). Seismic events may occur during the stimulation process or the reinjection of flowback or waste water fluidsor during the withdrawal process of subsurface fluids. Most seismicity that is associated with these events has historically been of low moment magnitude (from 1 to 4 on the Richter Scale), where it produces sensations, but little or no damage (Small *et al.*, 2014). Recent events involving waste water disposal in the US highlight the uncertainty of the induced magnitude of these events, the distance over which they may travel and their persistence over time (Keranen *et al.*, 2013, 2014).

Graham *et al.* (2014) observed that the risk of induced seismicity from deep-well injection is minimized by existing rules and industry practices. Such regulations restrict the volumes and injection rates to levels that result in pressures that are below fracture pressures. When observed, induced seismicity is considered relatively low in terms of the likelihood and magnitude of damages. Thus, induced seismicity is not a major concern for many oil and gas operations (Graham *et al.*, 2014). Such a view is supported by a 2012 report by the USA National Academy of Sciences on induced seismicity. It states "The process of hydraulic fracturing a well as presently implemented for shale gas recovery does not pose a high risk for inducing felt seismic events." Since the publication of the NAS report, recent earthquakes in Canada, including several of a magnitude greater than 4 on the Richter Scale, have shown that, with increased hydraulic fracturing in a region, induced seismic events may be more common although still rare.

Risk Mitigation

Most of the risks posed by unconventional resource development can be mitigated using conventional engineering or control techniques. Where novel technology is required or where performance cannot be forecast well, financial markets will discount the project (demand higher performance bonds and charge higher rates). In turn, this makes the project relatively less competitive or decision-makers will impose extra controls or conditions of performance.

For decision-makers, the option of changing project location may be attractive but requires confidence in estimates of challenged or problematic locations for drilling or development. There are likely to be signs that negative outcomes such as water contamination or microseismic events may be occurring, yet the scale or magnitude necessary to prompt any consequent action may be obscure or be subject to contentious and non-conclusive debate.

On a political time scale, the decision to intervene to stop or prohibit projects remains problematic. Since political preferences and discount rates may reflect very short time frames and uncertainty, combined with a perceived high risk this may result in termination or the denial of project approval. Other options include diminishing the scale of operation, shifting to other areas, demanding changes in technology or more frequent inspections or the requirement of full remediation performance bonds.

The role of financing agents is important in terms of indemnification from risk and also to adjust lending rates and insurance until investors are satisfied that the full contingency impact has been accounted for, given current models. The historical option of choice by regulators has been to use public agencies to provide additional insurance coverage or to legally exempt companies from overall liability from damages.

The outliers in the regulatory process are impacts that are persistent or that affect areas or people who were not included in the risk assessment or impacts that increase in intensity over time. In the case of water, chemical contamination can have serious health consequences and may not lend itself to remediation (spillage of the chemical MTBE has this characteristic[9]) over politically or socially accepted time frames. Induced seismicity may also fall into this category, where rules designed to respond to induced events by shutting down projects may come too late.

The risks implied by the potential responses are stark. If decision-makers are too risk averse or where uncertainty is deemed to be insurmountable, the tendency will be to either burden projects with unattainable safety controls, delay to the point where the economic returns cannot be justified or to turn down the project or switch projects to rely on older, less desirable, technologies.

Resolving the Issue

Power system development and delivery are risky ventures. Nonetheless, modern society cannot live without power. Thus, allowing risk perception that may be unfounded has the potential to impose very large costs, both monetary and environmental, on society.

It is possible to reconstruct and re-characterize the risks involved with unconventional fuel development. Taking into account a key need of regulators, namely to have confidence in outcomes from their decisions well past their term of office or the life of projects under consideration, there are three actions that would appear to respond to regulatory risk.

- First, develop new probability estimation techniques tied to communication systems or methods that instill clarity and perception of safety on the part of the public (for instance that not every strong-motion event is dangerous).
- Second, indemnify decision-makers so as to increase confidence in making decisions and to eliminate destabilizing delays in project or capacity approval activities.
- Third, increase enforcement for infractions of rules,[10] so that compliance with quality standards can be relied on by the public.

[9] See Johnson *et al.* (2000) and An *et al.* (2002) for a discussion of groundwater and chemical interactions.

[10] Zoback (2012) recommended a set of five basic practices that could be used by operators and regulators to safeguard an injection operation from inducing seismicity. (1) Avoid injection into active faults and faults in brittle rock; (2) formations should be selected for injection (and injection rates should be limited) to minimize pore pressure changes; (3) local seismic monitoring arrays should be installed when there is a potential for injection to trigger seismicity; (4) protocols should be established in advance to define how operations will be modified if seismicity is triggered; and (5) operators need to be prepared to reduce injection rates or abandon wells if triggered seismicity poses any hazard.

References

An, Y.-J., D.H. Kampbell, G.W. Sewell, 2002. Water quality at five marinas in Lake Texoma as related to methyl tert-butyl ether (MTBE). *Environmental pollution* 118 (3), 331–336.

Christopherson, S., N. Richtor, 2012. How shale gas extraction affects drilling localities; lessons for regional and city policy makers, *Journal of Town and City Management* 2(4); Henry Stewart Publications, London.

Cox, L.A., 2013. Improving causal inferences in risk analysis, *Risk Analysis* 33 (10).

Graham, J.D., J. Rupp, O. Schenk (2014). Unconventional gas development in the USA: exploring the risk perception issues. Conference paper, Harvard School of Public Health, Harvard Center for Risk Analysis.

Grasso, J.-R., 1992. Mechanics of seismic instabilities induced by the recovery of hydrocarbons. *Pure and Applied Geophysics* 139, 3–4, 507–534.

Johnson, R. *et al.*, 2000. MTBE – to what extent will past releases contaminate community water supply wells?, *Environmental Science & Technology* 34 (9), 210A–217A.

Keranen, K., H.M. Savage, G.A. Abers, E.S. Cochran, 2013. Potentially induced earthquakes in Oklahoma, USA: links between wastewater injection and the 2011 Mw 5.7 earthquake sequence, *GeoScienceWorld* 41 (6), 699–702.

Keranen, K., M. Weingarten, G.A. Abers, B. Bekins, S. Ge, 2014. Sharp increase since 2008 induced by massive wastewater injection. *Science* 345 (6195), 448–451.

Kim, W.-Y., 2013. Induced seismicity associated with fluid injection into a deep well in Youngstown, Ohio. *Journal of Geophysical Research: Solid Earth* 118 (7), 3506–3518.

Majer, E.L. *et al.*, 2007. Induced seismicity associated with enhanced geothermal systems. *Geothermics* 36 (3), 185–222.

Shrader-Frechette, K.S., 1991. *Risk and Rationality: Philosophical Foundations for Populist Reforms*. University of California Press.

Small, M.J. *et al.*, 2014. Risks and risk governance in unconventional shale gas development, environmental science and technology, *Special Issue: Understanding the Risks of Unconventional Shale Gas Development*, 48, 8289–8297, 1 July 1 2014.

Wiseman, H., 2014. The capacity of state institutions to govern shale gas development risks. *Environmental Science and Technology*, DOI: 10.1021/es4052582.

Zoback, M.A., 2012. Managing the seismic risk posed by wastewater disposal, *Earth Magazine*, April 2012.

12

Regulation of Unconventional Gas in Colombia

ANA CRISTINA SÁNCHEZ-THORIN AND ORLANDO CABRALES

Introduction

This chapter reviews recent changes in Colombian oil development and the corresponding changes in regulatory institutions and processes necessary to manage future operations and oversight.

Colombia is considered one of the South American countries with the most significant potential for shale gas, after Argentina and Brazil, taking into account its competitive energy framework and regime (Gómez, 2014). Colombia has also a strategic geographical advantage in terms of oil and gas exportation, owing to its double coastline in the Pacific and Caribbean. The Colombian government offered in 2012 the first unconventional reservoirs blocks, of which six were awarded in the bidding round "Ronda Colombia 2012" and one in "Ronda Colombia 2014". More than 15 blocks awarded prior to 2012 have the potential for unconventional oil and gas and in early 2014 were granted the fiscal terms developed in 2012. In ten years Colombia doubled its oil production, reaching in 2013 a million barrels per day and is intending to expand its energy portfolio by opening new opportunities in the offshore and unconventional realms.

In order to access and develop this resource intelligently, the Colombian Government took two years to prepare the regulatory framework for unconventional resources exploration and production by developing a Knowledge Management Programme. This regulation is based on strong scientific principles coupled with state-of-the-art technical standards to prevent and control potential environmental and social impacts.

Communication and coordination between four governmental bodies was of key importance during this process. These bodies were

- The Ministry of Mines and Energy (MME), which is in charge of adopting the national mining and energy policy and regulations.
- The National Hydrocarbons Agency (ANH according to its acronym in Spanish), which manages and promotes the use of hydrocarbon resources (oil and gas) and is overseen by the Ministry of Mines and Energy.

- The Ministry of Environment and Sustainable Development (MESD), which is in charge of adopting the national environmental policy and regulations.
- The National Authority of Environmental Licenses (ANLA, according to its acronym in Spanish) which is the authority in charge of evaluating and granting the environmental licenses for hydrocarbon exploration and production, which is ascribed to the MESD.

The first section of this chapter provides an overview of the prospects for unconventional gas development in Colombia using a 20-year forecast with three scenarios, those of scarcity, base case and high case.

This section is followed by a description of Colombia's efforts to overcome the technical challenges of uncovering the true technically based environmental challenges and articulating these to the government in order to develop the regulatory framework through the Knowledge Management Programme.

The regulations, which focused on freshwater use, groundwater protection, flowback water management and disposal, seismicity, emissions and chemical additives, are addressed along with a description of the key environmental and technical requirements for each issue.

The final section describes future challenges and builds on experience from the exploration process, including adjusting requirements designed to create play-based regulation while informing public opinion and political debates and strengthening the authority for regulatory enforcement.

Unconventional Gas Prospective Resources in Colombia

Oil represents 50% of the country's exports and about 25% of the national government revenues. By December 2012 Colombia had achieved historical average-oil production greater than 1000 thousand barrels per day (kbpd) and an average natural gas production of 1.2 billion cubic feet (bcf). By 2015 (January–April) oil production had increased to an average 1029 kbpd. In recent years Colombia has seen record levels in exploration activity (A3 wells and two- and three-dimensional (2D/3D) seismic programs) and in foreign direct investment. Oil price volatility in recent months has dramatically impacted the oil and gas sector, forcing the government to introduce fiscal incentives to keep production levels relatively high and to begin considering tax incentives for the exploration needed to maintain competitiveness.

Colombia has identified significant hydrocarbon potential in the Caribbean offshore, especially in the heavy oils of the Llanos basin and in onshore basins such as the Caguan-Putumayo, which remain largely unexplored. In this case there is increasing production potential, coupled with increased rates of recovery in existing fields, which now average 18%. According to the National University of Bogota (Vargas, 2012), the YTF ("yet-to-find") in the country is now generally acknowledged to be around 47 billion barrels of oil equivalent.

Most oil in Colombia is found in the shale belts dating from the Cretaceous and Palaeogene ages, concentrated in the Middle Magdalena Valley (VMM) and in the Eastern Cordillera (COR) and Catatumbo (CAT) basins (see Figure 12.1). These formations are considered the sources of Colombia's most prolific hydrocarbon deposits (Vargas, 2012).

As Figure 12.2 shows, Colombia has unconventional oil potential resources that range from 1000 to 10 000 million barrels in the base case and high case scenarios, respectively; the upshot is that over the next two decades unconventional oil may constitute from 11% to 26% of the total oil production in the country (UPME, 2012).

The unconventional natural gas prospective resources range from 2 Tcf in the base case scenario to 10 Tcf in a high case scenario (see Figure 12.3). This means that, in the next two decades, unconventional gas may constitute an impressive 32% to 60% of the total gas production in Colombia (UPME, 2012).

Meeting the Challenges

In spite of the promising prospects for unconventional resources, Colombia faces the need to incentivize investment, as competitors, such as Argentina, Mexico and Brazil, rank within the top ten countries worldwide with technically recoverable shale gas resources (Gómez, 2014). Additionally, internal challenges due to lack of experience induced governmental authorities to become technically robust, in terms of the environmental and technical challenges of unconventional exploration and production, in order to develop a strong and efficient regulatory framework.

Incentivising Foreign Investment

Following a decision made in 2011 by the Ministry of Mines and Energy and the National Hydrocarbons Agency offering the first unconventional oil and gas blocks in the 2012 bidding round (Ronda Colombia), the initial challenge was to make Colombia an attractive country for enhanced foreign investment. This challenge was addressed by three mechanisms.

- The Colombian Congress approved a royalty discount on existing rates of 40% for unconventional oil and gas;
- Exploration and production contracts were amended to incorporate incentives for the exploration and development of unconventional reservoirs. The main incentives included longer phases for exploration and pilot projects (from six to nine years) and for development (from 24 to 30 years). Additionally, the windfall profit fee was changed, increasing the trigger price from less than 50 dollars to 85 dollars per barrel.

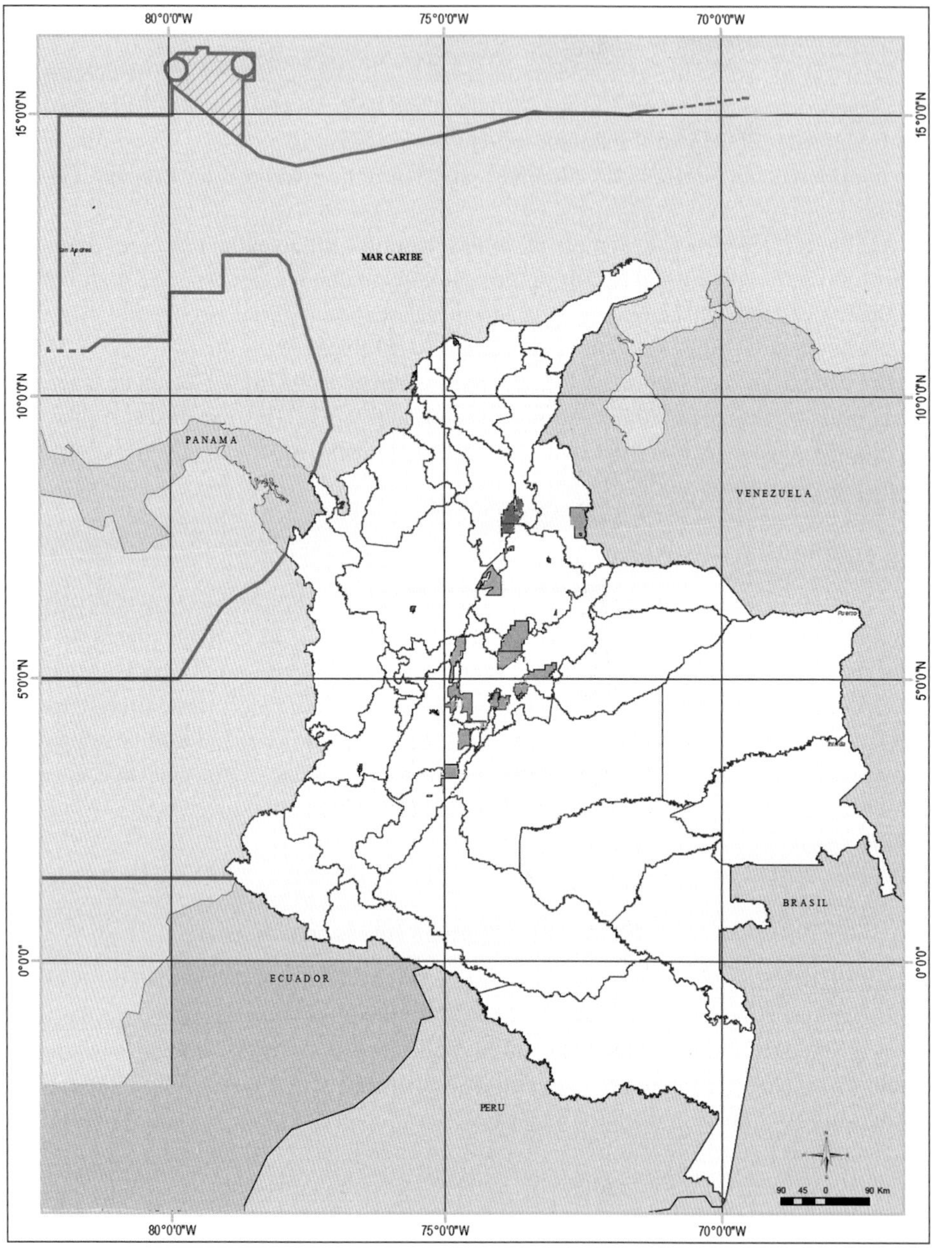

Figure 12.1 Blocks with shale prospectivity in Colombia. *Source*: ANH (2015).

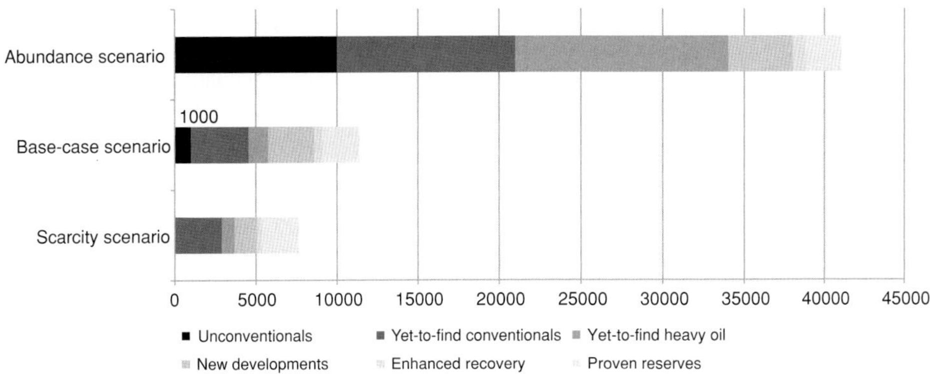

Figure 12.2 Scenarios of crude oil incorporation, in million barrels. *Source:* UPME (2012).[1]

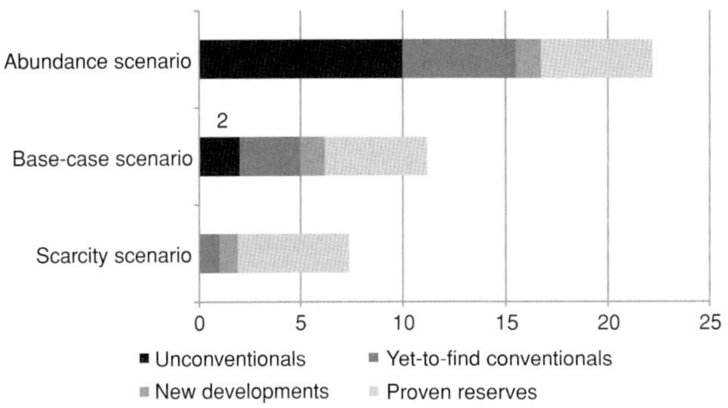

Figure 12.3 Scenarios of natural gas incorporation, in Tcf. *Source:* UPME (2012).

- Finally, in 2014, the Colombian government established rules for exploration, and production contracts signed prior to the bid round of 2012 and located in areas with unconventional potential could apply the 2012 incentives.

In addition, the ANH established, as a prerequisite for companies aiming to enter the tendering process for unconventional blocks, that they should exceed the thresholds of net profit, proven reserves and minimum production. This prerequisite gave assurance that companies exploring and producing in unconventional blocks had the economic capability to comply with stringent environmental and technical requirements.

[1] Scenario confirmation included testing assumptions based on a percentage of each of the project reserves, assuming declination rates which vary according to each type of resource; additionally, discovery sizes for each project and years for the start of each project were considered, on the basis of the available information from companies' investment plans (UPME, 2012). The term "none" means that unconventionals in an "scarcity scenario" would be assigned to a potential moratorium at the time of the study, since governmental institutions had not yet developed the regulatory framework. This was a challenge which was overcome through a Knowledge Management Programme implementation (see the next section, "Meeting the challenges").

These actions resulted in seven blocks being allocated:

- six in 2012, through the Ronda Colombia 2012, which were assigned to Exxon Mobil, Ecopetrol (National Oil Company), Shell and ConocoPhillips.
- one block in Ronda Colombia 2014, which was assigned to the Canadian company Parex Resources.

Knowledge Acquisition

By the end of 2012, when the first six blocks were assigned, various countries had declared moratoria and prohibitions on the hydraulic fracturing of shales in their own territories. This was in response to potential environmental risks, especially those associated to groundwater contamination. These actions raised a special concern within the Colombian environmental authorities (MESD and ANLA). Colombia did not follow suit and declare a similar moratorium; however, the Colombian governmental authorities (MME, MESD, ANH and ANLA) did decide to address this international debate by formulating and developing an academically based Knowledge Management Programme. It was constructed as an attempt to understand, from a technical point of view, each environmental issue associated with the activity and how to address it. This process took two years, which resulted in the publication of a new regulatory framework for exploration and production activities.

The Knowledge Management Programme aimed to acquire the best available technical information especially from the countries that had the most robust experience, such as Canada and the United States of America, as well as from countries and/or states which had declared an explicit opposition to the activity. The acquisition of knowledge was achieved through three different tools: technical workshops, site visits and discussions with international regulatory entities.

Technical Workshops

The Knowledge Management Programme involved four workshops, which took place in Bogotá Colombia and to which the national government brought 24 experts, as shown in Table 12.1. The four workshops were titled as follows:

- Environmental and social impacts of unconventional reservoirs E&P
- Regulatory framework and planning for unconventional reservoirs E&P
- Industry best practices of unconventional reservoirs E&P
- Unconventional natural gas development: social, economic and environmental implications

The main conclusions and recommendations of the workshops are shown in the following tables.

Table 12.1 *Experts involved in the Knowledge Management Programme workshops*

Issue	Topic addressed	Expert	Institution
Water protection and water use	Regulatory framework and water protection	John Deutch	MIT
	Groundwater protection and flow-back water management	David Yoxtheimer	Pennsylvania State University
Geology and seismicity	Induced seismicity associated with injection wells	Mark Zoback	University of Stanford
	Shale O&G geological principles	Thomas Grimshaw	University of Texas
Ecosystems, socio-economics and community disturbance	Ecological footprint and ecosystem management	Kathryn Mutz	University of Colorado, Boulder
	Socio-economic aspects of shale E&P	Iryna Lendel	Cleveland State University
	Workforce supply and demand and community issues	Jon Laughner	Pennsylvania State University
		Aviezer Tucker	University of Texas
	Emissions	Susan Stuver	Texas A&M University
Regulatory framework	IEA Golden Rules and shale E&P in world consumption projections	David Goldwyn	Former US Department of State and Department of Energy official
	Colorado case study	David Neslin	O&G Conservation Commission, Colorado

Table 12.2 *General unconventional exploration and production aspects*

- The "unconventionality" of a reservoir is indicated by the low permeability of the formation.
- Each reservoir is geologically unique.
- Hydraulic fracturing and horizontal drilling have been used for decades globally but the scale at which they are used in unconventional reservoirs is greater.
- In Colombia unconventional reservoirs are located within 1400–2300 metres from the surface, thousands of feet below fresh water aquifers.

Table 12.3 *Building a regulatory framework*

- Regulation should cover the full project lifecycle.
- Regulation should be target, risk prevention and performance based.
- Regulation should include "universal" applicable regulations and specific local needs.
- Regulation should not be based on concerns generated on estimations or errors assumed in calculations.
- Bad science leads to poor regulation.
- Exploration should be used as a learning regulatory phase for data acquisition that will set the basis for production.
- Regulation should have a technical rather than a political basis.
- Regulation should be subject to continual improvement based on field information and new technological advances.
- Regulation should be designed according to the region at a later phase.
- Authorities should have enough capabilities to conduct evaluation, compliance and monitoring activities.

Table 12.4 *Groundwater protection*

- Freshwater aquifers are usually shallow.
- Deeper forms of groundwater have greater salinity and are not considered a resource for human consumption.
- Aquifer depth information should be made available prior to hydraulic fracturing.
- Groundwater protection requires well integrity, where the surface casing integrity is critical, especially between 100–1000 metres.
- Fractures may extend 100–300 metres above the fracture well.
- Fracture modelling requires knowing the direction, length and vertical projection of a fracture.
- It is unlikely that hydraulic fracturing fluid may migrate to aquifers, owing to the large distance between fracking and aquifers.
- Regulation should determine the minimum safety distance to conduct hydraulic fracturing.
- Biogenic methane, CBM (coal bed methane) and thermogenic methane may be naturally occurring in aquifers.
- Pre-drilling groundwater methane testing should be conducted.
- Methane leaks due to hydraulic fracturing are unlikely but may be follow from well integrity loss.
- Chemicals such as arsenic and BTEX may be naturally present in reservoirs.
- Chemical water-quality baseline prior to drilling and hydraulic fracturing should be established.
- Groundwater monitoring programme should include the project lifecycle.
- Groundwater may be vulnerable to chemical or fluid surface spillages.

Table 12.5 *Flowback water, management and disposal*

- Flowback composition may include hydraulic fracturing fluid additives and naturally occurring substances within the reservoir and is generally rich in salts.
- Naturally occurring radioactive material (NORM) such as radium 226 or thorium 232 may be present in flowback water.
- NORM may be precipitated into sludge for its treatment and disposal.
- Flowback is generally rich in salts.
- No surface water discharges of flowback shall be conducted unless there is a strict treatment that fulfils set standards.
- Flowback should be recycled or reinjected into injection or disposal wells.
- Injection wells should be built to ensure that injections are hydraulically isolated.

Table 12.6 *Seismicity*

- Triggered seismicity involves natural movements that have speeded up in time due to processes such as injection.
- Triggered seismicity associated with hydraulic fracturing does not generate significant mass movements or subsidence.
- Seismic events associated with hydraulic fracturing are extremely rare and are never of enough magnitude to cause damage.
- Large earthquakes are caused by large faults which are easily identifiable; injection in or near active faults should be avoided.
- The Colombian national seismological network should be densified in unconventional reservoir regions in order to detect shallow seismic events.

Table 12.7 *Freshwater use*

- Water quantity used for hydraulic fracturing varies with the reservoir geology but may typically range between five to 12 millions of gallons per well.
- Non-potable water sources (e.g., those with high salinity) can be used as a hydraulic fracturing fluid base.
- Low flow and high flow water seasons are key to determining a project's water availability.

Table 12.8 *Air quality and sensitive ecosystems*

- Shale gas is a global alternative for greenhouse emissions reduction substituting coal by natural gas.
- Green completions should be implemented in the development phase.
- Multi-well platforms should be used to minimize the environmental footprint.
- Regulation should include an environmental zoning to exclude sensitive ecosystems.

Table 12.9 *Socioeconomic aspects*

- Demand in local labour is created by core shale-related industries and ancillary shale-related industries.
- Increase in gas supply can influence industrial growth and development.
- Increase in activity implies an increase in the labour force, especially during drilling. It takes a total of 420 people (13 full-time equivalents) to drill a well.
- Cooperation between academia, industry and government is essential to understand the expected labour demand and to generate short training programmes, forums and certifications.
- Shale gas may be a solution for 50 years but not necessarily for long-term needs. It is a temporary bridge technology.
- Technological development and alternative energy sources may supply future demand.

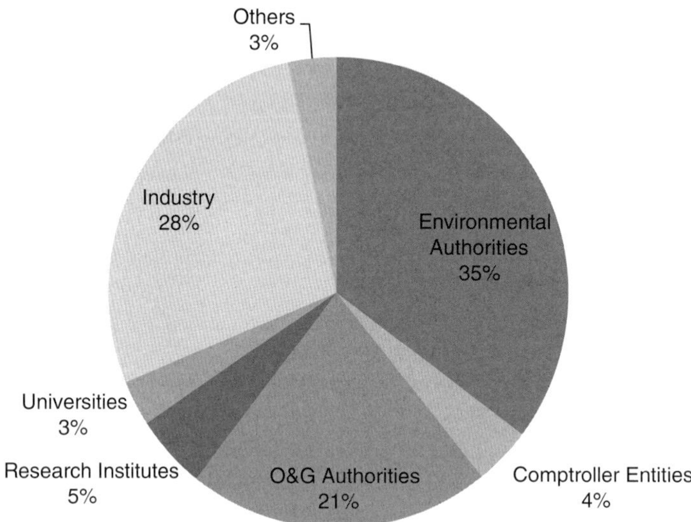

Figure 12.4 Knowledge Management Programme workshop participants.

The workshops were attended by 235 participants from a wide range of backgrounds and associations, as shown in Figure 12.4.

Site Visits and Meetings with Regulators

Following the workshops, a series of site visits to areas of operations was arranged in British Columbia and Texas. Here representatives of the Ministry of Mines and Energy, the Ministry of Environment and Sustainable Development, the National Authority of Environmental Licences and the National Hydrocarbons Agency visited well sites and witnessed hydraulic fracturing activities in remote and densely populated areas. These

Table 12.10 *Regulatory framework development principles*

Principle	Value
Apply and learn from lessons of experienced countries.	Regulation and project execution is benefited by proved technologies and best practice.
Define and treat separately the challenges of the exploratory phase and production phase.	Of special importance in the environmental regulatory framework, this principle allows the putting in place of requirements which may not be applicable in the exploration phase but are fully implementable in the production phase, e.g. green completions.
Provide technical answers for the public.	Regulation should be technically based so that information is acquired throughout the project lifecycle. Thus quantitative date will be available to respond adequately to public enquiries.
Apply prevention principle.	This principle prevents potential impacts that might happen if there is not enough information on the science behind the impact, e.g. surface water disposal might not be allowed owing to uncertainty whether naturally occurring radioactive material is present in the flowback fluid.
Continuously improve regulations.	As the activity develops, play-based and region-based regulations should be developed and adjusted to local needs.

activities were mainly supported by the Canadian and US Embassies, who provided mainly logistical support for site visits and meetings with governmental bodies.[2]

The Knowledge Management Programme had great value in bringing together a range of Colombian authorities who could benefit from a common and shared learning environment. A principal byproduct was a cooperative conversation that ultimately resulted in an appreciation of the fact that responsible environmental and social E&P of unconventional gas is possible and increases in value by having a basis in scientifically based regulations.

Current and Proposed Regulation

Once the technical bases were established, the approach adopted by the Colombian Government to develop regulations included the principles stated in Table 12.10.

[2] The visits that were accompanied with meetings included US federal agencies such as the Department of Energy, the Department of State, the Environmental Protection Agency, the Department of Commerce, the Department of the Interior, the Bureau of Land Management and entities in various US and Canadian states such as Texas, Alberta, British Columbia and New York, the last being of significant importance as this state declared a moratorium on directional drilling and high volume fracking as a result of the meeting held with the Department of Environmental Conservation at Albany.

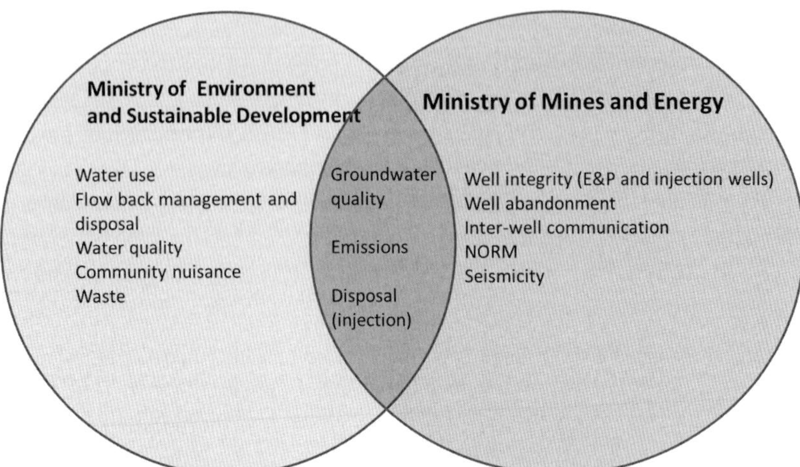

Figure 12.5 Inter-institutional framework on issues to be addressed by regulation.

Bringing Together the Energy and Environmental Authorities

Some of the main environmental concerns such as aquifer contamination and methane migration can be prevented by adequately constructing the wells and routinely conducting integrity tests. With this in mind, the Ministry of Environment and Sustainable Development (MESD) expressed concern regarding which environmental authorities should be responsible for the observance and protection of natural renewable resources. The suggestion was that the most appropriate body was the Ministry of Mines and Energy (MME). This in turn created a need for some inter-institutional regulatory framework that could determine the issues to be addressed by each Ministry in the future regulation. Figure 12.5 shows the development of this framework.

As Figure 12.5 shows, the issues to be included by each authority were clearly established and logically separate. The characteristics of groundwater quality following wastewater disposal through potentially overlapping injection wells was addressed by separating the prevention of incidents that may lead to the contamination of groundwater sources. This was later addressed by the Ministry of Mines and Energy by establishing requirements for the E&P wells and the injection wells in terms of construction and design, cement resistance, integrity testing and inter-well communication prevention. Meanwhile, the Ministry of Environment and Sustainable Development addressed the baseline monitoring of groundwater quality, establishing an area of review around the E&P wells in addition to the injection wells, and the parameters to be monitored. It also established the monitoring frequency rules for before, during and after well-drilling activities as well as the reporting requirements associated. In relation to emissions the MESD established the baseline air quality monitoring parameters as well as requirements for complete combustion and venting prohibition, whilst the MME established an emissions permit, historically its competence.

In addition to the inter-institutional framework a decision was made to hire an international consultant with sufficient experience in regulatory development to provide inputs into the regulations, both technical and environmental, to oversee the effort. Mr David Neslin, former Director of the Oil and Gas Conservation Commission for Colorado, was appointed with a team of specialists in order to contribute on the regulation development.[3]

Out of this process, the following regulation was developed:

- terms of reference for the Environmental Impact Studies for Hydrocarbon Exploration with an annex for unconventional reservoirs, which were adopted through Resolution 0421, 2014 (MESD and ANLA);
- Decree 3004 (2013) and Resolution 90341 (2014), which establish the technical requirements for the E&P of unconventional reservoirs (MME and ANH).

Addressing the Key Environmental Issues

A final regulation was developed based on the Knowledge Management Programme results as well as the inputs provided by Mr David Neslin and his team. In the process, three challenges were continuously faced on each of the following topics:

- maintaining a balance between performance-based and prescriptive regulation;
- distinguishing good practice from obligatory requirements; and
- tailoring the requirements to the Colombian context.

Ultimately each issue was addressed in the regulation trying to manage these challenges.

Below is a summary of the requirements for each issue:

Groundwater Protection

Well design and construction preventive measures:

- For all casings the pump and seal method should be used and should overlap the next casing by a minimum of five per cent.
- Requirements for conductor and surface casing include that they should be cemented up to the surface and 150 feet below the deepest freshwater aquifer encountered.
- Requirements for the intermediate and production casings include cementing 500 feet above the shoe and conducting CBL-type cement tests.
- The intermediate casing shall seal all horizons which may have a presence of oil and gas or corrosive substances.
- All casings shall have pressure tests.
- Cement shall be designed to resist 300 psi for 24 hours and 800 psi for 72 hours.

[3] Mr David Neslin was the former chair of the State Review of Oil and Natural Gas Environmental Regulations; (STRONGER), which published, in 2010, hydraulic fracturing guidelines and supported the development of FracFocus.org.

- If there is any sign of improper cementing, drilling activities shall be suspended and corrective actions implemented, and evidence of success in their implementation shall be sent to the authorities.
- Within ten feet of the surface casing an integrity test shall be applied to the geological formation to test for the initial rupture pressure at the shoe.

Additional requirements were established for existing wells to be converted into new E&P wells as well as into injection wells.

Hydraulic Stimulation Preventive Measures

- Even though Colombian shales are estimated to lie between 1500 and 2400 metres deep, a minimum distance was established between the base of freshwater aquifers and the reservoir where hydraulic fracturing would be conducted; this was set at five times the fracturing radius,[4] i.e. approximately 500 metres.
- For coal bed methane this distance shall be greater than twice the fracturing radius or ten times the vertical width of the stimulated interval.
- Prior to all hydraulic fracturing activities, pressure tests shall be conducted to all casings exposed.
- The operator must monitor the annular pressure of all casings permanently during hydraulic fracturing.
- If there is any increase above 200 psi, hydraulic fracturing activities shall be suspended immediately and notified.
- If pressures indicate that there may be communication between the hydraulic fracturing fluid and the annulus the operator shall suspend the hydraulic fracturing activities, notify the authorities, implement corrective actions and send evidence of their implementation and success.

Groundwater Baseline Monitoring

- Groundwater baseline information to be gathered in an 'area of review'. This area was defined according to the type of well:
    - The area of review for vertical and horizontal wells includes the area between the vertical and horizontal sections of the well or well arrangement. The horizontal radius shall be the longest lateral projection for the well or well arrangement.
    - The area of review for a vertical well (without horizontals) is the vertical area between the vertical section length of the well or well arrangement and three times the hydraulic fracturing radius.
- The regulation establishes that a conceptual aquifer model shall be produced for the area of review. The permeability of the stratigraphic units that separate the aquifer from

[4] The hydraulic fracturing radius is defined as the distance reached by the fracture as a result of hydraulic stimulation.

the target formation shall be determined as well as the conductivity and vertical and horizontal flow characteristics.

- The physico-chemical composition of the water shall be acquired with field-based collected samples within the area of review, in representative aquifer samples, for 49 basic parameters plus 13 additional parameters that have been associated with hydraulic fracturing activities such as barium 226 and thorium 232, BTEX, arsenic and formaldehyde, among others. Samples shall be taken in both the dry and wet seasons.
- Monitoring shall be done every quarter for all 62 parameters within the area of review.

Flowback Water Treatment and Disposal

This issue constituted a special sensitivity for designers as they considered regulatory requirements, since there is still a great deal of uncertainty about the potential contaminants that may be moved from unconventional reservoirs by the fracturing fluid. Ultimately, it was felt that only the exploratory phase could truly provide insights on the issue. Consequently, the authorities decided to create a principal regulation that would prohibit the storage of flowback water in pits above ground and the disposal of flowback fluid in surface waters.

As a result, flowback water and produced water can now only be disposed in injection wells, for enhanced oil recovery, or in disposal wells. Strict requirements were also addressed in the regulation to deal with injection well design, construction and integrity testing, as well as area of review, groundwater and surface water monitoring.

Flowback and produced water can also be disposed of by aspersion above ground, subject to prior treatment. Regulation for wastewater disposal above ground, necessary in order to define the thresholds, is still under development by the national environmental authorities.

Storage tanks for flowback and/or produced water are required to have a retention or containment barrier capacity up to 110% of the largest tank within the barrier. Storage tanks shall be frequently monitored to verify their integrity, as well as that of all tanks, hoses and any other storage or transportation structure, for both flowback and/or produced water and chemical additives for the hydraulic fracturing fluid.

Freshwater Use

Colombia has a wet–dry seasonality that may cause severe droughts to occur during the low-flow season in some areas in the country. According to the World Water Council, Colombia has a water stress indicator between 0 and 0.1, thus qualifying as a "no stress" country in terms of water scarcity.[5] However, during the dry season there could be some regions which could be significantly affected. For instance, during the first quarter of 2014 the Guajira region in the north and the Orinoco river region in eastern Colombia suffered a severe drought, causing the death of hundreds of animals, which led to the enhancement

[5] http://www.worldwatercouncil.org/library/archives/water-crisis. The water stress indicator has the following ranges: no stress, 0–0.1; low stress, 0.1–0.2; mid stress, 0.2–0.4; high stress, 0.4–0.8; very high stress, 0.8–1.0.

of regional forecasting alerts through permanent monitoring and implementation of local water availability measures for human and animal consumption. Colombia is also highly vulnerable to the effects caused by the El Nino Southern Oscillation (ENSO). This helps to explain why water availability was so thoroughly assessed during the baseline acquisition, which included understanding the specific flows for the surface and groundwater sources to be impacted and determining the potential conflict uses with communities within the area of influence.

For exploration activities, the regulation stated that the following elements were of highest priority:

- the reuse of flowback and produced water for fracturing other wells or for re-fracturing activities;
- the use for fracturing of non-potable water and residual municipal waste water;
- alternatives for low-flow seasons;
- measures and restrictions for low-flow seasons to sustain water use by communities present in the area of influence of the project.

Seismicity

Expert opinion has concluded there is a low but possible association of induced seismic events with hydraulic fracturing, flowback and produced water re-injection. Thus, as unconventional basins are located in a highly mountainous region of the Andes, preventive measures should be undertaken with regard to hydraulic fracturing projects.

With this in mind, requirements established by the Ministry of Mines and Energy and the Colombian Geological Service, include the following.

- The presence of geological faults shall be determined at any depth within an imaginary cylinder surrounding the E&P and injection wells, of which the cylinder dimensions may vary according to the depth of the well.
- A baseline seismicity monitoring for the region shall be conducted before, during and after hydraulic fracturing activities and for injection wells during the lifetime of the well.
- A seismicity network shall be installed by the project and shall be operated and interpreted by the Colombian Geological Service.
- Injection wells shall be installed to monitor the volumes and pressures at which fluids are injected.

In addition to these requirements, fallbacks have been established so that hydraulic fracturing cannot be undertaken within one kilometre of a large geologically active fault and no injection wells can be constructed within two kilometres of a large geologically active fault.

If a seismic event occurs that supersedes 4 on the Richter Scale and the epicentre coincides with the area of review of the project, activities shall be suspended and corrective actions adopted and their success demonstrated to the authority.

Emissions

In terms of emissions the regulation of the MESD prohibits the venting of any gas, and complete combustion must be implemented. Volatile organic compound reduction devices shall be installed in all storage tanks to achieve at least 90% emission reduction.

The following baseline information shall be installed prior to initiation of project activities:

- emissions inventory and potential receptors;
- air quality monitoring including parameters such as methane, hydrogen sulphide, BTEX and formaldehyde.

Chemical Additives

Information about all chemical additives used in the hydraulic fracturing fluid shall be published prior to each fracturing phase and adjusted after the phase has finished, if the fluid composition varied during the activity. The concentration shall be expressed as a percentage of the mass and the total volume of the fluid, and information on health, eco-toxicological and biodegradability characteristics shall be included in reports to be submitted to the ANLA for evaluation.

If an additive has a commercial secret, the family name shall be provided to the ANLA, who will undertake all measures to avoid disclosing the commercial secret to the public but will evaluate its toxicity, based on the information provided. However, if an unplanned event occurs, information shall be provided to the medical entity that requires it for clinical diagnosis or medical treatment.

The Future: Learning from Exploration and Regulation in order to Improve the System

Seven exploration blocks have been assigned for the exploration and production (E&P) of unconventional reservoirs in Colombia. Of these, there are unconventional prospects in about 15 so-called conventional blocks. It is estimated that in the next few years around ten exploratory wells will be drilled in the middle Magdalena region, depending on the commitments established within the contracts of the ANH with the companies, and on oil prices.

The exploratory phase will not only prove the potential of each basin but will also inform the government on the specific environmental and social challenges associated with each basin, so that regulation can be tailored to specific local conditions within the regions, aiming towards a future play-based regulation.

The exploratory phase will also provide clarity on seismicity issues associated with water injection and potentially with hydraulic fracturing within highly complex geological mountainous regions. In addition, the flowback water composition will be revealed in each basin, so that the potential presence in the surface of hazardous contaminants and especially of naturally occurring radioactive materials within the basins, is adequately qualified and quantified. This is expected to provide a better understanding of the scale of the environmental risks associated with the management and disposal of this fluid.

Water use is a special concern in some regions in Colombia during low-flow seasons. This is the reason why flow-back recycling and disposal and treatment through aspersion in the soil should be conducted.

The increase in the drilling of injection and disposal wells is of key importance in order to dispose of flowback water safely; however once water is disposed of in an injection well it is taken out of the hydro-geological cycle, and thus building wastewater treatment plants to adequately dispose water through soil aspersion becomes an important preferred option.

A key challenge for the Colombian government is also to prevent the exploration and production activities of unconventional basins to become a political rather than a technical issue. Since September 2014, a few months after the regulations were published, concerned national opinion leaders have been opposed to the decision of the Colombian government to undertake the change embedded in these rules on hydraulic fracturing. This attitude seems to be based on misinformation and misunderstandings of the proposed rules. The Colombian government has now reacted by bringing back the academics and former regulatory experts to talk to the public in a series of academic forums, trying to raise the debate to the level of technical and scientific discussion.

It would appear that the greatest challenge Colombia faces in the next few years is that of regulatory enforcement. The ANLA and the ANH, the national authorities responsible for enforcing the environmental and technical requirements respectively, need to strengthen the resources allocated to evaluating and controlling unconventional exploration and production projects. Competence within the authorities on best practice, performance standards and the best available technology will be of key importance to encourage responsible activities and improve regulatory requirements.

References

ANH (2014a). Estadísticas de Producción. http://www.anh.gov.co/Operaciones-Regalias-y-Participaciones/Sistema-Integrado-de-Operaciones/Paginas/Estadisticas-de-Produccion.aspx.

ANH (2014b). El ABC de los Yacimientos No Convencionales (unpublished document).

ANH (2015). Mapa de Tierras. http://www.anh.gov.co/Asignacion-de-areas/Paginas/Mapa-de-tierras.aspx.

Gómez, C. (2014). Shale gas development in Latin America. A working paper by the Americas Society and Council of the Americas Energy Action Group.

UPME (2012). Supply and demand scenarios for hydrocarbons in Colombia.

Vargas, C. (2012). Evaluating the total yet to find hydrocarbon volume in Colombia. *Earth Sciences Research Journal* 16.

13

Regulation of Unconventional Gas in India

VIJAY KELKAR AND RAHOOL PANANDIKER

State of the Oil and Gas Sector in India

According to the Energy Information Administration of the US (EIA), India is currently the fourth largest consumer of energy globally after the United States, China and Russia. But, according to BP Energy Outlook 2035 (BP 2014), India is likely to be by far the largest source of growth in energy demand, surpassing China, by 2035. By then India's energy demand is expected to have grown by 132%, compared with growth rates of 71%, in China and Brazil, and 20% in Russia. As against this impending growth in demand, there is an already evident deficit in domestic supply and thus India depends upon significant imports to meet its energy demand. If the current state of affairs and consequent lack of clarity relating to the regulatory and fiscal environment continues, there is every indication of a worsening of India's supply–demand gap.

In the financial year 2013–2014 (FY13–14), India imported 76% of its crude oil, 36% of its gas and 23% of its coal requirements (India PNG Statistics, Ministry of Petroleum and Natural Gas 2014, and Annual Report, Ministry of Coal 2014). The import shares are expected to increase to 85%, 51% and 30% respectively by 2030. The increasing value of imports imposes a significant burden on the Indian economy. In FY13–14, India spent over $115 billion on fossil fuel imports.

The increasing dependence on imports has exposed the country's energy sector to volatile energy prices in international markets coupled with threats of changes in regulations and/or geo-political conditions. Given the high level of imports, this situation has considerable macroeconomic implications. It is central to India's push for energy independence and security. As a result, managing energy security and reducing long-term imports are key concerns faced by the country.

India today faces the daunting challenge of meeting its growing need for energy resources in a cost effective, sustainable and environmentally friendly manner. Though the contribution of hydro, nuclear and other renewable energy is expected to rise in the coming years, India's dependence on coal, oil and gas is likely to continue. In recent years, natural gas has increasingly become the preferred option globally, as it offers clean extraction and relatively low prices compared with expensive liquid fuel.

Low domestic supply and high Asian oil-linked liquefied natural gas (LNG) prices have not allowed natural gas to increase its share in the overall energy mix in India. India is currently facing a shortage of natural gas. According to 'Vision 2030' (Report on natural gas infrastructure in India submitted to the Petroleum and Natural Gas Regulatory Board, 2013) the demand–supply gap for natural gas was 98 million standard cubic metres per day (mmscmd) in FY13–14 and is expected to increase to 272 mmscmd by FY2030.

In order to reduce this huge supply–demand gap it is imperative for the Indian government to look for alternative sources of supply. One major unexplored resource that India has is unconventional hydrocarbon resources. India has significant shale gas, coal bed methane (CBM) and gas hydrates reserves besides its plentiful coal reserves, but the development and extraction of these resources will require a positive policy environment.

Developing Unconventional Fossil Fuels

Shale oil and gas development in the United States has already changed the dynamics of the energy sector globally and has had significant economic and geo-political repercussions in the global markets. Following their successful development in the US, unconventional oil and gas, primarily shale gas and oil, have emerged as a potential key supply source. There are many different unconventional sources that merit attention. In this section, we discuss the different potential unconventional sources in India and the estimation of potential reserves, as well as presenting a brief status report on development of the resources, including regulatory and policy measures, if any.

Coal Bed Methane

Coal bed methane (CBM) is natural gas stored in coal seams, generated during the process of coalification.

CBM Potential Reserves

India is endowed with abundant reserves of coal, estimated to be the fifth largest in the world (BP statistical review). Consequently, CBM has the potential to be developed as a key resource for natural gas with projected reserves of about 92 Tcf (Directorate General of Hydrocarbons (DGH) estimates, 2014, p. 2).

CBM Development: Current Status

In order to expedite the process of CBM development in India, the government formulated a policy for the exploration of CBM in 1997. The Ministry of Petroleum and Natural Gas (MOP&NG) became the administrative Ministry, and the Directorate General of Hydrocarbons (DGH) became the implementing agency for CBM policy. Up to 2014, the DGH had awarded 30 blocks in four CBM rounds and three blocks on a nomination basis covering 17 000 sq. km. Only three of these blocks are producing so far: namely Raniganj (South), Raniganj (East) and Jharia. Current production from CBM is around 0.3 mmscmd but production is expected to reach 4 mmscmd by 2016–2017 (Working group report on petroleum and natural gas sector for the 12th Five Year Plan, 2012–2017, 2014).

However, increasing production faces many challenges in India. Out of the awarded 33 blocks, 14 are up for relinquishment. A large number of issues hamper the development of CBM resources around the country, such as inferior coal quality, low production from wells, regulatory issues, land acquisition issues, 'environmental delays', a lack of gas pipeline infrastructure and the requirement for a large number of wells to be drilled at low cost.

Coal to Liquids or Gas (CTL/G)

Coal to liquids or gas is essentially a technology to convert coal into liquid fuel. The liquid fuels produced through this process are made suitable for transportation by the removal of carbon or the addition of hydrogen, either directly or indirectly. The CTL process presents an important opportunity to improve India's energy security.

The government allocated coal blocks to private operators for the development of CTL projects (*Business Standard*, 12 June 2013). Jindal Steel & Power was allocated Ramchandi coal blocks in Odisha, and coal block north of Arkhapal Srirampur was awarded to Strategic Energy Technology Systems Pvt Ltd (SETSPL) – a joint venture between a consortium of Tata companies and Sasol of South Africa. However, years after allocations were made, companies had not even obtained the necessary clearances. The recent de-allocation of all coal blocks by the Supreme Court of India has put the projects further into limbo (*The Hindu*, 24 September 2014).

As an illustrative exercise, assume that one CTL (coal gasification project) can lead to an average incremental production of 36 million barrels of oil equivalent (boe). Further, assume that four CTL projects can be established during the period 2020 to 2030, i.e. a project can be launched every 2.5 years starting in 2020. On this basis, an average annual incremental production of 70–80 million boe between 2020 and 2030 is estimated (Kelkar Committee Report 2013–14). A clear policy push is required from the government if CTL is to play any role in the future energy mix of India.

Underground Coal Gasification (UCG)

Underground coal gasification (UCG) refers to the process of converting coal, especially reserves below mineable range or in a thin seam, to gas through the injection of air and water into coal seams. The UCG process can capture up to 80% of the resource potential by converting inaccessible coal deposits into useable natural gas. To exploit such coal resources it would be necessary to carry out *in situ* gasification. While a few pilot or demonstration facilities have been built worldwide, India needs to set up its own units to demonstrate how the available technology can be adapted for Indian coal deposits.

India has approximately 350 billion tons of potential coal reserves of which only one third is mineable. According to some estimates, if 5% of India's non-mineable coal reserves were successfully utilised for gasification, it could yield three trillion cubic metres of gas

Table 13.1 *The different blocks offered for CBM and their status as of April 2014 (Ministry of Petroleum and Natural Gas)*

No.	Block	State	Area (in sq. km.)	CBM Round/ Nomination	Contractor	Remarks
I. Under development/production phase (eight blocks)						
1	Raniganj (South)	West Bengal	210	FIPB	GEECL (100)	Commercial production started since 2007. Current production: 350 000 m^3/day.
2	Raniganj (North)	West Bengal	350	Nomination basis	ONGC (74)- CIL (26)	Block under development since 2013. Under production.
3	Jharia	Jharkhand	85	Nomination basis	ONGC (74)- CIL (26)	Block under development since 2013. Under production.
4	RG(E)-CBM-2001/I	West Bengal	500	CBM Round-I	Essar Oil Ltd (100)	Currently producing around 140 000 m^3/day.
5	SP(E)-CBM-2001/I	Madhya Pradesh	495	CBM Round-I	RIL (100)	Block has entered development phase. CBM produced is utilised for internal consumption. Currently producing CBM @387 SCMD.
6	SP(W)-CBM- 2001/I	Madhya Pradesh	500	CBM Round-I	RIL (100)	Development phase is up to December 2014. Pipeline up to Phulpur is being laid.
7	BK-CBM-2001/I	Jharkhand	95	CBM Round-I	ONGC (80)- IOC (20)	Block under development since 2013.
8	NK-CBM-2001/I	Jharkhand	340	CBM Round-I	ONGC (80)-IOC (20)	Block under development since 2013
II. Blocks under exploration phase (six blocks)						
9	SH(N)-CBM-2003/II	Chhattisgarh	825	CBM RoundII	RIL (100)	MWP of Ph-I completed. Entered Ph-II but no work carried out owing to non-receipt of EC from CECB.
10	SP(N)-CBM-2005/III	Madhya Pradesh	609	CBM Round-III	REL (45)- RNRL (45)- Geopetrol (10)	MWP of Ph-I completed. Ph-I expired on 17 February 2014. Ph-I transfer and entering into Ph-II is in progress.
11	SR(N)-CBM-2005/III	Madhya Pradesh	330	CBM Round-III	DIL (90)- Coal Gas (10)	Ph-1 completed. Phase duration to be regularised.
12	RM-CBM-2005/III	Jharkhand	469	CBM Round-III	Arrow (35)- GAIL (35)- EIG (15)- TATA Power (15)	MWP could not be completed owing to law and order problem (Force Majeure).

13	AS-CBM-2008/IV	Assam	113	CBM Round-IV	Dart Energy (60)-OIL (40)	MWP is yet to be started. Ph-I is up to September 2014.
14	MG-CBM-2008/IV	Tamil Nadu	667	CBM Round-IV	GEECL (100)	Out of CBM area of 667 sq. km, 295 sq. km overlaps with PEL/PML area of ONGC. Contractor and ONGC to resolve the impasse. Contractor reiterated its stand to have unrestricted access.
III. Contracts not effective due to non-grant of PEL by State Government (five blocks)						
15	GV(N)-CBM-2005/III	Andhra Pradesh	386	CBM Round-III	Coal Gas (10)-DIL (40)-Adinath (50)	PEL not granted by AP government due to objection raised by the Singarani Coalaries Company Ltd (SCCL) concerning overlap of CBM block with mining lease held by SCCL.
16	RM(E)-CBM-2008/IV	Jharkhand	1128	CBM Round-IV	Essar Oil Ltd (100)	PEL order issued. PEL deed is yet to be signed.
17	TL-CBM-2008/IV	Odisha	557	CBM Round-IV	EOL (100)	PEL not granted by Odisha Government.
18	IB-CBM-2008/IV	Odisha	209	CBM Round-IV	EOL (100)	PEL not granted by Odisha Government.
19	SP(NE)-CBM-2008/IV	MP & Chhattisgarh	339	CBM Round-IV	EOL (100)	PEL order for MP granted, but PEL for Chhattisgarh is not granted.
IV. Blocks under relinquishment/ relinquished (14 blocks)						
20	BS(1)-CBM-2003/II	Rajasthan	1045	CBM Round-II	RIL (100)	Under relinquishment. MWP of Ph-I completed.
21	BS(2)-CBM-2003/II	Rajasthan	1020	CBM Round-II	RIL (100)	Under relinquishment. MWP of Ph-I completed.
22	SK-CBM-2003/II	Jharkhand	70	CBM Round-II	ONGC (100)	Under relinquishment owing to poor prospectivity. MWP of Ph-I completed.
23	NK(W)-CBM-2003/II	Jharkhand	267	CBM Round-II	ONGC (100)	Under relinquishment owing to poor prospectivity. MWP of Phase-I remained unfinished with two test wells.
24	ST-CBM-2003/II	Madhya Pradesh	714	CBM Round-II	ONGC (100)	Relinquished owing to poor prospectivity.

(*cont.*)

Table 13.1 *(cont.)*

No.	Block	State	Area (in sq. km.)	CBM Round/ Nomination	Contractor	Remarks
25	WD-CBM-2003/II	Maharashtra	503	CBM Round-II	ONGC (100)	Relinquished owing to poor prospectivity.
26	BS(3)-CBM- 2003/II	Rajasthan	790	CBM Round-II	ONGC (74)- CIL (26)	Relinquished owing to poor prospectivity.
27	BB-CBM-2005/III	West Bengal	248	CBM Round-III	British Petroleum (100)	Relinquished owing to poor prospectivity. Ph-I MWP of five test wells and two core holes not competed.
28	MR-CBM- 2005/III	Chhattisgarh	634	CBM Round-III	Arrow (35)- GAIL (35)- EIG (15)- TATA Power (15)	Contractor opted to relinquish without completing MWP owing to poor prospectivity.
29	TR-CBM-2005/III	Chhattisgarh	458	CBM Round-III	Arrow (35)- GAIL (35)- EIG (15)- TATA Power (15)	Relinquished owing to poor propspectivity. MWP of Ph-I completed.
30	BS(4)-CBM-2005/III	Rajasthan	1168	CBM Round-III	Geopetrol	MWP has not been initiated owinge to non-granting of Consent to operate/ Consent to establish.
31	BS(5)-CBM-2005/III	Rajasthan	739	CBM Round-III	Geopetrol	MWP has not been initiated owing to non-granting of Consent to operate/ Consent to establish.
32	KG(E)-CBM-2005/III	Andhra Pradesh	750	CBM Round-III	Geopetrol	Under relinquishment as Petroleum Exploration License (PEL) not granted by AP Government.
	ST-CBM-2008/IV	Madhya Pradesh	714	CBM Round-IV	Dart Energy (80)- TATA Power (20)	Under relinquishment after completing Ph-I.

equivalent (Implementing Clean Coal Technology Through Gasification And Liquefaction, 2013).

UCG Resource Estimation

The Oil & Natural Gas Corporation (ONGC) signed an Agreement of Collaboration (AOC) with Skochinsky Institute of Mining (SIM), Russia, for the implementation of an underground coal gasification (UCG) program in India (*Mining Weekly*, 26 February 2013). As a follow-up, memoranda of understanding were signed with various coal companies for accessing coal/lignite blocks for evaluating their suitability for UCG. After evaluating a number of coal/lignite blocks, Vastan Mine block, belonging to GIPCL in Surat district, Gujarat, was found suitable for UCG. This site has been taken up by ONGC as an R&D project to establish UCG technology. All the groundwork and inputs for pilot construction have been finalised for the implementation of a UCG pilot at Vastan. In parallel, other sites have been studied for their suitability for UCG. The ONGC and Neyveli Lignite Corporation Ltd (NLC) jointly identified Tarkeshwar in Gujarat and Hodu-Sindhari and East Kurla in Rajasthan. One more site that was also jointly identified by ONGC and Gujarat Mineral Development Corporation Ltd (GMDC) is Surkha in Bhavnagar district, Gujarat.

Shale Gas

Shale gas refers to natural gas that is trapped within shale formations. Shales are fine-grained sedimentary rocks that can be rich sources of petroleum and natural gas. Over the past decade, the combination of horizontal drilling and hydraulic fracturing has allowed access to large volumes of shale gas that were previously uneconomical to produce.

In less than a decade, shale gas has transformed the US natural gas sector, opening it up as potentially a major exporter to major importers.

Shale Gas: Resource Potential

Currently, there are few technical data available to make proper shale gas or oil estimates with any degree of confidence. However, varying, estimates of shale resource in India do exist.

The EIA, in its 2013 global assessment of shale gas reserves, estimated that India has 96 Tcf of technically recoverable shale gas reserves, which could meet India's estimated demand for 26 years (EIA-Technically Recoverable Shale Oil and Shale Gas Resources, 2014).

Technically reliable estimation of shale gas or oil resources as well as the nature of these resources should be one of the priority areas for the Indian Government before contemplating the formation of a policy framework for its exploration and extraction. The geological conditions in India are very different from those in the US, so our Government needs to have a comprehensive understanding of requirements and associated costs before creating its shale gas policy.

Key prospective basins for shale gas exploration in India are the Cambay Basin in Gujarat, the Assam-Arakan Basin in northeast India, the Krishna-Godavari, Cauvery, Vindhyan and Gondwana basins in Central India.

Policy Framework

The delay in coming out with a clear policy on shale gas exploration has clearly dampened the interest of investors in the exploration of the resource. However, in October 2014, the Government notified policy guidelines for the exploration and exploitation of shale gas and oil by national oil companies (NOCs) in their onshore Petroleum Exploration License (PEL) or Petroleum Mining Lease (PML) block. The policy marks a first step and covers only the onshore acreage that was awarded on a nomination basis to the state-run explorers ONGC and Oil India Ltd. Royalties and taxes remain the same as for the conventional oil and gas in those blocks.

However, private operators are still awaiting a policy framework that provides them with an opportunity to explore the shale resources. As of now, there is a non-level playing field, which reduces the attractiveness of the sector for private players.

Current Status

The Indian national oil company Oil & Natural Gas Corporation (ONGC) drilled the first shale gas well in the country in October 2013, in the Cambay basin, which is expected to have a shale gas potential of 20 Tcf. The company plans to explore 30 additional shale gas wells over the next two years with committed investments of INR 600 crore (approximately USD 89 million) for the project (*Mint*, 26 November 2013).

The ONGC has also signed a memorandum of understanding with Conoco-Phillips to explore for shale gas. The two have undertaken joint studies in Cambay, Krishna Godavari, Cauvery and Damodar basins, on the basis of which a drilling programme was agreed in the Cambay Basin area in Gujarat.

Shale Oil

Shale source rock could be rich in natural gas or oil. The oil produced from shale formations is called shale oil. Eagle Ford Shale and Bakken shale in the US are examples of shale formations that are rich in oil. Production in Eagle Ford shale oil increased from 0.1 million barrels per day (bpd) in 2011 to almost 1 million bpd in 2014.

Shale Oil: Reserve Estimates

According to the 2014 EIA report on resource assessment of shale gas and shale oil, India has 87 billion barrels of risked shale oil resources in place. The risked technically recoverable shale oil resources are estimated to be 3.8 billion barrels.

Gas Hydrates

Gas hydrate is a crystalline solid consisting of a gas molecule surrounded by a cage of water molecules. In India, gas hydrate research is being steered by the Ministry of Petroleum and Natural Gas under the NGHP (National Gas Hydrate Programme), initiated in 1997. The NGHP draws participation from the Directorate General of Hydrocarbons (DGH), NOCs and national research institutes.

Gas Hydrates: Resource Potential

Earlier studies have foreshadowed that there might be gas hydrate resources of 1894 trillion cubic metres (Tcm) in India. In February 2012, the US Department of Energy (USDOE) announced that around 933 trillion cubic feet (Tcf) is the concentration of gas hydrate in sands within the gas hydrate stability zone (Indian Minerals Yearbook, 2012). This estimate is encouraging, though the estimated amounte of sand present is an approximation based on gross geological depositional models.

Under the National Gas Hydrate Programme (NGHP), the National Institute of Oceanography (NIO) completed a study around the site NGHP-01–10, where ~128 m thick gas hydrate has been recovered. The studies indicate a resource estimate of ~16.5 million cubic metres from a gas-hydrate-bearing sediment over an area of 0.98 km^2.

Based on the findings of NGHP Expedition-01, the Krishna Godavari deep water basin and the Mahanadi deep waters are considered potential areas where large tracts of turbidity sand channel systems can be expected in the delta sequence accumulations. Based on the geophysical studies carried out so far, in over 5000 km^2 in the Krishna Godavari and Mahanadi offshore deep water areas, more than 50 sites have been identified for NGHP Expedition-02. These locations are being prioritised in consultation with NGHP and US scientists. Geo-scientific studies are continuing to identify more locations in the area.

Gas Hydrate: Current Status

The exploitation of methane from gas hydrates is still at a research stage globally. Various factors, such as the characteristic nature of gas hydrates, their dissociation and stability and environmental factors have to be thoroughly understood before pilot production testing for the extraction of methane is contemplated. The NGHP has taken several initiatives in this direction, on the basis of the global R&D trends.

Existing Regulations and Other Policy Measures in Place

As discussed above, in 1997 the Indian Government formulated a CBM policy covering the exploration and production of CBM in India and started awarding CBM blocks through international competitive bidding (ICB) rounds in 2001 (Reliance Power Website, 2014). However, the growth in production has been very slow. Regulatory issues and pricing uncertainty have impacted the growth of CBM in India. For example, Arrow Energy bought two CBM blocks in India and was itself later acquired in a 50 : 50 joint venture of Shell and PetroChina. The company's international assets, including those in India, were hived off into DART Energy, a new entity. DART has sought to amend its production sharing contract to reflect the new name but is yet to get clearance. As mentioned above, a key issue impacting growth is prices. Initially, the government delayed the fixing of prices for CBM, which led to significant delays in projects. However, the Government has now decided on a formula for domestic natural gas prices based on the average of specific global benchmarks (taking into consideration prices at Henry Hub, the National Balancing Point (Europe) and the Russian and Canadian gas hubs), which will be revised every six months. The gas price was increased to $5.61 per million British thermal units (mmbtu) from the earlier $4.2 per mmbtu from November 2014. There is also the possibility to add a premium over and above this estimate, especially for gas from more complex fields. Whether this price will be sufficient to evince interest in the sector (especially when CBM players were selling gas in the range of US$5.1 to US$22.8/unit) still remains to be seen.

In the case of shale gas, the Government has made public a draft regulation, which is still under consideration. The draft policy suggests a new revenue-sharing model with no cost recovery. The companies will be asked to quote the percentage of output that they are willing to share with the Government at different production slabs. The main motive is to minimize government intervention and eliminate the current issues in accounting and the "supposed" incentive for gold plating. The proposed policy for shale oil and gas is along lines that already exist in CBM contracts, where the government gets royalty- and production-linked payment (PLP). *Ad valorem* royalties (taxes based on value) at the prevailing rate for crude oil and natural gas would be applicable to shale oil and gas respectively and would accrue to the State Governments, whereas the production-linked payment on an *ad valorem* basis will be made to the Central Government.

The Government is currently considering the creation of a unified licensing policy (ULP), aiming to unify and simplify the policies for oil and gas, coal bed methane and shale. The policy aims at ending the current 'cost recovery', which has been a main issue of contention.

The key features of the proposed policy are:

- Revenues from the awarded block will have to be shared from the first day of production; in the current system, expenditure is recovered before profits are shared with the Government.
- No approval required to drill wells
  The oil and gas companies will not need approval to drill if they find oil and gas. The policy eliminates a long-winded procedure in which the Directorate General of Hydrocarbons had to ratify discovery,
- Minimise government interference
  The policy limits the role of bureaucrats in managing the fields.

Roadmap for the Development of a Regulatory and Policy Framework for the Exploitation of Unconventional Resources in India

Despite the fact that unconventional oil and gas resources can play an important role in India's energy mix, there is no cohesive existing framework for the exploration and production of such resources in India. We propose a framework and initiatives that we believe are critical to the fruitful development of the sector. Such a framework requires:

1. a commensurate fiscal regime that balances risks and rewards, backed by transparent administrative and regulatory policies with processes and practices for operational efficiency;
2. a move towards market-determined pricing;
3. an institutional framework that builds confidence in the investor community;
4. the creation of market conditions to incentivise the exploitation of unconventional resources:
5. environmental preparedness;
6. development of technical capabilities.

We recommend some key levers towards instituting a policy and regulatory framework to exploit the potential of unconventional resources.

Fiscal and Regulatory Issues

1. The main driver for the development of unconventional resources in India will be the creation of a stable and investor-friendly policy and regulatory regime with clear roles for governments, regulators and other stakeholders.
2. The government must adopt a fiscal regime that responds to the risk–reward paradigm of the exploitation of unconventional resources. It is critical to establish a risk sharing contract that balances the risks of unconventional exploitation and production. A production sharing agreement is one such regime, which would align the interests of the state and the operators. To be noted is the fact that the 'biddable' revenue sharing contract (RSC) model has an inherently misaligned risk–return structure, which leads either to lower levels of production, owing to the resulting reduced exploration efforts, or to lower recovery ratios to high windfall gains to operators, encouraging contract instability due to political economy factors. It is thus no surprise that a 'biddable' revenue sharing model does not find much favour across the world.
3. Most importantly, a revenue sharing regime, as has been proposed by the government, is cost insensitive and in a country such as India – where cost escalation could also occur on account of regulatory delays as well as technical, logistic and market factors – a pure revenue sharing regime which does not factor in costs would be too rigid a structure to attract investors.
4. To improve investor confidence, it is necessary that contract stability and sanctity is maintained and retrospective changes to contracts are avoided.

5. Given the long-term nature of the sector, it is necessary to have contracts based on the economic life of the resource base rather than arbitrary tenures.
6. Some fiscal reforms could also be taken up to provide additional incentive.
   a. Including oil and gas under the proposed goods and services tax (GST) framework would facilitate the development of markets by simplifying and standardising taxation norms and ensuring similar prices of oil and natural gas across the country. Even though the government expects to introduce GST in the fiscal year 2016–2017, the inclusion of petroleum and natural gas in the structure still seems a distant dream. The reason is that the Central Government would need the concurrence of 20 States out of the 29 to impose GST on petroleum. The Constitutional Amendment Bill, tabled in the Lok Sabha, seeks to bring petroleum under GST but gives States the power to impose value-added tax (VAT), while the Central Government will levy excise duty on it. This was the middle path since States did not want to impose GST on petroleum as they earn a huge amount from it.
   b. Extending the definition of 'mineral oil', as used in Oilfield Regulation and Development (ORD) Act 1948, to the Income Tax Act, 1961.

Move Towards Market Based Pricing

7. One of the main issues impacting the growth of domestic natural gas is the low domestic prices paid to producers, which in turn affects production. Multiple regimes, akin to Administered Pricing Regimes, introduce an element of arbitrariness to the setting of prices thus creating a sense of instability in estimating long-term price movements, which is critical for investment decisions.
8. A rational and fair pricing policy for gas, an exhaustible resource, is vital in recognition of the principle of intergenerational equity. As discussed in the Kelkar Committee Report, the principle implies that the natural resource should be priced at the highest price possible in the market, i.e. the pricing should be market-determined. A gas molecule consumed by the current generation is, in effect, denied to future generations. In other words, the present generation is essentially borrowing these resources from their children and grandchildren, and equity requires that future generations be fairly compensated. At the very minimum, the resource price cannot be less than the maximum opportunity value of that resource. Hence, the fair price can only be the best price gas can command either in the domestic or international market, i.e. the price that is market-determined in an environment where exchange is conducted in a transparent manner in an arm's length basis. So, one of the most important incentives will be a complete deregulation of prices, allowing buyers and sellers to negotiate prices at arm's length. Producer prices for natural gas should be unfettered of any government intervention, allowing for gas pricing by producers on a market-determined basis through transparent arm's length transactions. Market-determined gas pricing should apply to all forms of gas (conventional and unconventional) irrespective of the source. The decision for a transition to market-determined producer prices should be taken and communicated by the government at

the earliest opportunity, which, given the long-term planning required for exploration and production (E&P) activity, will enable operators to plan their investments upfront.

Institutional Framework

9. It is necessary to clarify the role and responsibilities of the government and the regulatory bodies in order to ensure a smooth adoption of policy as well as administration and management of the contracts with the operators.
10. It is recommended that the Directorate General of Hydrocarbons (DGH) play the role of regulator, taking on the fiduciary responsibilities of ensuring the optimal exploitation of resources and leaving the fiscal and treasury aspects to other entities. At present, the multiple roles of government as policy maker, regulator and operator lead to conflicts of interest and dampen investor confidence in the sector. This creates the need for an independent and transparent regulatory mechanism, so the DGH should move from its current role of being an advisor to being an independent regulator for the upstream oil and gas sector.
11. To ensure that the DGH can operate effectively, it should be empowered with an independent financing and staffing mechanism. The funds required for day to day operations must be made available automatically, on a formulaic basis, through the Oil Industry Development (OID) process. The DGH should be established as a multi-member, multidisciplinary body with professional teams that have expertise in legal, environmental, financial and technical and other domains. The DGH should further have the flexibility in its charter to access global experts.
12. **Expediting the approvals process:** A large number of E&P projects in India get delayed owing to delays in different regulatory clearances, such as land and environment. The process of attaining approvals and clearances should be streamlined. Also, approvals not received within the set timeframe should be deemed obtained. For the statutory clearances required on blocks given out for exploration – such as clearances from the environment or defense ministries – the best practice is to obtain these clearances before the acreages are put on offer. This may delay the process of offer but would enable the Government to put its own house in order before 'trapping' investors into blocks which they are then unable to work. It is important that a timely and transparent system is established in order to speed up the approvals process, so that the development of unconventional resources is not hampered by these issues.

Developing Market Conditions

13. Given the nascent and controlled nature of the gas market in India, it will be extremely challenging to attract sufficient private sector interest in the development of gas infrastructure. The onerous task of building a gas pipeline infrastructure should therefore be taken up by the government though a public sector enterprise (PSE) that has experience in constructing and managing gas pipelines. The PSE should be given responsibility for

developing the plan or blueprint, securing funding and constructing the infrastructure by getting the appropriate partners on board and operating the pipelines.

Environmental Preparedness

14. **Strong environmental norms and procedures:** Unconventional oil and gas developments also require strict environmental laws to be in place, so that the development does not harm the environment. India can use the studies under way globally to access information about potential harm and formulate appropriate polices. Production from shale involves the fracking of rocks and also uses a lot of water. With India already facing a water crisis, it is important to promote the optimal usage of water. Given the controversy surrounding the exploitation of unconventional resources, it is critical to embark on a campaign to educate the polity on the costs and benefits of this important venture. This will go a long way in ensuring that post-award issues are minimised.

Developing the Technical Capabilities

15. **Investment in R&D:** The development of unconventional resources, especially shale gas, is currently at a nascent stage. For rapid development, it is important to prove that the resources exist and can be recovered at an economically viable cost. Government's investment in finding innovative techniques to extract shale gas from Indian basins in a cost effective manner will give a major boost to the sector. Once resources and viability are proven, investments will automatically follow given the huge potential of the Indian market.
16. Create 'Missions' of national importance for specific unconventional resources such as gas hydrates. The recent success of the 'Mars Mission' is an example to be emulated.
17. It is necessary to facilitate the NOC's pursuit of technology alliances with companies and institutions, both nationally and internationally, that have already developed intellectual property in the exploitation of unconventional resources, in order to expedite the application of new technologies and expertise. Indian companies, especially the NOCs, should be encouraged to form joint ventures and alliances that leverage existing global technology expertise to optimally explore unconventional resources.

Concluding Remarks

India has significant amounts of unconventional resources but requires the right policy direction to create any material impact. The industry is still at a very nascent stage and requires clarity and promotion from the Indian Government.

India is among the largest producers of coal in the world and accounts for one of the largest coal reserves. Coal gasification can provide an important alternative non-conventional source of oil and gas to meet our growing energy needs. Initiatives that can encourage the sustainable exploitation of non-conventional sources of energy, by the development of a suitable contract model, policies and administrative support for coal mining, by streamlining the approval and clearance processes and by a transition to market-determined pricing of natural gas, would go a long way in increasing hydrocarbon production domestically. An average annual incremental production of 70–80 million boe between 2020 and 2030 is estimated.[18] Estimates for the entire gamut of potential production from unconventional petroleum sources could be more than twice this number. This could translate to savings in the import bill of USD 25–30 billion annually, thus reducing import dependence by approximately 20%–25% of net imports of oil and gas.

Additionally, it is imperative that the Indian Government and companies (NOCs) start to build their technical capabilities in the sector. Technology alliances – R&D joint ventures with global players having exposure to unconventional resources – may help in expediting the extraction process. Some Indian companies, such as Reliance and GAIL, have already invested in shale gas assets in the US and may look to use that expertise in the Indian basins.

Unconventional resources could come to be a significant part of India's energy mix in the long term. Putting together the right regulatory and policy framework will go a long way in developing these resources.

References

BP Energy outlook, 2014. http://www.bp.com/content/dam/bp/pdf/Energy-economics/Energy-Outlook/Energy_Outlook_2035_booklet.pdf (Accessed in August 2014)

EIA, 2014. Technically recoverable shale oil and shale gas resources: an assessment of 137 shale formations in 41 countries outside the United States, http://www.eia.gov/analysis/studies/worldshalegas/pdf/chaptersxx_xxvi.pdf (Accessed in October 2014)

Implementing clean coal technology through gasification and liquefaction – the Indian perspective, 2014, http://sphinxsai.com/2013/conf/PDFS%20ICGSEE%202013/CT=43(824–830)ICGSEE.pdf (Accessed in September 2014)

India seeks collaboration with SA on coal gasification technology, 2013. *Mining Weekly*, 26 February 2013 (Accessed in September 2014)

Indian Minerals Yearbook, 2012, http://ibm.nic.in/IMYB_2012_Petroleum%20and%20Natural%20Gas.pdf (Accessed in October 2014)

Indian PNG Statistics – Ministry of Petroleum and Natural Gas and Annual Report – Ministry of Coal, 2014. (Accessed in August 2014)

Odisha holds up processing of CTL coal blocks, 2014, *Business Standard*, June 12, 2013 (Accessed in August 2014)

ONGC begins shale gas exploration in India, 2014. *Mint* , 26 November, 2013 (Accessed in October 2014)

Reliance Power, 2014. http://www.reliancepower.co.in/business_areas/fuel_business/coal_bed_methane.htm (Accessed in October 2014)

Report of the Committee on the roadmap for reduction in import dependency in the hydrocarbon sector by 2030, parts I and II, Kelkar Committee, Ministry of Petroleum and Natural Gas, New Delhi, 2014. (Accessed in September 2014)

Report of the working group on petroleum and natural gas sector for the 12th Five Year Plan (2012–2017), 2014. http://petroleum.nic.in/docs/reports/wgreport.pdf (Accessed in August 2014)

Status of CBM blocks awarded to PSUs and private/JV companies, 2014. petroleum.nic.in/docs/cbmblocks.pdf (Accessed in August 2014)

Supreme Court quashes allocation of 214 coal blocks, 2014 *The Hindu*, 24 September. 2014 (Accessed in September 2014) 'Vision 2030', 2014. Natural gas infrastructure in India: http://www.pngrb.gov.in/newsite/Hindi-Website/pdf/vision-NGPV-2030–06092013.pdf (Accessed in August 2014)

14

Failure to Frack: Pitfalls of Governance and Risk in Polish Shale Gas

MICHAEL CARNEGIE LABELLE

Introduction

The shale gas revolution, which reshaped America's energy landscape, built up expectations in resource-rich Poland. On the basis of earlier assessments, Poland's geology was perceived to be rich in shale gas and oil. Initially high estimates were given credence when the Polish Geological Institute published a report in 2012 estimating shale gas deposits at 1920 bcm with recoverable shale gas at 346–768 bcm. The report stated, 'These resources are therefore 2.5 to 5.5 times higher than documented conventional gas fields in Poland (145 bcm)', which is equivalent to 35–65 years of cumulative domestic gas consumption (Polish Geological Institute, 2012). Politically, the game was on before the report was even finished. In the footsteps of the US, it seemed that shale gas offered to boost Poland's energy security and competitiveness and to offset both imported gas and domestic coal as current energy sources. By 2015, however, after a number of companies pulled out from Poland, these estimates, or at least the ability to access the trapped gas, proved elusive.

The associated above-ground political and commercial activities surpassed any underground activities, despite the move of oil and gas companies to explore in Poland. This chapter focuses on these above-ground activities and the surrounding environmental and investment risks noted by the established governance structure in Poland. It does not assess the direct environmental effect posed by hydraulic fracturing but assesses, by primary and secondary sources, how companies and state institutions implement procedures that ultimately translate into protection of the environment. Examined are the governance structures overseeing and facilitating shale gas development and the risk assessment and reactions to these of companies and government units. Ultimately, these activities led to decision-making on deploying more investment cash and equipment into Poland. The findings indicate that sustained investments into shale gas exploration and extraction failed because of the geological and administrative complexity and the overall short- and long-term costs associated with exploration in Poland (see Goldthau and LaBelle, forthcoming, for a comparative study of the acceptance and rejection of shale gas in Bulgaria and Poland). Geology was only one element preventing extensive exploration.

The governance of shale gas development in Poland rests on a strong division between state authorities; this contributes to the complexity of introducing hydraulic fracturing technologies to extract shale gas. There are five state ministries involved in overseeing shale gas developments and they each attempt to adhere to a political mandate to develop the sector. At the same time, they each maintain their professional competencies. This chapter will define and explore 'governance' as a means to explain the interactions of state agencies and ministries. The specialization of each state entity leads to conflicting or confusing administrative procedures for private companies attempting to explore for shale gas.

Risk mitigation is central to the shale gas story in Poland. On the basis of the US experience, significant caution is exercised by state agencies in reviewing applications for shale gas exploration. Risk assessments are also made by companies looking to invest large sums of money into exploration and possible extraction. In this chapter two different types of risks are recognized, 'commercial' and 'governance'; they apply to the rolling out of new energy technologies and reflect both the commercial side of contracts and the policy environment (LaBelle, 2012). Joining together an assessment of the governance system and of the risks present in Poland provides a novel, but structured, type of account of the country's attempt to create a domestic shale gas industry. As will be described, in spite of failure to launch in the short term the door may still be open for long-term development of the industry.

This chapter first frames a discussion of the governance of an emerging shale gas industry and the associated risks for both state and private actors. The deployment of new energy technologies into a market is discussed as a way to understand the engrained difficulty that state institutions and private actors face when using a previously unavailable technology in a country. This risk–governance framework provides a background to understand the turbulent attempt in Poland to roll out a sufficiently large shale gas industry, in a country with an active hydrocarbon sector. A description is provided in the third section of Poland's efforts to encourage and assist shale gas exploration. By early 2015 these efforts were clearly not enough, as many oil and gas companies withdrew from Poland. The final section will address why the country's earlier attempts failed to encourage sufficient levels of activity to build a foreign-led shale gas revolution.

Risk–Governance Framework

A risk–governance framework (LaBelle, 2012) was originally developed to assess the transition to a low-carbon EU energy sector. This provides the risk–governance categories necessary to highlight the interaction of private companies with state institutional efforts to roll out new energy technologies. There are two categories, 'commercial risks' and 'governance risks' (Figure 14.1). Commercial risks concern factors influencing short-term decisions about whether a project should proceed. Generally, these are uncontrolled risks, which investors examine to determine pricing and rates of return on investments. Some of these risks can be hedged, thereby mitigating them, while others, such as the 'environmental legal compliance risk', call for a significant amount of upfront work for their mitigation, although the risk always remains. Investor perceptions are balanced between financial returns and risks (Wüstenhagen and Menichetti, 2012). The second category, governance

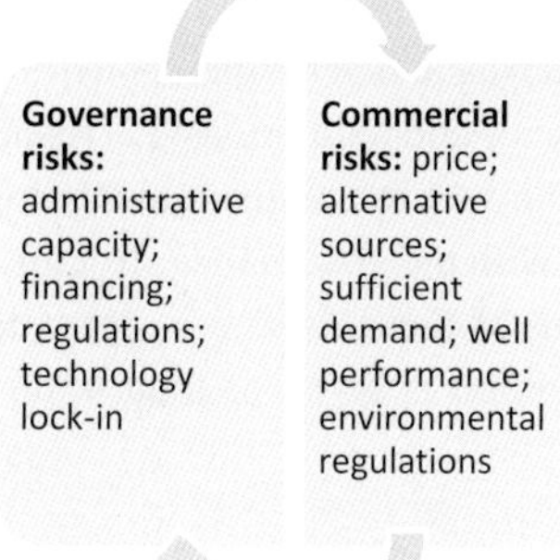

Figure 14.1 Governance and commercial risk categories.

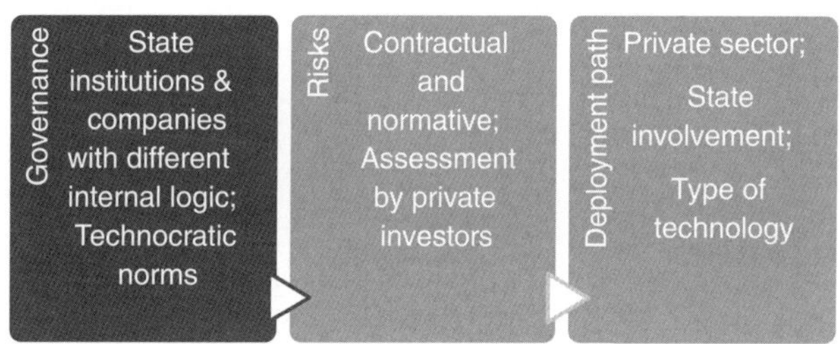

Figure 14.2 Deployment process for new technology.

risks, draws from longer-term policies and institutional (re)actions impacting investments and the ability of a technology to scale up. Often, administrative capacity and state policies are focused on supporting older technologies or strategies, thereby supporting engrained rules and procedures and conflicting with more rapid advances in technologies (Unruh, 2002).

Governance and the level of risk inform the ability of a country to foster innovation in energy technologies. Figure 14.2 combines the two conceptual frameworks. First, 'governance' represents deep-seated national socio-economic and political conditions conforming to a particular energy system – thereby influencing the speed, style and manner of entrance of any new energy technology. Second, risks associated with a country determine how and when a technological development path emerges. These are placed within two categories of *governance risks* and *commercial risks,* which hold deeper normative beliefs and reflections of measurable contractual risks. Third, a technology's *deployment path* is shaped by governance methods, risks and policies that determine whether push or pull incentives are used to commercialize a product (e.g. subsidies or tax breaks). New energy technologies, whether carbon or non-carbon based, challenge existing technological regimes; therefore, for both renewable energy technologies (RETs) and carbon-based energy technologies

(CETs), overcoming institutional and social obstacles is the first task in gaining market entry.

There are three phases to fostering and permanently deploying new energy technologies: (1) governance, (re)shaping policies and regulations; (2) risk perceptions influencing private and public actors willingness to invest in the given technology; and (3) deployment of the technology and how it interacts with the governance structure and between different actors. To create a better understanding of these three concepts, definitions are briefly provided and later a more detailed discussion is given of their application to the Polish shale gas story.

Governance

First, *governance* is defined along the lines of self-organization, involving state institutions, local, regional and national . . . '(such as economic, political, legal, scientific, or educational systems), each of which has its own complex operational logic such that it is impossible to exercise effective overall control of its development from outside that system' (Jessop, 1998, p. 30). The reason for choosing this definition is that the components of the state, such as political and legal entities, may be unable to coordinate their actions fully – even if attempts have been made – and inserting new technologies requires the development of new regulations, laws, taxes, etc. Consequently the effort becomes fraught with obstacles within each state, for private and social systems alike.

Governance proceeds along sectoral lines (Bulmer, 2007): expertise is deployed in specific policy or technical realms. This can also include the separation of responsibilities and actions, along a vertical axis, between state agencies and ministries, such as environmental protection (agency) and economic development (ministry), and horizontal separations between state agencies (concession licensing and environmental protection) (Christensen and Lægreid, 2013, p. 362). The division of tasks creates 'organizational complexity' (Christensen and Lægreid, 2013, p. 362) and important questions for the state that must be resolved if the deployment of a technology is to be successful.

It is through the interaction of groups of actors that a particular technology gets final approval. The approval in essence 'emerges from the depths of the horizontal and vertical organizational units of the state', finally to see the light of day. As will be described below, the administrative delays in Poland lend credence to the idea of the 'depth and breadth of the state apparatus'. The governance of shale gas and its connected hydraulic fracturing techniques are now highly controversial, leading to significant precautions and time delays within all governance units (see *Action Coalition Framework* by Sabatier and Jenkins (1988), for further considerations on policy making in controversial areas).

Risk

Second, perceived systemic *risks* challenge established procedures engrained in state and private actors. Notably, risk perceptions influence energy investment decisions (e.g. Unruh, 2000, 2002; Wiser *et al.*, 2004; Woerdman, 2004; Bekkers and Thaens, 2005;

Jamison *et al.*, 2005; Wiser and Bolinger, 2006; Hoffmann *et al.*, 2009). Defining and identifying risks and mitigation strategies in the energy sector influences the course of action (Engau and Hoffmann, 2009; Hoffmann *et al.*, 2008; LaBelle, 2012; Tsoutsos and Stamboulis, 2005; Unger, 2010; Walsh, 2012). For our purposes here, these risks may be split between institutional (Levi-Faur, 2009; Levi-Faur and Jordana, 2005; Unruh, 2002, 2000) and normative (Hoffmann *et al.*, 2009). For the energy sector, the portfolio management of generation assets and network design are central in risk management (Wüstenhagen and Menichetti, 2012), at the most basic level to ensure security of supply. Different risks apply depending on the perspectives of actors and chosen technologies. As the literature above demonstrates, the interaction – over time – of risks, governance practices, perspectives and technologies drive forward investment and the renewal of energy systems.

Long-term investments into new technologies require stability and predictability; R&D efforts emerge in an environment where technologies are commercialized (Burer and Wustenhagen, 2009; Walsh, 2012; Weyant, 2011). An unpredictable environment does not exclude commercialization, but influences the timing and scale of deployment within a country. Normative influences, such as political and social demands, push forward investments and can demonstrate the deployment potential of energy technologies (such as solar photovoltaic). But, for new technologies to become self-perpetuating and embedded in a national energy system, deeper structural changes, to, for example, regulations or financial incentives, are necessary. These may not occur in markets marked by instability and shifting political and economic perspectives. Risk perception can hold back the significant deployment of a technology.

Deployment Path

The deployment of new energy technologies must address existing technologies and governance systems previously put in place to operate a national energy system. As Christensen and Lægreid (2013, p. 362) state, there is 'organizational complexity' within the division of administrative tasks. Complexity also marks the wider social and commercial environment in which technologies necessarily exist. Therefore, the deployment of energy technologies must also address a wider environment of state–technology–social relations. A better understanding of this interaction is seen through the concept of the techno-institutional complex (TIC). The TIC provides an explanation of the 'complex systems of technologies embedded in a powerful conditioning social context of public and private institutions' (Unruh, 2000, p. 818). The horizontal and vertical division of the state administration (described above) underpins the TIC and the protection of specialization within existing administrative processes. 'TIC[s] develop through a path-dependent, co-evolutionary process involving positive feedbacks among technological infrastructures and the organizations and institutions that create, diffuse and employ them' (Unruh, 2000, p. 818).

The deployment of a new technology relies on market and government push–pull factors. Government or social acceptance is key to scaling up. Technological familiarity and the

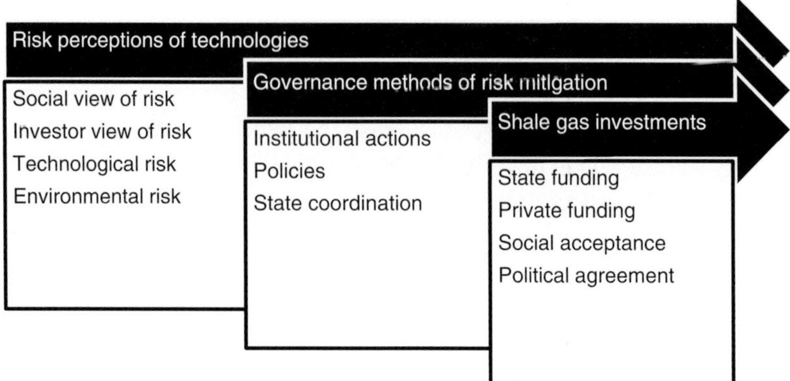

Figure 14.3 Diagram of innovative flows from state institutions to shale gas investments.

common practices of society, business actors and state institutions exert a strong influence in determining success (see LaBelle and Goldthau, 2013). The two general methods for deployment are: (a) tax incentives or regulation *pushing* a technology into the marketplace; conversely, (b) consumers can *pull* out a technology, based on their own requirements, from the marketplace (Walsh, 2012). The general approach discussed in this chapter is a top-down push policy environment, where a government seeks to roll out a particular technology.

Summarizing the three phases involved in rolling out new energy technologies (governance, risk perceptions, and deployment) rests on building an understanding of how path dependency perpetuates administrative actions, shaping interactions with the owners of new technologies. Deployment is dependent on the risk perceptions of owners and on the policy environment created by the complex interactions forming the governance system.

Figure 14.3 outlines the flow of innovative technologies from broad governance indicators to administrative support. The diagram moves from risk perceptions and governance methods to the 'social context of public and private institutions' (Unruh, 2000, p. 818) and the push–pull policies utilized for the deployment of selected energy technologies. Each energy technology must overcome institutionalized and socially entrenched barriers that maintain traditional energy technologies.

Methodology

As mentioned carlier, the methodology of the study comprises qualitative methods of interviewing stakeholders and reviewing secondary sources. These are presented as personal communications (Goldthau and LaBelle, forthcoming). The diverse range of stakeholders interviewed, and the consistent – and similar – messages received from participants indicate that a sufficient sample size was achieved. Interviewees were selected on the basise on their involvement in five categories; this enabled a diverse set of views to be heard while the central concept, of deploying fracking technology in Poland, was always kept in mind. There were two rounds of interviews, conducted in Poland in September 2012 and May 2013; further clarifications were sought in 2015. A total of 27 interviews were carried out. The list of interviewees can be broken down into five categories: government officials or executives (6), company officials (7), academic and legal personnel (5), an elected parliamentary official (1) and representatives of non-governmental organizations (8). Twenty-two of those interviewed were seen as representing a national level of interest and five a local or regional level, the latter being interviewed in the shale gas region of Lublin. The interviews were semi-structured and the interviewees represented companies, organizations and individuals who were active or otherwise concerned with the Polish shale gas sector. Included in the list were officials from foreign affairs, environmental, industry and social groups.

Polish Shale Gas: Governance of Shale Gas Exploration

'We want to have renewable energy sources, but hard coal and lignite – and soon shale gas – will remain our principal energy sources. That's where the future of the energy sector lies', Tusk [Poland's Prime Minister] told reporters. '. . . we've decided that renewable energy sources, which are an important complement to Poland's energy sector, will be limited as much as EU rules will allow', he said (PHYS. ORG, 2013).

Poland has put its weight behind fossil fuels, including shale gas. 'Shale gas offers benefits for economic development of country. Oil and gas sector, chemical industry, metallurgy, tourism . . . all will benefit from shale gas' (Foreign Ministry Official A, personal communication, 2012). The power sector, as noted, is very dependent on coal. Only three per cent of Poland's electricity is produced by gas (Polish Information and Foreign Investment Agency, 2013). Poland plans to build a nuclear power plant to provide lower carbon emissions. However, construction has not begun. In the European Union, Poland has vetoed the EU Low Carbon Roadmap 2050 and also the EU Energy Roadmap 2050, blocking low carbon goals. Renewable energy makes up seven per cent of Poland's energy mix but, as the Prime Minister of Poland has stated, this will be kept to a minimum (PHYS. ORG, 2013). The renewal of coal fired power plants is a priority, as 40 per cent of generation plants are over 40 years old, with 15 per cent over 50 years old (Polish Information and Foreign Investment Agency, 2013, p. 7).

Poland places a priority on revitalizing and providing continuity in its energy infrastructure. This section of the chapter will analyse government efforts to foster a shale gas industry. It will do this by assessing key factors that interviewees identified as essential to developing the shale gas industry. These include: (a) a consideration of the environmental risks; (b) investment incentives; (c) business strategies; (d) R&D cooperation between firms; and (e) market preparation. A discussion of these key factors is followed by a review of how interviewees identified both the short- and long-term risks in Poland and how these may prevent, or even encourage, shale gas extraction. Assessing these influential factors provides a structured approach to describing Poland's efforts and provides a method of critically analysing, through the risk–governance framework, why Poland has faltered in boosting shale gas exploration and production.

State Organization

The history of interaction between the private sector and the state in the energy sector is informative for describing the experiences on both sides. In the energy sector, this can best be seen in the privatization of energy companies, which occurred in waves. In 1990, 33 distribution companies were formed out of one vertically integrated company, 14 of these being partially privatized by 2004 (Radzka, 2006). However, the Polish Treasury still maintains a significant ownership of shares in these energy companies, leading one Polish economist to state that 'this is not privatization but rather state ownership with private management controlled by the Polish state' (The Economist, 2011). This leaves foreign investment mostly confined to the generation sector, as state ownership is maintained in distribution and transmission, thus maintaining a state monopoly on many energy services. Importantly, this translates to limited interaction with private investors for state administrators.

In the area of shale gas, the Polish government also maintains a strong presence. There are five state ministries involved: Economics, Environment, Finance, Foreign Affairs and Treasury. Conflicts can develop, such as the economics ministry being involved in taxation while the treasury owns energy firms, which are affected by the tax level – a lower tax level results in higher profits for the treasury. There are a number of state owned firms active in the oil and gas sector and now involved in shale gas, including PGNiG, PKN Orlean and Grupa Lotos. The Environment ministry oversees the drilling and extraction process by granting concessions and licences to companies, although much of this is conducted through regional offices in cooperation with municipalities.

Poland emerged onto the shale gas scene in 2007 when initial estimates were very high. However, these were revised downwards and placed at 146 trillion cubic feet of shale gas (Energy Information Administration, 2013). Nonetheless, estimates are again being adjusted as more research is conducted. The current process is attempting to determine the amount of resources in the country, estimated in 2012 by the Polish Geological Institute and National Research Institute (PGI-NRI) at approximately 346–768 bcm. (The next report was due by the end of 2015.) That would equate to 35 to 65 years of projected cumulative gas consumption in Poland, or 120 to 200 years of current domestic gas production (Polish Geological Institute, 2012). Despite a lot of publicity and political support, Poland at the start of 2015 had only 70 wells, which are mainly for research; commercial production is still some way off.

Notable is a study (funded by the Polish oil and gas firm, PKN Orlean) analysing employment and economic development in three scenarios (Cylwik *et al.*, 2012; Czyzewski *et al.*, 2012). Under the most intensive scenario, with high foreign direct investment (FDI) there is the potential to create 27 000 direct jobs and 483 000 indirect jobs and a contribution of 0.8% to the growth of the Polish GDP per year over the period 2019–2025 (Czyzewski *et al.*, 2012, pp. 30–31). Studies prepared by the Kosciuszko Institute and the Polish Ministry of Treasury have assessed the socio-economic and taxation impact of shale gas (Albrycht *et al.*, 2012a, 2012b) with pragmatic and favourable outcomes. Despite these serious considerations of the potential, there remains only limited signs of an industry being established. The focus should shift to a deeper level of analysis and ask 'Why?'

Environmental Risks

Environmental risks are generally proclaimed as the top reason why shale gas cannot be extracted in Europe. However, in Poland there is general agreement, on both the impact and best approaches to protect the environment, from all stakeholders including the Polish environmental movement, the oil and gas companies and the state administration. A representative for an umbrella group of environmental NGOs saw shale gas as one component in the country's energy mix, with an effective use of shale gas going to local communities where it is extracted. 'We believe this local use of shale gas, of course following all the strict tools of best available technologies, could be a transition fuel that could complement the use of renewables' (Representative for Polish Environmental Organizations, personal communication, 2013). While this Green Party leader was not directly opposing shale gas, some in his party do. Rather, his efforts are focused on showing alternatives – in renewable energy and through energy efficiency measures. Overall, anti-fracking activities, noticeably present in other countries, are more subdued in Poland, with a moderate and nationally contextualized approach dominating (see Materka, 2001, 2015).

Oil and gas companies do not oppose the existing environmental regulations, but clarity and predictability are sought. 'Polish environmental regulations are very strict. We are fine with this, as long as the regulations are clear' (Oil and Gas Company Official A, personal communication, 2012). The task of protecting the public and educating them about the environmental aspects of fracking, by oil and gas companies active in Poland, is viewed as the government's job. While companies are able to put forward their own information and experts, they know people are skeptical about information coming from them (Country manager for an oil and gas services firm, personal communication, 2012). Nevertheless, as a company official noted, a company should not be greedy and should spend money on acquiring 'very objective environmental scientific knowledge and [consultation with] individually respected professors' (Country manager for an oil and gas services firm, personal communication, 2012). Regulatory approval in Poland does not require public consultation by companies. Despite this, many shale gas companies in Poland voluntarily hold meetings with local citizens, in cooperation with state agencies.

The environmental ministry of Poland views a plurality of information as important for local stakeholders. 'We believe this is the most important factor to provide information so it [should be] coming from at least two, if not three sources' (Advisor to Ministry of Environment, personal communication, 2012). Public input and regulatory approval remains at the local level. Local governments must inform citizens of pending applications and provide necessary documents. Public meetings with companies and other stakeholders can be held at the instigation of the local governments. Generally, public involvement is encouraged by authorities (Regional Deputy Director, personal communication, 2013).

The mitigation of environmental risks depends on the administrative procedures in place. In Poland there are numerous local, regional and national government agencies and ministries involved in approving any type of resource extraction – including shale gas. The administration of the environmental assessment process is broadly cited as an area needing to be streamlined to enable faster extraction of shale gas. At the current pace the Polish state audit office predicts that it will take until 2026 to drill 200 wells in order to provide a credible picture of the shale gas reserves (Supreme Audit Office of Poland, 2014).

The above criticisms resulted in legal revisions. Poland revised the regulatory and oversight process for comprehensive Environmental Impact Assessments (EIAs) in 2013. Changes impacted drilling above 5000 meters and in areas outside environmentally sensitive zones (such as in Natura 2000 zones). Despite – or because of – these changes, the European Commission started an infringement procedure in 2014. The legal alterations cancelled the need for a *comprehensive* EIA. Instead, a simplified EIA can be done. However, the comprehensive EIA has not gone, as the initial simplified review can prompt a more comprehensive review (Communication with Energy and Natural Resources Lawyer, 2015). From the European Commission's point of view, the lack of a comprehensive EIA review contravenes the Environmental Impact Assessment Directive (Neslen, 2014). Under the new legal structure, the local authority still maintains veto powers over any exploration or extraction activities. Different interpretations of what constitutes an EIA, and at what stage it is done, are a source of discord between the Commission and Poland. Local authorities maintain a strong hand in the process by working with state-level scientists. The differences over EIA procedures between Poland and the Commission, and efforts to balance demands on the shale gas industry and environmental protection, characterize the long-term complexity of bringing hydraulic fracturing to Poland.

Technologically, the extraction of shale gas relies on unconventional techniques which have limited procedural precedent. Existing administrative procedures are applied, but they prove slow and vague for fracking. Applications for shale gas research (drilling) took on average 132 days, despite a the law stating 30 days for approval and the Ministry of Environment repeatedly emphasizing shale gas as a priority (Supreme Audit Office of Poland, 2014). For final approval for a new research well, within the Environmental Protectorate Office (excluding local authorities), agreement must be made not just between regional and central offices but with offices in other regions as well, in order to create a uniform decision (Regional Deputy Director, personal communication, 2013). The attempt to simplify the EIA approval process constitutes an attempt to remedy the time of overall assessment.

Even from a Polish environmentalist perspective, the time for approval also indicates deeper administrative problems. The lack of prescribed procedures accounting for the characteristics of fracking technology, on the one hand, and incredibly protracted administrative procedure times on the other hands (such as 132 days), caused an environmentalist to expresses sympathy: 'I agree [that]the companies have to wait [a] very long [time] for a decision.' However, the answer, for her, is not less environmental regulation but a raised 'quality of administration' to produce a sufficient solution (Representative for Polish Environmental Organizations, personal communication, 2013). Reliance on past procedures drawn up for conventional drilling inhibits the rapid roll-out of fracking technology envisioned by the government.

Incentivizing Investments

Tax incentives, or even a stable tax environment, can prompt companies to invest in new energy technologies. Most recently, Poland is pursuing a 'tax holiday' to give companies an incentive to invest quickly. Poland's tax structure has often been a point of contention for foreign investors in the country. Traditionally, the tax and royalty rates were set low, because the gas and oil sector was state owned. Any company profits were placed back in the state treasury; taxes, therefore, just shifted the money from one state pocket to another. Considerable debate took place with even the drafting of a law to increase the tax paid on shale gas extraction. In 2012 and 2013, this was envisioned to be a total of 40% of profits, including Poland's standard 19% corporate tax rate. Notable was the raising of the extraction fee of USD 1.50–1.811 per 1000 cubic metres of gas to USD 6.15–7.38 (Business New Europe, 2012, based on 2012 USD/PLN exchange rate). One interviewee, echoing others, stated, 'The taxes are too low and companies know it' (Advisory to Ministry of Environment, personal communication, 2012). There was a general expectation that rates would increase, but it was uncertain by how much. However, subsequent to the state audit office report issued in 2014, the Prime Minister of Poland stated that the new law would not be as far-reaching, 'Today we understand that in order to count money from shale gas, we must first of all begin to extract it' (Bernat, 2014). The proposed legislation grants a six-year tax break for shale gas projects (Natural Gas Europe, 2014a) and allows deeper drilling without additional approval (Ciobanu, 2013).

Business Strategies: Pulling-Out, Independence, Cooperation

Ambiguity over both geology and the legal framework causes companies to act in three ways. First, the lack of a legal framework and poor test results has already caused companies like Marathon Oil, Talisman Energy, Exxon Mobil and Chevron, to withdraw from Poland (Bernat, 2014). Second, some foreign firms remain independent (San Leon or BNK). Third, joint ventures emerge as a risk reduction strategy. State owned oil and gas group Lotos stated in 2012 that they were looking for foreign partners, with technical expertise as a way of reducing the technical risks (Warsaw Voice, 2012). Another perspective sees local partnerships helping foreign firms to navigate local social and administrative challenges. Chevron has shifted from an independent direction to a joint venture strategy.

The Zurawlow region is marked by a strong environmental movement against the exploration efforts by Chevron. In 2012, environmental protestors successfully argued in court that Chevron was contravening the EU's Bird Directive. In June 2013, further protests erupted at the same site, preventing the installation of a fence in preparation for drilling work. This environmental movement was often cited by interviewees as the 'only' strong opposition force in Poland (private geologist, personal communication, 2012), thus increasing the environmental compliance risk for Chevron. In December 2013, Chevron and the state owned oil and gas company PGNiG signed a strategic agreement to cooperate in the region where the environmental protests had occurred. Cooperation between Chevron and PGNiG was expected to reduce the financial and exploration risks, including development costs, while PGNiG expected to benefit from Chevron's global experience in shale gas (United Press International, 2013). At the beginning of 2015, despite these trial efforts, Chevron gave back their awarded concessions and said they would leave Poland.

Reflecting on the attempt at cooperation between PGNiG and Chevron, there was indeed an opportunity for Chevron to deal with local opposition while offering PGNiG access to their expertise and knowledge. Even before then, the view was held that state owned oil and gas companies active in shale gas exploration had to rely on foreign expertise and technology. 'We don't have the knowledge in Poland' (Chief Economist, state owned oil and gas company A, personal communication, 2012). The present writer, on a flight from Warsaw to Chicago, met an American drill-hand who was on a 10-day break from working in northern Poland on a pneumatic drilling rig. The company was American, the rig was imported from Romania and the crew were all non-Polish (Gas drilling rig crew leader, personal communication, 2013). Thus cooperation and partnerships were already an established part of the oil and gas industry. The high visibility of signed joint-cooperation agreements may only be for public or political consumption, while significant exchanges of expertise between foreign and domestic companies are carried out on a regular basis.

R&D, Innovation and FDI

Fostering cooperation between state owned companies also emerges as a challenge equal in magnitude to streamlining state institutional efforts to upgrade administrative procedures. State owned companies are affected by non-cooperation both internally and externally. According to one executive of a state owned oil and gas firm, he had trouble cooperating with the semi-independent upstream portion of the same company. In addition, he cited the organization's inability to foster cooperation with other state owned firms over equipment, such as drilling rigs; from his perspective, this clearly inhibited the use of the best available technologies (Chief Economist, state owned oil and gas company A, personal communication, 2012). In July 2012, five Polish firms agreed to spend €415 million on the joint exploration and extraction of shale gas. However, in January 2014 this cooperation was cancelled as no agreement could be reached on how it would actually be achieved (Natural Gas Europe, 2014b). So, while the Polish Treasury is the owner of these firms, there is an inability to build a coordinated response even within and between state firms.

The Polish government is using other mechanisms to stimulate R&D in order to foster a domestic shale gas industry which may also result in export opportunities. In a push to increase R&D in the shale gas technology applicable for Poland's geology, the government announced a cooperative R&D grant program for the commercialization of technologies. The government will provide PLN 500 million for the 'Blue Gas project', with companies investing an equal amount. It is expected, according to an official Embassy statement, that 'Scientists and businesses will cooperate on ways of identifying areas worth exploring, on exploration technology and on limiting the impact of extraction on the environment' (Embassy of Poland in Cairo, 2012). The expectation exists to develop Polish expertise and technology and export these (Chief Economist, state owned oil and gas company A, personal communication, 2012; Czyzewski *et al.*, 2012; Embassy of Poland in Cairo, 2012; Foreign Ministry Official A, personal communication, 2012). This is also relevant for Ukraine, which has similar geology.

Emerging from the interviews was the commonly held view that Poland needs FDI in the shale gas sector to reach a critical mass fully exploiting the country's reserves. With FDI, a fully developed shale gas industry, specializing in innovative fracking technologies, can emerge and be globally competitive. 'The technology has all the elements of innovation. It is composed of many different businesses' (Foreign Ministry Official A, personal communication, 2012). Buy-in by multi-national companies was viewed as necessary to fund R&D, which in turn could be exported. 'Without international capital the Polish shale gas project will never happen' (Country manager for oil and gas services firm, personal communication, 2012).

Preparing the Market

The commercialization strategy of the Polish state and of Polish companies rests on current R&D and commercialization projections, sitting within the national and international gas markets. The integration of shale gas into the marketplace requires new infrastructure and market mechanisms, including: (a) developing a transport network from well-head to a national gas network; (b) local distribution systems connected to a national gas grid, eliminating the necessity for households to burn coal; (c) scaling up of the wholesale gas market, such as building new gas fired power plants or increasing demand from industry (large consumers could help offset the cost of building the gas network); and (d) a gas trading platform, with sufficient volumes of gas traded that a market price can be established. These are all contractual and governance risk issues that market participants must balance in their investments.

The identified hurdles for distribution and marketing are one reason why an environmental leader says shale gas should be used for local communities. Keeping gas close to its source requires much less infrastructure to be built. For example, if roads are used to transport gas then the roads will require additional investment (Representative for Polish Environmental Organizations, personal communication, 2013). In addition, liberalization of the gas market will send a price signal to the market, clarifying whether it is profitable to drill in Poland (Economic Development Representative A, personal communication, 2012).

The success of a new shale gas sector in Poland rests on how well risks are mitigated. The risks range from commercial risks, involving performance or price, to governance risks involving the administrative capacity in state institutions and the clarity of environmental regulations. Companies have left Poland because these risks were deemed too high.

Discussion: Risk Governance of Shale Gas

Encouraging shale gas extraction in Poland is not isolated from other economic and social activities. Governance mechanisms for technological innovation, from the R&D stage to market policies, influence how fracking is introduced into a country. The social acceptance of carbon based energy sources such as coal and gas is high in Poland. The roll-out of hydraulic fracking technology, from this point of view, should be possible, particularly when the situation in Poland is compared with that of other countries having significant environmental movements challenging the use of shale gas. But the story and the success of shale gas extraction goes beyond social acceptance; technologically and institutionally there must be a forward momentum.

Seismic technology, horizontal drilling and water injection are new technologies that follow a typical national innovation path. Using R&D and the assessment and mitigation of risks for both companies and the environment, state institutions provide the policy framework to encourage or discourage deployment. Engrained institutional procedures are being used with difficulty to bring to market a technology perfected for American soil. Adapting this to the Polish situation requires flexibility both in the field and in the approving state institutions. The examination of risks provides a lens to view the intersection of governance challenges and private sector investments. Reviewing the governance and commercial risks and how they affected the deployment of shale gas technology in Poland between 2009 and 2015 provides a cautionary tale about the rolling out of a politically supported technology.

Governance Risks

Governance risks in Poland are closely connected with perceptions held by company officials in the oil and gas industry. The blurred progression of institutional decisions around shale gas, from taxation to environmental compliance, holds back investors from fully committing. This is reflected in the minimal drilling activity by operators, which is merely sufficient to maintain their concession licence. They refrain from any additional activity that would speed up the discovery and extraction of shale gas.

An initial governance definition highlighting the inability to coordinate different state institutions (Jessop, 1998) underpins the lack of joined-up state actions – despite all positive intent. Institutional risk aversion, expressed through engrained administrative procedures, low staffing levels and vagueness in environmental assessment criteria, inhibit innovation in this area of the energy sector – despite stated government and social support. This results in foreign firms avoiding the country or scaling back operations, and in turn a higher dependence on state owned companies to adopt and exploit shale gas technologies. State institutions must deal with the 'vertical and horizontal' internal split of agency and ministry responsibilities and they do attempt to streamline procedures (like environmental assessments) or create incentives (like tax holidays), but these either encounter other institutional constraints (from the European Commission) or are too late as companies are already seeking to go elsewhere.

State institutions, whether the Environmental Ministry or the State Treasury, each attempt to fulfil their role within the state (environmental protection and revenue generation respectively) but, owing to the complexity of the system established for a previous Techno-Institutional Complex (Unruh, 2000), it becomes bogged down in institutional risk mitigation and conflicting institutional governance logic. Even when laws are amended in order to simplify deployment (as in 2013), other levels of governance (from the EU) emerge to muddy the waters. In addition, new technologies may be adopted by state owned companies but they lack the financial or expert capacity to fully deploy a technology. Thus even when technological change is demanded, a lack of experience and institutional inertia may undermine the ability to change. Socially, if there is little pressure to change then the energy system as a whole remains locked in to older technologies and systemic paths.

Commercial Risks

There are three main commercial risks emerging from interviews with Polish stakeholders. These are: price risks, demand risk and environmental compliance. The interviews indicated that the uncertainty associated with these risks is less about agreements between parties than about short-term government actions (the realm of governance). However, one key market factor is that the price differential between domestic shale gas and LNG or Russian piped gas rests both with global and continental market forces and also with geopolitics.

Price and demand risks are central, and environmental compliance risks emerge as a key category. Both are viewed as important to stabilize and clarify whether foreign investments could play a role in Poland's efforts to develop shale gas. Interestingly, it is not less environmental regulation that interviewees requested but clarity and sufficient administrative resources. Commercial risks influence companies' decision-making processes about which investments to make and at what scale. The working assumption is that both the market and the demand will be developed enough when full commercial production begins. Therefore, it is interesting to note that investments proceed even under this uncertainty.

The commercialization strategies of companies are informed by the risks present in each country. This includes the financial mechanisms for new investments and bureaucratic hurdles that must be overcome. In the case of Poland, uncertainty in the tax regime influences the speed of investments. Environmental risk mitigation on the part of the state, stemming from unreformed processes used for conventional gas extraction by state owned firms, raises the risk level for foreign firms. Chevron's entrance, exit, re-entrance and then a second exit despite a joint venture with state owned PGNiG point to shifting business strategies. The split between companies that have withdrawn from Poland and those independent companies that have stayed also highlights local challenges.

Deployment

Poland pursues shale gas for three reasons: (1) to reduce its reliance on Russian gas; (2) to prompt industry to locate to Poland; and (3) to develop an export industry based on fracking technology. The state says 'go', society says 'go' and firms say 'go', yet little goes. Understanding why the shale gas revolution stays in the starting blocks requires a deeper look into the path dependency of technologies and the institutional structure established in another energy era for conventional oil and gas. Geologically, Poland's ground is proving to be a challenge for companies to extract shale gas – despite the institutional encouragement.

Experience with the private sector, and how investors balance risks, is also lacking, as evidenced by the on-and-off privatization of energy companies. This factor is combined with a constitutional mandate to maintain control over state owned energy firms, meaning there are inherent factors affecting how institutions deal with the private sector, narrowing the scope for private investment into the energy sector including the speed at which FDI can infuse into the shale gas sector. Without considerable change affecting how energy innovation and risk mitigation occurs, the institutional processes, encumbered by the geology, will continue to slow down not just shale gas development but in addition other energy technologies, such as renewables, that also need institutional reforms and state efforts to align with private efforts. In both areas, that of carbon based technologies and that of renewable technologies, failure to break through rigid institutional structures will inhibit the deployment of new energy technologies, and thus traditional forms of coal technology will remain dominant.

Conclusion

This Polish case study regarding the fostering of shale gas technology is important for other countries pursuing the new American dream of cheap fracked gas. Poland's intention to develop the shale gas sector has been surrounded by extensive political and media hype. However, institutional blockages emerged as another stumbling block. The shale gas story in Poland can serve as a cautionary tale for other countries seeking to deploy shale gas technologies or other new energy technologies. Institutional alignment within the state must match the physical and investment environment of a country. Wishful political thinking does not build an energy system. Administrative flexibility, investment incentives and clarity are needed for all energy technologies. Governance and commercial risks influence the speed of shale gas development. Importantly, these risks are reflected in broader political, social, business and even environmental considerations.

Acknowledgement

The author acknowledges the support of the EU FP7 large-scale integrated research project GR:EEN Global Re-ordering: Evolution through European Networks, Number: 266809.

References

Albrycht, I., Garpiel, R., Łaszczuk, A., Łazarski, A., Matyka, M., Pokrywka, Ł. *et al.*, 2012a. The impact of shale gas extraction on the socio-economic development of regions – an American success story and potential opportunities for Poland (Report abstract). The Kosciuszko Institute, Warsaw.

Albrycht, I., Matyka, M., Kotala, W., 2012b. Win–win strategy for taxation on shale gas extraction in Poland (policy brief). The Kosciuszko Institute, Warsaw.

Bekkers, V., Thaens, M., 2005. Interconnected networks and the governance of risk and trust. *Information Polity: Int. J. Government & Democracy in the Information Age* 10 (1/2) (March 2005), 37–48.

Bulmer, S., Dolowitz, D., Humphreys, P., Padgett, S., 2007. *Policy Transfer in European Union Governance: Regulating the Utilities*. Routledge.

Bernat, P., 2014. UPDATE 1 – Polish PM sees approval of new shale gas law within two weeks. Reuters.

Burer, M.J., Wustenhagen, R., 2009. Which renewable energy policy is a venture capitalist's best friend? Empirical evidence from a survey of international cleantech investors. *Energy Policy* 37, 4997–5006. doi:10.1016/j.enpol.2009.06.071.

Business New Europe, 2012. Poland eyes shale gas boost as it reveals tax plans [WWW Document]. URL http://www.bne.eu/story4104/Poland_eyes_shale_gas_boost_as_it_reveals_tax_plans (accessed 17 October 2012).

Christensen, T., Lægreid, P., 2013. The new regulatory orthodoxy: a critical assessment. In *Handbook on the Politics of Regulation*, p. 712. Edward Elgar.

Ciobanu, C., 2013. Poland's shale gas bubble 'bursting' [WWW Document]. Inter Press Serv. URL http://www.ipsnews.net/2013/07/polands-shale-gas-bubble-bursting/ (accessed 29 March 2014).

Cylwik, A., Pietka-Kosinska, K., Lada, K., Sobolewski, M., 2012. Economic potential for shale gas production in Poland in 2010–2025. Scenario analysis. Centre for Social and Economic Research, Warsaw, Poland.

Czyzewski, A.B., Bodnari, E., Kozieja, G., 2012. Gas (r)evolution in Poland: which way to success? PKN Orlen.

Embassy of Poland, 2012. Government to pump PLN 500m into shale gas R&D projects – Trade & Investment Promotion Section in Cairo [WWW Document]. URL http://cairo.trade.gov.pl/en/aktualnosci/article/y,2012,a,27820,Government_to_pump_PLN_500m_into_shale_gas_RD_projects.html (accessed 30 March 2014).

Energy Information Administration, 2013. EIA/ARI world shale gas and shale oil resource assessment.

Energy and Natural Resources Lawyer. 2015. Personal communication, April 3.

Engau, C., Hoffmann, V.H., 2009. Effects of regulatory uncertainty on corporate strategy – an analysis of firms' responses to uncertainty about post-Kyoto policy. *Environ. Sci. Policy* 12, 766–777. doi:doi: 10.1016/j.envsci.2009.08.003.

Goldthau, A., LaBelle, M. (forthcoming). The power of policy regimes: explaining shale gas policy divergence in Bulgaria and Poland. *Rev. Policy Research*.

Hoffmann, V.H., Trautmann, T., Hamprecht, J., 2009. Regulatory uncertainty: a reason to postpone investments? not necessarily. *J. Manag. Stud.* 46, 1227–1253. doi:10.1111/j.1467–6486.2009.00866.x.

Hoffmann, V.H., Trautmann, T., Schneider, M., 2008. A taxonomy for regulatory uncertainty – application to the European Emission Trading Scheme. *Environ. Sci. Policy* 11, 712–722. doi: 10.1016/j.envsci.2008.07.001.

Jamison, M.A., Holt, L., Berg, S.V., 2005. Measuring and mitigating regulatory risk in private infrastructure investment. *Electricity J.* 18 (6) (July 2005), 36–45. doi:10.1016/j.tej.2005.06.002.

Jessop, B., 1998. The rise of governance and the risks of failure: the case of economic development. *Int. Soc. Sci. J.* 50 (155), 29–45.

LaBelle, M., 2012. Constructing post-carbon institutions: assessing EU carbon reduction efforts through an institutional risk governance approach. *Energy Policy* 40, 390–403. doi:10.1016/j.enpol.2011.10.024.

LaBelle, M., Goldthau, A., 2013. Escaping the valley of death? Comparing shale gas technology policy prospects to nuclear and solar in Europe. *J. World Energy Law & Business* 7 (2) (18 December 20013): jwt020. doi:10.1093/jwelb/jwt020.

Levi-Faur, D., 2009. Regulatory capitalism and the reassertion of the public interest. *Policy Soc.* 27, 181–191. doi:10.1016/j.polsoc.2008.10.002.

Levi-Faur, D., Jordana, J., 2005. The rise of regulatory capitalism: the global diffusion of a new order. *Ann. Am. Acad. Pol. Soc. Sci.* 598 (1): 200–217.

Materka, E., 2001. End of transition? Expropriation, resource nationalism, fuzzy research, and corruption of environmental institutions in the making of the shale gas revolution in northern Poland. *Debate: Journal of Contemporary Central and Eastern Europe* 19, no. 3: 599–631.

Materka, E., 2015. Poland's quiet revolution: of shale gas exploration and its discontents in Pomerania. *Central European Journal of International and Security Studies* 9 (3).

Natural Gas Europe, 2014a. Poland tries to speed up shale gas industry [WWW Document]. URL http://www.naturalgaseurope.com/poland-indigenous-shale-gas-fracking (accessed 29 March 2014).

Natural Gas Europe, 2014b. Co-operation in Poland's shale gas sector crumbles [WWW Document]. URL http://www.naturalgaseurope.com/poland-shale-gas-co-operation (accessed 4 April 2014).

Neslen, A., 2014. EC serves notice to Poland over shale gas defiance. *The Guardian*, 30 July 2014. http://www.theguardian.com/environment/2014/jul/30/ec-serves-notice-to-poland-over-shale-gas-defiance.

PHYS. ORG., 2013. Poland anchors energy strategy in coal, shale gas: PM [WWW Document]. URL http://phys.org/news/2013-09-poland-anchors-energy-strategy-coal .html (accessed 3 March 2014).

Polish Information and Foreign Investment Agency, 2013. Energy sector in Poland.

Polish Geological Institute-National Research Institute (PGI-NRI) in Warsaw, 2012. Assessment of shale gas and shale oil resources of the Lower Paleozoic Baltic-Podlasie-Lublin Basin in Poland, March 2012.

Radzka, B., 2006. Liberalisation, privatisation and regulation in the Polish electricity sector.

Sabatier, P., Jenkins-Smith, H., 1988. An advocacy coalition framework of policy change and the role of policy-oriented learning therein. *Policy Sciences* 21, 129–168.

Supreme Audit Office of Poland, 2014. NIK on shale gas search – Supreme Audit Office [WWW Document]. URL http://www.nik.gov.pl/en/news/nik-on-shale-gas-search .html (accessed 16 January 2014).

The Economist, 2011. Privatisation in Poland: Overcoming miner obstacles. *The Economist*. www.economist.com/blogs/schumpeter/2011/07/privatisation-poland.

Tsoutsos, T.D., Stamboulis, Y.A., 2005. The sustainable diffusion of renewable energy technologies as an example of an innovation-focused policy. *Technovation* 25, 753–761. doi:10.1016/j.technovation.2003.12.003.

Unger, H., 2010. Innovation and market entry in the energy industry: lessons for fuel cells and new technologies. *J. Bus. Econ. Res.* 8, 63–71.

United Press International, 2013. Chevron teams up with Poland's PGNiG to explore Poland's shale gas [WWW Document]. UPI. URL http://www.upi.com/Business_News/Energy-Resources/2013/12/13/Chevron-agrees-to-explore-Polish-shale-gas-potential/UPI-40791386934916/ (accessed 13 December 2013).

Unruh, G.C., 2000. Understanding carbon lock-in. *Energy Policy* 28, 817–830. doi:doi: DOI: 10.1016/S0301–4215(00)00070–7.

Unruh, G.C., 2002. Escaping carbon lock-in. *Energy Policy* 30, 317–325. doi:doi: DOI: 10.1016/S0301–4215(01)00098–2.

Walsh, P.R., 2012. Innovation Nirvana or Innovation Wasteland? Identifying commercialization strategies for small and medium renewable energy enterprises. *Technovation* 32, 32–42. doi:10.1016/j.technovation.2011.09.002.

Warsaw Voice, 2012. Lotos seeks partner for shale gas projects in Poland [WWW Document]. Warswavoice.pl. URL http://www.warsawvoice.pl/WVpage/pages/article.php/21383/news (accessed 16 October 2012).

Weyant, J.P., 2011. Accelerating the development and diffusion of new energy technologies: beyond the 'valley of death'. *Energy Econ.* 33, 674–682. doi:10.1016/j.eneco.2010.08.008.

Wiser, R., Bachrach, D., Bolinger, M., Golove, W., 2004. Comparing the risk profiles of renewable and natural gas-fired electricity contracts. *Renewable and Sustainable Energy Reviews* 8 (4) (August 2004), 335–63. DOI: 10.1016/j.rser.2003.11.005.

Wiser, R., Bolinger, M., 2006. Balancing cost and risk: the treatment of renewable energy in western utility resource plans. *Electricity J.* 19 (1) (January 2006), 48–59. DOI: 10.1016/j.tej.2005.11.012.

Woerdman, E., 2004. *Path Dependence and Lock-in of Market-Based Climate Policy*. Developments in Environmental Economics, Vol, 7. Elsevier. http://www.sciencedirect.com/science/article/B8G5W-4NYSVF2-5/2/454f119a804ed7f5641fa82ec786d28e.

Wüstenhagen, R., Menichetti, E., 2012. Strategic choices for renewable energy investment: conceptual framework and opportunities for further research. *Energy Policy* 40, 1–10. doi:10.1016/j.enpol.2011.06.050.

15

Unconventional Gas Regulation in Australia and the US: Case Studies of Four Jurisdictions

IAN CRONSHAW AND R. QUENTIN GRAFTON

Introduction

The four jurisdictions examined in this chapter, in terms of unconventional gas extraction, include, in Australia, Queensland and New South Wales and, in the United States, Pennsylvania and New York. Although they each have resources of unconventional gas within their own States, they display sharply differing approaches to development.

All four State jurisdictions exist within their own respective Federal (Government)–State frameworks. In both countries, States regulate natural resource developments within their boundaries, but Federal Governments do have important, potentially overriding, powers. Local rules and regulations can also affect development. Thus, gas producers face a complex Federal–State local regulatory environment. A comparison and evaluation of the history of approaches in each jurisdiction provides valuable insights about 'best practice' regulation for unconventional gas extraction.

Queensland

Commercial production of coal bed methane (CBM) began in Australia in 1996, in Queensland's Bowen Basin (see Figure 15.1). Production remained at relatively low levels, until 2006 when the Surat Basin commenced production, with output quadrupling in the six years to 2012, reaching around 6 bcm, equivalent to around one third of Eastern Australia's gas demand or 13% of Australia's gas consumption. Over the decade to 2012, proven and probable reserves expanded tenfold, to around 900 bcm. The overwhelming majority of this expansion has been in the Surat Basin. In addition, a number of promising basins, notably the Galilee Basin, hold substantial CBM resources.

Almost all Australia's unconventional gas output is CBM from Queensland, with a very small amount from New South Wales. As of late 2012, some 1200 wells were producing CBM in Queensland (QCA, 2014). The major companies involved, until quite recently, were Australian and include Origin Energy, Arrow Energy, Santos and the Queensland Gas Company (now a subsidiary of BG). Furthermore, AGL, an Australian company, also has interests in fields operated by Origin and Arrow.

Gas development in Queensland, and particularly CBM, was stimulated by the State's Gas Scheme, which stipulated that electricity retailers source a percentage of their power

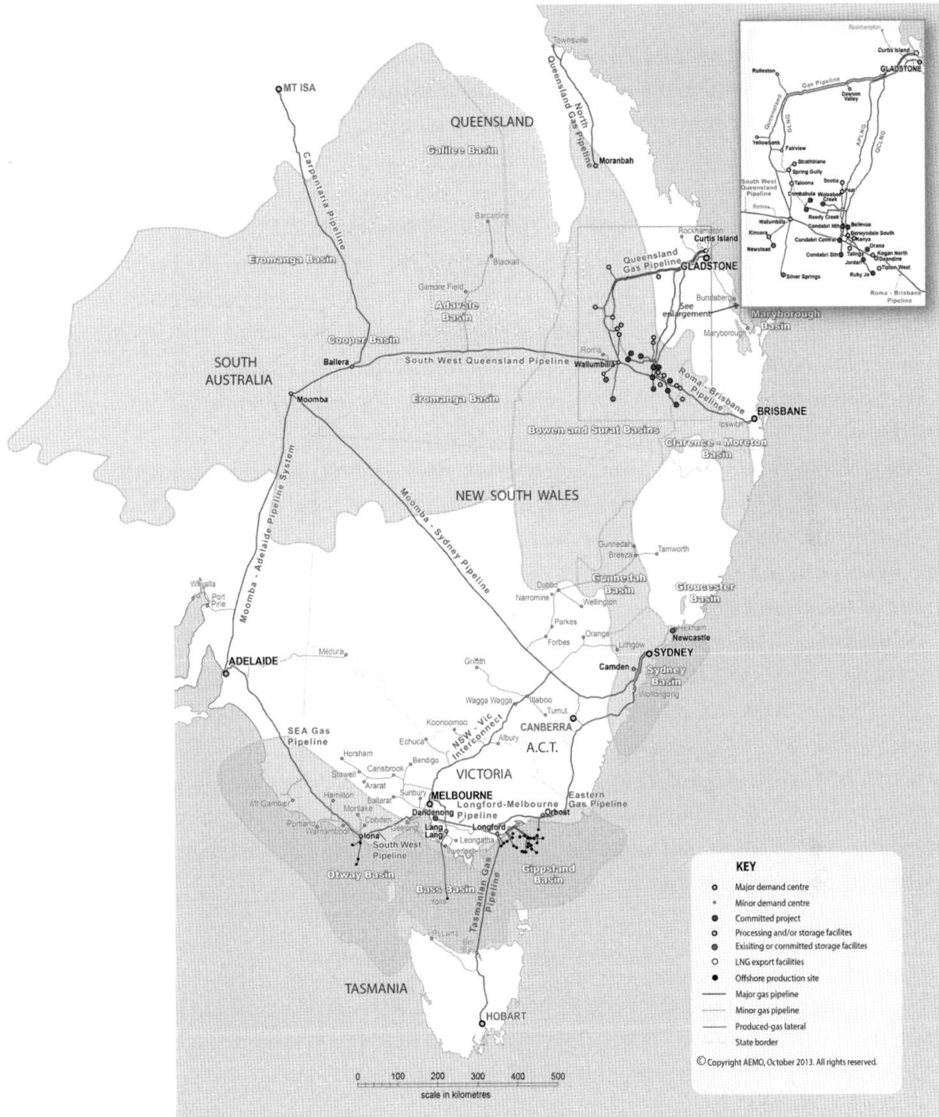

Figure 15.1 2015 Gas Statement of Opportunities map for eastern and south eastern Australia. *Source:* AEMO (2013).

from gas fired sources (15% in 2013). The scheme, designed along analogous lines to those of the Federal Mandatory Renewables Energy Target, boosted the State's gas production industry and diversified the power sector. The Gas Scheme supplemented the advantages of gas fired electricity, despite the relatively high cost (compared with coal) and the initially lower availability of gas. The scheme was terminated at the end of 2013.

Gas Demand Driven by LNG Expansions

The development of a major LNG export industry in Queensland will lead to a rapid increase in CBM output. The three plants on Curtis Island adjacent to Gladstone should be exporting some 33 bcm well before 2020, around 5% of globally traded natural gas, or around 7% of projected global LNG trade in that year. The first plant started exports in late 2014 and a second plant joined it in the second half of 2015. While large in terms of LNG exports, this will represent less than 1% of total global gas demand in 2020.

Gas supply to the plants of around 1500 PJ per year (close to 40 bcm) will entail a six-fold increase over 2012 production levels. Local industry participants have been joined, in the more than $60 billion invested in these LNG plants (BREE 2013), by a number of international players in gas production, trading and purchasing, including BG, ConocoPhillips, Petronas, Total, Kogas, and Sinopec. By mid 2015, nearly 5500 wells were producing CBM in Queensland (APPEA 2015).

A possible fourth LNG project, involving Shell, PetroChina and Arrow, could have added more than 500 PJ to these demands but in early 2015 the project was shelved, in the face of rising costs and lower oil (and therefore LNG) prices. It is possible that gas from the fields designed to feed this fourth project may eventually feed the first three, or new trains at those projects.

Queensland domestic gas demand also seems likely to grow over the decade to 2020, to around 7 bcm (AEMO 2013). Thus, gas production seems set to expand by a factor of eight over the next few years. Such rapid growth is by no means unique globally, with clear precedents in the western United States and in the province of Alberta in Canada, where CBM has expanded from virtually nothing in 2002 to 10 bcm a decade later. The rapid expansion of the Queensland industry has posed challenges for an industry intent on raising gas production and constructing three plants in the same location, as well as for governments, regulators and communities.

Rapid Expansion Poses Problems

The challenges of rapid expansion in unconventional gas production in Queensland differ from those surrounding the large-scale shale gas development seen in the United States. In Queensland, the differences include:

(i) the treatment, disposal or re-use of the large quantities of water extracted from coal seams, and the potential impact on existing aquifers through such extraction;
(ii) land access by CBM producers, and the impact on land of drilling and ongoing operations;
(iii) to the extent that hydraulic fracturing techniques are used (to date fracture stimulation has been used in only about 8% of production wells (APPEA 2015)) the type of chemical additives employed and the possible contamination of aquifers; and
(iv) other water impacts, such as the cross contamination of underground aquifers.

To these challenges associated directly with the CBM expansion must be added those related to the simultaneous construction of three LNG plants on Curtis Island, such as the social impacts on Gladstone. These include the rapid rise in housing costs as well as environmental issues, for example, the rapid and massive dredging of Gladstone Harbour and the disposal of the spoil and its possible negative impacts on the Great Barrier Reef. Local environmental concerns have arisen despite the fact that environmental approvals for additional dredging and development, obtained in December 2013, contained a large number of conditions (Department of Environment 2013; Davies *et al.* 2013). These port developments, however, have been primarily driven by coal exports and related coal port activities.

Key Regulatory Developments

Queensland has a long history of conventional oil and gas production, from the Cooper Basin in the far west to the Roma area in the Surat Basin. Initially, these developments formed the basis of regulatory approaches, as set out and modified in more recent legislation such as the State Petroleum and Gas (Production and Safety) Act 2004. Water impacts of petroleum operations were also recognised in the Water Act 2000, which requires petroleum lease holders to assess water levels and quality prior to commencing operations.

The rapid growth of the petroleum industry in recent years and its massive expansion have led to a change from this petroleum-based regulation to one more specifically tailored to CBM issues, with a large number of regulatory instruments specific to the industry having been introduced in recent years. These include a Code of Practice for Constructing and Abandoning Coal Seam Gas Wells (2011), Coal Seam Gas Water Management Policy 2012 and Coal Seam Gas Recycled Water Management Plan 2013. In 2010, Queensland specifically banned BTEX chemicals used in the CBM sector (Queensland Government 2010).

One important initiative, reflecting the current and future importance of the Surat Basin, was the declaration of the basin as a cumulative management area (CMA) under the Queensland Water Act 2000. The role of the key gas-producing formation in the Surat Basin, the Walloon Coal Measures, as a geological formation of the Great Artesian Basin was also a key factor in this decision. This led to the production of an underground water impact report over 2011 and 2012. As a result, a large-scale regional water monitoring network of almost 500 monitoring points is being established. The report predicted that of some 21 000 bores in the CMA, 85 were likely to see water levels decline below the trigger threshold in the following three years. Petroleum tenure holders are required to make good any water losses, and to carry out baseline assessments before production begins in order to underpin that obligation.

At the beginning of 2013, the Office of Groundwater Impact Assessment (OGIA) was established. It has taken over previous functions of the Queensland Water Commission and is responsible for monitoring and reporting annually on the implementation of the report. The office is an independent entity under the Water Act 2000, with its head a statutory officer, and is independently and fully funded by an industry levy on petroleum tenure holders, currently raising some $7.5 million per year (QCA 2014). This levy is a funding arrangement that provides continuity to OGIA, allowing it to pursue longer-term research and fulfil its mandate of linking science and policy. At present, the total annual costs of regulatory provision in Queensland, estimated at around AU$29 million, are only partially met by industry revenues, estimated at AU$14 million (QCA 2014).

Surat Underground Water Impact Report (UWIR)

In an area of concentrated CBM development, the impacts on water levels caused by individual CBM projects can overlap. The Queensland Government may declare such an area to be a 'cumulative management area' (CMA). The area of planned concentrated CBM development in Queensland has been declared as the 'Surat CMA'.

When a CMA is declared, OGIA is required to prepare a cumulative assessment of impacts of CBM water extraction, and to develop integrated regional management arrangements. These assessments and management arrangements are to be set out in an underground water impact report (UWIR). When prepared, a UWIR is submitted for approval to the Chief Executive of the Department of Environment and Heritage Protection. The first UWIR was approved for the Surat CMA in July 2012 (Queensland Water Commission 2013). A regional groundwater flow model was developed to support the development of the UWIR and is used to predict future water level impacts in the coal seams as well as in adjacent aquifers.

The Surat UWIR includes:

- maps of predicted water level impacts;
- a water monitoring strategy;
- a management strategy for springs that could be affected by falls in water levels; and
- an assignment to individual CBM operators of responsibilities to carry out activities such as specific parts of the water monitoring strategy.

The maps of predicted water level impacts support the continuation of water supplies for bore owners affected by CBM water extraction. If a bore supply is impaired by CBM water extraction at any time, the CBM operator has a responsibility to find a solution to the problem. This framework requires that in areas where the UWIR predicts that water levels will fall by more than a trigger threshold within three years ('immediately affected areas'), CBM operators must enter into agreements as soon as possible with bore owners about arrangements to maintain water supplies. This promotes the setting up of sound and workable arrangements before any impairment occurs.

The OGIA is carrying out research activities in collaboration with other bodies and CBM operators to further improve understanding of the groundwater flow system. The regional groundwater flow model will be redeveloped to incorporate the new knowledge, and the Surat UWIR will be revised over 2015.

In addition to the OGIA, a number of other government agencies are involved in the oversight of various stages of the industry, notably the Queensland Department of Environment and Heritage; since 2012 a number of more specialised agencies have been established to deal with CBM issues more specifically.

The Coal Seam Gas (CSG) Compliance Unit (formerly the LNG Enforcement Unit) monitors CBM operators on an ongoing basis and ensures that they comply with the current laws and policies. The Unit, which had 28 inspectors in mid 2013, inspected some 369 CBM wells in the year ending 30 June 2013, plus 154 drilling rigs (Queensland Government 2013). The Unit also provides an integrated whole-of-government approach to managing complaints. It includes staff from across government, including environmental and groundwater experts, petroleum and gas safety specialists and staff specialising in land access issues.

This Unit, managed by the Department of Natural Resources and Mines (NRM), works closely with other government agencies such as the Department of Environment and Heritage Protection (EHP) and the GasFields Commission Queensland (see below) to regulate the CBM industry and build community confidence in the management of the industry. The Unit is responsible for a range of activities including:

- monitoring compliance relating to CBM activities;
- managing and investigating complaints;
- inspecting sites where CBM activities are conducted; and
- sampling 300 groundwater bores per year to monitor the impacts of groundwater quality from CBM activity and to verify the monitoring data that are being supplied by CBM companies.

In addition to these developments, in 2013 the Queensland Government established the GasFields Commission to manage the co-existence of rural landholders, regional communities and the CBM industry in Queensland. The Commission is specifically designed to give local communities a more direct say in the responsible development of the CBM industry. Although relatively newly established, its creation seems to have made a strong positive contribution to improving landholder and industry relations.

In March 2014, the Queensland Government passed a new regional planning act, intended to recognise and protect four key areas of interest:

- priority agricultural areas;
- priority living areas;
- strategic environmental areas; and
- strategic cropping areas.

Resource development can only occur in these areas if proponents reach agreement with a landholder, or if a Regional Interest Development Approval is granted. The regional planning act has mandated the Gasfields Commission with a new statutory role to provide advice to the Government.

Landholders welcomed the new Act as a means to redress the power imbalance between landholders and potential resource developers, by giving landholders a greater say in what will actually occur on their land.

In sum, the regulatory regime in Queensland is relatively complex and evolving, with differing responsibilities among agencies at various stages of the project life cycle. Newly

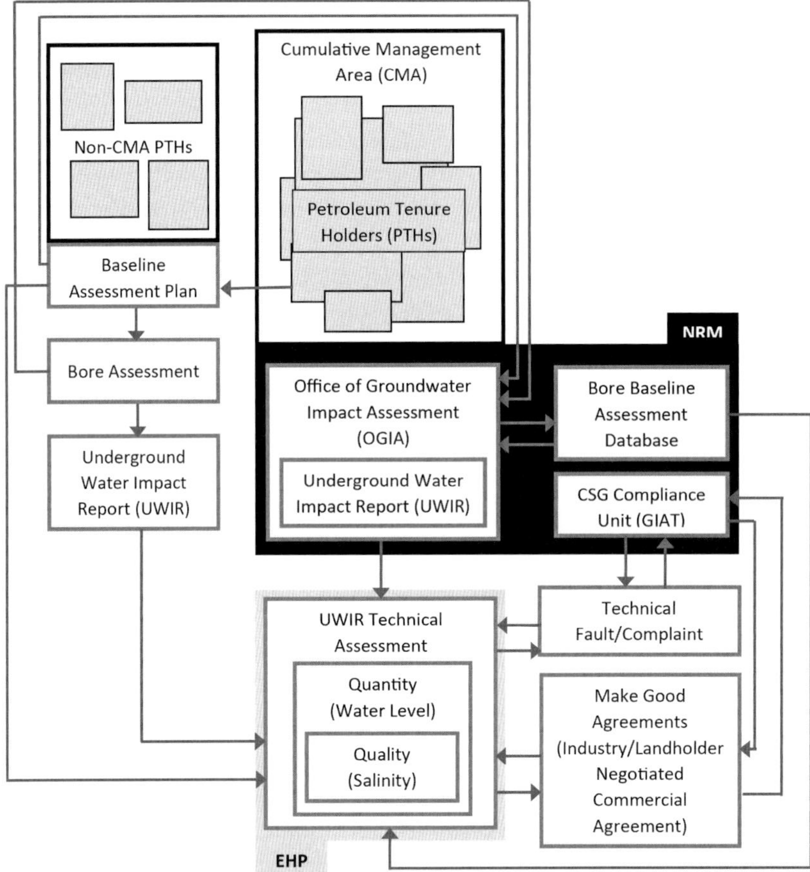

Figure 15.2 Responsibilities for CBM water management in Queensland. *Source:* Queensland Competition Authority (2014).

created organisations have been given a mandate to focus on CBM regulation. Water management, which is at the core of the regulation, and the subject of most stakeholder concern, is shown in more detail in Figure 15.2, which highlights the roles of OIA, CSGCU, NRM and EHP.

The Australian Federal Government, through the Environment Protection Biodiversity Conservation Act (EPBC Act) 1999, can also regulate matters of national environmental significance, such as those relating to international conventions on endangered species. In June 2013, the then Federal Government passed amendments to the EPBC Act, effectively rendering water resource issues of national significance, reflecting community concern over water issues associated with CBM development. In October 2013 the newly elected Federal Coalition Government signed a Memorandum of Understanding with the Queensland Government to create a one-stop shop for environmental approvals covered by the EPBC

Table 15.1 *Queensland – simplified regulatory responsibilities for CBM-LNG project life cycle*

Activity		Responsibility
Pre-exploration		Chiefly NRM, environmental approvals EHP, GasFields Commission
Exploration/Planning	Development plan approval	NRM[c]
	Baseline water assessment	NRM, OGIA, EHP[b]
	EIS	EHP
	Underground water impact reporting	NRM, OGIA, EHP
Development/Construction	Development planning and construction (dams, pipelines, LNG facility)	NRM, EHP, DSDIP[d]
	CBM recycled Water Management Plan	DESW, DSDIP
Production	Well production, water monitoring	NRM
	Water disposal/reuse	DEWS[a] EHP
	Transport, LNG processing	DEWS, EHP, NRM
Compliance and Monitoring	Water monitoring	EHP, NRM
	Safety, drilling, reporting	NRM (CSGCU)
	Final reporting, closure, rehabilitation	EHP

[a] Department of Energy and Water Supply
[b] Department of Environment and Heritage Protection
[c] Department of Natural Resources and Mines (includes OGIA and CSGCU)
[d] Department of State Development, Infrastructure and Planning

Act, accrediting Queensland Government processes; this action was designed to remove any Federal–State duplication of process, subject to a formal bilateral agreement between the two governments within a year.

Rapid Developments Now Under Review

In summary, the evolutionary path to regulatory oversight in Queensland started with the general use of existing petroleum legislation and then moved quickly to more specific and complex legislation and oversight, with more purpose-designed institutions. The speed of development has been dramatic, with some 7000 wells active in mid 2015, and the proliferation of legislation and regulatory bodies and guidelines has been equally rapid.

Seven major pieces of legislation are involved, run by six different State Departments. The response of the Queensland Government to the situation was to commission a

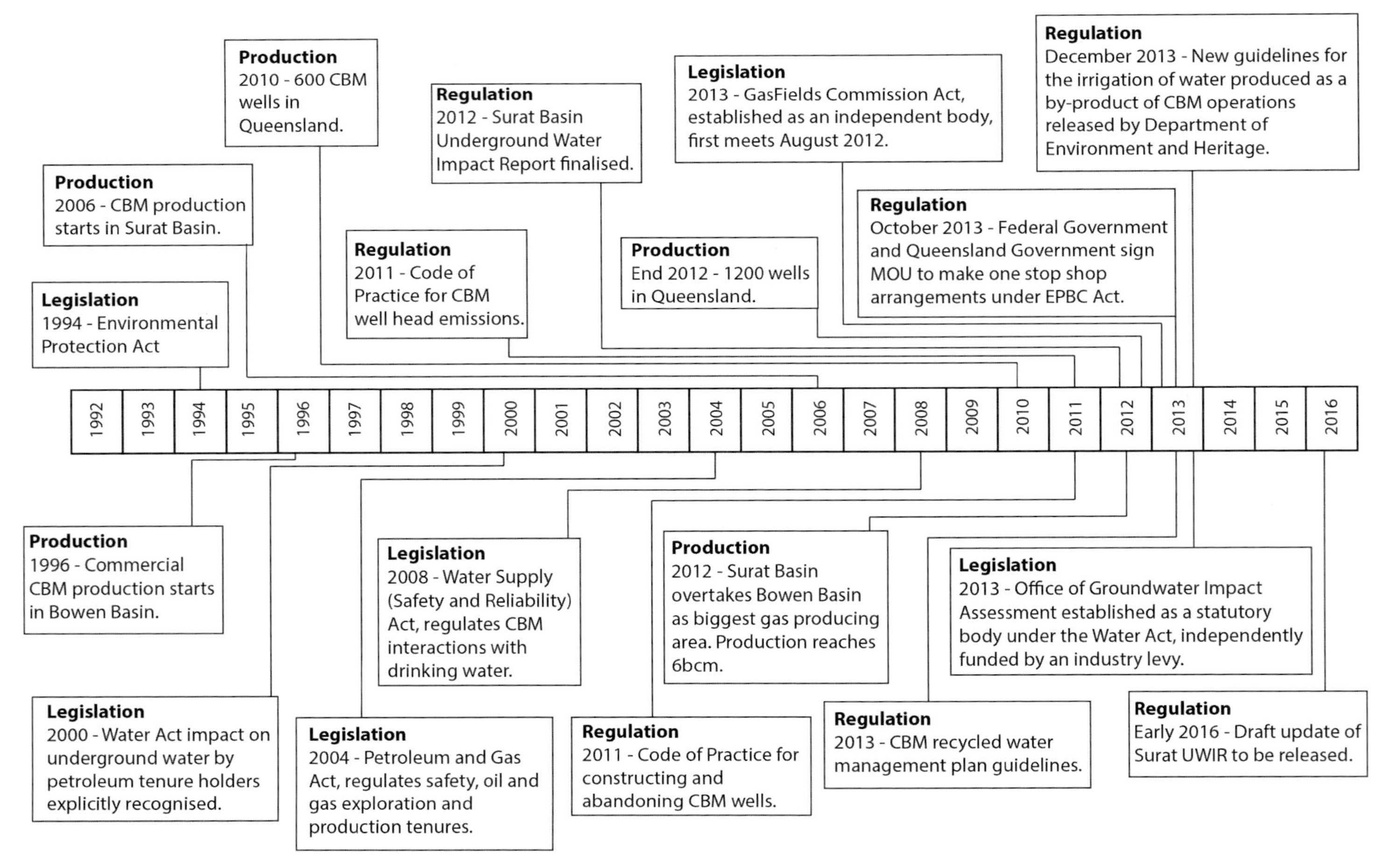

Figure 15.3 Timeline summarising the development of CBM in Queensland.

comprehensive review of the regulation of the CBM industry, to be performed by the Queensland Competition Authority, which reported in February 2014. The report summarised the complexity, costs and overlap in regulation and made significant proposals for reforms designed to streamline the regulatory process without compromising environmental or other public policy goals. Other reviews are also under way, including one of the Strategic Cropping Land Act 2011, which was designed to protect the State's best cropping land from development that would otherwise harm its productive capacity.

The approach to regulation of the industry in Queensland has evolved rapidly from that pertaining to conventional petroleum extraction to the more specific needs of CBM production, notably in water extraction and disposal. The review of beneficial-use approval processes, however, still remains one of the most challenging issues. A number of regulatory instruments and institutions have been purpose-built, and in many ways these approaches embody facets of global best practice. The present regime that has evolved is now complex, with multiple agencies involved.

New South Wales

By contrast with Queensland, the State of New South Wales (NSW) has little history of oil and gas exploitation, although small-scale CBM production has taken place. Commercial production dates from a development in 2001 based near Camden in the Sydney Basin. Gas plays a relatively smaller role in the New South Wales energy economy than in the Queensland energy economy, with gas use at 3.6 bcm/annum, around two-thirds that of Victoria and less than that of Queensland, despite the fact that NSW has the largest population of any Australian State (AEMO 2013). In part, this is so because of the dominant role of coal fired power, the lack of local production and the relatively late construction of gas delivery infrastructure via pipeline from the Cooper Basin in south Australia and more recently from Bass Strait in Victoria. Nevertheless, gas remains a key energy input in a number of industries, including petrochemicals, fertiliser and cement, and is also an important input to the power sector. Together, large industrial users and the power sector account for about half the NSW gas use.

New South Wales Faces a Difficult Transition

Any gas production developments in NSW take place against a challenging short- to medium-term market situation. Gas users in NSW have generally benefitted from long-term relatively stable contract-based prices. This, by global standards, has led to low prices, as the Australian east coast gas market has not been exposed to international price developments. Nevertheless, a significant proportion of these wholesale contracts are set to expire soon; around 80% of contracted wholesale supply will expire before 2018. While a well-functioning eastern Australian gas market could be expected to provide additional gas, the costs and prices of new gas supply are certain to rise especially given competition for the gas supply needed to meet the expansion of Queensland LNG export facilities in the same period before 2020.

Overall, the development of policy surrounding the eastern Australia gas market, including key network connections, remains a work in progress. A deficiency of pipeline capacity, in particular, does not provide complete confidence that the market will deliver gas supplies efficiently in the necessary timeframe. The market also lacks transparency in several key areas, such as contracts and pricing.

The situation in New South Wales on CBM development has many similarities with Queensland. In particular, landowners do not own or control mineral rights on their properties; these are vested in the Crown, namely the State Government. Water quality and quantity issues, as everywhere in Australia, are highly socially and politically sensitive. Further, population densities in rural areas tend to be higher in the proposed developments in NSW than in Queensland. Such locations also have little history of oil and gas development. Thus, despite its experience of coal mining and transport, the NSW population has relatively little understanding of the impacts or, indeed, the costs and benefits, of such production.

Legislation governing CBM extraction is rather general, as might be expected from the immature stage of development. For example, the Petroleum Onshore Act 1991, the Protection of the Environment Operations Act 1997 and the Water Management Act 2000 are the major pieces of relevant legislation.

NSW Reserves are Substantial, with Three Projects

New South Wales CBM reserves are estimated at nearly 75 bcm, with substantial additional contingent resources of 160 bcm (RLMS 2013). Currently, there is one small project under way, with three larger projects at planning and assessment stages. The small Camden project just to the south of Sydney, operated by AGL, produces around 5% of NSW current gas supply. Production started in 2001, from around 150 gas wells, of which 117 have been fracture stimulated (AGL 2015); 96 are still in production. Proposals to expand this to the north were first advanced in 2010 but in early 2013, responding to community concerns, the project developer, AGL, requested that the NSW Department of Planning and Infrastructure suspend its assessment of the proposed expansion, including deferring public hearings. In July 2015, AGL announced it would not proceed with this expansion project.

The Metgasco project, based in the north east of NSW, in the Clarence–Moreton Basin, aimed at producing 0.5 bcm per annum, includes over 10 bcm of reserves plus some 60 bcm of proven, probable and possible resources. In 2012, a gas sales agreement was signed with a large local dairy firm. Some 50 appraisal wells have been drilled. In March 2013, Metgasco suspended its Clarence–Moreton Basin field operations because of uncertainty surrounding NSW government support and regulations. Field operations and subsequent development and production may recommence when the regulatory environment is judged to be more certain, although public opposition to gas developments in this relatively more populated region is strong. The NSW Government offered some AU$25 million in compensation to buy back and cancel the project's three petroleum exploration licences (PELs), effectively ending the project.

The company AGL has a second project at planning stage, the Gloucester project, located about 100 km north of Newcastle. The project envisages 110 gas wells, producing from formations between 200 and 1000 metres deep, for a total output of 0.4 bcm per annum that should approach one-fifth of NSW gas demand, on the basis of proven reserves of 17 bcm. The project would benefit from proximity to one of the main gas trunklines, at Hexham, near Newcastle.

An Environmental Assessment for the Gloucester project was submitted in 2010, following some years of community consultation, and the project was approved by the NSW Planning Commission in early 2011. This decision was challenged in the Land and Environment Court but the original approval was upheld late in 2012. Commonwealth approval under the EPBC Act was obtained in early 2013, subject to some three dozen conditions relating to matters of national environmental significance. The construction of processing facilities and pipeline interconnections was intended to allow the first gas to be delivered in 2015, with full output by 2018. But, in January 2015, the detection of BTEX chemicals in the flowback water associated with a pilot hydraulic fracturing program led to the suspension of work on the project. Flow testing of the four pilot wells continued in the second half of 2015. Lower forward gas prices are expected to affect the project's viability. A final investment decision is possible in 2016.

An additional project, the Hunter Gas project, was effectively abandoned in 2015, with relevant Petroleum Exploration Licences included in the NSW Government's buy-back scheme.

The largest CBM project proposed in NSW is the Santos' Narrabri project, intended to ultimately produce some 1.8 bcm per annum from 850 production wells, based on proven reserves of 60 bcm. Such production levels could meet half the NSW gas demand and, when connected to the existing gas delivery infrastructure, offer some transport cost savings over interstate supplies. The first stage of the development involved the drilling of 15 appraisal wells and received Commonwealth approval in October 2013. On 3 April 2014, the company lodged its preliminary environmental report. The report highlighted the benefits of the project, including the creation of 1200 jobs during construction with up to 200 ongoing jobs, plus a regional community benefit fund, estimated to reach $160 million over 20 years, in addition to royalties over the project's lifetime of up to $1.6 billion. The company indicated that all development would occur outside the two kilometre radius from residential zones, future growth areas and critical industrial clusters, as defined by State legislation. Water extraction at 1.5 Gl per annum, while substantial, should be seen in the context of the 300 Gl of water drawn by other users in the Namoi River catchment. Conditional approval for produced-water processing facilities was given in August 2015, with site works starting in late 2015. A full Environmental Impact Assessment was expected in late 2015.

In March 2014, Santos and AGL adopted a set of principles in NSW that recognised the companies' long standing position that they will not undertake drilling activities on private land without the voluntary consent of the landholder (Santos, 2014). The Principles were signed by Santos, AGL, NSW Farmers, Cotton Australia and the NSW Irrigators Council.

Progress is Slow, and Approvals Processes Remain to Be Clarified

The ongoing concerns regarding CBM developments of a number of groups in NSW prompted the State Government to take a number of regulatory interventions.

In November 2012 the Government issued the Strategic Regional Land Use Policy, which was designed to give greater protection to valuable agricultural land and water resources from the impacts of mining and CBM developments. This was to be achieved by identifying strategic agricultural land and implementing a new Gateway process. The Gateway assessment is an independent, upfront, scientific assessment of the impact of new, significant, mining and CBM State proposals on strategic agricultural land and its associated water resources.

Strategic agricultural land comprises:

- biophysical strategic agricultural land – land which has the best-quality soil and water resources and is capable of sustaining high levels of productivity; and
- critical industry cluster land – a concentration of significant agricultural industries potentially affected by coal seam gas or mining developments.

The Mining and Petroleum Gateway Panel, comprising independent scientific experts, was established to oversee the Gateway assessment process, which must occur **before** an applicant can submit a development application. The Gateway process applies to the two million hectares of strategic agricultural land which was mapped in the Upper Hunter and New England north west regions of the state in 2012 and to the equine and viticulture clusters in the Upper Hunter region.

In February 2013 the NSW Government commissioned an independent review of CBM activities from the Chief Scientist and Engineer. The interim report of this review, released in July 2013, advised the NSW Government to commit to a vigilant, transparent and effective regulatory and monitoring regime. As a first step, the need for a strong policy on data gathering and handling was identified. Land owner compensation, company insurance and operator penalties, along with the need to address knowledge gaps, were also highlighted as areas to be strengthened. A final report was delivered in September 2014.

In March 2013, the NSW Government appointed a Land and Water Commissioner to provide independent advice to the community regarding exploration activities on Strategic Agricultural Land throughout the State. The intention is for the Commissioner to provide advice to the community on applications for mineral and petroleum rights, taking into consideration CBM, landowner rights, access agreements and compensation, and to oversee land access agreements; this would include publishing general remuneration information.

In the latter months of 2013, the New South Wales Government moved to clarify all these studies and approaches by:

(i) imposing a moratorium on CBM exploration and extraction in Sydney's drinking water catchment;

(ii) prohibiting CBM development in residential areas and within 2 km of such residential zones. Vineyards and horse studs are also included in these provisions, as are future residential growth areas; and
(iii) starting up the Gateway Panel (see above) for pre-development approval for CBM development. This panel will work with the Commonwealth Independent Expert Scientific Committee (see below) in its deliberations.

The agreement between the States and the Commonwealth in November 2013 may be a positive step for improving the investment climate for coal seam gas (CSG, also known as coal bed methane (CBM)) in NSW. Overall, the degree of uncertainty inherent in the NSW Government regulatory approaches, as they existed in mid 2015, seemed set to slow the development of the industry in the State. Similarly, widespread public concerns related to access by gas companies to land, lack of compensation and lack of transparency and information about most aspects of company activity and performance and to potential water contamination have not been fully resolved.

In this regard, the Final Report of the NSW Chief Scientist, delivered at the end of September 2014, marked an important contribution to the development of regulatory policy in NSW. The report set out succinctly ways to address community concerns, with many recommendations aimed at adopting best practice approaches seen elsewhere. Recommendations included:

- land access arrangements, including measures to compensate landholders and others affected;
- the full cost of regulation of the industry to be borne by the industry through fees and charges;
- clear designation of those areas where CBM can and cannot occur, using a strong evidence base, notably for water and other social and environmental impacts;
- much greater levels of data transparency and access; and
- development of a purpose-built regulatory regime with a single regulator to support it, in recognition that the current approach was incremental and ad hoc.

Given these major modifications to current approaches, the Chief Scientist concluded that the industry could be effectively managed.

In November 2014, the NSW Government responded to this Final Report, with its Gas Plan, and committed itself to accepting all the Chief Scientist's recommendations. The Environment Protection Authority (EPA) was appointed as the lead regulator for gas exploration and production. Landholders and others affected by development would receive independent advice on compensation benchmarks from the Independent Pricing and Regulatory Tribunal (IPART), and a Community Benefits Fund would be established, based on both industry and government contributions, to ensure that the communities most directly affected by gas development would share at least some of the benefits.

As part of its Gas Plan, the State Government also committed to implementing a much more selective approach to designating areas for CBM development; subsequently, a number

Table 15.2 *New South Wales – simplified regulatory responsibilities for CBM project life cycle*

Activity		Responsibility
Exploration	Petroleum prospecting, exploration	
	Production licensing	NSW Minister for Resources and Energy through Division of Resources and Energy, DTIRIS[a]
	CBM exploration, assessment, production	OCSG[b]
	Implementation of Land Access Agreements	Land and Water Commissioner
Development	Pre-development	Gateway process
	Assessing development applications for major projects, strategic regional land use policy, planning for CBM	DPI[e]
	Water Licensing	NSW OW[c]
	Water Impact Statement	NSW OW
	Environmental Protection Licence	NSW EPA[d]
Production	Well integrity	OCSG
	Groundwater monitoring	NSW OW
Compliance	Environmental/Health	NSW EPA (compliance with EPL)
	Compliance with non-environmental issues	OCSG
	Site rehabilitation	OCSG

[a] Department of Trade, Investment, Regional Infrastructure and Services
[b] Office of Coal Seam Gas, within DTIRIS
[c] NSW Office of Water
[d] Environmental Protection Authority
[e] Department of Planning and Infrastructure

of petroleum exploration licences (PELs) have been bought back and cancelled with modest compensation for the licence holder. Sixteen PELs were bought back under this program, ending in late 2015, reducing the area under petroleum title from 60% of the State to 8.5%. The State Government also announced a Strategic Release Framework to ensure that only selected areas would be released for gas production; the total area is likely to be significantly reduced compared with that licensed for exploration prior to the Gas Plan announcement. Further, the Groundwater Baseline Project was to be expanded, to provide the knowledge base to underpin these decisions.

Many of these features directly address community concerns and incorporate best practice within the regulatory design framework. The key test of their effectiveness will lie in their implementation, which will only be capable of assessment as legislation is enacted and scientific, regulatory and other competencies are developed and applied. Despite these

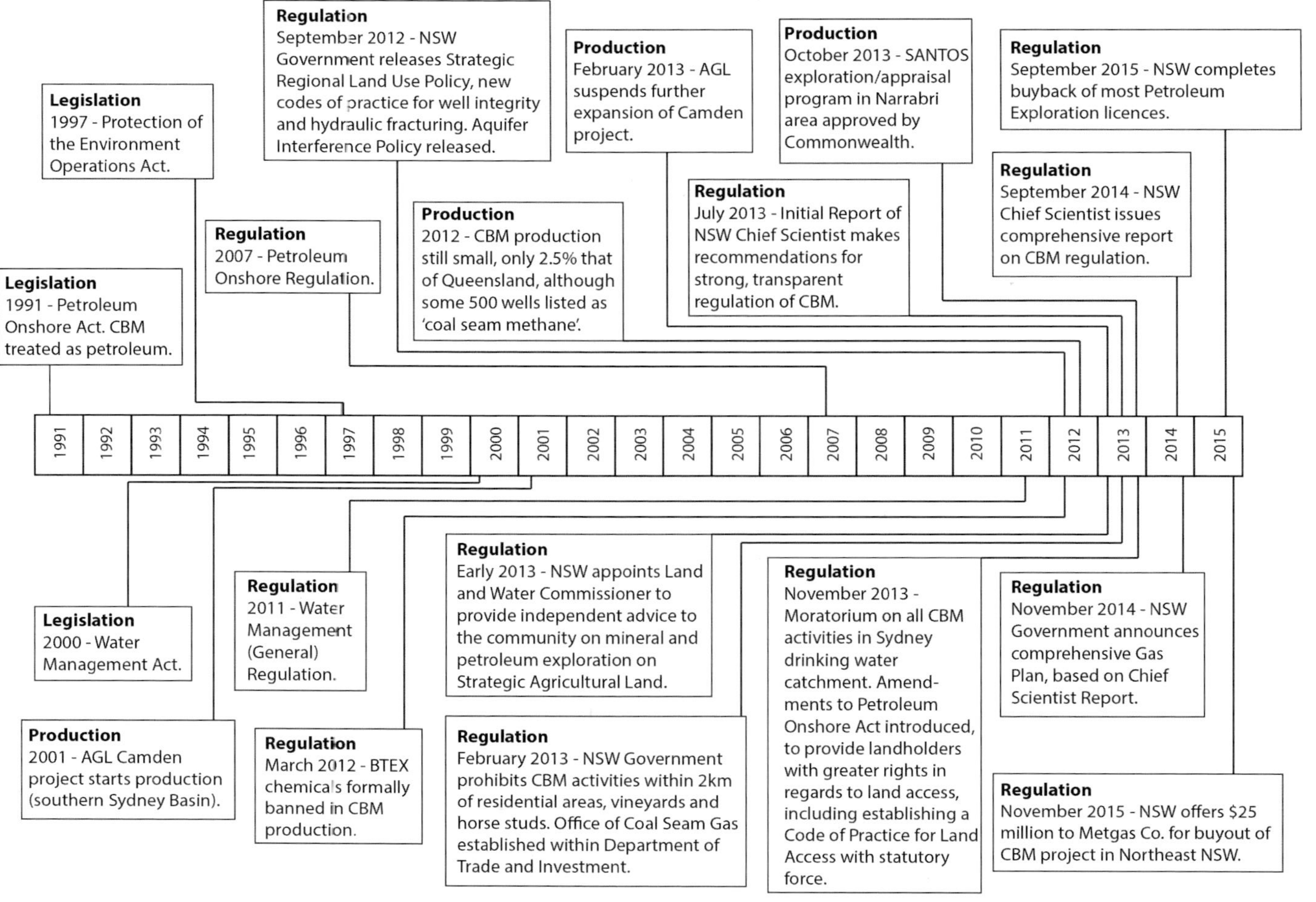

Figure 15.4 Timeline summarising the development of CBM in NSW.

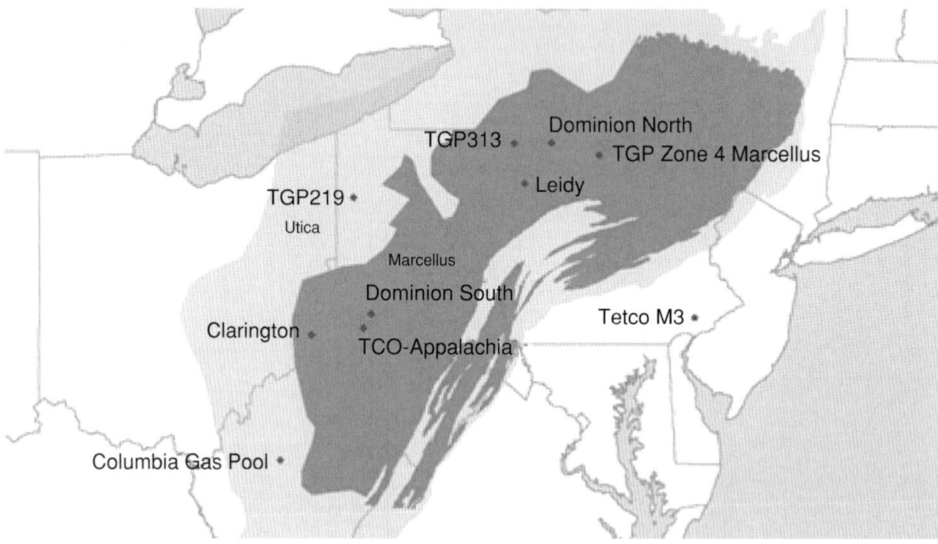

Figure 15.5 Map of Pennsylvania and New York States, showing the Marcellus Basin. *Source:* US Energy Information Administration (December 2015).

initiatives, many opponents of CBM developments remain unmollified by the announced measures.

Pennsylvania

With an area of only 120 000 square kilometres, Pennsylvania is only about 15% the size of NSW but with a greater population, of about 13 million. Nevertheless, some parts of the rural north of the State have relatively low population densities, below 10 persons per square mile (4 per km^2). Pennsylvania has a long history of petroleum exploration, dating back some 150 years, although not always in the regions where shale gas is now being produced.

Pennsylvania is a very mature petroleum province, and recent advances in drilling techniques have allowed the rapid expansion of production in the Marcellus Shale formation (see Figure 15.5). This is a geological formation underlying much of West Virginia, Pennsylvania, Ohio and New York States. Currently, Pennsylvania accounts for almost 90% of the Marcellus output. The formation underlies about two-thirds of Pennsylvania, but not the more populous south east region of the state.

Output from Marcellus Shale Has Expanded Dramatically

The Marcellus Shale formation, at depths of 1800 to 3000 metres, is generally much deeper than the coal bed methane formations tapped in Australia. The shale gas resource had been considered uneconomic, owing to its low permeability, but advances in horizontal drilling, hydraulic fracturing and advanced seismic techniques have overturned this view. As a result, gas output in Pennsylvania has increased from 0.3 bcm in 2008 to nearly 60 bcm in 2012, a 200-fold increase over the period.

The State's current production level accounts for nearly 7% of United States gas production. Estimates for 2013 indicate a nearly 50% increase in output over 2012 (Figure 15.6), and data for 2014 indicate that growth continued at remarkable rates, so that by the end of 2014 the Marcellus output was around 150 bcm per year, with the US total gas output at a record annual rate of over 750 bcm, despite declining gas and oil prices and falling numbers of active drilling rigs. Subsequently, in late 2015, the active gas rigs in the region fell to barely 60 (from 150 in 2012), output peaked at around 165 bcm on an annualised basis and wholesale prices in the region were around $1 per Mbtu lower than the already low Henry Hub value, $2.30.

Reserve estimates have grown from zero in 2007 to 180 bcm in 2009 and to 650 bcm in 2011, with large and rapid ongoing growth since then. In recent years, between 1500 and nearly 2500 wells have been drilled each year; the vast majority are horizontal wells. Productivity per well has risen sharply, both in terms of gas output per well and reduced drilling times and costs. The size of the Marcellus output is already 'redrawing' the North American gas map. The region has lowered its demand for gas imported from other regions and countries and is sending exports into southern Ontario. Shale gas production is also reducing the eastward flow of Alberta gas to both Canadian and US users. A major LNG export project recently approved by the US Government will draw heavily on Marcellus gas, exporting through existing port facilities in Maryland that were originally designed for LNG imports.

Environmental Concerns Have Been High

While the depth of unconventional gas production is well below that of formations from which underground water is extracted, considerable public concerns have arisen over the use of fracturing technologies and the possible contamination of groundwater, either by fracturing fluids or through cross contamination. The quantities of water used per fracturing stage (and there may be up to 20 fracturing operations per horizontal well) require between 2 and 4 Ml, and sometimes as much as 8 Ml. By comparison, an Olympic-sized swimming pool contains about 2 Ml of water. Concerns exist about water availability and the treatment and disposal of any water produced with the gas. In contrast with the coal bed methane output in Australia and elsewhere, large-scale dewatering and the disposal of extracted water is not required, but the treatment of flowback water remains an issue. Land disturbance through drilling operations is also a concern, although the technique of drilling multiple wells from a single drilling pad can reduce noise and disturbance.

A key difference between the Australian and United States regulatory environment is that, typically, US landowners own the subsurface mineral rights. Hence, successful gas extraction provides an ongoing revenue stream to the landowner throughout the production phase, long after drilling operations are completed. Consequently, the regulation of mineral

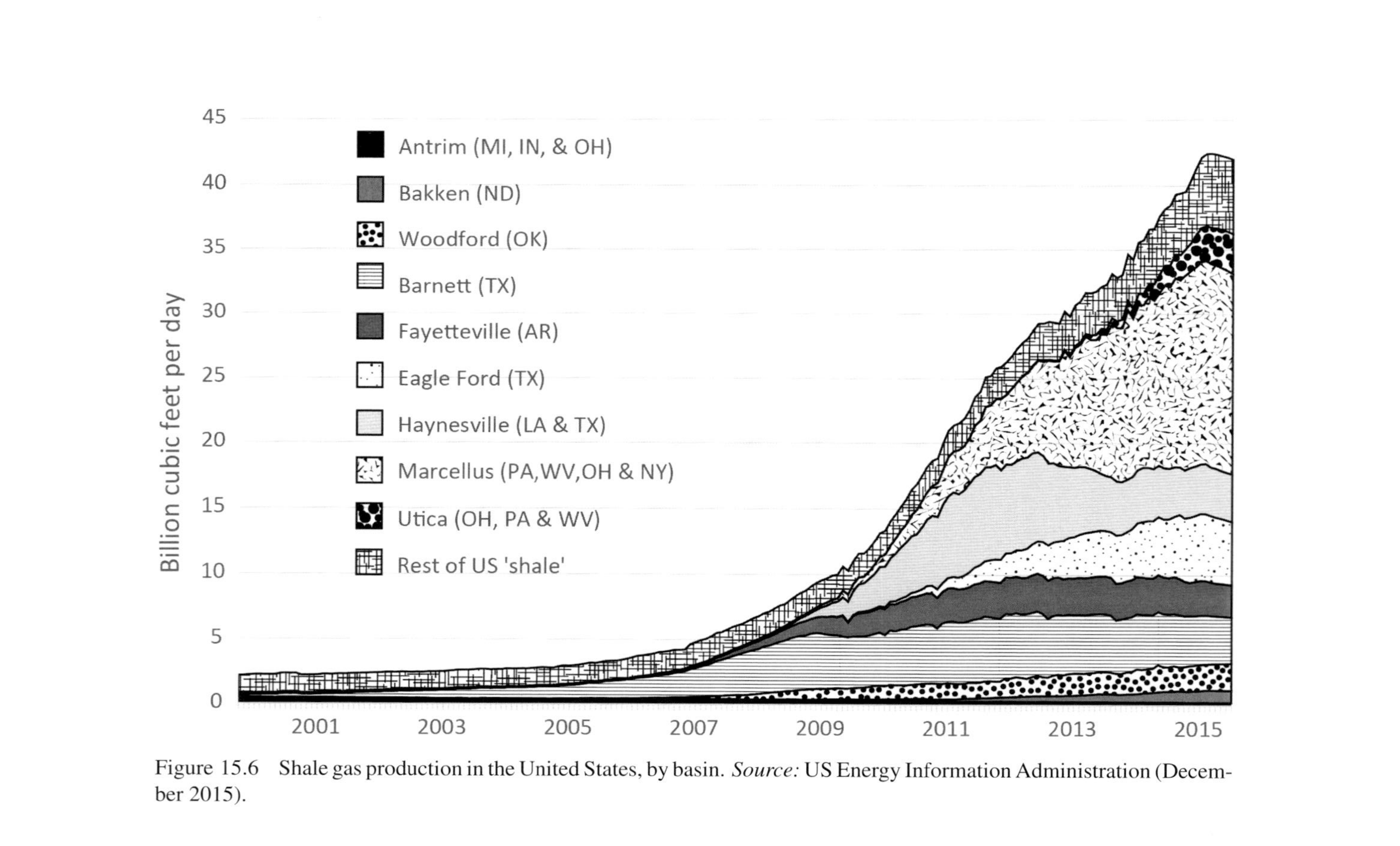

Figure 15.6 Shale gas production in the United States, by basin. *Source:* US Energy Information Administration (December 2015).

rights and access is a private contractual arrangement between landholders and producers. Governments generally have only a small role to play in terms of the right to extract from a specific parcel of land, although local governments can, and do, regulate access, for noise and traffic reasons. Rights of 'eminent domain' do not extend to gas pipelines, so that delays in extending the pipeline network have occurred, slowing the expansion of the gas output.

The Existing Regulatory Regime Has Been Heavily Modified in Recent Years

The oil and gas industry in Pennsylvania has been strongly regulated since 1956, through the State-based Oil and Gas Act. Subsequent laws (the Coal and Gas Resource Co-ordination Act, the Oil and Gas Conservation Law) have intensified this regulation but, notwithstanding the long history of oil and gas production, regulations accounting for the novel technology of hydraulic fracturing have had to be addressed. For instance, the oil and gas code was extensively modified over the period 2009 to 2011 to cope with the widespread use of new technologies and the sharp increase in gas output. Prior to the finalisation of these changes, in early 2011, there was a moratorium on unconventional gas development for a few months and a year of public meetings followed by a report on developments on the Marcellus Shale. The report became the basis for Act 13 (Murtazashvili 2015), which amended the Oil and Gas Act of 1984; thereafter, gas output increased dramatically.

Managing Water Impacts is a Special Focus for Regulation

Water impacts are regulated under the Clean Streams Law and Water Resources Planning Act. Regulation and enforcement are located in the Department of Environmental Protection (DEP), within the Office of Oil and Gas Management, which has a strong focus on the rapidly growing output from the Marcellus Shale. The rapid growth in gas output has been mirrored by an equally rapid growth in the Office's permitting, inspection and enforcement staff, from around 60 in 2010 to more than 200 by 2013. Costs are recovered from the industry through a new drilling permit fee, currently raising more than $15 million per annum.

Operators seeking a drilling permit must specify proximity to surface water and water supplies. Landowners should be automatically notified of drilling permit applications. Natural gas wells must be cased and cemented through all fresh water aquifers, before intersection with the gas-producing zones. Since late 2013, operators have been required to report bi-annually on well status, production, waste handling and water management. Drilling is, generally, not permitted within 500 feet of buildings or watercourses, a 'setback' limit that is amongst the highest in the United States (in Texas it is 200 feet while in neighbouring Ohio it is between 100 and 200 feet). This setback limit was recently increased from the previous 200 feet. Pennsylvania, in common with New York and Ohio, has additional setback restrictions for major water sources, such as private water wells, lakes and major streams, extending to 1000 feet.

On water management, which is a key regulatory focus, drilling companies, as part of the permitting process, must develop a Water Management Plan. This Plan should cover the entire life cycle of the water used and should include the identification of quantities and sources of water, impact analysis on the water source and options for the storage and disposal of waste and formation water. For water from hydraulic fracturing, a re-use plan must be included.

River Basin Commissions Play a Key Role

Rules for water quality and use have been developed in conjunction with the relevant river Basin Commissions, namely the Delaware River Basin Commission and the Susquehanna River Basin Commissions (RBCs).

These two bodies must take into account all developments within their respective basins, including interstate impacts. Permits are required for any water withdrawals, mirroring the approach taken in the much drier western States, and are significantly more demanding than those of most eastern States. While there are, in fact, a number of River Basin Commissions in the United States, these Commissions appear to be the only two with authority to regulate shale gas development and with the power to shut down operations if water impacts exceed agreed levels. The Delaware RBC has used its authority over water permitting to impose an indefinite moratorium on shale gas development in the basin. Two of the four States covered by the Delaware RBC have moratoria of their own (New York and New Jersey). Shale gas developments in the other two (Pennsylvania and Maryland) lie outside the Delaware River Basin; hence the Delaware RBC moratorium has little practical impact at present on shale gas development.

From the perspective of the DEP, adverse water impacts in the vicinity that are observed within a year (previously six months) are assumed to be the driller's responsibility. The radius of concern has increased in recent years, with the limit now set at 2500 feet of the drill site. This provision provides a strong incentive to undertake predrilling water testing, although Pennsylvania does not mandate this (unlike many other jurisdictions). In practice, about 90% of water wells within 1000 feet of a gas well were tested before drilling.

Disclosure Has Improved

Disclosure of the composition of fluids used in hydraulic fracturing operations has proved controversial. Federal Safe Drinking Water legislation provided for this until a 2005 amendment excluded fracturing fluids from its scope. Pennsylvania, in common with almost all gas-producing States, requires that all chemical use must be fully disclosed to the DEP and that this information is available to landowners, local governments and other interested parties, although there are exemptions.

The Pennsylvania State Government has moved to restrict the ability of local and municipal governments to control oil and gas development. A specific State Act requires local governments to allow oil and gas development and associated facilities in most zones. Other States give local governments much more control over local regulations, such as

Table 15.3 *Pennsylvania – Simplified regulatory responsibilities for unconventional gas development for LNG exportation*

Activity		Responsibility
Pre-exploration		Drilling companies enter into lease with landowners, a private contractual agreement
Exploration		DEP[b] issues drilling permits, including Water Plan, for accessing, storing and disposing of water. Some municipalities have banned hydraulic fracturing within their city limits
Development	Drilling oversight	DEP
	Water extraction	Authorisation required from Delaware River Basin Commission, or Susquehanna River Basin Commission with DEP, assessed on a regional basis
	Water disposal	DEP
	Export authorisation	Federal DOE[c] for non-FTA countries (for FTA countries it is virtually automatic)
	Environmental approvals for LNG plant	US FERC[a]
Production		DEP mandates mechanical integrity testing for all operating wells. Semi-annual reporting for all wells
Compliance and Monitoring		DEP – Oil and gas compliance reporting system available online

[a] The Federal Energy Regional Commission has a strong regulatory presence in all interstate pipeline transmission issues.

[b] Pennsylvania Department of Environmental Protection, Office of Oil and Gas Management

[c] Department of Energy (Federal)

those concerning traffic movements and noise control. Nonetheless, some municipalities, including Pittsburgh in 2010, have banned certain operations, notably hydraulic fracturing, within their boundaries.

A Federal EPA investigation into links between fracking operations and drinking water (discussed further below) contained two case studies from Pennsylvania: Bradford and Susquehanna counties. The study found some private water wells had been damaged by methane and ethane migration caused by nearby fracking. However, there was no evidence of contamination by waste water or fracking fluids (EPA 2015).

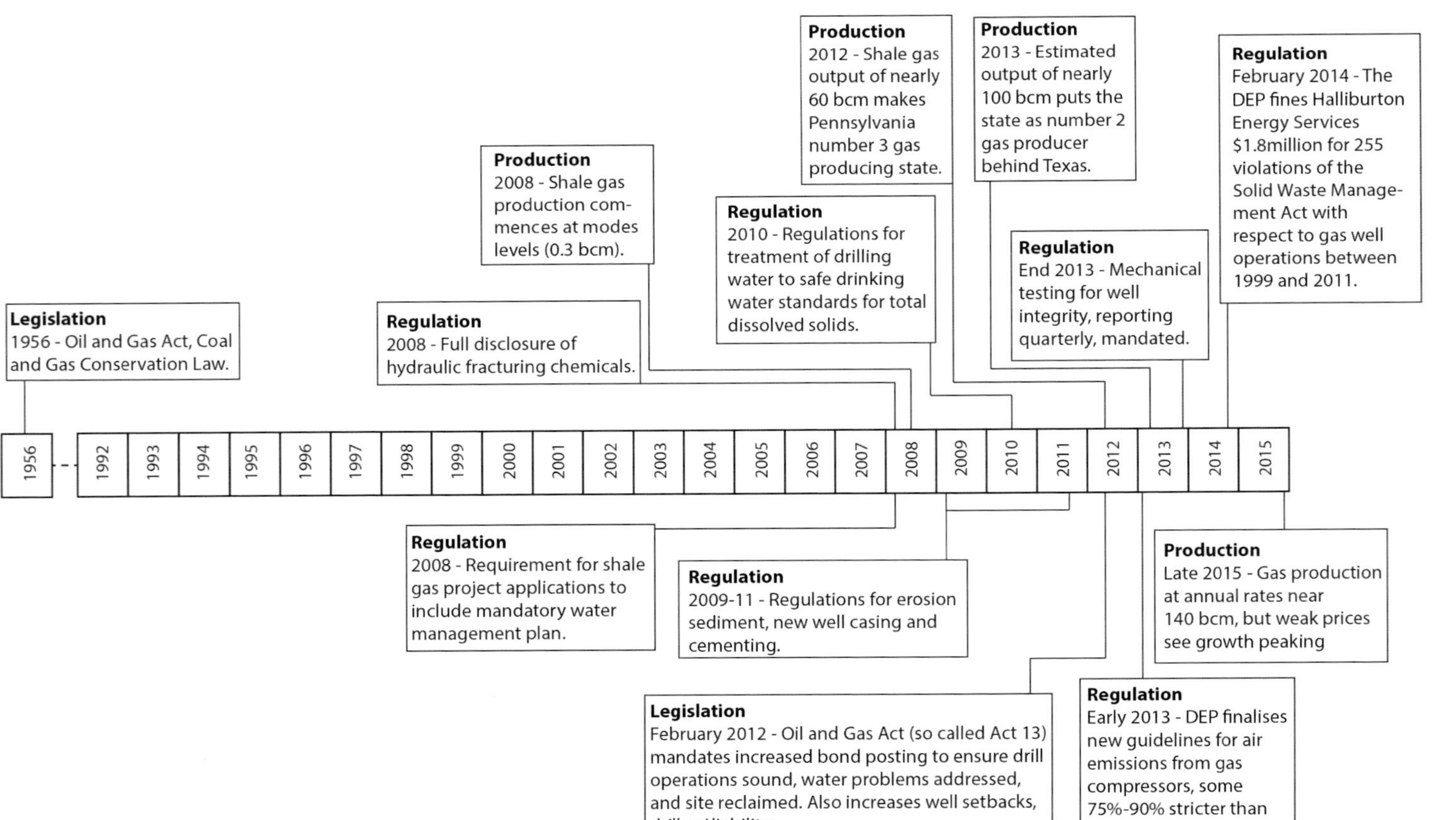

Figure 15.7 Timeline summarising the development of shale gas in Pennsylvania.

Summary

The regulatory approach in Pennsylvania reflects the rapid growth in the shale gas industry. Regulation and the structures and institutions supporting it have been very dynamic, notwithstanding the State's long-standing petroleum industry. Regulations have been changed in scope and direction, for example, those covering the radius of concern over drilling activities. The scope of regulation is comprehensive, covering permitting, drilling operations, well integrity and water sourcing use and disposal, and incorporates regional approaches. Regulatory standards are used, often with minute detail, including, for example, the cementing types and practices to be used to ensure well integrity.

More recently, incentive-based rather than compliance-based regulation is being developed (for example, regarding predrilling water monitoring or more flexible cementing materials to improve well integrity). Transparency is less than ideal, although it is improving in areas such as fracturing-fluid disclosure. Outright prohibitions on use of particular compounds, for example BTEX, are rare although some practices are banned (for example venting, although not flaring). Overall, the State has undertaken a well-resourced response to the challenges posed by the very rapid expansion of the shale gas industry, but popular discontent remains high.

New York

New York State lies immediately north of Pennsylvania, and the Marcellus Shale formation extends across the state border into the south western portion of New York.

A Moratorium Has Been In Place Since 2008

The state is somewhat larger in area than Pennsylvania. Its population of nearly 20 million is heavily urbanised, although the area underlain by the Marcellus Shale is much less densely populated. There is a history of oil and gas production dating back to the nineteenth century and, although some 14000 wells are still active, its oil and gas industry is less familiar than in Pennsylvania. Its gas output is less than 1% of that of Pennsylvania, and there is no unconventional gas production. In large part, this can be explained by a much more cautious approach to the regulation of new gas output, with a *de facto* moratorium in place since 2008 (made *de jure* in 2010), now likely to extend into the indefinite future.

Petroleum development in the State is overseen by the Bureau of Oil and Gas Regulation of the Department of Environmental Conservation (DEC). The Bureau oversees the permits necessary for any oil or gas drilling, which must conform to requirements established in a generic environmental impact statement (EIS). In 1992, the Bureau adopted a generic impact statement covering natural gas production but, although there were references to hydraulic fracturing, it contained no actual regulations on the subject. In 2008, the Bureau was instructed to remedy this gap by preparing a supplemental generic EIS. A draft was posted for public comment, with first recommendations appearing in 2010. Further revisions have been made to this and presented in 2011 and 2012.

The ongoing nature of the process of developing a generic EIS has meant that no permits for hydraulic fracturing could be issued, effectively stopping unconventional gas production. In late 2010, the New York legislature imposed a legislative moratorium on hydraulic fracturing which was supplanted by an Executive Order from the Governor that imposed a moratorium on hydraulic fracturing until the supplemental EIS was concluded. Over the following four years, Governor Andrew Cuomo took no decision as to whether to accept the hydraulic fracturing rules developed by DEC or to restart the process. Finally, in December 2014, Governor Cuomo definitively banned hydraulic fracturing in the State. In doing so, he accepted the recommendation of the New York State Department of Health. The Health Department cited potential health risks, noting groundwater contamination in Wyoming and increased traffic deaths in Pennsylvania. The option of allowing some limited gas exploitation in the Southern Tier region, adjoining the border with Pennsylvania, and underlain by part of the Marcellus formation, was rejected. The ban appears to be open ended, having no defined time limit.

Municipalities Have Successfully Delayed Shale Gas

In New York State, local municipalities have tried to stop shale gas production, relying on powers that allow them to control traffic and noise. This happened in Binghamton, just north of the Pennsylvania border in south central New York State, and also further to the north, in the small towns of Dryden and Middlefield. While these municipalities are small, their rules have the potential for State-wide precedents. For instance, Binghamton enacted a two-year ban on hydraulic fracturing in late 2011, but this was challenged by a group of landowners keen to take advantage of potential gas production on their properties. The landowners' lawsuit was upheld by the State Supreme Court but the town appealed this decision. While the appeal was dropped early in 2014, the other towns, including Dryden, which was the first community to ban hydraulic fracturing, remained in the State Court of Appeals having had their bans upheld by lower courts. In large measure, these local- and community-based actions explain the difference in unconventional gas development in New York State compared with Pennsylvania (Arnold and Holahan 2012).

The impact of these production bans on New York gas supplies, and on its gas prices, has been muted by the massive expansion in nearby Pennsylvania. Historically, New York prices had been $1–2 per Mbtu higher than those at the Henry Hub, located in Louisiana, reflecting the cost of transportation from that region, as the north east region of the United States was previously a relatively small gas producer (see Figure 15.8). The growth of the Marcellus gas output saw the hub price for Pennsylvania gas fall below the Henry Hub price over 2013, and the trend continued through 2014 and 2015. The availability of Marcellus gas has meant that gas to New York and Boston has recently traded around Henry Hub prices for long periods. Nevertheless, occasional very severe cold weather (as seen in December 2013 and in early 2014) has caused significant price spikes, so that city gate prices, averaged over 2013, have been around $1.37 and $3.18 per Mbtu higher than

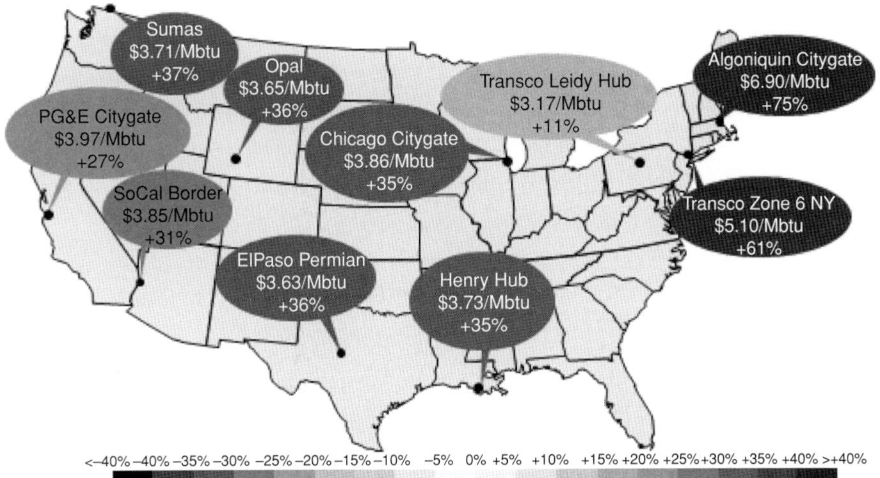

Figure 15.8 Percentage changes in spot natural-gas prices at major trading locations through the 31 December 2013 delivery date. *Source:* US Energy Information Administration (December 2015).

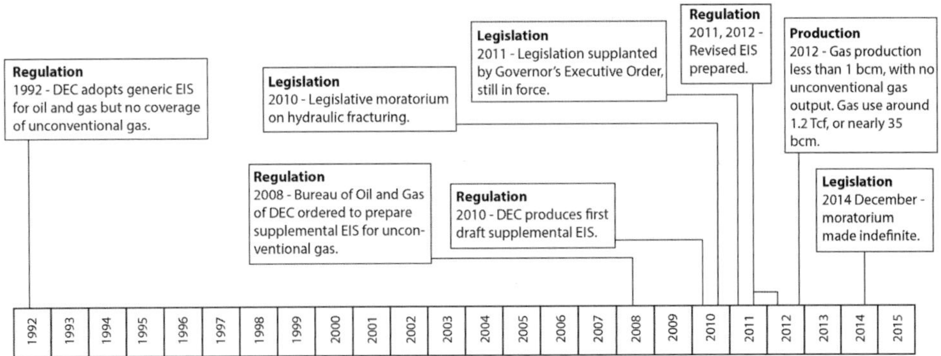

Figure 15.9 Timeline summarising the development of shale gas in New York.

Henry Hub for New York and Boston, respectively. This highlights that pipeline, storage and deliverability constraints can be as significant as gas production in price formation.

The Role of Federal Governments

In both Australia and the United States, the federal system of Government means that States are generally the locus of resource development and environmental standards. Nonetheless, in both countries Federal Governments have taken a number of initiatives to promote best practice in regulation and to provide a certain amount of investment certainty and uniformity, especially where production basins straddle state borders or where pipeline systems inevitably cross those same borders. Water issues, arguably the most intense point of concern in both countries, inevitably fail to respect State boundaries. Given the shortage of regulatory resources, both financial and manpower, and the generally lower level of State resources in general, there is a case for greater, not less, Federal involvement, even if it is merely in a coordinating or facilitating role. This is required, even if only to avoid *ad hoc* policy making and 're-inventing the wheel' in each state jurisdiction.

Australia

Federal–State coordination

A key player in the coordination of efforts in Australia has been the Standing Council on Energy and Resources (SCER), a Federal–State Ministerial Council operating under the authority of the Council of Australian Governments (COAG), which brings together the Heads of the State Governments.

The Council's terms of reference include ensuring the safe, prudent and competitive development of the nation's mineral and energy resources and markets, in order to optimise long-term economic, social and environmental benefits to the community, by:

- facilitating national oversight and coordination of governance, policy development and programme management to address the opportunities and challenges facing Australia's energy and resources sectors into the future;
- providing national leadership on key strategic issues and effectively integrating these strategic priorities into Government decision-making in relation to the energy and resources sectors; and
- enhancing national consistency between regulatory frameworks to reduce costs and improve the operation of the energy and resources sectors.

It has discharged these responsibilities with respect to CBM by undertaking a number of major nationwide initiatives.

While there are many common features of regulation across States (for example both Queensland and NSW require the full disclosure of hydraulic fracturing chemicals to State regulators, and BTEX chemical use has been banned in most States for some years), it is clear that Australia does not yet have a nationally consistent approach to CBM regulation. Bearing in mind the differences already seen between NSW and Queensland, the State of Victoria has had a moratorium in place since 2012, reaffirmed after the election of a new State Government in late 2014 and not likely to end before 2016, following a wide-ranging Parliamentary Inquiry. In early 2015, Tasmania extended a similar moratorium on hydraulic fracturing for five years.

To address this inconsistency across states, SCER produced a National Harmonised Regulatory Framework for Coal Seam Gas. The Framework was finalised after a year-long process including public and industry consultation. It focuses on four main areas:

well integrity; water management and monitoring; hydraulic fracturing; and chemical use including disclosure. It incorporates many features of international best practice described by the International Energy Agency (2012). For instance, it strongly urges greater transparency and greater disclosure of all chemicals used in CSG activities.

A second key activity of the SCER that is also relevant to CBM activities is the National Multiple Land Use Framework. The SCER anchors its approach to CBM within the framework for multiple and sequential land use and development, most notably focussing on the principles that communities and landholders should participate in decision-making on land use change and that that involvement should be timely and genuine.

Independent Scientific Advice is Essential for Evidence-Based Policy

An important Australian Government initiative was the formation in 2012 of the Independent Expert Scientific Committee (IESC) to provide advice on the impact of CBM and coal mining projects on water resources. The committee is backed by some $150 million in funding to undertake regional assessments. The formation of this committee, and its level of resourcing, is an acknowledgement of, and a response to, knowledge gaps. These gaps include an understanding of inter-aquifer connectivity and potential leakage, of gas and water flows within coal seams and of how the desorption of methane from coal seams may alter the properties of surrounding formations.

Sufficient federal funding for research provides some assurance that the advice provided will be complete, objective, regionally based, allows scope for full cumulative assessments and avoids duplication. As a result, there would seem little merit in a State's funding additional work in, for example, the impacts of fracturing fluid chemicals, if this is being adequately addressed by the IESC. Indeed, the State Governments of Queensland, New South Wales, Victoria and South Australia can seek the advice of the IESC on CBM issues under the National Partnership Agreement.

The IESC would appear to be an ideal entity for dealing with trans-border issues, such as those associated with the Great Artesian Basin (GAB), which includes regions of three States. An integrated programme of hydrological research was to be conducted on the GAB in 2015–2016, comprising four inter-related components: the ecohydrology of groundwater-dependent terrestrial vegetation; the ecological values of baseflow and surface-water–groundwater connectivity regimes in non-perennial streams; subsurface ecology; and the composition and resilience of GAB-spring communities. The research will improve understanding of ecological responses to the changes in water quantity and quality associated with coal seam gas extraction and coal mining.

On the advice of the IESC, the Australian Government provided some $4.2 million for the national assessment of the chemicals associated with CBM extraction in Australia. This assessment is examining human health and environmental risks from the chemicals used in the drilling and hydraulic fracturing of CBM in Australia. It will provide an evidence base for management of these chemicals and will improve public access to information on these chemicals. The project started in mid 2012 and completion is likely in 2015–2016 (NICNAS 2015).

The national assessment of chemicals involves a collaboration between the national industrial chemicals regulator, i.e. the National Industrial Chemicals Notification and Assessment Scheme (NICNAS, the lead agency), the Commonwealth Scientific and Industrial Research Organisation (CSIRO), the Federal Department of the Environment and, in an advisory role, Geoscience Australia. The assessment will:

- identify chemicals used in drilling and hydraulic fracturing for CBM in Australia;
- characterise potential short- and longer-term adverse effects on human health and on the surface and near-surface environment caused by the chemicals used in drilling and hydraulic fracturing;
- assess the public, occupational and environmental risks arising from their use, based on current Australian work practices, considering
  - the surface handling of drilling and hydraulic fracturing chemicals at the well site (this will include activities such as handling, storage, transport, mixing, injection, surface spills, spills into natural water courses), and
  - the surface handling of water from CBM wells (including activities such as transport from the well site, storage, treatment, and waste disposal);
- develop models and simulators to better predict soil, surface water and shallow groundwater concentrations of the chemicals used in CBM extraction due to transport and site spills, overflows, runoff and leaks from surface ponds.

The Gas Industry Social and Environmental Research Alliance (GISERA) is an industry–government (CSIRO) body researching the socio-economic and environmental impacts of unconventional gas production, with an initial focus on Queensland's CBM industry. Groundwater and surface water impaction is an important focus, but social and community issues are being studied also. The work of GISERA has confirmed important community benefits, not just in income but in the reversal of rural decline and the retention of younger people.

United States

As in Australia, the individual States are the major locus for regulation of unconventional gas development and, as a result, some 31 different State-based regulatory systems can be identified (27 States actually produce shale gas). Differences in geology, hydrology and population density can explain much of the variation in regulation.

The Federal Government exerts authority over Federal lands, encompassing substantial areas of the western United States, plus control over air and surface water quality and endangered species. Shale gas operations were exempted from the key provisions of Federal environmental laws in 2005. To date, efforts in the Congress to reverse these exemptions have been unsuccessful.

Pavillion, Wyoming

In response to complaints by domestic well owners about problems with well water quality, in 2008 the United States EPA initiated an investigation into the ground water quality near Pavillion, Wyoming. The Draft Report (DiGiulio *et al.* 2011) noted that domestic water wells overlay the 169 vertical gas-producing wells, which produced gas from the shallow Wind River Formation, a complex shale and sandstone formation, as well as from the deeper Fort Union Formation. It observed that hydraulic fracturing had taken place at depths as shallow as 372 metres although generally deeper than 500 metres, and with associated surface casing as shallow as 110 metres. By comparison, domestic water wells were as deep as 244 metres. Initial investigations indicated that high concentrations of certain organic compounds in shallow monitoring wells, near surface pits and used for the disposal of drilling wastes and produced and flowback waters, were a possible source of the shallow groundwater contamination. While there were anomalies in the deeper groundwater, determining the source of these was more difficult. The EPA released its 2011 findings for public and scientific review, noting at the time of release that the circumstances in Pavillion were atypical. In particular, hydraulic fracturing was taking place in and just beneath the drinking water aquifer. These were production conditions unlike those in most shale gas production areas in the United States.

The findings of the Draft EPA Report were strongly contested by ENCANA, the current field operator, and other industry bodies. In 2012, the State appropriated some $750 000 for the construction and installation of residential cistern systems and a water loading station in the town of Pavillion, with completion for late 2013.

In 2013, the State of Wyoming announced it would investigate further the drinking water quality in Pavillion. The work, to be conducted by the Wyoming Oil and Gas Conservation Commission and the Wyoming Department of Environmental Quality, would be supported by the EPA. The key focus of this work was the 14 water wells, all with at least one constituent above safe water drinking standards, located within 1320 feet (a quarter mile, about 400 metres) of some 50 oil or gas wells around the water wells, dating in some cases back to the 1950s. The integrity of these oil and gas wells, plus the historic disposal pits in the area, would also be carefully studied in the investigation.

Even in the absence of final conclusions from this work, it appears that the case of Pavillion demonstrates the importance of baseline water data along with proper water quality information before, during and after oil and gas development.

EPA Takes the Lead

The United States Environmental Protection Agency (EPA) has been actively involved in monitoring the unconventional gas industry (see the fact box on Pavillion, Wyoming). It is undertaking a major investigation into the relationship between hydraulic fracturing and drinking water (EPA, 2012, 2014). Some 18 separate research projects are under way, designed around the five stages of the fracturing cycle, including:

(1) water acquisition, notably the impacts of the withdrawal of large water volumes;
(2) surface spills of fracturing fluid;

(3) the possible impacts of injection and fracturing processes on drinking water;
(4) flowback and produced water; and
(5) wastewater treatment and waste disposal.

The EPA study is examining the toxicity of some 1000 compounds employed in hydraulic fracturing from 25 000 wells drilled in 2009 and 2010. Five case studies have been chosen by the EPA for in-depth evaluation, generally responding to evidence of impaired water quality, health and environmental concerns. These include:

(1) two counties in Colorado, under study for potential drinking water impacts from CBM activities;
(2) Dunn Country in North Dakota, where impacts of a well blowout during hydraulic fracturing for oil in the Bakken formation can be seen;
(3) Bradford County in north eastern Pennsylvania, on the border with New York, where the intent is to evaluate Marcellus Shale gas production, in a region of naturally occurring methane in the groundwater;
(4) Washington County in south western Pennsylvania, an area of around 800 square miles, with a high population density of around 240 per square mile and where more than 200 new gas wells have been drilled every year since 2009; and
(5) Wise County Texas, from the perspective of drilling for shale gas from the Barnett Shale formation.

A general progress report was issued in December 2012 (EPA, 2012) and a preliminary report for public comment and peer review was released in June 2015. Research in the area is underpinned by a collaborative agreement between the US Department of Energy and the US Geological Service. The Draft Report concluded that there were a number of potential mechanisms by which hydraulic fracturing could affect drinking water sources. The report found that there were specific instances where such mechanisms had led to the contamination of drinking water, but their number was small relative to the number of hydraulically fractured wells.

A key conclusion of the draft EPA report was that no evidence was found that hydraulic fracturing has led to widespread, systemic, impacts on drinking water resources in the United States. The report will be peer reviewed, and public comment sought. When finalised, the report could create a global standard for high quality in-depth research on impacts on drinking water, with national and most likely global implications.

Most States are moving to greater levels of transparency with respect to requirements concerning hydraulic fracturing fluids, but there is considerable variation in those requirements. As a result, in 2012 EPA announced a proposed rulemaking, under the Toxic Substances Control Act, for more uniform and widespread disclosure of these chemicals.

The EPA has been more successful, under the Clean Air Act, in reducing methane venting through Reduced Emissions Completions (so-called green completions), which became mandatory from 2015. Nevertheless, industry has resisted this implementation date, citing difficulties of supplying the necessary equipment. In relation to wastewater from hydraulic fracturing and drilling operations, EPA announced in 2011 a plan to develop standards for discharges, citing the inability of many water treatment plants to treat this type of wastewater adequately, the first proposals to be advanced in 2015.

The EPA is leading an effort to reduce methane emissions more broadly, with a focus on the growing oil and gas industry. Methane accounts for an estimated 10 per cent of US greenhouse gas emissions, of which nearly one-third is from the oil and gas sector. Pipelines and new technologies for detecting and monitoring leaks will be enhanced alongside the existing efforts at oil and gas production and processing facilities.

Overall, the US Federal Government role in regulating unconventional gas is modest but it is growing in some key areas. Given the high degree of heterogeneity in US unconventional gas production, and the resultant diverse array of legislative and regulatory approaches, arriving at uniform federal standards is probably unlikely and potentially undesirable, but guidance on minimum standards and transparency across states would be desirable.

Cooperative Approaches Are Worth Examining

The State Review of Oil and Natural Gas Environmental Regulations (STRONGER), a non-profit multi-stakeholder body, and the Ground Water Protection Council (GWPC) are two organisations that seek to promote cooperative approaches across states. These two organisations work to share information to improve the quality of regulatory policy and practice in the various States. Previously, federal agencies (DOE and EPA) and the American Petroleum Institute provided funding for STRONGER and GWPC, but this has ended.

The most recent STRONGER review of the Pennsylvania regulatory regime, in 2013, identified areas of strength, including rapid adaptation in the previous four years, but also highlighted areas for improvement, notably in standardised public data, radioactive waste disposal, surface casing depths and pre-drill water sampling. The GWPC is a similar non-profit association of State water regulators. This Council has reviewed State regulatory practices for groundwater protection. In conjunction with the Interstate Oil and Gas Compact Commission, the Council has established FracFocus, an online information system disclosing chemical use in hydraulic fractured wells. As of late 2015, 23 states now use this site as a means of ethical disclosure, covering some 106 000 wells. This is already an important and growing initiative in both the US and Canada (where five provinces use a licensed version of the system).

A third body has been established recently to address regional issues in the Appalachian area. In March 2013, a group of environmental organisations, philanthropic foundations and major energy companies established the Center for Sustainable Shale Development (CSSD) at the urging of the DOE Shale Gas Advisory Panel. The Center, based in Pittsburgh in western Pennsylvania, is designed to provide shale gas producers throughout the Appalachian region with independent certification of performance standards for shale gas production, and to promote higher and more uniform standards throughout the industry. The CSSD has established 15 performance standards, covering air and water pollution, that include areas such as green completions, flaring and venting and waste water recycling; It opened formally at the beginning of 2014, with four companies seeking certification including Shell and Chevron.

Selected Incidents, Breaches and Infraction

New South Wales

In March 2010, Eastern Star Gas was fined by the NSW EPA for two cases of discharging water containing high levels of total dissolved solids. The lease is now operated by Santos, and legal proceedings are continuing with respect to the reporting of wastewater discharges in 2010 and 2011. In January 2014, Santos was fined $52 000 in respect of contaminated saline wastewater discharges in the Pilliga State Forest, at a time when Eastern Gas Star was the operator.

In 2013, AGL was fined for excessive nitrogen oxide emissions from gas compressor engines south of Sydney. The incidents were self-reported, and the causes fixed.

In 2012, it was found that Metgasco had disposed of excess produced water in the Casino sewage treatment plant (in north eastern NSW), a process outside the licence terms. The NSW EPA issued a formal warning to Richmond Valley Council, the sewage plant operator, that such treatment was in breach of their Environment Protection Licence.

Pennsylvania

In 2009, 15 families from Dimock, a small town in Susquehanna County with contaminated water, filed a federal lawsuit against a major gas-producing company, Cabot. Cabot insisted that the methane migration in Dimock water wells was naturally occurring, pointing to tests taken after drilling had been halted in the area. Nevertheless, the DEP fined the company $120 000 for the methane migration incidents, barred it from drilling within the Susquehanna County community and ordered it to pay for a water pipeline to Dimock residents. Cabot agreed to pay for temporary water supplies at the affected homes.

In February 2014, DEP fined the company Halliburton some $1.8 million for repeated breaches of hazardous waste disposal laws regarding waste from gas wells.

Queensland

In the first half of 2010, the Queensland Department of Employment, Economic Development and Innovation investigated reported gas leaks near Tara, around 300 km west of Brisbane. Some 58 CBM wells were examined, with one found leaking methane well above the lower explosive limit and four leaking at or above 10 per cent of the lower explosive limit; 21 minor to very minor gas leaks were also found. The operator, QGC, now a subsidiary of BG, was issued with three compliance directions to repair the leaking wellheads. These directions were subsequently issued to all ten CBM operating companies to inspect and verify wellhead integrity. Random field inspections to monitor compliance were also increased (DEEDI 2010).

In 2012, the LNG Enforcement Unit, DNRM, investigated gas bubbling in the Condamine River near Chinchilla, in the vicinity of CBM wells. The investigation showed that the gas was biogenic methane, but was unable to establish the source although the gas composition was consistent with gas from the Surat Basin geological foundations. While gas migration to the surface of the water had been observed historically, the current activity appeared more vigorous than that previously observed. No environmental harm at the seep areas was detected (DNRM 2012).

Both Countries Show a Wide Diversity of Approaches

Both Australia and the US show a wide diversity of approaches to regulating unconventional gas. In Australia the State of Victoria, with large but diminishing offshore gas resources and much less onshore oil and gas history, imposed a moratorium on hydraulic fracturing, effectively blocking unconventional gas production. South Australia, where onshore gas production has a long history that includes the use of advanced stimulation techniques, has encouraged potential shale gas production.

In the United States there is a wide variety of regulatory approaches and widely differing geological formations, exploitation techniques and water issues. Texas, with many decades of deployment of hydraulic fracturing techniques, was the initial focus for unconventional gas output in the Barnett Shale, but has kept regulation within the purview of the traditional oil and gas regulator, the Texas Railroad Commission. North Dakota, which has seen an almost overnight revolution in unconventional hydrocarbon production, based on hydraulic fracturing, has had to adapt much more rapidly to the regulatory challenges. Oil output from the Bakken formation in North Dakota has leapt from little more than 100 000 b/day in 2007 to approach 1 Mb/day at the end of 2013, peaking in early 2015 at a little over 1.2 Mb/day, which is an amount second only to that of Texas. Given the emphasis on oil, and lacking existing gas-gathering infrastructure, nearly a third of North Dakota's gas output is flared although this proportion is falling.

In summary, the recent regulatory histories of the two countries offer many lessons, particularly for those regions that have opted to prohibit or defer potential development.

Victorian Gas Market Taskforce

In the Australian State of Victoria, the report of the Gas Market Taskforce was released in October 2013. Chaired by former Federal Minister Peter Reith, the Taskforce was charged with providing policy options to the Victorian Government to improve the eastern States' gas market, including increasing gas supply in a market context of rising demand and increasing prices. The State of Victoria currently has no unconventional gas production.

In the course of making its recommendations, the review addressed a number of aspects of potential unconventional gas production, including recommending the lifting of the State moratorium on hydraulic fracturing and new CBM exploration licences. The review also detailed clear recommendations for applying the highest standards of regulatory and scientific oversight of this industry. Among its most important findings in this area were the need to:

- implement the SCER National Harmonised Regulatory Framework, including aligning Victorian legislation with the 18 leading practices embodied in the framework;
- appoint a Gas Commissioner, to engage with landholders and communities and build confidence with those groups in unconventional gas;
- establish an independent water science programme to undertake comprehensive baseline data studies;
- establish an independent Water Science Committee under the Gas Commissioner;
- subject the gas industry to similar licensing requirements as other water users; and

- remove the moratorium on hydraulic fracturing, subject to permanent bans on BTEX chemicals and requiring the disclosure of all chemicals used and the independent monitoring of impacts.

To address the asymmetry between rewards and risks, the review recommended a portion of State royalties be awarded to local communities, with local input into decision-making plus higher upper limits on compensation for loss of amenity.

As in other areas of Australia, water issues formed the major area of stakeholders' concerns, covering possible water contamination, loss of supply, equitable water use and, in general, a lack of information. The review highlighted some wider issues with respect to gas markets in eastern Australia, notably lack of transparency, lack of liquidity and forward markets for gas and weaknesses in transmission markets, issues still persisting in late 2015. In general, the review recommendations are consistent with contemporary regulatory thinking in many OECD countries. However, as of late 2015, the Victorian Government have not followed these recommendations and a ban on hydraulic fracturing remains in force.

Key Insights

Most jurisdictions where unconventional gas development has occurred have had to cope with a rapid increase in gas production. Some jurisdictions have responded by outright bans or open-ended moratoria. Others have had to advance rapidly up a regulatory 'learning curve', with often under-resourced regulators and regulatory instruments not necessarily designed with the new technologies in mind. Other jurisdictions have long experience in oil and gas regulation, with large well-resourced regulators, so that development has been able to occur relatively rapidly.

In most producing jurisdictions there is a move to intensify regulation and also to create more purpose built, better-resourced, institutions, and to simplify, clarify and streamline legislation. Greater regulatory scrutiny can be seen, for example, in the increase in the radius of concern for drilling, greater disclosure of drilling and fracturing chemicals and more attention to regional cumulative impacts, notably on water resources. While greater regulatory intervention will entail some increase in developers' costs, recent research shows such increases to be relatively modest, of the order of seven to ten per cent (IEA 2012).

Water is a Common Concern: Basin-Wide Approaches are a Key Tool

In the case of shale gas large quantities of water are needed in the production process and obtaining that water, treating it and re-using or disposing of it are major concerns. While water issues differ for CBM in Australia, it is clear that they are seen as vital to many stakeholders. The interaction between water regulation and petroleum, while not new, has become much more intense. Increasingly, basin-wide approaches are being taken, for example the Surat Basin study in Queensland, and the Susquehanna Basin study in the US, with more stringent procedures for water permitting, withdrawals and disposals that are at least as rigorous as those applying to other users. Such basin-wide approaches provide a useful basis to study the cumulative impacts of large, if dispersed, unconventional gas developments. Basin-wide approaches have existed in North America for some decades and are becoming more common in Australia. Overall, it would seem that closer interaction between water regulators and oil and gas regulators is important.

Australia and the United States Have Important Differences

One important difference between the two countries concerns the ownership of mineral rights. In North America landowners generally own the subsurface rights, while in Australia royalty-type revenues accrue to States. Hence, US landowners have considerable incentives to push for gas development, as seen for example in attempts by landowners in south central New York State to overturn bans on shale gas drilling. Such incentives are absent in Australia and, given that local communities must tolerate a very considerable increase in for example road movements and noise, strong local opposition is unsurprising. Potential partial remedies for this might include upfront access or other payments to landowners and some means of distributing royalty revenues back to affected areas, as proposed in the Reith Review to the Victoria Government in 2013.

Generally, areas of low population density have seen an easier introduction of new gas and oil output, as for example in North Dakota or central Queensland. In regions of higher population density, such as New York State and areas close to Sydney, development has been much more difficult, or impossible, to initiate.

In the United States, the unconventional petroleum industry was led by relatively smaller companies, with backing from federal research and information dissemination programmes. More recently, larger companies, such as Exxon Mobil, have bought into the unconventional hydrocarbon developments. In Australia, early work on CBM was led by companies seeking to improve coal mine safety, but later the work was picked up by local companies Santos, Origin, Arrow and AGL. More recently, these companies have been supplemented for LNG developments by much larger international gas producers and buyers, such as BG, Petronas, Conoco Phillips, Total, Sinopec, CNOOC and Kogas.

The generally smaller, more innovative, US companies have driven the rapid industry output and productivity gains but at the potential expense of environmental impacts, notably in the water sphere. A combination of larger, better-resourced, companies in tandem with much stronger regulation could provide more sustainable outcomes.

Pre-Drilling Data Should Be Required

The rapid increase in drilling activity, and the generally higher intensity of such activity for a given gas output, has meant that regulators have not been properly resourced, at least for initial developments. The rapid take-off of the unconventional gas industry has meant that high quality pre-drilling data on, for example, water quality, may be missing. The large number of wells being drilled may also mean that regulators need higher levels of compliance activity, the costs of which should be recovered from industry, in order to maintain community support and high standards.

To ensure a social licence for its activities, the industry should respond positively to greater community demands for more transparency on all aspects of operations, including on chemical use, in order to reconcile understandable commercial desires to protect valuable intellectual property with very high levels of community concern. Regulators also need to create the appropriate environment to encourage innovation with regard to the use of lower volumes of hydraulic fracturing chemical and less harmful additives, and the recycling of fracturing fluids. A strong consensus is also emerging that certain classes of chemicals seem to have little place in the industry, especially BTEX compounds (Queensland Government 2010, Victorian Government 2013).

Striking the Right Federal–State Balance Can Be Tough

Nationwide approaches have an important role in unconventional gas developments. They include providing adequate resources for research, establishing basic regulatory frameworks and avoiding reinvention of' the 'regulatory wheel' by states or regions. Such approaches could be initiated by Federal Governments, or by cooperative activities funded by groups of States and industry. The US example of cooperative industry-funded regional arrangements to promote best practice, including the involvement of universities, is an approach worth considering elsewhere. Nevertheless, maintaining community confidence in the independence of such work may prove difficult. Exchanges of regulatory experience at an international level can also be very useful, notwithstanding obvious basic differences in, for example, geology, social and economic contexts.

In regions where there is little experience in a given area of onshore oil and gas production, caution can be justified. Time to examine others' experience and to build public trust through much higher levels of engagement, baseline data collection and appropriate modification of regulatory instruments and institutions may be needed if unconventional gas development is to have a long-term future.

Properly Resourced, Purpose Built Regulatory Bodies Can Be Justified

Purpose built institutions, such as Queensland's Gas Commission, the CSG Compliance Unit and the Office of Groundwater Impact Assessment, provide noticeable improvements in stakeholder engagement and regulatory efficiency. Regular review is, however, a necessary step to ensure that regulatory objectives continue to be met efficiently and effectively.

In summary, unconventional gas has delivered substantial benefits to the United States and is about to do the same in Australia. Nevertheless, it is an industry that requires a high and ongoing commitment to levels of regulation, compliance and verification. These regulatory commitments, and theireffectiveness, are sure to test the competence of authorities over coming years.

Conclusions

Potential gas production areas, where no unconventional gas production has taken place, for regulatory or other reasons, have the opportunity to build regulatory regimes that incorporate best practice (QCA 2014, Victorian Government 2013, O'Kane 2013, IEA 2012). Key data from some federal studies (EPA 2015, NICNAS 2015) have become available over 2015 and will do so over the coming years, and it will provide an important additional opportunity to improve the quality of regulation. Some key features of such a best practice regime, plus areas for existing regulators that require further development, are summarised below.

Pre-Drilling

Prior to drilling, a number of steps are necessary. Pre-drilling information baselines, especially for water levels and quality, basin-wide and at individual well level, are an essential prerequisite for ensuring that water quality and availability are not affected by extraction. High levels of understanding of complex water systems are required, as well as of underground fault systems, if seismic events are to be avoided.

Access to Drilling Sites

Land access arrangements require a much higher level of community consultation than at present; this should be backed by institutions that can 'level the playing field' between companies and landowners. Simple no-fault arrangements for compensation when, and if, there is pollution of air or water need to be in place prior to drilling. Risk and reward arrangements need to be developed that clearly recognise that most costs and risks are borne locally while the benefits, often substantial, can be much more diffuse and regional and national in scope.

Drilling Regulation

Transparency around all aspects of the industry needs to be greatly improved, including about the chemicals used in hydraulic fracturing, with appropriate outright bans, based on global evidence. Well integrity, throughout the production process, from 'cradle to grave' and into the hereafter, needs to be an ongoing regulatory focus, reflecting the greater stresses where wells are hydraulically fractured on multiple occasions. Some jurisdictions have used environmental bonds, or funds collected over the lifetime of mining projects, to ensure safe and ongoing decommissioning. Emissions of methane and other organic compounds need to be eliminated or reduced to very low levels. In fact such emissions should be eliminated during well completion and production, and flaring or venting should be prohibited through green completions. These policies need to be enforced through real-time monitoring.

Water Issues

Water concerns need to be addressed using regional approaches to measure cumulative impacts, including catchment or basin-wide approaches. Extracted water or wastewater must be treated to accepted standards. Guidelines for the beneficial use of water should be developed and accepted by all stakeholders in the industry. Clearer guidelines should be developed for the disposal of wastewater streams, such as saline wastes, solids and all drilling wastes.

Regulatory Agencies

Regulatory and compliance agencies should be fit for purpose and adequately resourced, with industry levies providing continuity and certainty of resourcing. Funding from industry levies should underpin the basic science, the ongoing development of regulation, high level monitoring and compliance.

Ongoing reviews are necessary to ensure effective and efficient regulation, backed by evidence, and exchanges with other jurisdictions. Regulators will continue to learn from the industry and each other. Cross-jurisdiction learning should promote best and improving practice, using a mix of outcome-based regulation and more prescriptive approaches.

This chapter draws material from Cronshaw and Grafton (2016).

References

AEMO (2013). Gas statement of opportunities: for eastern and south-eastern Australia. Australian Energy Market Operator.

AGL (2015). Groundwater Management Plan for the Camden Gas Project. www.agl.com.au/~/media/AGL/About%20AGL/Documents/How%20We%20Source%20Energy/Camden%20Document%20Repository/Water%20Plans/20151030_Camden%20Gas%20Project%20%20%20Groundwater%20Management%20Plan.pdf (accessed 22 December 2015).

APPEA (2015). Industry statistics. Australian Petroleum Production and Exploration Association. www.appea.com.au/industry-in-depth/industry-statistics/ (accessed 21 December 2015)

Arnold, G. and Holahan, R. (2014). The federalism of fracking: How the locus of policy-making authority affects civic engagement. *Publius* 44: 344–368.

BREE (2013). 2013 Australian Energy Statistics, Bureau of Resources and Energy Economics, Canberra, July.

Cronshaw, I. and Grafton, R.Q. (2016). A Tale of Two States: Development and Regulation of Coal Bed Methane Extraction in Queensland and New South Wales, Australia. *Resources Policy* 50: 253–263.

Davies, J.D., McCormack, C.V., and Rasheed, M.A. (2013). Port Curtis and Rodds Bay seagrass monitoring program: biannual western sasin & annual long term monitoring. November 2012. Report, TropWATER, James Cook University, Cairns.

Department of Employment, Economic Development and Innovation (DEEDI) (2010). Investigation report: Leakage testing of coal seam gas wells in the Tara 'rural residential estates' vicinity. Queensland Government www.dnrm.qld.gov.au/__data/assets/pdf_file/0011/119675/tara-leakage-csg-wells.pdf (accessed 22 December 2015).

Department of Environment (2013). Abbot Point and Curtis Island projects approved, New safeguards to protect the long-term future of the Great Barrier Reef. Media Release 10 December 2013. www.environment.gov.au/minister/hunt/2013/mr20131210.html (accessed 21 December 2015).

Department of Natural Resources and Mines (DNRM) (2012). Summary Technical Report – part 1: Condamine River gas seep investigation. Queensland Government. www.dnrm.qld.gov.au/__data/assets/pdf_file/0005/119669/condamine-river-gas-seep.pdf (accessed 22 December 2015).

DiGiulio, D.C., Wilkin, R.T., Miller, C. and Oberley, G. (2011). Investigation of ground water contamination near Pavillion, Wyoming (Draft). US Environmental Protection Agency, Denver. www.epa.gov/sites/production/files/documents/EPA_Report OnPavillion_Dec-8-2011.pdf (accessed 22 December 2015).

Environmental Protection Agency (EPA) (2012). Study of the potential impacts of hydraulic fracturing on drinking water resources (Progress Report). www.epa .gov/sites/production/files/documents/hf-report20121214.pdf (accessed 22 December 2015).

Environmental Protection Agency (EPA) (2014). EPA's study of hydraulic fracturing for oil and gas and its potential impact on drinking water resources (External Review Draft). http://cfpub.epa.gov/ncea/hfstudy/recordisplay.cfm?deid=244651 (accessed 22 December 2015).

Environmental Protection Agency (EPA) (2015). Assessment of the potential impacts of hydraulic fracturing for oil and gas on drinking water resources. EPA/600/R-15/047a, June 2015.

International Energy Agency (IEA) (2012). *Golden Rules for a Golden Age of Gas: World Energy Outlook Special Report on Unconventional Gas*. OECD Publishing, Paris.

National Industrial Chemicals Notification and Assessment Scheme (NICNAS) (2015). National assessment of chemical associated with coal seam gas extraction in Australia: frequently asked questions. Department of Health, Australian Government. www.nicnas.gov.au/communications/issues/fracking-hydraulic-fracturing-coal-seam-gas-extraction/faqs (accessed 22 December 2015).

Murtazashvili, I. (2015). Origins and consequences of state-level variation in shale regulations: the cases of Pennsylvania and New York. In (eds. W.E. Hefly and Y. Wang) *Economics of Unconventional Shale Gas Development Case Studies and Impact*. Springer.

O'Kane, M. (2013). Initial report on the independent review of coal seam gas activities in NSW by NSW Chief Scientist and Engineer. Sydney, Australia: NSW Government.

Queensland Competition Authority (QCA) (2014). Coal Seam Gas Review: Final Report. Queensland Competition Authority, Brisbane.

Queensland Government (2010). Government bans BTEX use in coal seam gas sector. Media Release 4, August 2010. http://statements.qld.gov.au/Statement/Id/70932 (accessed 21 December 2015).

Queensland Government (2013). CSG compliance program exceeds targets. Media Release, 24 July 2013. http://statements.qld.gov.au/Statement/2013/7/24/csg-compliance-program-exceeds-targets (accessed 21 December 2015).
Queensland Water Commission. (2013). Queensland Water Commission Final Report. www.parliament.qld.gov.au/Documents/TableOffice/TabledPapers/2013/5413T2898.pdf (accessed 22 December 2015).
RLMS (2013). Eastern Australia gas reserves and resources at 31 December 2012. Resources and Land Management Services, Brisbane.
Santos (2014). Santos enters into a set of principles for land access in NSW. Media Release, 28 March 2014. www.santos.com/media-centre/announcements/santos-enters-into-a-set-of-principles-for-land-access-in-nsw/ (accessed 21 December 2015).
Victorian Government (2013). Gas Market Taskforce: Final Report and Recommendations. www.energyandresources.vic.gov.au/__data/assets/pdf_file/0003/1143417/Victorian-Gas-Market-Taskforce-Final-Report-October-2013.pdf (accessed 22 December 2015).

16

Regulation of Unconventional Hydrocarbons in Alberta, Canada

MICHAL C. MOORE

Introduction and Chapter Design

This chapter explores the unique nature of Canada's and Alberta's regulatory systems, which have been created and have evolved to deal with a range of unconventional fuels, standards of extraction and transport and with the environmental externalities associated with industry operations. Alberta can truly be called unique, since its economy is dependent on a principal resource product – hydrocarbons – with tremendous reserves ranging from conventional to unconventional. The province's regulatory and tax structures, in particular, have changed in tandem with the growth and diversification of the energy industry; as a consequence the expertise and experience of staff and regulators tends to be more focused than that of other provinces or states, simply as a reflection of the relatively narrow set of resources at the heart of regulatory applications.

Alberta is landlocked and must work closely with external agents, including two adjacent provinces, the Government of Canada and the principal export client, the United States, in terms of access and tariffs. Alberta's role in energy regulation is linked closely to the responsibilities and issues faced by the National Energy Board as well as by counterparts in the United States. Because the industry is so dominant in the overall economy, swings or changes in fuels and finished energy products can dramatically affect the province and its economy and in turn lead to unusually large impacts on the rest of the nation.

The chapter reviews the nature of the unconventional resources in Alberta, discusses the core nature of regulation for hydrocarbon resources and then presents an overview of the regulatory institutions and processes at both the National Energy Board and the Alberta Regulator (AER).

Canadian Provinces and Energy

Energy resources are available throughout Canada, although not always in sufficient quantities to offset the need for foreign imports (such as natural gas imports to eastern Canada). Most hydrocarbon resources, both oil and gas, are located in western Canada, and largely they are found in Alberta. From the early days of the development of oil and gas in Alberta, a key export relationship has dominated trade with the United States, over and above serving demand in the balance of Canada.

In part this is due to the distances involved, but primarily it stems from the fact that western Canadian production capacity exceeds Canadian demand, with the result that Canada is a net exporter. A key aspect of the Canadian confederation is that each province is deemed to own, and given the responsibility to manage, its own natural resources. Federal oversight and control is only apparent when specific environmental issues arise from resource development or when goods such as oil, gas or electricity cross provincial boundaries.

As pointed out by Deyholos and Cuschieri (2013), the oil and gas regime in Canada is "concession-based and the Crown does not normally co-own or directly participate in oil and gas projects".[1] The owner of the mineral rights, whether the Crown or the owners of freehold estates, will typically grant a company a lease that gives the lessee the right to explore and drill for and remove and dispose of minerals for a set term, in exchange for rental fees and royalty interests.

In Alberta the property rights to the surface of the land and the property rights to the minerals under the land are mostly separate and distinct.[2] Consequently, one party can hold title to the surface rights while another party holds title to the oil and gas beneath the surface. In the process of deciding on project proposals, the regulator must take these ownership interests into account not only in terms of the resources but also in terms of the externalities that might emerge from project development.

The regulator is responsible for project oversight through the project's entire life cycle and, when a project reaches its useful economic end, companies have a responsibility to continue to mitigate or eliminate environmental threats by abandoning or decommissioning wellsites in an appropriate fashion. According to Deyholos and Cuschieri (2013), if a *Reclamation Certificate* is required, the company will have to ensure that the worksite has met 'equivalent land capability', which is the ability, after reclamation, to support the various land uses that existed prior to oil and gas activities.

The Nature of Energy Regulation

Energy, from primary sources to processed and delivered fuels and electricity, is vital for any modern society. Since the demand for energy resources is continuous (with allowance for seasonal or other variations), ensuring adequate supplies, affordable prices and predictable quality is a key role for governments. Direct responsibility for overseeing the operations of the energy industry is typically delegated to a special commission or regulatory body with the ancillary tasks of monitoring market power, assuring adequate capacity and implementing mandates such as renewable energy investment and maintaining policy-established

[1] There are exceptions to this, as evidenced by recent investments by Alberta in an upgrader project (for example the Sturgeon Refinery developed by the North West Redwater Partnership) within the Alberta industrial heartland.

[2] There are still descendants of homesteaders that were living in Alberta before it became a province that received both surface and mineral titles and continue to hold them today.

levels of environmental quality. In the case of Canadian provinces,[3] the oversight duties include the holding of hearings and the giving of permissions for new capital investments, the development of reserve capacity for oil and gas, the monitoring of ongoing field operations for compliance with project approval conditions and finally consideration of the broad issues of public health and safety and environmental compliance in plants, pipelines, drilling and storage facilities.

Regulators typically derive their authority and assigned roles from a legislative grant or licence. Within this, most regulatory institutions are required to maintain independence from policy-makers in decision-making but not in the case of the structure or design and expected performance standards of the systems they oversee. They often have authority to monitor economic or environmental violations, but operate within budget and with performance oversight from the legislative branch.

It is convenient to group the key responsibilities for energy regulators into the following main categories:

- ensuring adequate energy supplies (for domestic and export consumption);
- conducting public hearings on energy project proposals;
- adjudicating and reviewing utility ratemaking[4] for regulated utilities in electric systems;
- ensuring that policy goals are met;
- maintaining and distributing data and information resources for the public, policy-makers and investors is an optional or additional role.

The Nature of Unconventional Gas and Regulation

Unconventional gas (as well as oil) operations owe much of their success to the implementation and use of new technology, which typically uses directional drilling to reach smaller pockets of trapped resource and to gain access to horizontal formations where gas molecules are located in tight sand and shale formations with limited permeability. Whereas conventional and traditional gas wells use capillary action to attract gas to migrate to lower-pressure zones (such as a well), unconventional wells must inject a fluid to crack or fracture the rock having oil or gas resources and then extract the liquid–gas solution and separate the constituents, finally recovering the gas itself. Horizontal drilling and hydraulic fracturing are not new technologies; horizontal drilling dates as far back as the late 1940s (hydraulic fracturing actually goes as far back as the late 1860s, when nitroglycerine was used to fracture shallow, hard-rock, oil reserves).

This process is by its very nature land and resource intensive, involving large volumes of fracking fluids such as water, butane or even CO_2 and sand or other proppants, the transport of which ultimately requires the use of roads that may not be purpose built for

[3] Only Alberta and the offshore boards in Newfoundland and Nova Scotia are true independent and quasi-judicial regulators. British Columbia is arguably a hybrid entity and in the other provinces the regulator is a branch or division within the responsible ministry. As well, in those jurisdictions, typically the regulator is responsible for royalty policy and administration.

[4] Most utility pricing issues are considered in electricity hearings, although marginal cost pricing objectives are present in oil and gas deliberations as well. For more on this topic see Gilbert (1991).

these operations, and the final disposal of used fluids. The range of land-use, economic and environmental concerns is large and creates the need for unique regulatory requirements, hearing processes, data presentation, oversight and enforcement and finally remediation.

Unconventional oil and gas development is regulated throughout its life cycle, although not consistently from jurisdiction to jurisdiction. The range can include applications from preconstruction land leasing and seismic testing, to ongoing operations and water or fluid use, to end-of-life restoration and abandonment.

Regulators familiar with oil and gas operations are typically charged with the development of rules and standards for the industry, and they maintain specially trained staff to conduct hearings and oversight and enforcement actions.

Canadian Energy Regulation and Supporting Agencies

Canada is a federation of 10 provinces and three territories and, with the exception of Quebec, is a common law jurisdiction. The federal constitution divides legislative authority between the federal parliament (the *federal Crown*) – which has jurisdiction over matters of interprovincial, national and international scope – and the provincial legislatures (each a *provincial Crown*) – which have jurisdiction over matters of a more regional or even local nature.

According to Deyholos and Cuschieri (2013) "Participants in the oil and gas industry are often subject to both federal and provincial regulators because these levels of government have overlapping or shared legislative authority in the areas of natural resource development, transportation, marketing and the environment".

The key federal statutes and regulations governing oil and gas exploration include the following:

- the National Energy Board Act (R.S.C. 1985, c. N-7) (NEB Act), which established the National Energy Board and which deals with energy-related issues under the authority of the federal Crown, including: the administration of oil and gas interests; production and conservation; construction and operation of pipelines; traffic, tolls and tariffs on pipelines; export and import of oil and gas; and the interprovincial trade of oil and gas.
- the Canadian Environmental Assessment Act (S.C. 1992, c. 37) (CEAA), which sets out rules intended to ensure that the review of a project is sufficient to avoid adverse environmental effects and to encourage sustainable development. This includes the process for the environmental assessments that are triggered under the CEAA if a federal authority proposes a project, provides funding for a project, grants an interest in land to a project or exercises regulatory jurisdiction over a project.
- the Canadian Environmental Protection Act (S.C. 1999, c. 33) (CEPA), the primary policy document, which addresses the protection of the Canadian environment as well as public health, including the prevention and management of risks posed by hazardous substances.

The Canadian federal government has two primary *agencies* that deal with energy policy and regulatory issues such as conventional and unconventional natural gas development. They are the National Energy Board (NEB), which carries out regulatory responsibilities, and Natural Resources Canada (NRCan), which is responsible for policy. Both the NEB and NRCan report directly to the Federal Minister of Natural Resources.

In addition, the Canadian Environmental Assessment Agency (CEA Agency) is involved in oversight for unconventional resources by providing environmental assessments that support "informed decision making in support of sustainable development", CEAA (2015); the Department of Fisheries and Oceans (DFO), which is under the authority of the Fisheries Act (R.S.C. 1985, c. F-14), delivers programmes and services that support the sustainable use and development of Canadian waterways and aquatic resources, including monitoring pollution in waterways and conducting environmental assessments. (The NEB has been delegated responsibility for CEAA oversight in the energy sector.)

The National Energy Board (NEB)

The NEB is the principal federal agency involved in day to day regulation of the energy industry when its projects cross provincial boundaries or where export products are part of the project in question. It conducts hearings and collects much of the basic data used by other federal and provincial agencies in evaluating the state of the industry within Canada.

The main responsibilities of the NEB are established in the National Energy Board Act (NEB Act) and include regulating:

- the construction, operation and abandonment of pipelines that cross international borders or provincial boundaries, as well as the associated pipeline tolls and tariffs;
- the construction and operation of international power lines and designated inter-provincial power lines; and
- imports of natural gas and exports of crude oil, natural gas liquids, natural gas, refined petroleum products and electricity.

Additionally, the NEB has regulatory responsibilities for oil and gas exploration and production activities on lands owned or under the stewardship of the federal government (Canada Lands) not otherwise regulated under joint federal–provincial accords. These regulatory responsibilities are set out in the Canada Oil and Gas Operations Act and the Canada Petroleum Resources Act.

The NEB conducts an environmental assessment (EA) during its review of applications for facilities and activities under its jurisdiction. For certain projects, an EA is also required by federal legislation such as the Canadian Environmental Assessment Act 2012 (see the comment above on NEB oversight), the Mackenzie Valley Resource Management Act, the Inuvialuit Final Agreement or the Nunavut Land Claims Agreement. Certain Board inspectors are appointed as Health and Safety Officers by the Minister of Labour to administer Part II of the Canada Labour Code as it applies to NEB-regulated facilities and activities.

The NEB has an important information role, in tandem with that undertaken within NRCan, to monitor and publish data and to evaluate overall aspects of energy supply, demand, production, development and trade that fall within the jurisdiction of the federal government.

Natural Resources Canada (NRCan)

Natural Resources Canada (NRCan), is the agency of the federal government responsible for the oversight of natural resources, energy, minerals and metals, forests, earth sciences and mapping and remote sensing. It was created in 1995 by amalgamating the now-defunct Departments of Energy, Mines and Resources and Forestry. According to its website and various publications about its mission, the agency works to ensure the *responsible development* of Canada's natural resources, including energy, forests, minerals and metals. Natural Resources Canada (NRCan 2015) also uses its expertise in earth sciences to build and maintain an up-to-date knowledge base of the Canadian landmass and resources.

Because resource use and development have international impacts and implications for economic development, the NRCan agency works with American and Mexican government scientists, along with the Commission for Environmental Cooperation, to help produce the North American Environmental Atlas, used to characterize and evaluate environmental issues on a continental basis over time.

As pointed out above, in the Canadian constitution the responsibility for natural resources is left to the provinces, not the federal government. However, the federal government has jurisdiction over off-shore resources, trade and commerce in natural resources, statistics, international relations, and international boundaries.

The department is governed by the Resources and Technical Surveys Act, RSC, c. R-7 and the Department of Natural Resources Act, SC 1994, c. 41. The department is "structured along business lines according to types of natural resources and areas of interest". The agency currently has the following sectors that deal primarily in the energy area:

- Earth Sciences Sector
- Energy Sector
- Innovation and Energy Technology Sector
- Minerals and Metals Sector

Alberta Energy Regulation and Supporting Agencies

The Alberta Energy Regulator (AER) exists to ensure the "safe, efficient, orderly, and environmentally responsible development of hydrocarbon resources over their entire life cycle. This includes allocating and conserving water resources, managing public lands, and protecting the environment while providing economic benefits for all Albertans", AER (2015).

The AER succeeds the Energy Resources Conservation Board, combining regulatory functions from the then Ministry of Environment and Sustainable Resource Development (now Environment and Parks) that relate to public lands, water and the environment. In this way, the AER provides full-life-cycle regulatory oversight of energy resource development in Alberta – from application and construction to abandonment and reclamation, and everything in between.

The AER is authorized to make decisions on applications for energy development, monitoring for compliance assurance, decommissioning of developments and all other aspects of energy resource activities (activities that must have an approval under one of the six provincial energy statutes). This authority extends to approvals under the public lands and environment statutes that relate to energy resource activities.

Authority for Acts and Regulations

Alberta's Acts and Regulations pertaining to Alberta Energy are derived from the laws of Alberta which direct the management of Alberta's resources. They are supplemented by a series of guidelines, effective from January 2009, that describe the principles involved in assessing, levying and collecting Alberta's Crown royalty share of natural gas and gas products (those produced from lands subject to a Crown lease agreement). These also describe the special procedures that will be used to carry out the assessment, levy and collection of royalties when resources are extracted and transported from the field.

The Gas Royalty Operations Information Bulletins highlight any proposed changes to legislation, regulations or operating procedures. They are published on a monthly basis and provide industry clients and interested parties with updated information regarding reporting requirements for the determination and calculation of natural gas royalties.

AER Governance

According to its mandate, the AER's governance structure is designed to achieve the benefits of both strong corporate-style oversight and independent adjudication.

For corporate governance, the AER is headed by a chair, who leads a board of directors, all appointed by the Lieutenant Governor in Council (the provincial Cabinet) and all part-time directors of the AER. These directors set the general direction of the regulator's business affairs and are charged with approving regulatory change and with setting performance expectations for the AER and its chief executive officer.

The CEO, who reports directly to the chair, oversees day-to-day operations, which include receiving and making decisions on applications, developing regulatory standards and guidelines that monitor and investigate energy resource activities for compliance and closing energy developments through the processes of remediation and reclamation. The AER has significant discretion to use their hearings as part of the decision-making process.

Hearings are the responsibility of commissioners, who report to a Chief Hearing Commissioner. The commissioners, who are appointed by the Lieutenant Governor in Council, are responsible for the AER's adjudicative functions, acting as the decision makers on major applications and conducting hearings.

As with many other jurisdictions, Hearing Commissioners are independent adjudicators; their decisions may be overturned only by the Court of Appeal of the province.

Role in Unconventional Resources

The development of unconventional resources has presented a challenge for an industry and public regulator with a long history of managing abundant conventional resources. The Alberta Energy Regulator points out that advancements in technology, such as horizontal wells and multi-stage hydraulic fracturing, have enabled the economic development of previously unattainable oil and gas resources. The use of this advanced technology creates new risks and opportunities. For example, unlike conventional hydrocarbon pools, unconventional resource development requires a greater intensity of infrastructure (wells, roads and other facilities). This difference, and others, is driving the AER's work to introduce a new *play-based regulatory framework*[5] for the development of unconventional resources.

The AER (2012) developed a discussion paper titled "Unconventional Oil and Gas in Alberta". The paper outlined a new framework that considers the advancements in technology, the greater scale, duration and intensity of development and the cumulative effects on water, air quality, surface disturbance and the public.

Density of Operations

Unconventional gas drilling and operations can be land intensive, even with the use of multiple well pads and directional wells. A key role of the AER is to regulate the density and location of these well pads, on a location-by-location basis and by determining the number of wells per section or per approximately one square mile.

Shale gas can be developed using various drilling and completion methods depending on the reservoir's characteristics, but "... generally, due to its low permeability, 8 to 16 subsurface drainage wells per section are required to effectively recover the underground gas in a shale play. The production from one horizontal wellbore with 10 fracture stages may in some cases be equivalent to drilling 10 vertical wells with a single fracture each" AER (2014).

The Regulation of Shale Gas Development in Alberta

Shale gas is currently regulated under the same legislation, rules and policies as conventional natural gas. The AER points out that "although shale gas development in Alberta has not been using horizontal multi-stage fracturing extensively, Alberta does have considerable experience with hydraulic fracturing".

[5] Play-based regulation involves implementing a single application and decision-making process for multiple wells, pipelines and facilities presently under different pieces of legislation. It requires all the operators in a planning or activity area to collaborate and jointly bring a single application for a single regulatory approval, which will be used for regulating all their unconventional oil and gas activities in that area in the future.

Wellbores are required to have cemented casings in place which meet stringent requirements and which are also set to depths far below any fresh water aquifers, to ensure an impenetrable barrier between the shale gas formation and the wellbore that might contaminate groundwater.

The fluid used in hydraulic fracturing operations has raised concerns about toxicity and groundwater contamination, both during and after the fluid has been injected into the resource reservoir.

Alberta does not allow produced fluids to be sent to municipal wastewater treatment systems. Fluids that cannot be treated and recycled must be disposed of in approved disposal wells, where the fluids are injected deep underground in geologically isolated reservoirs for permanent disposal.

AER Directive 083: Hydraulic Fracturing – Subsurface Integrity mandates that licensees demonstrate that operational risks have been considered in the selection and design of the wellbore construction and that monitoring and testing to ensure that well integrity is maintained have been carried out.

To address risks of inter-wellbore communication, whereby fluid and/or pressure from hydraulic fracturing operations impact a nearby oil or gas well, Directive 83 requires licensees to carry out a risk assessment and to prepare a well control plan to manage unintended inter-wellbore communications and reduce the impacts if a communication does occur.

Current regulatory requirements for shallow fracturing operations, outlined in Directive 027, focus on shallow zones up to 200 metres below the surface. Zones between 200 and 600 metres (depths that are above the base of groundwater protection in many areas) may be subject to future development, and current requirements may be extended to these zones. Under Directive 83, all licensees carrying out hydraulic fracturing operations in this zone must:

- conduct a risk assessment;
- observe the prescribed setbacks for water wells and top of bedrock;
- use environmentally friendly chemical additives or fluid compositions above the base of groundwater protection.

In addition, the AER requires the disclosure of hydraulic fracturing fluid composition, water source and volume data on a well-to-well basis. Once reported, information about what chemicals are being used can be obtained from a special website created to educate the public and maintained by Fracfocus (2015). It is functionally a chemical disclosure registry specifically for hydraulic fracturing operations.

Any fluid that is returned to the surface from hydraulic fracturing operations must be handled under strict guidelines mandated and enforced by the AER. Even if the fluid is treated, it cannot be released back into a natural water body. The AER restricts fracturing within a 200 metre lateral distance of water wells, to reduce the potential impact on or interference with the domestic use of aquifers or water wells.

For other specific directives created to address unconventional gas development, see the Appendix to this chapter.

Induced Seismicity

Recent events, involving injected wastewater or fracking fluids in Alberta, Ohio, Texas and Oklahoma, have increased public interest and scrutiny in the issue of induced seismicity. Induced seismicity refers to small earthquakes (seismic events) that are assumed to be tied to active well injection. The AER site suggests that "typically, induced seismic events have been low in magnitude and have rarely been felt at the surface. Historically, induced seismic events have been associated with some aspects of energy development, particularly oil and gas extraction activities and the deep-well disposal of wastewater. As the use of hydraulic fracturing has increased, it has been linked as a potential cause of induced seismicity in North America and elsewhere in a limited number of cases. The AER notes that there has been no evidence that harm from induced seismicity has occurred to the public, workers, property, structures, surface and/or groundwater in Alberta".[6]

Regulatory Projects

Additionally, the AER is testing a new framework called the Play-Based Regulation Pilot to govern unconventional oil and gas development. Play-based regulation (PBR) designs regulatory requirements and processes to suit the risks and desired outcomes for a specific resource play. It incorporates a unique regional assessment of projects, which is a departure from the regulation of conventional oil and gas development. The limited scale PBR pilot project is a test designed to improve the broader scale evaluation of:

- surface infrastructure needs and impacts;
- wellbore integrity;
- water impact; and
- air impacts.

The key element of the process is to have more transparent and timely disclosure of development plans, improving opportunities for stakeholder input and collaboration among operators to optimize infrastructure development.

Information and Statistics

The use of data is a key element of ongoing regulatory activity in the Province of Alberta as well as Canada-wide. The Oil and Gas Policy and Regulatory Affairs Division (Oil and

[6] According to the agency, they "monitor seismic activity across Alberta using the Regional Alberta Observatory for Earthquakes Studies Network (RAVEN) and networks operated by Natural Resources Canada, the University of Alberta, University of Calgary, Montana Bureau of Mines and Geology and the University of Western Ontario. Data collected from these stations are also used to document natural and induced earthquakes which are compiled into a comprehensive earthquake catalogue, or seismic database for Alberta". Bulletin 2015–07: Subsurface Order No 2: Monitoring and Reporting of Seismicity in the Vicinity of Hydraulic Fracturing Operations in the Duvernay Zone, Fox Creek, Alberta, mandates that operators must monitor for seismic activity within five kilometres of their wells if hydraulic fracturing operations are being conducted. In addition, operators must have a response plan in place to address potential events and must follow a "traffic light" process with staged action thresholds.

Gas Division) of Natural Resources Canada (NRCan) provides an annual review of, and summaries of the trending of, the crude oil, natural gas and petroleum product industry in Canada and the United States. In Alberta, the regulator collects information on a routine basis from all operators within its boundaries, as well as the voluntary reporting of industry performance from NGOs such as CAPP. The Province continues to depend on national geological surveys for geophysical data and mapping.

Conclusions

The presence of a strong and diverse hydrocarbon industry in Alberta has prompted sophisticated regulatory institutions and rules in order to maximize resource development potential, to increase overall economic activity and public revenues and to ensure that public health and safety standards have been met. There have been several iterations of the regulatory agency, typically reflecting changes in political or legislative intent. The latest version – a single regulatory agency – is being challenged by the newly elected government. They have stated a goal of gaining more efficiency and accountability to the public. While the outcome of such a proposal is unknown at the time of writing, the role of the AER and its specialized focus on unconventional fuel development serves as a model for other jurisdictions seeking to develop this type of resource.

Appendix A

Additional directives governing natural gas operations in Alberta

Directive 008: Surface Casing Depth Requirements
Directive 009: Casing Cementing Minimum Requirements
Directive 017: Measurement Requirements for Oil and Gas Operations
Directive 020: Well Abandonment
Directive 031: REDA Energy Cost Claims
Directive 035: Baseline Water Well Testing Requirement for Coalbed Methane Wells Completed Above the Base of Groundwater Protection
Directive 038: Noise Control
Directive 040: Pressure and Deliverability Testing Oil and Gas Wells
Directive 044: Requirements for Surveillance, Sampling, and Analysis of Water Production in Hydrocarbon Wells Completed Above the Base of Groundwater Protection
Directive 050: Drilling Waste Management
Directive 051: Injection and Disposal Wells – Well Classifications, Completions, Logging, and Testing Requirements
Directive 055: Storage Requirements for the Upstream Petroleum Industry
Directive 056: Energy Development Applications and Schedules
Directive 058: Oilfield Waste Management Requirements for the Upstream Petroleum Industry
Directive 059: Well Drilling and Completion Data Filing Requirements

Directive 083: Hydraulic Fracturing – Subsurface integrity measurement and reporting of used water volumes as well as disclosure and reporting of chemicals used

Appendix B

Alberta Statutes

The key Alberta statutes and regulations governing oil and gas exploration include the following:

- the Mines and Minerals Act (R.S.A. 2000, c. M-17), which applies to all mines and minerals owned by the provincial Crown and deals with exploration and drilling, oil and gas leases and royalties;
- the Petroleum and Natural Gas Tenure Regulation (Alta. Reg. 263/1997), which details the procedures for obtaining oil and gas concessions from the provincial Crown;
- the Oil and Gas Conservation Act (R.S.A. 2000, c. O-6) (OGCA), which sets out rules intended to prevent the waste of oil and gas and provide for the economic, orderly and efficient development of oil and gas through licensing and approval requirements for drilling and operating facilities and the regulation of oil field and pool developments;
- the Oil Sands Conservation Act (R.S.A. 2000, c. 0–7) (OSCA), which has a similar purpose and effect as the OGCA, but applies directly to oil sands;
- the Surface Rights Act (R.S.A. 2000, c. S-24) (SRA), which establishes the Surface Rights Board and outlines the rules for obtaining land access rights required for exploiting oil and gas;
- the Environmental Protection and Enhancement Act (R.S.A. 2000, c. E-12) (EPEA), which sets out rules intended to support and promote the protection, enhancement and wise use of the environment through sustainable development, including with respect to environmental impact assessments, the release and storage of hazardous substances and the remediation of contaminated lands;
- the Public Lands Act (R.S.A. 2000, c. P-40) (PLA), which deals with the administration of public lands held by the provincial Crown, including outlining the process for the disposition of public lands, access to public lands and the enforcement of penalties for violations; and
- the Water Act (R.S.A. 2000, c. W-3), which applies to any action that may disturb water resources, including the consumption, use and storage of water for any purpose.

References

Alberta Energy, 2014. Policy and regulations, http://www.energy.alberta.ca/OurBusiness/3718.asp.

Alberta Energy, 2015. Sturgeon upgrader, http://www.energy.alberta.ca/3444.asp, 2015.

Alberta Energy Regulator, 2012. Regulating unconventional oil and gas in Alberta, 2012, https://www.aer.ca/documents/projects/URF/URF_DiscussionPaper_20121217.pdf.

Alberta Energy Regulator, 2014. Play-based regulation pilot project. www.aer.ca/about-aer/spotlight-on/pbr-pilot-project.

Alberta Energy Regulator, 2015. Who is the AER and what is its role? https://www.aer.ca/about-aer/spotlight-on/unconventional-regulatory-framework/who-is-the-aer-and-what-is-its-role.

Alberta Industrial Heartland, 2015. Key oil and gas project updates. http://industrial-heartland.com/index.php?option=com_content&view=article&id=130%3A;project-status&catid=53&Itemid=160.

Canadian Environmental Assessment Agency, 2015. Basics of environmental assessment, https://www.ceaa-acee.gc.ca/default.asp?lang=En.

Deyholos, R. and Cuschieri, D., 2013. Canada – oil & gas: a comparative guide to the regulation of oil and gas projects. Torys LLP. http://www.torys.com/insights/publications/2013/01/canada-oil-gas-a-comparative-guide-to-the-regulation-of-oil-and-gas-projects.

Fracfocus, 2015. A chemical disclosure registry. http://www.fracfocus.ca.

Gilbert, R.J. ed., 1991. *Regulatory Choices, A Perspective on Developments in Energy Policy*, University of California Press.

National Energy Board, 2015. The role of the NEB. https://www.neb-one.gc.ca/index-eng.html.

National Resources Canada, 2015. Web page. http://www.nrcan.gc.ca.

17

When Unconventional Becomes Conventional: Regulation of Natural Gas Development in British Columbia, Canada

PAUL JEAKINS

Introduction

The shift in British Columbia (BC) from conventional to unconventional natural gas development has been more evolutionary than revolutionary. While local, national and international media highlighted a steady stream of management issues related to unconventional development, the province had the opportunity to anticipate issues arising in other unconventional jurisdictions during the modernization of its oil and gas sector legislation.

British Columbia has had a history of heightened concern for the environment stretching back decades. The 'war in the woods' that took place in the 1980s and 1990s changed the face of forest management and raised citizens' expectations for care and attention of the natural environment. As articulated by Dr Fred Bunnell (2013) 'although no one compiles statistics, British Columbia appears to have hosted more frequent drama, colour and noise around social licence than anywhere else in the world.'

As natural gas development and liquefied natural gas (LNG) rose in prominence in the province over the past decade, the focus in British Columbia shifted from forestry to ensuring that the hydrocarbon sector is also analyzed, planned and managed in such a way that a sustainable economy does not come at the expense of environmental or social well being.

The British Columbia Oil and Gas Commission was established in 1998 as the provincial regulator of the oil and gas industry for the province. Since then, it has driven innovation and provided a responsive approach to a rapidly changing sector that includes new tools and conditions to address many of the challenging issues raised by unconventional development.

At its inception the Commission was deliberately structured, from initial planning through to reclamation, on the basis of the regulatory life cycle of an oil and gas activity, with all the necessary internal expertise; regulation-making and statutory decision-making powers were housed in one agency.

Between the years 2005 and 2010 – at the time that the industry was moving from conventional to unconventional in the province – BC's regulatory approach was changing as well. The development of the new Oil and Gas Activities Act (the OGAA) brought together the Pipeline Act (1950s), components of the Petroleum and Natural Gas Act (1970s) and the Oil and Gas Commission Act (1990s).

Therefore, when it came to adopting regulations and processes for unconventional development, the Commission only needed to re-invent part of the wheel, allowing for a comprehensive approach to regulating unconventional natural gas.

As much as the shift to unconventional natural gas development raises new management challenges, it also presents opportunities to align the regulatory response to society's values and to manage the gas resource in more innovative and holistic ways. This in turn has allowed for a shift in management thinking from an operational or individual-application basis to a more tactical analysis level.

The Institutional and Regulatory Environment

Federal Responsibilities in Canada

In Canada, policy for the development of natural resources is the jurisdictional responsibility of the provincial level of government.

The federal level of government has jurisdiction over the import and export of oil and natural gas and over interprovincial oil and gas pipelines, First Nations reserves and values that are in the national good, such as species at risk. The Federal Ministry of Natural Resources has a well-developed Energy Division that:

- undertakes research;
- monitors energy markets;
- promotes energy efficiency;
- assists in technology development;
- coordinates with provinces and territories on topics of mutual interest.

The National Energy Board is Canada's regulator for the following:

- the construction, operation and abandonment of pipelines that cross international borders or provincial boundaries, as well as the associated pipeline tolls and tariffs;
- the construction and operation of international power lines and designated interprovincial power lines;
- imports of natural gas and exports of crude oil, natural gas liquids, natural gas, refined petroleum products and electricity;
- oil and gas activities for offshore lands and the Territories, except where that role has been devolved (Yukon and Northwest Territories) or there is a resource management accord (Newfoundland and Nova Scotia offshore).

In addition, federal agencies have a lead role in regulating aspects of species at risk, migratory species and navigable waterways that have an effect on natural gas development.

At the national level, but outside both the federal and provincial governments directly, the Canadian Standards Association (CSA) facilitates the development of national standards for some hydrocarbon development activities. This includes: oil and gas pipeline systems; the storage of hydrocarbons in underground formations; liquefied natural gas systems (LNG);

and emergency and security management (CSA 2014). Some provincial jurisdictions have adopted the CSA standards as part of their regulatory framework. For example, BC's Pipeline Regulation states '. . . a pipeline permit holder must not design, construct, operate or maintain any of the following except in accordance with CSA (Standard) Z662 (Pipeline Regulation, 2010 (BC) s3).

Provincial Regulatory Responsibilities: BC

Regulating natural gas in BC starts from the premise that the provincial government has already set policy and laws allowing development. This includes policy and regulation for tenure dispositions, exploration, development, production and intraprovincial transportation. Most of those issues are dealt with under the province's Petroleum and Natural Gas Act.

Regulatory responsibility is delegated to the Commission through the Oil and Gas Activities Act. The Commission is an agent of government overseen by a Board of Directors, which includes: the Ministry of Natural Gas Development's Deputy Minister as Chair; the Commissioner of the Commission as Vice Chair; an independent director as defined in the OGAA; and several Board advisors. Under the OGAA, the Commissioner has the ability to designate a person as a decision-maker under the Act or Regulations. The Commissioner can also delegate the exercise of any power conferred or imposed on the Commission under the OGAA, other than powers and duties of the Board (Oil and Gas Activities Act s7).

As defined in the OGAA s4, the job of the Commission is to ensure the resource is regulated in a manner that:

(i) provides for the sound development of the oil and gas sector, by fostering a healthy environment, a sound economy and social well-being;
(ii) conserves petroleum and natural gas resources;
(iii) ensures safe and efficient practices;
(iv) assists owners of petroleum and natural gas resources to participate equitably in the production of shared pools of petroleum and natural gas.

From its inception, the Commission was set up as a single-window, full life-cycle regulator with responsibilities for overseeing oil and gas operations in BC, including exploration, development, production, pipeline transportation and reclamation.

The Ministry of Natural Gas Development (MNGD) oversees the development of legislation for the province's oil and gas sector. Other ministries also provide policy guidance through their legislation (e.g., the Water Act,[1] Land Act, Forest Act). As part of the single-window setup, the Commission has authority embedded directly in the OGAA to administer some portions of BC's Land, Forest, Heritage Conservation, Environmental Management and Water Acts[1] on behalf of government. As such, the Commission administers the

[1] The Water Sustainability Act came into force on 19 February 2016 – updating and replacing the Water Act. More information is available at http://engage.gov.bc.ca/watersustainabilityact/act/.

portions of these Acts affecting oil and gas activities, known as 'specified enactments' or 'specified provisions'.

As the OGAA states, 'For the regulation of oil and gas activities, the Commission, instead of the official named in a specified provision, has all the powers relating to a discretion, function or duty referred to in the specified provision, including, without limiting this, the powers in the specified enactment relating to the administration and enforcement of an authorization, and is charged with all the responsibilities pertaining to that discretion, function or duty' (Oil and Gas Activities Act s8). Other jurisdictions in Canada and abroad are either investigating this approach or actively moving toward a single-window structure.

The Canadian Constitution recognizes and affirms the existing Aboriginal and treaty rights of the Aboriginal peoples of Canada. The courts have directed Crown agencies to consult with First Nations when contemplating decisions that could potentially impact treaty or Aboriginal rights. In BC approximately 200 distinct First Nations make up almost five per cent of the provincial population. Virtually all upstream activity occurs within Treaty 8 in the northeast portion of the province. While the Commission can rely on some aspects of engagement being undertaken between an applicant and a First Nation, the Commission has the primary responsibility to consult and, where appropriate, accommodate First Nations in relation to potential impacts to their treaty or Aboriginal rights, as part of the decision-making process.

Starting the Commission in this holistic way has certainly assisted with application efficiencies. But it has also meant that the Commission has had the opportunity to build up the internal expertise necessary for all aspects of natural gas development – from initial planning through to reclamation. Decision-makers can draw upon that expertise during the assessment of application strategies. However, the cross-pollination of all that expertise within one shop also allows for a comprehensive approach in the formulation of cumulative activity and tactical-management strategies for a number of issues across larger spatial and temporal scales.

How BC Moved from Conventional to Unconventional Thinking

Modernizing Legislation While Anticipating Change

As with many North American jurisdictions, industry in BC switched from the development of conventional natural gas resources to the development of unconventional natural gas resources in a very short period of time. In 2006, 90 per cent of BC's natural gas activity was in conventional plays; by 2011, 90 per cent of proposed development projects were targeting unconventional formations. Production is following that trend, as shown in Figure 17.1.

In 2005, the BC government was actively promoting coal bed methane (CBM) as a development opportunity, culminating in the first wells being produced in late 2008. At the same time, the tenuring of shale and tight sand formations was skyrocketing. In 2005–2006, the price of gas was around $13 USD per Mbtu (million British thermal units), work in Texas' Barnett formation was showing producers how to unlock shale gas economically

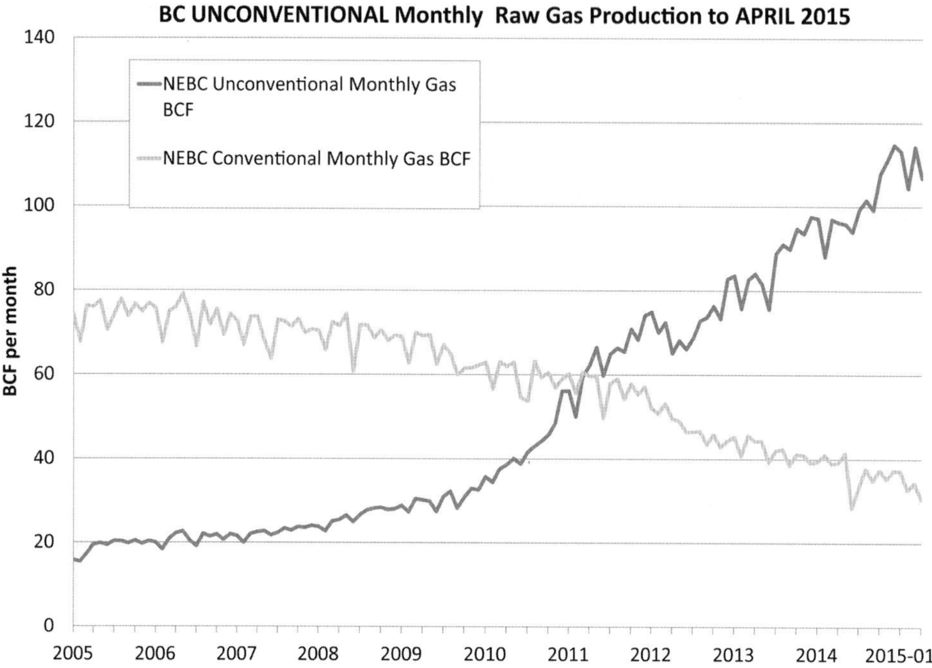

Figure 17.1 British Columbia's Unconventional monthly raw gas production in billion cubic feet (bcf). *Source:* BCOGC (2014b).

and Canadian and US companies began to take up tenure in the Horn River and Montney basins in BC (see Figure 17.2). The focus on shale and tight sands shifted money and resources away from CBM, and BC's producing CBM wells were shut-in in 2010.

As the industry has changed its technology, so too have regulators changed their approach to regulating unconventional development. It is critical to keep legislation and regulations related to fast-moving unconventional development updated and understandable, in order to successfully manage environmental and social values over the long term (Beach *et al.*, 2014; Cleland, 2011; STRONGER, 2013; Wiseman, 2014).

A number of regulatory changes specific to unconventional development were made initially as part of the OGAA's development and – as part of the Commission's commitment to continuous improvement – more changes are under way. These include changes to the Drilling and Production Regulation, to the Water Regulation and to the OGAA, to allow the Commission to take on the regulation of oil refineries and gas-processing plants.

Developing New Legislation

The new statute was designed to merge the existing relevant legislation into a new modern Oil and Gas Activities Act. As unconventional activity began to rise steeply, Commission

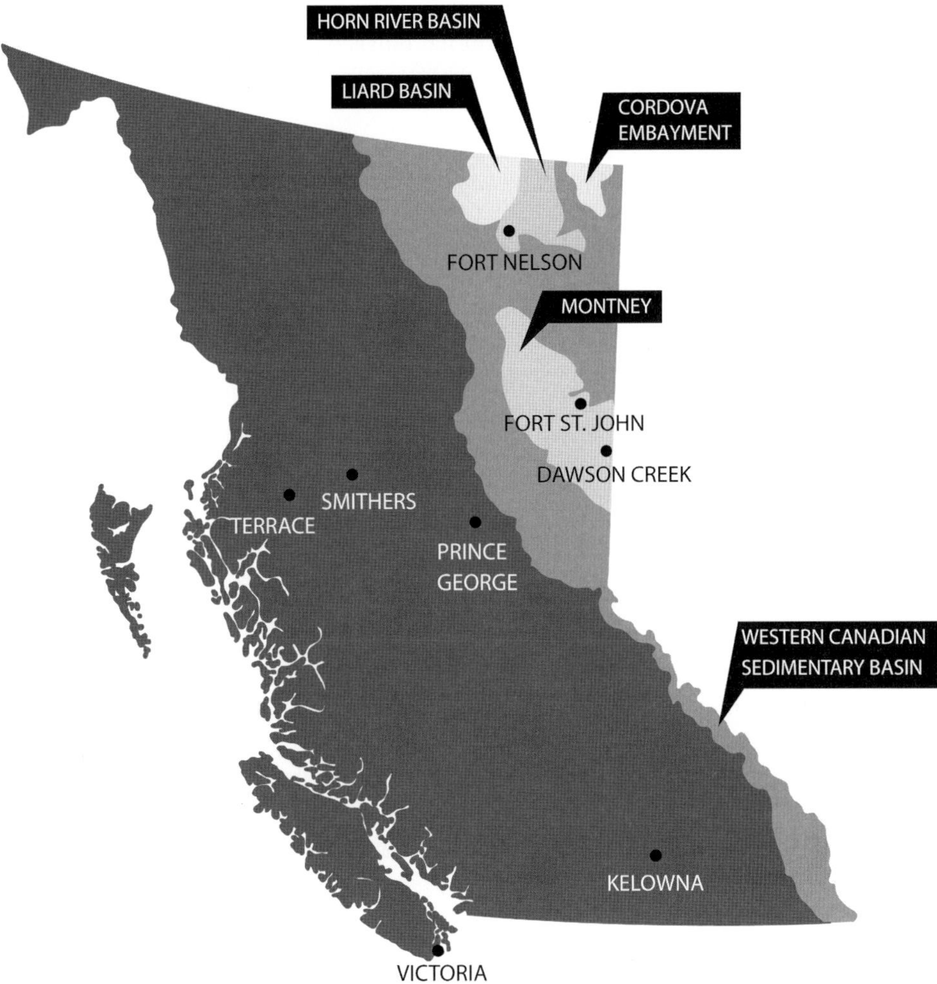

Figure 17.2 Shale and tight sand basins in British Columbia. *Source:* BCOGC (2014b).

and MNGD staff recognized the opportunity to build innovation into the Act and Regulations, knowing that the rapid shift from conventional to unconventional activities was going to change how natural gas was developed in BC. The OGAA was passed in 2008 and brought into force in October 2010 with the completion of all associated regulations.

The development of the OGAA and its regulations attempted to balance prescriptive and results-based approaches. The OGAA was also set up to allow for innovation by other government agencies (proposed through the Commission) and by industry through the designation of special projects (e.g., OGAA s75). Section 75 projects can include the development of new technologies or methods or new completion techniques to enhance recovery. However, during the development of the OGAA, Section 75 was also contemplated as an opportunity to propose innovative practices, for the management of other values such as species at risk, that could be contrary to the provisions set out in the OGAA or its regulations.

New Regulatory Structure

Regulations under the OGAA are enacted by the Lieutenant Governor in Council or by the Commission's Board, depending on the subject matter. The Lieutenant Governor in Council – the Cabinet of the sitting elected government – is given broad regulation-making authorities to establish a framework within which the Commission operates. Generally, this broad regulation-making authority includes:

- establishing policies and operational procedures of broad application (i.e., integration with Acts that apply across the natural resource sector) to be followed by the Commission;
- environmental protection and management;
- establishment of administrative penalties.

An efficient feature is that the Board of the Commission is given regulation-making powers generally for matters related to the specifics of carrying out oil and gas activities. This includes deciding what information is required in permit applications, the procedures around consultation and notification, the carrying out of oil and gas activities and the adoption of national or international codes, standards or rules.

Commission experts have the opportunity to provide input into the regulations developed by Cabinet, but the policy is researched and recommended by MNGD and subsequently approved by Cabinet. The Cabinet-level regulations under the OGAA are:

- the Administrative Penalties Regulation;
- the Environmental Protection and Management Regulation (EPMR);
- the Exemption Regulation;
- the Oil and Gas Activities Act General Regulation;
- the Pipeline Crossings Regulation.

Regulations enacted by the Commission are informed by MNGD and other government agencies, are approved by the Commission's Board and consist of:

- the Consultation and Notification Regulation (CNR);
- the Drilling and Production Regulation (DPR);
- the Emergency Management Regulation;
- the Geophysical Exploration Regulation;
- the Liquefied Natural Gas Facility Regulation;
- the Oil and Gas Road Regulation;
- the Pipeline Regulation.

While developing or changing either type of regulation goes through the same rigorous consultation and legislative drafting processes at the Ministry of Justice, the more technical Board regulations rely on the specific expertise of the Commission's staff and Board, and can usually be changed with a shorter turnaround. This approach has been particularly effective in the evolving unconventional regulatory environment. The Board deals only with the oil and gas sector whereas government must consider the effects on other Acts and regulations. This singular focus results in shorter policy processes for technical regulations, shorter drafting timelines and fewer political and strategic considerations to be accommodated.

The OGAA provides authority for ministerial oversight on how provisions of the Act are being implemented by the Commission. The Minister for Natural Gas Development can order an independent audit of the Commission related to any aspect of the Act's implementation. The Minister responsible for the Wildlife Act can also order an independent audit of the Commission's performance related to the protection and effective management of the environment (OGAA s10).

This oversight works in two ways. It allows the relevant policy ministries an opportunity to audit the Commission, in order to ensure that it is implementing the policy as originally intended. It also means that the Commission's experts must work closely with other agencies in the development and implementation of policy and regulations to ensure a common understanding across agencies with respect to their implementation.

Researching Change

A number of regulatory changes specific to unconventional development were made initially as part of the OGAA's development, and more are going through the system at the time of writing. These changes have come about as a result of learning from the experiences of other mature unconventional regulatory jurisdictions, as a result of staff and other agency input and consultation with the public and First Nations and as a result of filling research gaps. Building on the successes and mistakes seen in other jurisdictions can assist in generating effective and efficient policy approaches (Wiseman, 2014), but local values and issues will drive how other approaches are interpreted and ultimately implemented.

Anticipating what would happen as a result of the transition from conventional to unconventional development was important. For years the Commission has been an international affiliate of the United States Interstate Oil and Gas Compact Commission (IOGCC). While the IOGCC has some measure of advocacy for the sector in its mandate, it largely brings together oil and gas regulators to track, evaluate and disseminate information on state innovations and best practices.

As part of the Commission's switch to regulating unconventional development, Commission staff conducted a thorough regulatory review of North American jurisdictions as well as incorporated information available through the IOGCC and other jurisdictions such as Australia (Productivity Commission, 2009). Each jurisdiction offered what it had learned, and this contributed to the Commission's understanding of how regulating unconventional gas differs from regulating conventional gas.

Making Changes

Strategically, the basics of how the Commission regulates activities have not changed with increasing shale gas development. Paramount within the Commission's statutory decision-making authority to regulate oil and gas activities is protecting health, safety and the environment. Among other things, it ensures that water permitting and licensing is done in an environmentally sustainable manner, fugitive methane leaks are minimized and contamination is regulated.

All aspects of the regulatory framework continue to adapt and evolve to address unconventional gas development. Examples of regulatory changes subsequent to the move to unconventional gas after the OGAA was brought into force include:

- making disclosure of hydraulic fracturing fluid components mandatory (2012);
- implementation of an Oil and Gas Road Regulation (2013);
- implementation of an Emergency Management Regulation (2014);
- implementation of a Liquefied Natural Gas Facility Regulation (2014).

In addition to regulatory changes, operational and procedural changes implemented between 2010 and 2014 include:

- footprint analysis for unconventional development;
- the use of FracFocus.ca as both a vehicle for companies to submit hydraulic fracture fluid data and a site for the general public to review the data;
- permit conditions related to induced seismicity (this will be moving to regulatory requirements in the near future);
- increased oversight for water, including longer-term licences and short-term permits, and requirements for quarterly and annual public disclosure of actual water use;
- the development and implementation of BC's North East Water Tool (NEWT) and prior to that, the modelling of streams and water courses;
- the development and implementation of an Area-based Analysis Tool (ABA);
- deployment of a Mobile Air Monitoring Laboratory (MAML);
- an exclusion zone policy prohibiting new drilling activity within one kilometre of schools and an enhanced review process for any well permits within two kilometres of a school.

From Operational to Tactical

As with conventional development, the regulation of unconventional natural gas development must recognize linkages across strategic, tactical and operational levels of policy and management. Planning at each level must be related to the other two levels, as follows.

- For Commission purposes, the strategic level encompasses legislation, policy and provincial- or regional-scale planning.
- The tactical level consists of a number of tools operating across a region or, in the case of unconventional gas, a basin or a number of basins.

- Operational planning reacts to individual or grouped (i.e., multi-activity) natural gas applications and activities (i.e., post-approval). A statutory decision on an application for an oil and gas activity is considered an operational decision, and relies on strategic- and tactical-level direction and guidance.

The current dominance and consolidation of BC's natural gas development into shale and tight sand basins has allowed for a shift in thinking about regulatory management from an operational or individual application basis to a broader tactical level.

The MNGD is responsible for the disposition of tenures in BC. The size of a natural gas tenure in the province ranges between 250 hectares (ha) and 10 000 ha; between 30 and 70 companies hold tenure in each of the province's unconventional basins. The Commission and other government agencies must manage the cumulative effects across a number of different natural gas-company tenures (as well as across other sectors) for a variety of environmental and social values.

The Province of British Columbia defines cumulative effects as 'changes to environmental, social and economic values caused by the combined effect of present, past and reasonably foreseeable human actions or natural events' on the land base (Government of British Columbia, 2014).

An individual company can certainly operate in a manner that may protect a piece of land that has a given environmental or social value, but its focus is constrained to a specific and limited operational footprint. The overall approach to the definition of economic, social and environmental values must come from government. Companies generally focus on establishing social licence for their operations where it can be defined as 'demands on and expectations for a business enterprise that emerge from neighbourhoods, environmental groups, community members, and other demands of the surrounding civil society' (Gunningham *et al.*, 2004).

Government agencies such as the Commission, however, manage the impact on those values within the policy and legislation set out by the elected government. However, 'where the community is sufficiently determined, social licence can be denied to government . . . when enough of the community believes the government is not shepherding Crown lands as well as they should' (Bunnell, 2013).

Within that context, the Commission's role is to regulate the province's unconventional gas resources for provincially identified environmental and social values, utilizing all available data and input to ensure the protection of public safety and conservation of the environment while respecting those affected by oil and gas activities.

Given the multi-sectoral use of a limited landbase, decisions on individual applications are often not true balances between social, environmental and economic values at all scales for all time. The balancing of values can occur at the tactical or strategic levels where thresholds may be employed (see 'Area-Based Analysis' below).

Because the focus of natural gas extraction activities is concentrated in four unconventional basins in BC, the opportunity to manage values more holistically at a tactical level presented itself. Working with other government agencies, the Commission began both proactive and reactive regulatory shifts to deal with the change in natural gas development and began seeking tactical management opportunities.

Tactical Approach to Issues

To design tactical tools and plans for appropriately and effectively managing and regulating unconventional natural gas development in BC, the Commission's in-house experts work closely with policymakers from various ministries. They used jurisdictional scans, literature reviews, a study commissioned through the University of BC (Harshaw, 2012), input from First Nations and local residents, expertise from other government agencies, industry experts and reports and the Commission's own internal expertise to determine priorities and to analyze potential approaches.

The Commission's tactical approach has focused on these initial values and issues:

- consultation and notification;
- water;
- induced seismicity;
- waste water injection;
- current and future footprint;
- area-based analysis;
- compliance, enforcement, reporting and audits,

Some tools, analyses and reports developed specifically for issues and values related to BC's natural gas sector follow below.

Consultation and Notification

One of the regulations added when the OGAA came into force was the Consultation and Notification Regulation (CNR). The OGAA and CNR require oil and gas applicants to conduct consultation and/or notification with residents and stakeholders within specified distances before submitting an application for activity. This regulation is separate from the Commission's First Nation consultation and engagement.

The CNR process provides those potentially affected by activities related to natural gas development with the opportunity to discuss concerns with the proponent prior to submission of the application to the Commission. The recipient may also make a submission stating concerns directly to the Commission within the timeframe specified in the regulation.

Beyond company contact information, proponents must provide a description of the following to residents/stakeholders:

- the location of proposed activities;
- the proposed oil and gas activities and any significant structures and equipment to be added (constructed or used) for the activity;
- any roads that will be constructed to carry out the proposed activities;

- if and how the proposed activities relate to any existing oil and gas activities being carried out within the notification or consultation distances;
- the approximate order in which the proposed activities will be carried out;
- for multi-well pads, the entire schedule of activities over various years, where applicable.

If recipients are not satisfied with the response by the Commission to the information supplied, landowners on whose land oil and gas activities may occur have the option to appeal to a separate Oil and Gas Appeal Tribunal (OGAT). The Tribunal is an independent quasi-judicial body under the OGAA.

The Tribunal hears appeals from the Commission's decisions relating to, for example, orders, findings of contravention, administrative penalties and permitting decisions in relation to an oil and gas activity. The Tribunal may also hear appeals from decisions that have undergone a review by a review official under s71 of the OGAA.

The Commission also consults with First Nations, in respect of gas exploration and development activities, on behalf of the government. The Commission seeks to uphold the honour of the Crown through meaningful engagement and consultation with First Nations regarding potential impacts on treaty rights from proposed oil and gas activities and, where appropriate, to achieve accommodation with respect to impacted rights.

As stated, upstream development is largely concentrated in Treaty 8. The Commission has engaged in consultation process agreements with the Treaty 8 First Nations. These agreements spell out the consultation process and timing for the oil and gas sector, provide consultation capacity funding to the communities and define a meaningful consultation process. First Nations are continuing to have discussions with Government on issues of cumulative effects across sectors, revenue sharing and effects on their Treaty rights.

Water

Concern for water use in natural gas development is a constant theme in discussions with First Nations, local residents and the media and wider public. In a 2012 University of British Columbia study, a relative trade-off question (using Thurstonian analysis) found that the public in BC differentiated between concerns about drinking water and about the use of water for development.

As shown in Figure 17.3, sustaining clean drinking water was relatively more important than minimizing the amount of water used. The caveat to that finding is that other parts of the survey clearly showed that the use of water must be undertaken in an environmentally appropriate and sustainable manner.

The geologies of the Horn River and Montney basins are very different. As a result, the amount of hydraulic pressure needed to fracture the rock is dramatically different – thus the amount of water used per hydraulic fracture is dramatically different. The average volume of water used per well in the Horn River Basin is approximately 76 900 m^3, whereas the average volume is 8200 m^3 for Montney wells (BC Oil and Gas Commission, 2014b).

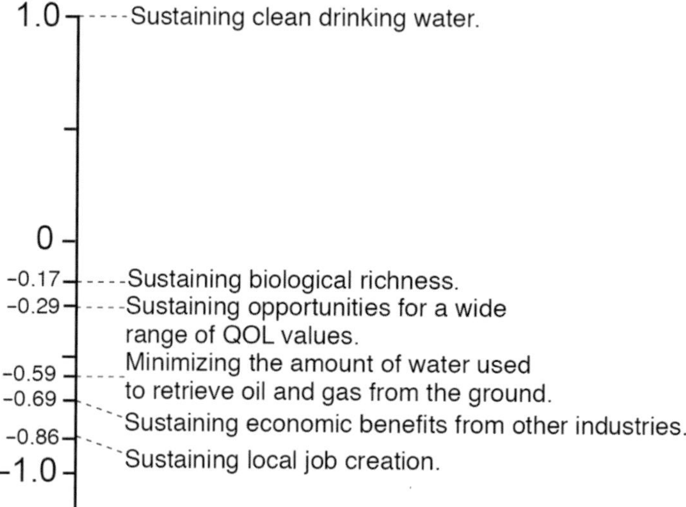

Figure 17.3 Relative priorities for sustaining natural resource values. The highest preference is at the top. *Source:* Harshaw (2012).

The increase in water use related to unconventional development has resulted in a number of new processes. While the primary source of water for most companies is currently fresh-water rivers, lakes and streams, companies also use water from other sources. This includes: recycling water as much as possible; waste water from a community; water from private land sources and deep saline aquifers. The shift to unconventional development has also resulted in an increase in the ways in which companies manage water; these include installing sizeable water storage pits, leading to a new footprint on the landbase.

The Commission is responsible for the review of applications and decisions about the water use from Crown land that is related to natural gas development. This is done through short-term water permits and longer-term water licences. The Commission works closely with the Ministry of Environment and the Ministry of Forests, Lands and Natural Resource Operations (FLNRO) on the overall management of water for the sector.

The OGAA and EPMR describe the results that a company must achieve related to water. There are more than 100 references to water in the EPMR relating to:

- spillage (prevention, reporting, containment and remediation);
- quality, quantity and flow (no material adverse effects, setback from waterworks and ground water recharge areas);
- riparian values (conservation of habitat and water values, protection of reserve zone, protection of trees and vegetation for temperature-sensitive streams);
- aquifers (no material adverse effects);
- operations within wetlands (maintenance of natural flow).

The Commission gives the ecosystem and communities top priority in allocating water use. It considers total water use in the ecosystem before allocating water access to a company. Community and ecological needs must be able to be sustained before a permit is issued and conditions may be attached to the permit. In fact, the Commission has suspended water use in northeast BC three times in the past five years owing to drought conditions.

To support the decision-making process for upstream water-use approvals, in light of the increase in water use during unconventional gas development, the Commission initiated the development of the North East Water Tool (NEWT), partnering with the FLNRO and Geoscience BC.[2] The NEWT has an underlying hydrological information database, developed through a hydrology modelling project (Chapman *et al.*, 2012).

It is designed to query locations on rivers or lakes throughout northeast BC to determine the monthly and annual average runoff at that location. These hydrology data represent the 30-year average (or 'normal') runoff (BC Oil and Gas Commission A). In addition, NEWT is designed to query all short-term water-use approvals and all water licences issued pursuant to the BC Water Act, in order to quantify how much water has already been allocated.[3]

An 'environmental flow' concept, based on environmental flow guidelines from Alberta (Locke and Paul, 2011) is included in NEWT. The Alberta methodology is being applied to northeast BC for two reasons:

- the similarity of river conditions between northeast BC and adjacent areas in Alberta;
- the absence of a comparable methodology available for BC at the present time.

The basic output from NEWT is guidance on natural water supply and availability, to assist statutory decision-makers with water-allocation decisions. This is just one piece of information that can be considered by the Commission in making a water-allocation determination (BC Oil and Gas Commission B, 2014). This tool is available online for anyone to query, at http://geoweb.bcogc.ca/apps/newt/newt.html.

As with any tool that relies on models and assumptions, NEWT has some limitations, as follows.

- The median error in the modelling was 3.7 per cent, and 78 per cent of the basins used for model calibration were modelled within ±20 per cent of their observed flow (Chapman *et al.*, 2012).
- The short-term water-use information available at present from the Commission's database and the water licence information available from the FLNRO database are limited, and the result is that total water allocations on a monthly time scale may be overestimated.

[2] Geoscience BC is a non-profit organization that receives funding from the provincial government. Its mandate is to attract mineral and oil and gas investment to British Columbia through generating, interpreting and publicly distributing geoscience data in partnership with First Nations, communities, governments and industry.

[3] FLNRO has released a northeast water strategy that helps manage water tactically across sectors. The Commission's NEWT is a cornerstone tool in that strategy.

- Many water licences have seasonal withdrawal approvals (i.e., a higher rate during the spring high flow and a lower rate during the winter low flow). The available digital data in the FLNRO database show only the maximum rate, which, when applied in NEWT, overestimate the amount of water licensed (BC Oil and Gas Commission A, 2014).

In addition, the Commission has developed a water portal – a map-based water information tool designed to provide public access to a wide range of water-related data and information in northeast BC. The data are displayed with flexible charts and analytical tools to assist users in understanding and using the data: http://waterportal.geoweb.bcogc.ca/.

The Commission publishes quarterly and annual reports on water-related approvals and use. These tools enable the Commission to ensure that decisions support the legislative and regulatory requirement to enforce the protection of water resources.

Induced Seismicity

The concept of induced seismicity is relatively new for BC and is leading to a regulatory response for unconventional development. In 2011, Natural Resources Canada (NRCan) made the Commission aware of a number of anomalous low-level seismic events in the Horn River Shale Basin. Working with NRCan and industry staff, the Commission undertook an investigation which culminated in an August 2012 report (BC Oil and Gas Commission, 2012).

Recommendations from the 2012 report led to considerable improvements in seismic monitoring in northeast BC that included:

- an expansion from two regional monitoring stations to 10 stations at the time of writing, to collect detailed seismic data;
- a mandatory shut down for industry operations if seismic activity reaches a certain threshold (magnitude 4.0).

Since that time, other events in the Montney formation have been observed and, in 2014, the Commission undertook another investigation and reported further findings regarding seismic activity (BC Oil and Gas Commission B, 2014).

The Commission tracks northeast BC seismic events through the NRCan website and industry-owned dense seismographic arrays. Events reported by the public are also investigated.

Figure 17.4 shows seismicity event locations within the Montney shale gas trend.

The investigation found that the recorded events did result from fluid injection during hydraulic fracturing, and staff subsequently made seven recommendations to minimize such events, all of which were accepted:

1. Improve the accuracy of the Canadian National Seismograph Network in northeast BC.
2. Perform geological and seismic assessments to identify pre-existing faulting.

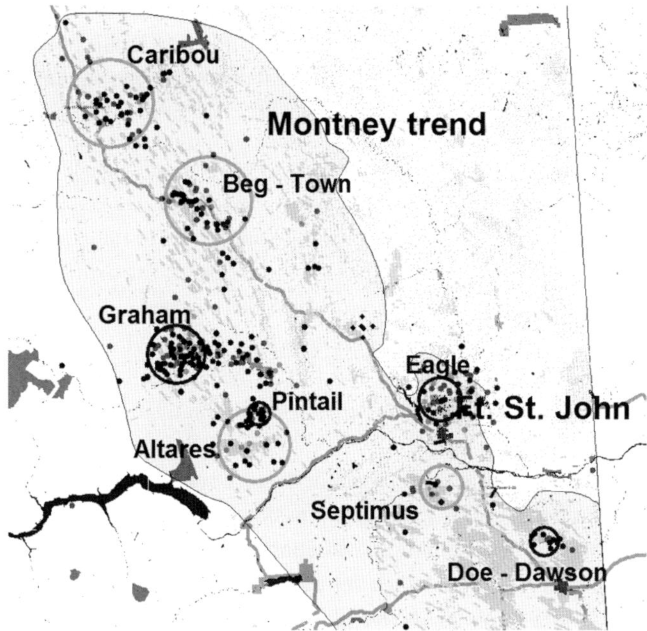

Figure 17.4 Seismicity event locations within the Montney shale gas trend. *Source:* BCOGC A (2014).

3. Establish induced-seismicity monitoring and reporting procedures and requirements.
4. Station ground-motion sensors near selected northeast BC communities to quantify the risk from ground motion.
5. Study the deployment of a portable dense seismograph array to selected locations where induced seismicity is anticipated or has occurred.
6. Require the submission of micro-seismic reports to monitor hydraulic fracturing for the containment of micro-fracturing and to identify existing faults.
7. Study the relationship between hydraulic fracturing parameters and seismicity (BC Oil and Gas Commission, 2012).

The Commission is acting on all additional recommendations from the 2014 report, which were to:

- increase regulatory scrutiny for disposal wells;
- encourage the deployment of high-resolution dense arrays;
- continue to improve regulations to address induced seismicity;
- assess the use of hydraulic fracturing buffer zones to protect sensitive infrastructure and subsurface projects.

Commission staffs are currently recommending changes to the DPR related to the above recommendations and conditions. As an interim measure, as a result of the 2012 study and recommendations, the following conditions are added to every permit until they are placed in the regulations.

'1. During fracturing operations on this well, the Permit holder shall immediately report to the Commission Emergency Contact 1–800–663–3456 any seismic event that was:
   (a) recorded by the Permit holder or any source available to the Permit holder as being magnitude 4.0 or greater and within a 3 km radius of the drilling pad, or
   (b) felt on surfaces within a 3 km radius of the drilling pad.

'2. In the event that a pad well is identified, either by the Permit holder or the Commission, as being responsible for the seismic event described in section 1(a) above, the Permit holder, subject to section 3 below, will suspend fracturing operations on this well immediately.

'3. Fracturing operations at this well, suspended under section 2 above, may continue if:
   (a) the Permit holder presents to the Commission a plan for mitigation aimed at reducing the seismicity or eliminating well operations related to the induced seismicity,
   (b) the Commission is satisfied with this plan, and
   (c) the Permit holder implements this plan.'

Additionally, the Commission takes the following specific steps related to seismic activity in northeast BC.

- It responds to seismic events as soon as they appear on the NRCan website or a private dense seismograph array. Reports from the public are also investigated;
- These events are compared with the locations of oil and gas operations
- If there is a temporal and geographic similarity between the seismicity and oil and gas activities, the operators are contacted with a request for more data. These data are analyzed.
- All data are used by the Commission to monitor seismic activity closely in areas of oil and gas operations. Actions are taken if and when required.
- Further steps may include the deployment of dense arrays, which are used to study seismic activity in greater detail, or the moderation of hydraulic fracturing parameters, which can include limiting well pressures or suspension of operations.

Waste Water Injection

The surface discharge of produced water is not allowed in BC. Natural gas producers must dispose of produced water into a subsurface formation via an approved and regulated disposal well, as per the Oil and Gas Waste Regulation, which is under the Environmental Management Act (BC Oil and Gas Commission C). This applies to the disposal of water associated with hydrocarbon production, flowback fluids or a combination of both, as produced water is defined to include recovered fluids from well completion or workover operations (including flowback fluids from fracture stimulations).

The regulation of waste water is a combination of prescriptive and results-based requirements. The aim is to prevent waste-water injection contaminating freshwater – as required in the EPMR and the DPR. Prescriptive requirements are outlined in both the regulations and the guidelines associated with subsurface disposal (BC Oil and Gas Commission D 2014).

The Commission recognizes that full-cycle water management is an important component of increased development activity. This includes fluid recycling, treatment and disposal. The Commission is working with industry in each of these fields. Moreover, significant work has taken place to achieve an improved understanding of disposal reservoirs and appropriate disposal well operating, monitoring, measurement and reporting requirements to mitigate potential issues.

Current and Future Footprint

An understanding of the current surface land use taken up by oil and gas activities combined with a prediction of what the land use will look like in the future can assist in determining whether setting development and other value thresholds is warranted.

Operators involved in shale gas development often point out that unconventional development, using horizontal drilling methods, leads to a smaller disturbance footprint on the surface. In 2013 the Commission analyzed historic surface land use by well sites, pipelines, roads, geophysical exploration programmes, facilities and associated oil and gas activities in northeast BC in part to track what the shift from conventional to unconventional looks like.

In addition, the shift to unconventional development has created opportunities to predict the interplay between subsurface resource development and surface land use. Shale basins that cover a large area allow the surface development to be planned differently from when one is attempting to explore and develop finite conventional pools. Horizontal drilling also helps in that surface activity can be planned to take place in such a way that the disturbance to environmental and cultural values is minimized.

Figure 17.5 shows an example of the shift in land-use patterns between past conventional natural gas development and current unconventional natural gas development. Furthermore, the shift from large-scale geophysical projects utilizing bulldozers to low-impact geophysical techniques over the past decade has been even more dramatic. Usually, the shift to unconventional development has also led to a drop in geophysical projects.

The northeast portion of BC, where almost all the oil and gas development has occurred and will occur for the foreseeable future, comprises approximately 16.3 million ha. Until 2006, most of the natural gas development activity occurred outside the main shale basins in the Western Canada Sedimentary Basin that extends into Alberta, Saskatchewan and Manitoba.

Unconventional development has led to shifts in land use, including an increase in the number of pits used for water storage and increased well-pad size, shifting from a conventional well-pad average size of 1.4 ha to an unconventional well-pad average size of

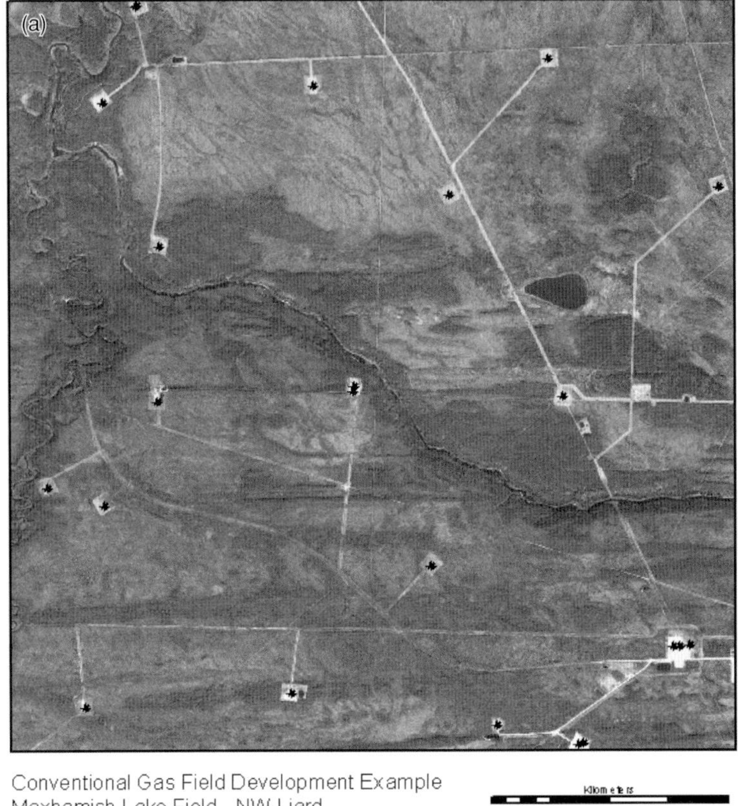

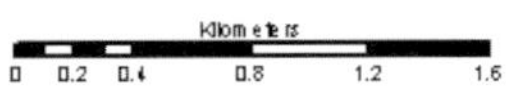

Figure 17.5 (a) Conventional and (b) unconventional development patterns. *Source:* BCOGC (2013).

4.5 ha. Table 17.1 highlights the amount of surface disturbance by activity for each basin. The current disturbance by formation ranges between 1.28 per cent and 4.08 per cent.

The ‘hunting’ nature of conventional development did not allow for accurate predictions on where and when development would occur. However, the ‘farming’ nature of unconventional shale development has allowed the Commission to put a bookend on where future development will likely occur. The potential for LNG export from the province has allowed planners to put another bookend on natural gas development, predicting when development is likely to occur.

With that in mind, early on in the planning for LNG the Commission and MNGD forecast that it was likely that natural gas production that would contribute to LNG exports (Figure 17.6). The forecast assumed five LNG plants exporting 82 megatonnes per year (11.6 bcf/day) by 2020, and thus is an upper-end forecast. The forecast also assumed that 25 per cent of the gas will come from outside BC (MNGD and BC Oil and Gas Commission 2014b). This forecast will be updated as it becomes clearer how many LNG plants and

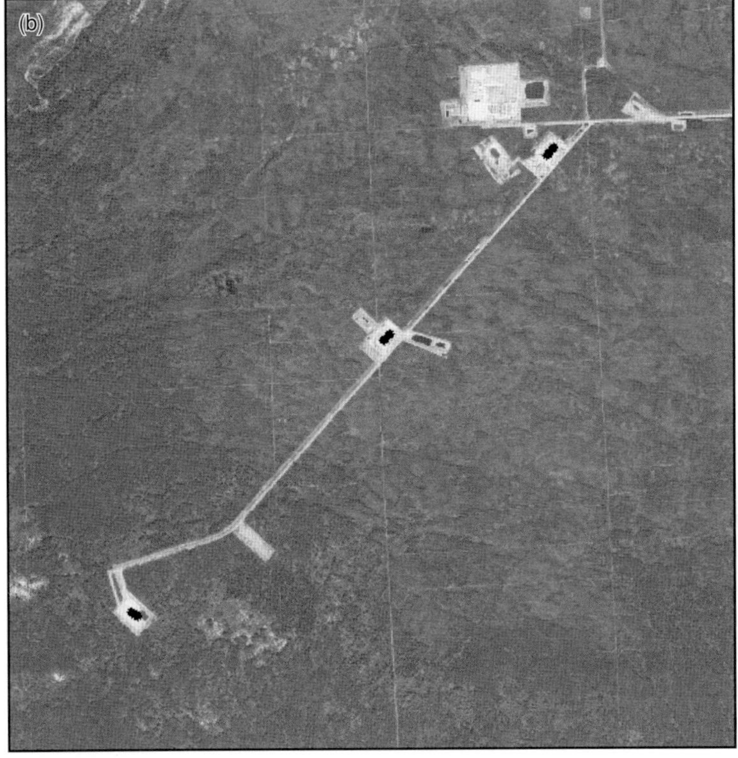

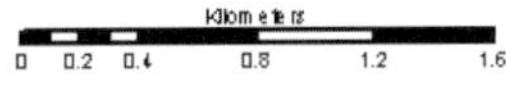

Figure 17.5 (*cont.*)

pipelines will be built over the next few years. As an upper threshold, the 2014 forecast indicated that the province can sustain environmental values, even with a maximum build out.

Linking production to the number of wells then allows further forecasting for the number of well pads and related disturbance, water use, proppant needs, lengths of pipelines, road disturbance and number of facilities. Figures 17.7 through 17.10 highlight some of these forecasts.

Area-Based Analysis

In addition to forecasting likely development, the Commission has developed an area-based analysis (ABA) approach as a framework for managing the potential environmental and cultural impacts of oil and gas development at a tactical level. The approach integrates strategic direction from statutes, regulations and existing land-use plans with localized management guidelines for identified environmental and cultural values into a framework for

Table 17.1 *Surface area used for oil and gas activities, by geological formation (BCOGC 2013)*

Activity	Western Canada Sedimentary Basin (ha)	Per cent of Western Canada Sedimentary Basin	Liard Basin (ha)	Per cent of Liard Basin	Horn River Basin (ha)	Per cent of Horn River Basin	Cordova Embayment (ha)	Per cent of Cordova Embayment	Montney Play Trend (ha)	Per cent of Montney Play Trend
Wells**	30 207	0.22	367	0.04	1240	0.11	499	0.16	13628	0.46
Roads**	83 454	0.62	3613	0.39	6490	0.57	2974	0.94	25352	0.85
Facilities**	1 543	0.01	45	0.005	286	0.02	16	0.01	739	0.02
Pipeliness**	43 785	0.33	1177	0.13	2630	0.23	1182	0.37	16749	0.56
Other Oil & Gas Infrastructures**	12 683	0.09	507	0.05	3838	0.33	347	0.11	5356	0.18
Geophysical Exploration (seismic lines)	227 497	1.69	7208	0.78	23 941	2.09	5634	1.78	67381	2.26
Basin Area	13 450 458		934 304		1 145 989		315 867		2 985 906	
Total Area Used for Oil and Gas Activities**	399 169	2.97	12 989	1.39	38 425	3.35	10 651	3.37	129 205	4.33
Net Area* Used for Oil and Gas Activities	375 111	2.79	11 939	1.28	36 474	3.18	9735	3.08	121 950	4.08

* The net area is that when the area shared by overlapping permit types is removed from the total area.

** The total area includes the area shared by overlapping permit types.

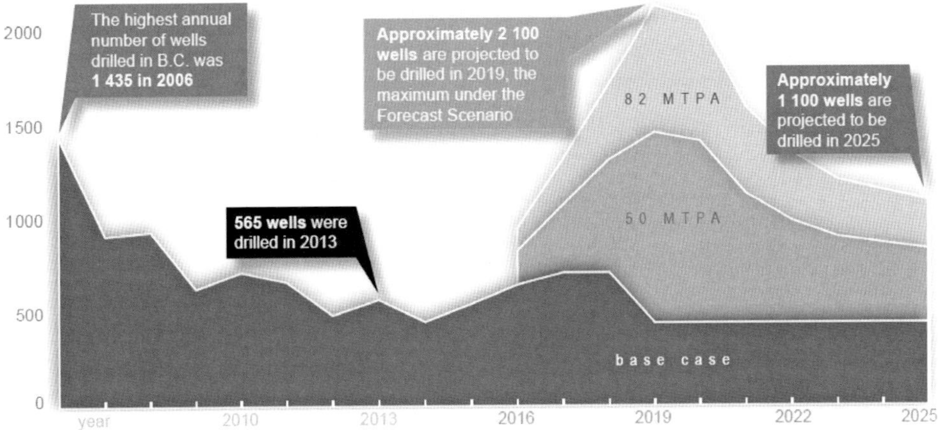

Figure 17.6 Current and predicted wells drilled by year. *Source:* BCOGC (2014b).

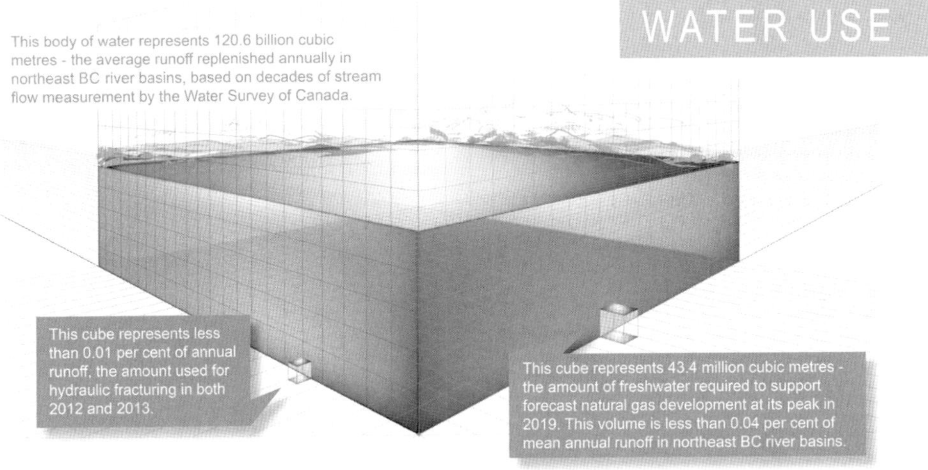

Figure 17.7 Current and predicted water use. *Source:* BCOGC (2014b).

assessing oil and gas development across a defined landbase. The ABA project includes the development of two ABA status tools: the ABA WebMap Tool and the Impact Assessment Tool.

Both are decision-support tools to assist in the review of individual activity applications as well as in the overall measurement and management of the potential cumulative effects of natural gas development on the landscape in the overall context of all industrial activity.

The area-based analysis covers the full extent of the Western Canada Sedimentary Basin in northeast British Columbia. This includes the key development basins: the Horn River, the Cordova Embayment, the Montney and the Liard Basin.

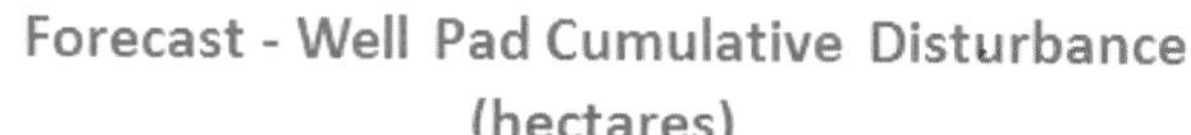

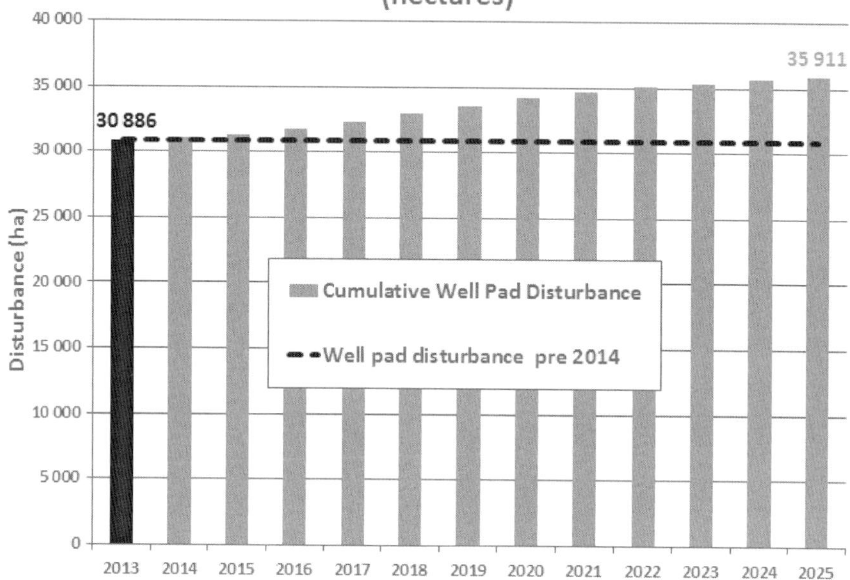

Figure 17.8 Predicted well pad surface disturbance. *Source:* BCOGC (2014b).

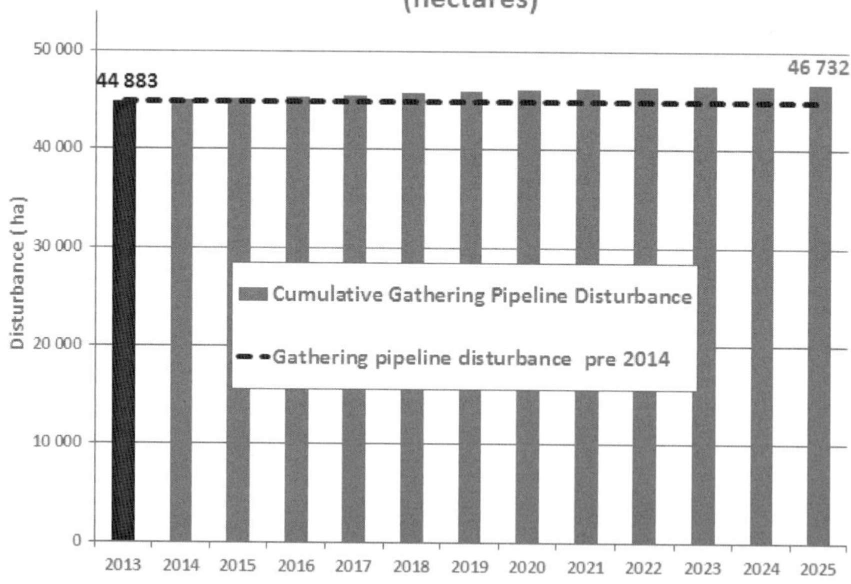

Figure 17.9 Predicted gathering pipeline surface disturbance. *Source:* BCOGC (2014b).

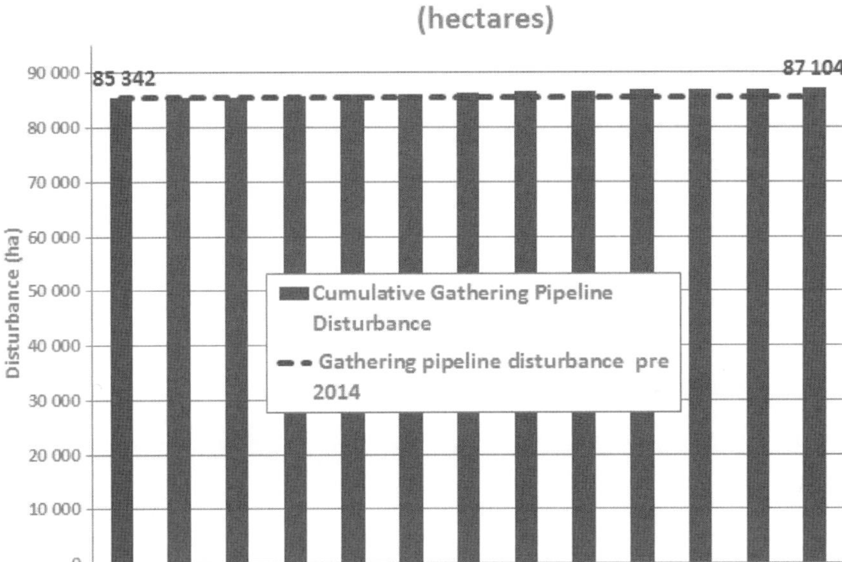

Figure 17.10 Predicted road surface disturbance. *Source:* BCOGC (2014b).

The initial values (Figure 17.11) being assessed focus on the biophysical components of the ecosystem, covering 650 000 ha of hydro-riparian reserves and 3.5 million ha of old forest.

The starting values are:

- old forest (as of January 2015);
- hydro-riparian ecosystems (riparian habitat, water quantity) (as of January 2015);
- high-priority wildlife habitat (expected fall 2015);
- resource features (in development);
- cultural heritage resources (in development).

The principle behind ABA is that the management response escalates as the impact on a value increases owing to industrial build out by all activities, not just oil and gas (refer to Figure 17.13).

The ABA permitting process allows all new development applications received by the Commission to be measured and assessed for cumulative effects and, where applicable, can inform approval decisions where the proposed impact exceeds a defined trigger.

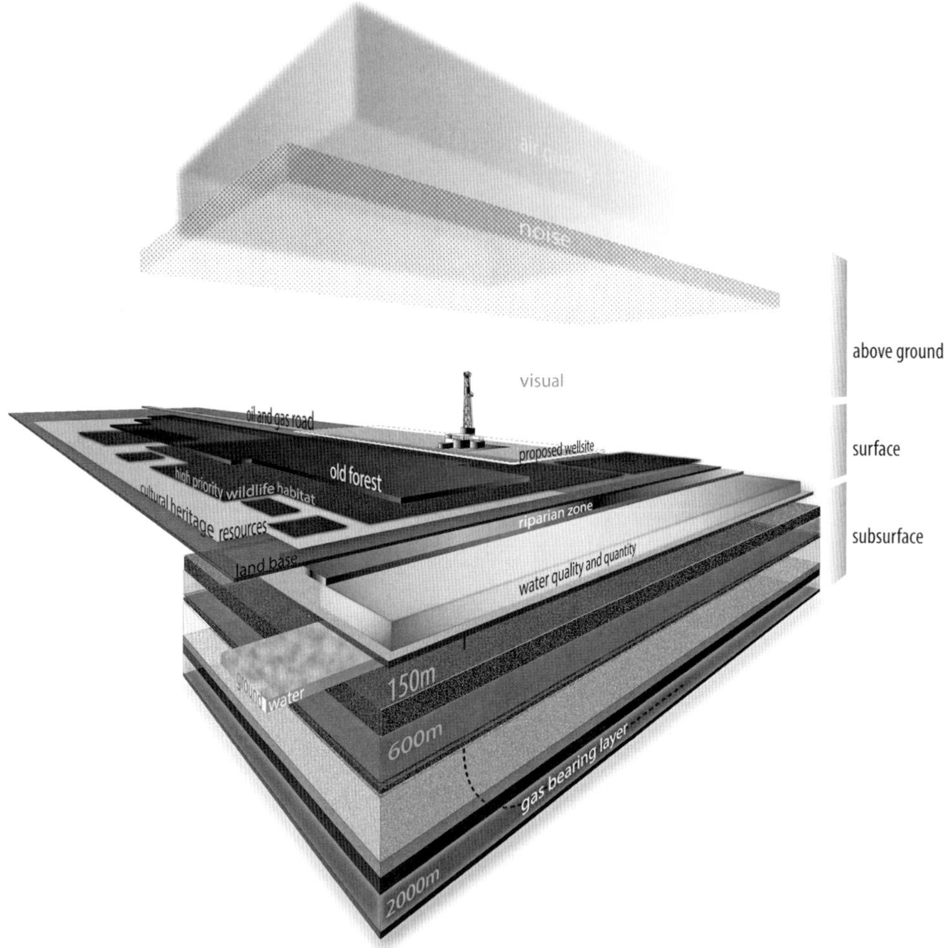

Figure 17.11 Area-based analysis schematic. *Source:* BCOGC A (2014).

An assessment framework is developed for each value (see Figure 17.12) and the information generated during the assessment is provided to the statutory decision maker (SDM) for use in their decision.

Area-based analysis evaluates cumulative impacts on the landscape at an ecological level, using ecological assessment units (such as Water Management Basins and Natural Disturbance Units). Where development has reached or exceeded a scientifically defined level in a given ecological assessment unit, that area is assigned an ABA status of 'enhanced management' or 'regulatory policy' (Figure 17.13).

In applying the area-based analysis to a permitting decision, SDMs will consider whether a proposed oil and gas activity impacts an ABA value dataset with ABA status 'enhanced management' or 'regulatory policy'.

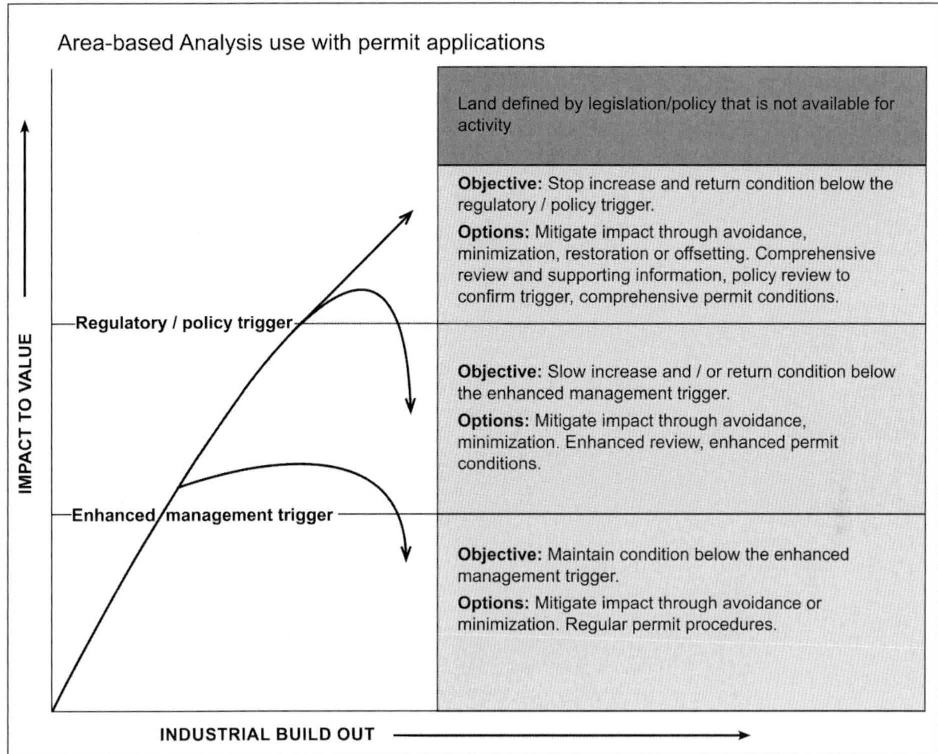

Figure 17.12 Area-based analysis use with permit applications. *Source:* BCOGC A (2014).

The goal of the analysis is to maintain conditions in the bottom bar, where applications are subject to routine reviews and operating procedures. When the enhanced management trigger has been reached and the current condition of the value is determined to be in the middle bar, the permitting process is subject to additional review and supplementary operating procedures will be considered.

Where oil and gas activity results in an ABA status that is 'enhanced management' or 'regulatory policy'the Ministry of Environment's Environmental Mitigation Hierarchy (Figure 17.14) must be followed to minimize disturbance to the landscape.

If a proposed oil and gas activity will push the status into enhanced management or regulatory policy, mitigative steps are required to move the impact below the trigger.

In the event that the regulatory policy trigger has been reached and the current condition of the value is determined to be in the top bar, the application review process is subject to additional senior or regional staff review and additional operating procedures will be considered.

The objective is to restore conditions below the regulatory policy trigger and ideally below the enhanced management trigger. This may include suspending permitting, confirming policy direction and implementing innovative approaches to mitigate the impact.

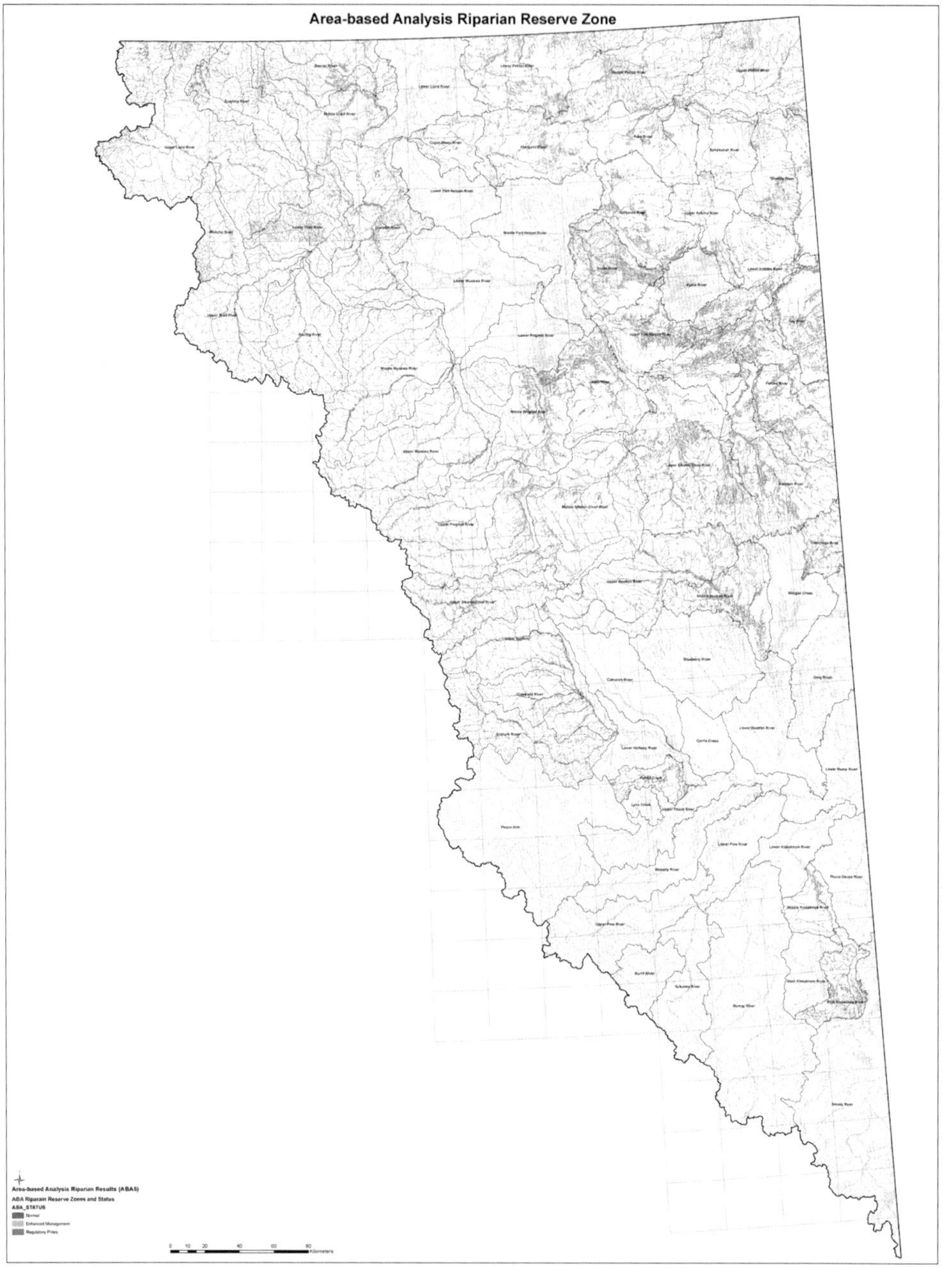

Area-based Analysis Riparian Reserve Zone

Figure 17.13 Area-based analysis for northeast BC riparian detailed map. *Source:* BCOGC A (2014).

Figure 17.14 Ministry of Environment Environmental Mitigation hierarchy.

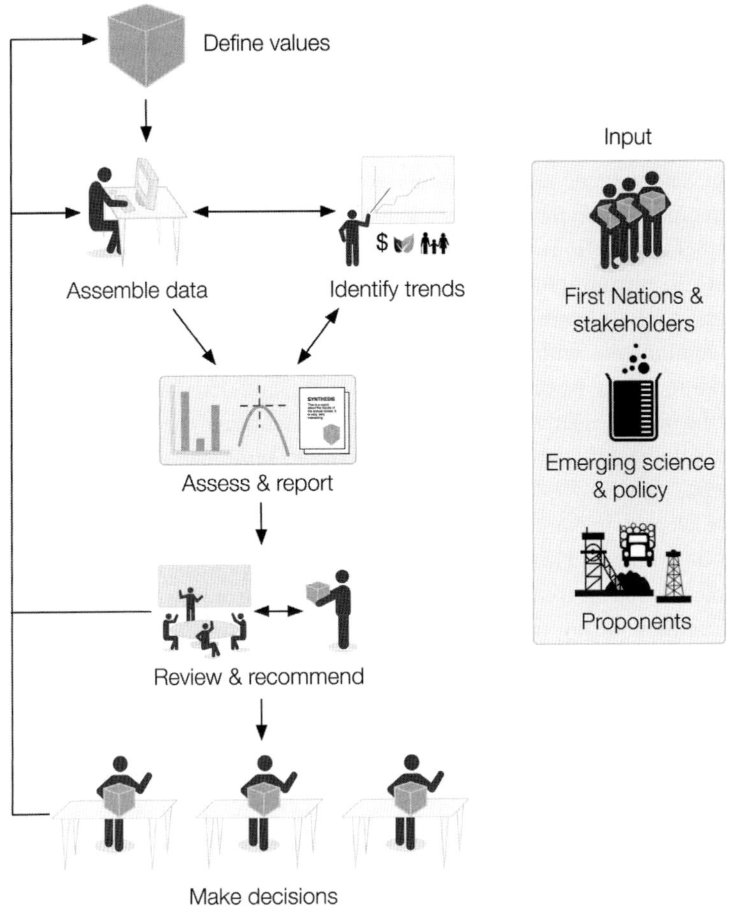

Figure 17.15 ABA workflow within the Commission. *Source:* BCOGC (2015).

Area-based analysis brings together all available strategic direction, tactical and operational data and research and input from others, including First Nations and area residents, for use by Commission statutory decision-makers. Figure 17.15 summarizes the utilization of ABA in operational decision-making within the Commission.

Since 5 January 2015, all applications have been assessed by the Commission for cumulative effects using ABA. Statutory decision-makers of the Commission will use ABA in their evaluation of applications to gauge the potential impact of applications on defined ABA values. And as of 1 September 2015, all applications for oil and gas activity must show that ABA has been considered and what mitigative strategies are in place if triggers are reached.

Area-based analysis reports will be completed periodically to summarize surface land-use disturbance against the defined values. The reports will be posted on the

Commission's public website. Information from the Commission's ABA reports will feed into the province's overall approach to cumulative effects analysis.

Compliance, Enforcement, Reporting and Audits

For both conventional and unconventional oversight, 'equally as important as having well-developed regulations on the books is adequate enforcement by staff both in the office and conducting field inspections' (Grimshaw and Groat, 2012).

The move to unconventional gas development in BC has led to changes in how the shifting regulatory structure is enforced by the Commission. The increase of horizontal drilling in combination with hydraulic fracturing as a means of wellbore completion has increased the focus on aspects of well-site development. This includes wellbore integrity (in view of the high pressures involved in hydraulic fracturing) and fluids handling (in view of the large volumes of water used, and the desire to minimize the use of fresh water and ground water).

Wellbore integrity is evaluated in various ways. The Commission requires the submission of well logs and is moving to perform periodic audits on these well-log submissions. As was standard practice during conventional development, the Commission continues to conduct on-site inspections during periods of casing and cementing to ensure that the required practices are followed.

As discussed above, the use of hydraulic fracturing has increased public scrutiny on both water use and management, and fluids handling is a compliance focus. In coordination with other regulators, the Commission has developed new water handling and storage standards that apply to all oil and gas industry proponents. These standards define how water should be transported and stored when it is used for hydraulic fracturing.

The standards are imposed at the time the activity is approved. During on-site inspections, Commission inspectors ensure these standards are put into practice. Additional requirements include specific construction practices to be followed during water storage site construction, and specific monitoring and data reporting for storage sites.

The Commission requires the submission of data from monitoring wells, as well as from all required leak-detection devices. These data sources are then subject to a Commission audit.

As the OGAA and regulations have a blend of prescriptive and results-based approaches, oversight has to take both into account. In addition to the review of prescriptive requirements, the Commission has instituted an audit programme to look at results-based requirements – i.e., the results that must be achieved without prescribing how.

The Commission audits industry performance relative to these requirements against internationally recognized standards. Audits are conducted on activities where on-site inspections will not provide sufficient confirmation of results. The combination of inspections (on site) and audits (measuring outcomes) is utilized to maximize the efficacy of Commission activities in ensuring regulatory compliance.

Overall, field inspections and enforcement actions are an extension of the approach to safety throughout the entire life cycle of an oil or gas project – from the application stage, to permitting, to construction and ultimately to reclamation and remediation. Each year, the Commission continues to target higher compliance rates to ensure the safety of British Columbians and the protection of the environment.

As part of its mandate, the Commission must report on its approach to decision-making, which includes contraventions. In 2013, the Commission began posting a quarterly overview of enforcement actions online.

Continuous Improvement: Knowledge Gaps and Research

As part of its own shift from conventional to unconventional, the Commission has garnered knowledge from jurisdictions that have already shifted from conventional to unconventional natural gas development, hired a number of new experts in-house and established a modern legislative framework that is continually improved; but it recognizes that some knowledge gaps still need to be filled.

The commission engaged Ernst Young (EY) in fall 2014 to conduct a focused assessment of the regulatory framework governing hydraulic fracturing. Specifically, EY was asked to:

- assess the Commission's current regulatory framework, including legislation, regulation, guidance, leading practices, policies, permit conditions, and industry standards;
- develop a detailed map of the relationship between existing regulatory instruments and the key issues presented by hydraulic fracturing;
- conduct a high-level scan of six selected jurisdictions with an industry and geology similar to BC;
- identify opportunities to improve the framework;
- on the basis of a set of guiding principles, develop leading-practice recommendations.

Ernst Young adopted an issues-based approach to the assessment. Three primary issue-groups were identified: water use and protection (water life cycle), induced seismicity and quality of life disturbances. The key issues within those groups were identified, and an analysis of the existing regulatory instruments was conducted to determine whether there were gaps or opportunities.

Ernst Young determined that the issues presented by hydraulic fracturing are being effectively managed by the Commission, within the current regulatory framework.

In looking at the six similar jurisdictions (Alberta, Saskatchewan, Texas, North Dakota, Pennsylvania and Colorado) the EY review also 'found that the key hydraulic fracturing issues that were identified were identical across all jurisdictions, varying only in the level of importance'.

The Commission also partners with universities to ensure that it has the best science around oil and gas development, including the University of British Columbia's Faculty of Applied Science, and researchers at the University of Victoria and the University of Northern BC. The Commission has provided funding, perspective and data on a number of projects as it uses science to fill any potential knowledge gaps. Projects are suggested by either university researchers or Commission staff and include:

- assessment of green additives and chemical compounds for hydraulic fracturing;
- modelling coupled interactions of proppant and shale during post-fracturing fluid flow-back and fracture closure;
- estimating the life-cycle water footprint of hydraulic fracturing;
- geohazard assessment and methane detection using low-cost drones;
- failure and risk analysis of BC oil and gas pipelines;
- developing numerical models of hydraulic fracture propagation;
- cementing fluid mechanics causes and solutions for surface casing vent flow;
- investigating hydro fracture and stress field interactions on critically stressed faults and induced seismicity using a FEMDEM modelling approach;
- assessment of groundwater baseline sampling criteria;
- evaluation of gas migration (GM) near production wells and abandoned wells associated with unconventional gas developments;
- acquiring and analysing multi-sensor airborne data.

The Commission will continue to strengthen its relationship with universities in order to utilize emerging science in the development of regulations and processes. The Commission strives as well to stay fully apprised of the latest technological breakthroughs and independent worldwide scientific research on possible industry effects on human health, social structures, climate, environmental and cultural values. The Commission continues to deepen and share its knowledge of BC's hydrocarbon resources and the local effects of oil and gas development.

Conclusion

Unconventional natural gas development has risen to prominence over the past decade, and in British Columbia it has become the primary means of producing natural gas. The mechanics are complex, and so are the regulatory requirements to address and protect values while staying on top of change. Society is changing as well, and its knowledge and expectations will continue to require the impartial balancing of values.

British Columbia enhanced its regulatory regime at the same time that natural gas development moved from conventional to unconventional. That approach provides the flexibility to continually adapt to industry innovations. The BC Oil and Gas Commission, as the provincial regulator of the industry, has regulation-making powers, which ensures a timely response to emerging issues. In addition, the Commission is actively working with universities to address any potential knowledge gaps.

Unconventional gas development has become the convention in BC, and the Commission is positioned to anticipate, adapt and continually improve its regulatory oversight.

Legal Disclaimer in Reference to BC Legislative References

These materials contain information that has been derived from information originally made available by the Province of British Columbia at: http://www.bclaws.ca/ and this information is being used in accordance with the Queen's Printer License – British Columbia, available at: http://www.bclaws.ca/standards/2014/QP-License_1.0.html. They have not, however, been produced in affiliation with, or with the endorsement of, the Province of British Columbia and thus **THESE MATERIALS ARE NOT AN OFFICIAL VERSION.**

References

Argonne National Laboratory, University of Arkansas (2014). Overview of regulatory requirements governing natural gas development in the Fayetteville shale region. <http://lingo.cast.uark.edu/LINGOPUBLIC/reg/index.htm>

Arthur, D.J., Bohm, B.K. and Coughlin, B.J. (2010). Summary of environmental issues, mitigation strategies, and regulatory challenges associated with shale gas development in the United States and applicability to development and operations in Canada. In *Proc. Canadian Unconventional Resources and International Petroleum Conference*. Society of Petroleum Engineers.

BC Ministry of Natural Gas Development, BC Oil and Gas Commission (2014). LNG forecast scenario. https://www.bcogc.ca/node/11277/download.

BC Oil and Gas Commission (2012). Investigation of observed seismicity in the Horn River Basin. http://www.bcogc.ca/node/8046/download.

BC Oil and Gas Commission (2013). Oil and gas land use in Northeast British Columbia. http://www.bcogc.ca/node/11039/download.

BC Oil and Gas Commission (2014a). Investigation of observed seismicity in the Montney Trend. http://www.bcogc.ca/node/11291/download.

BC Oil and Gas Commission (2014b). Water use for oil and gas activity 2013 Annual Report. http://www.bcogc.ca/node/11262/download.

BC Oil and Gas Commission (2015). ABA workflow within the Commission. Internal Presentation.

BC Oil and Gas Commission A (2014). ABA northeast BC riparian detailed map. ftp://ftp.bcogc.ca/outgoing/OGC_Data/Area_Based_Analysis/ABA_RRZ_Detailed_Map_NEBC.pdf.

BC Oil and Gas Commission B (2014). About NEWT. http://www.bcogc.ca/node/8137/download.

BC Oil and Gas Commission C (2014). BCOGC application guideline for deep well disposal of produced water. https://www.bcogc.ca/node/8206/download.

BC Oil and Gas Commission D (2014). Legislation. http://www.bcogc.ca/Legislation.

Beach, S., Wilkins, A., Winter, J. (2014). The future of energy regulation and policy development: a summary paper. University of Calgary. The School of Public Policy. *SPP Communique* 6 (5).

Bunnell, F. (2013). Social licence in British Columbia: some implications for energy development. *Journal of Ecosystems and Management* 14 (2).

Chapman, A. and Kerr, B. (2011). Development of a hydrology decision support tool. Unconventional Gas Technical Forum. Victoria, BC, 7 April 2011 (Slide presentation).

Chapman, A., Kerr, B. and Wilford, D. (2012). Hydrological modelling and decision-support tool development for water allocation, northeast British Columbia. Geoscience BC Summary of Activities 2011. *Geoscience BC* 2012 (1), pp. 81–86.

Cleland, F. M. (2011). *Seismic Shifts: The Changing World of Natural Gas*. Canada West Foundation.

Cleland, M. (2014). *From the Ground Up: Earning Public Support for Resource Development*. Canada West Foundation.

Council of Canadian Academies (2014). Environmental impacts of shale gas extraction in Canada. http://www.scienceadvice.ca/uploads/eng/assessments%20and%20publications%20and%20news%20releases/Shale%20gas/ShaleGas_fullreportEN.pdf.

Canadian Standards Association (CSA Group) (2014). About CSA Group. http://www.csagroup.org/us/en/about-csa-group.

Freeman, J., and Kolstad, C.D. (2006). *Moving to Markets in Environmental Regulation: Lessons from Twenty Years of Experience*. Oxford University Press.

Government of British Columbia (2014). What is the cumulative effects framework? *Cumulative Effects News*. 01, p. 1.

Grimshaw, T.W., and Groat, C.G. (2012). *Fact-Based Regulation for Environmental Protection in Shale Gas Development*. The Energy Institute, University of Texas at Austin.

Gunningham, N., Kagan, R.A. and Thornton, D. (2004). Social licence and environmental protection: why businesses go beyond compliance. *Law and Social Inquiry* 29, pp. 307–341, 308.

Harshaw, H.W. (2012). British Columbia Oil and Gas Resource Management Public Opinion Survey. University of British Columbia.The Science and Community Environmental Knowledge (SCEK) Fund. http://www.scek.ca/sites/default/files/documents/RA2010-02_Public_Opinion_Survey_final-report_public_Feb_8_12.pdf.

Interstate Oil and Gas Compact Commission (2014). What we do. http://iogcc.ok.gov/what-we-do.

Locke, A. and Paul, A. (2011). A desk-top method for establishing environmental flows in Alberta rivers and streams. Alberta Environment and Alberta Sustainable Resource Development. http://www.environment.gov.ab.ca/info/library/8371.pdf. Accessed 19 March 2015.

Oil and Gas Activities Act, Queen's Printer (SBC 2008). Retrieved from the Queen's Printer website: http://www.bclaws.ca/civix/document/id/complete/statreg/08036_01.

Penn State (2014). Issues and Impacts of Marcellus Shale and natural gas drilling for landowners, communities, and businesses including financial, environmental, economic and legal issues.http://extension.psu.edu/natural-resources/natural-gas/issues.

Productivity Commission (2009). *Review of regulatory burden on the upstream petroleum (oil and gas) sector*. Government of Australia. http://www.pc.gov.au/__data/assets/pdf_file/0011/87923/upstream-petroleum.pdf.

Sakmar, S.L. (2011). Shale gas developments in North America: an overview of the regulatory and environmental challenges facing the industry. In *Proc. North American Unconventional Gas Conference and Exhibition*. Society of Petroleum Engineers.

STRONGER, State Review of Oil & Natural Gas Environmental Regulations (2013). Hydraulic fracturing guidelines. http://www.strongerinc.org/sites/all/themes/stronger02/downloads/2013-STRONGER-Hydrualic-Fracturing-Guidelines.pdf.

Wiseman, H.J. (2012). *Regulation of Shale Gas Development, Including Hydraulic Fracturing*. Centre for Global Energy, International Arbitration and Environmental Law, University of Texas School of Law.

Wiseman, H.J. (2014). Regulatory islands. *New York University Law Review*, 89, 1661.

18

Leading Practice Regulation for Unconventional Reservoir Development in South Australia

BARRY GOLDSTEIN, MICHAEL MALAVAZOS AND BELINDA HAYTER

Introduction

Leading practice regulation starts with well-considered legislated objectives that drive the behaviour of both industry and regulators. Experience demonstrates that the foundations for regulation that consistently meet community expectations are the following.

1. Legislated objectives for protecting social, natural and economic environments from significant risks are needed
2. Regulatory principles should have certainty, openness, transparency, flexibility, practicality and efficiency.
3. Legislated requirements are needed for:
   (a) information on potential environmental impacts of operations, long before companies apply for approval for on-ground activities; the information needs to be sufficient to inform people and enterprises potentially affected by the relevant regulated activities in a timely manner;
   (b) engagement with people and enterprises potentially affected by the relevant regulated activities;
   (c) regulation to be contestable but without support for vexatious objections to land access; fair and efficient dispute resolution processes to prevail with all regulatory decisions ultimately open to resolution in court;
   (d) fair compensation to affected land-users for deprivation or impairment of the use and enjoyment of the land or for damage or losses;
   (e) project operators regularly to report on their environmental outcomes, and describe innovation to mitigate any reccurrence of incidents.
4. Lead regulatory agencies (or a one-window-to-government approach) should be used as an effective way to coordinate co-regulatory decisions in parallel, rather than in series.
5. Regulators should have the competence and capacity to deploy regulations effectively while adhering to the principles of certainty, openness, transparency, flexibility, practicality and efficiency, without taint of capture.
6. Compliance enforcement policies are to be published and robustly implemented to prevent, persuade or direct operations in order to avoid undesired outcomes.

Regulators can prevent and stop operations, require restitution, levy fines and cancel licences.

7. Project operators must monitor and report on the efficacy of their risk management systems to deliver environmental protection, while regulators probe to prevent and detect mishaps.
8. Fitness-for-purpose assessments of facilities and risk management systems should be undertaken periodically, whether or not non-compliant activities arise.
9. All compliance records should be public, so that the efficacy of regulation is transparent.
10. Because resource development technologies will evolve, regulation must be kept evergreen.

The introduction of new energy development technologies is inevitable, so best practice regulation will remain an aspiration. Expeditious and welcomed access to land for safe, compatible, multiple uses is the metric for regulatory performance. Public and investor trust is the most valuable starting point and the most desirable outcome of a virtuous exploration and production life cycle (Goldstein *et al.* 2007).

The South Australian Approach

Relevant Legislation

Petroleum exploration and development activities in South Australia are administered by the Department of State Development (DSD) under the South Australian *Petroleum and Geothermal Energy Act 2000* (PGE Act, onshore), the Commonwealth *Offshore Petroleum and Greenhouse Gas Storage Act 2006* (offshore) and the South Australian *Petroleum (Submerged Lands) Act 1982* (offshore). The PGE Act was proclaimed on 1 October 2009 and supersedes the *Petroleum Act 2000*. The PGE Act has a number of aspects that are considered a comparative advantage without precedent in other Australian legislation.

The high-level objectives of the PGE Act are to:

- sustain trusted practical, efficient, effective and flexible regulation for upstream petroleum, geothermal and gas storage enterprises and the construction and operation of transmission pipelines in the state;
- encourage and maintain competition in the upstream petroleum and geothermal sectors;
- minimise environmental damage from activities and protect the public from risks inherent in petroleum and geothermal operations;
- sustain effective consultation processes with people affected by regulated activities and the public in general;
- ensure as far as reasonably practicable the security of supply of natural gas.

These objectives drive certainty for business, by providing clarity in terms of regulatory requirements and for investment timelines, and for the public so that the community can expect their interests to be protected. The objectives refer to the protection of the public's interest in the sustainability of the natural, social and economic environments. It is important in this discussion to highlight that in the context of the PGE Act the definition of environment includes:

- land, air, water (including both surface and underground water);
- organisms and ecosystems – this includes native vegetation and fauna;
- buildings, structures and cultural artefacts;
- productive capacity or potential;
- the external manifestations of social and economic life, which include aspects such as human health and wellbeing;
- the amenity values of an area.

This definition of environment is consistent with that in the Environment Protection Act 1993 (EP Act) and is broad enough to ensure that potential impacts on all natural, social and economic aspects of the environment are identified, considered and appropriately addressed through the environmental assessment and approval provisions of the PGE Act.

A key lesson learnt from post-event investigations of significant incidents is that regulators must have relevant and up-to-date capabilities (competence and capacity) to be trusted to act in the interests of the public in protecting the natural, social and economic environments during upstream petroleum industry activities. Additionally, the risks of regulatory capture must be managed effectively. As the regulator of upstream petroleum and geothermal energy activities in South Australia, administering the PGE Act, DSD strives to maintain a one-window-to-government or -lead-agency approach, working closely with local co-regulatory agencies such as the South Australian Environment Protection Authority, the Department of Environment, Water and Natural Resources (DEWNR), SafeWork SA, the Department of Health, the Department of Planning, Transport and Infrastructure (DPTI) and the Aboriginal Heritage Branch of the Aboriginal Affairs and Reconciliation Division, in order to deliver an efficient application of all relevant laws and regulations applicable to the petroleum and geothermal industries in South Australia.

Properly resourced lead agencies transparently facilitate the delivery of all co-regulatory objectives and requirements and hence earn trust from the industry, co-regulatory agencies and the public. A one-window-to-government approach enables the management of approval processes in parallel rather than in series.

The PGE Act has been designed to enable a one-window-to-government approach such that, in complying with the objectives of the Act and its processes, upstream petroleum operations' compliance with obligations under other legislation will also be facilitated. These concurrent legislation and requirements include:

- The Commonwealth's *Environmental Protection, Biodiversity and Conservation Act 1999* (EPBC Act), which protects and manages nationally and internationally important flora, fauna, ecological communities and heritage places – defined in the EPBC

Act as matters of national environmental significance. The Commonwealth Government Department of Sustainability, Environment, Water, Population and Communities (SEWPaC) provides stewardship for the EPBC Act, and the legislation is a key part of the co-regulatory process across South Australia.

- South Australia's EP Act, and relevant policies that provide the regulatory framework to protect South Australia's environment, including land, air and water. This legislation was the result of the streamlined integration of six Acts of Parliament and the abolition of the associated statutory authorities. South Australia's Environmental Protection Agency (EPA) provides stewardship for this Act.
- South Australia's *National Parks and Wildlife Act 1972* (NP&W Act), which is the cornerstone for protecting natural environments within parks and regional reserves in the state. The Department of Environment, Water and Natural Resources (DEWNR) provides stewardship for this Act. The NP&W Act is significant as it is a key part of the co-regulatory approval regime for minerals and energy (including unconventional gas) resource exploration and production in South Australia.
- The *Work Health and Safety Act 2012 (SA)* (WHS Act), the state's lead legislation to protect people in the workplace. SafeWork SA provides stewardship for this Act.
- The South Australian *Native Vegetation Act 1991* (NV Act), administered by DEWNR.
- The South Australian *Natural Resources Management Act 2004* (NRM Act), administered by DEWNR.
- The South Australian *Development Act 1993*, administered by DPTI.
- The South Australian *Public and Environmental Health Act 1987*, and specifically the Public and Environmental Health (Waste Control) Regulations 2010, as administered by HealthSA.
- The South Australian *Native Title (South Australia) Act 1994*, administered by the Attorney-General's Department.
- The Commonwealth *Native Title Act 1993* (NT Act) administered by the Commonwealth Attorney-General's Department.
- The South Australian *Adelaide Dolphin Sanctuary Act 2005*, administered by DEWNR.
- The South Australian *Aboriginal Heritage Act 1988*, administered by the state's Aboriginal Affairs and Reconciliation Division.
- The South Australian *Marine Parks Act 2007*, administered by DEWNR.
- The South Australian *River Murray Act 2003*, administered by DEWNR.
- The South Australian *Arkaroola Protection Act 2012*, administered by DEWNR.

Compliance with these pieces of legislation is facilitated through collaborations and working arrangements between the DSD and the government agencies that administer them to ensure that the Statements of Environmental Objectives (SEO) that must be complied with for specific activities are consistent and in keeping with the relevant objects of each Act. The SEO and the collaborative relationships between DSD and co-regulatory agencies, including consultation arrangements, are described further during a consideration of the approval processes under the PGE Act.

Principles for Best Practice Regulation

The PGE Act was developed on the basis of the following six principles for regulatory best practice (Malavazos 2001):

1. **Certainty** The regulatory objectives are uniform, clear and predictable for all stakeholders.
2. **Openness** Stakeholders are appropriately consulted on the establishment of the regulatory objectives and information on outcomes is publicly available
3. **Transparency** The regulatory decision-making processes are visible and comprehensible to all stakeholders and industry performance in terms of compliance with the regulatory objectives is clear to all stakeholders
4. **Flexibility** The level of regulatory scrutiny, surveillance and enforcement needed to ensure compliance is determined on the basis of individual company compliance capability and the outcomes to be achieved
5. **Practicality** The regulatory objectives are achievable and measurable. Hand in hand with the flexibility principle and the objective-based legislation this also means that licensees are able to innovate to use the most effective technologies and practices to achieve the best outcomes.
6. **Efficiency** The compliance costs imposed on both government and the licensee by the regulatory requirements are minimised and justified. Negative impacts on communities are minimised, and licensees remain liable for the cost of their impacts. Furthermore, an appropriate rent (royalty) is paid to the community from the value realised from the development and production of its natural resources.

The above-listed Regulatory Principles can be achieved through the following regulatory strategies.

- Regulatory objectives and assessment criteria for those objectives are developed through broad stakeholder consultation involving industry, government agencies and the community to ensure acceptance and credibility in the environmental objectives to be achieved.
- Regulators and licensees maintain trustworthy capabilities (competence and capacity).
- Effective, informative stakeholder consultation by both project operators and regulators is initiated well ahead of land access. This drives operators to explain their planned activities and any potential risks, seek feedback on areas of interest or concern for the community and establish relationships and terms for land access with stakeholders well before applying for activity approval from the DSD, e.g. before any particular activity 'gets personal'.
- Public access should be provided to the details of risks through the Environmental Impact Reports (EIRs), regulatory objectives through the SEOs and reliable research to reduce key uncertainties and support risk management strategies, so that the basis for regulation is contestable. Department policies for compliance monitoring and enforcement and

the reporting of industry compliance should also be transparent and publicly available online.
- Timely notice of entry should be given prior to activities taking place, with sufficient operational details to inform stakeholders effectively.
- Potentially affected people and organisations should be able to object to land access – nevertheless the regulator, and the prescribed dispute resolution processes, do not support and hence will minimise vexatious objections.
- There should be fair and expeditious dispute resolution processes.
- There should be fair compensation to affected land users for costs, losses and deprivation of land use due to operations.
- Risks should be reduced as low as reasonably practicable (ALARP), while also meeting community expectations for overall outcomes.
- Licensees should monitor and report (to the regulator) on the efficacy of their risk management processes, and the regulator should probe this.
- The regulator can prevent and stop operations, require restitution or rehabilitation, levy fines and cancel licences.
- Industry compliance is publicly reported, so the efficacy of regulation is transparent.

Licensing and Approval Processes

In the context of the definition of environment under the PGE Act, and the principles of best practice regulation, the approval processes under the PGE Act comprise three key stages, as detailed in Figures 18.1, 18.2 and 18.3 and described below.

Stage 1: Licensing Approval

The first stage relates to the licence application and approval process, where a proponent applies for the appropriate licence to give them the right to undertake regulated activities within a licence area. A licence granted under this stage is not a right to do any on-ground activities; rather it is simply an exclusive right to an area within which the licensee can then apply for approval to undertake activities. The regulated activities are defined in section 10 of the PGE Act and include exploration for regulated resources, operations to establish the nature and extent of a discovery of that resource and the potential commerciality of its production and the production, construction and operation of transmission pipelines for carrying regulated substances. Such activities can only be and undertaken subsequent to approvals granted under Stages 2 and 3, which address the environmental and operational aspects of activities.

Only parties with the demonstrated capacity to invest in and safely conduct regulated activities are eligible to become PGE Act licence holders. Licences are available for exploration, for the retention of explored areas to conduct assessments of commerciality, for production, for pipelines, for preliminary and speculative surveys, for associated activities and for special facilities relevant to the regulated activities.

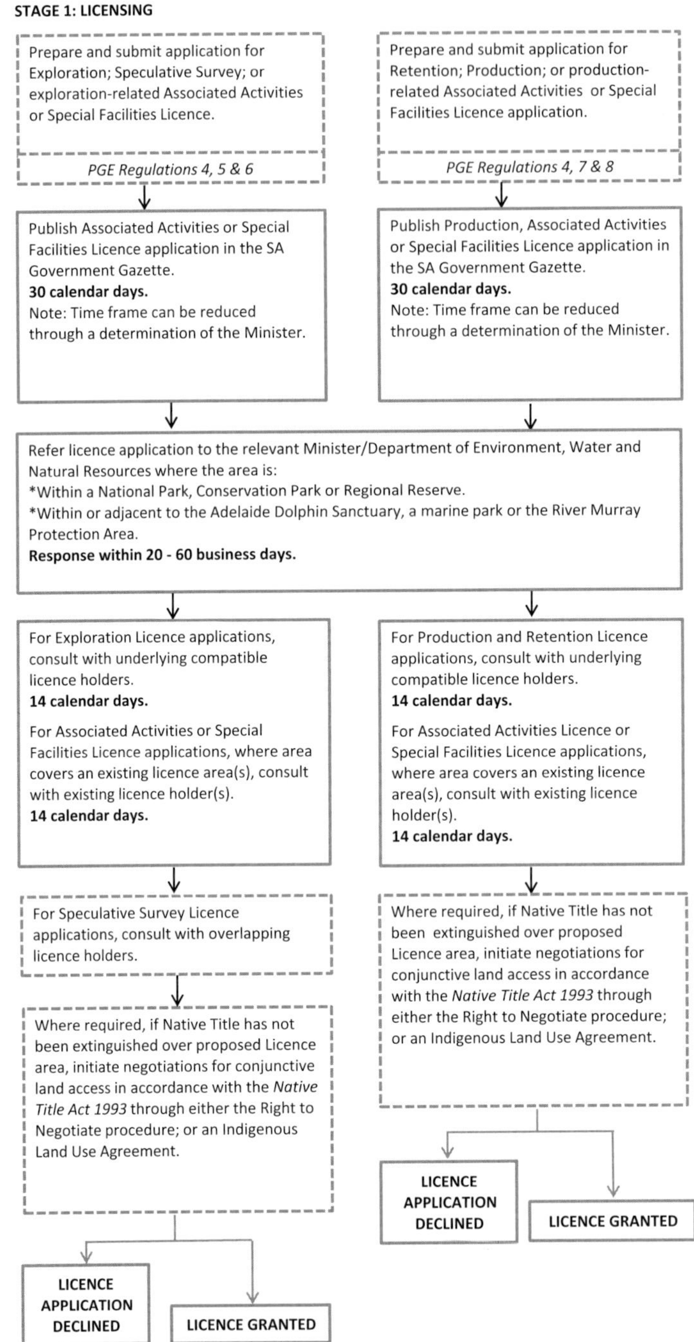

Figure 18.1 Stage 1 in the licensing and approvals process for exploration, retention, production and associated activities pursuant to the South Australian *Petroleum and Geothermal Energy (PGE) Act 2000*. (Dashed-line boxes, items initiated by proponent or licensee; solid-line boxes, items initiated by the Department of Statement Development.)

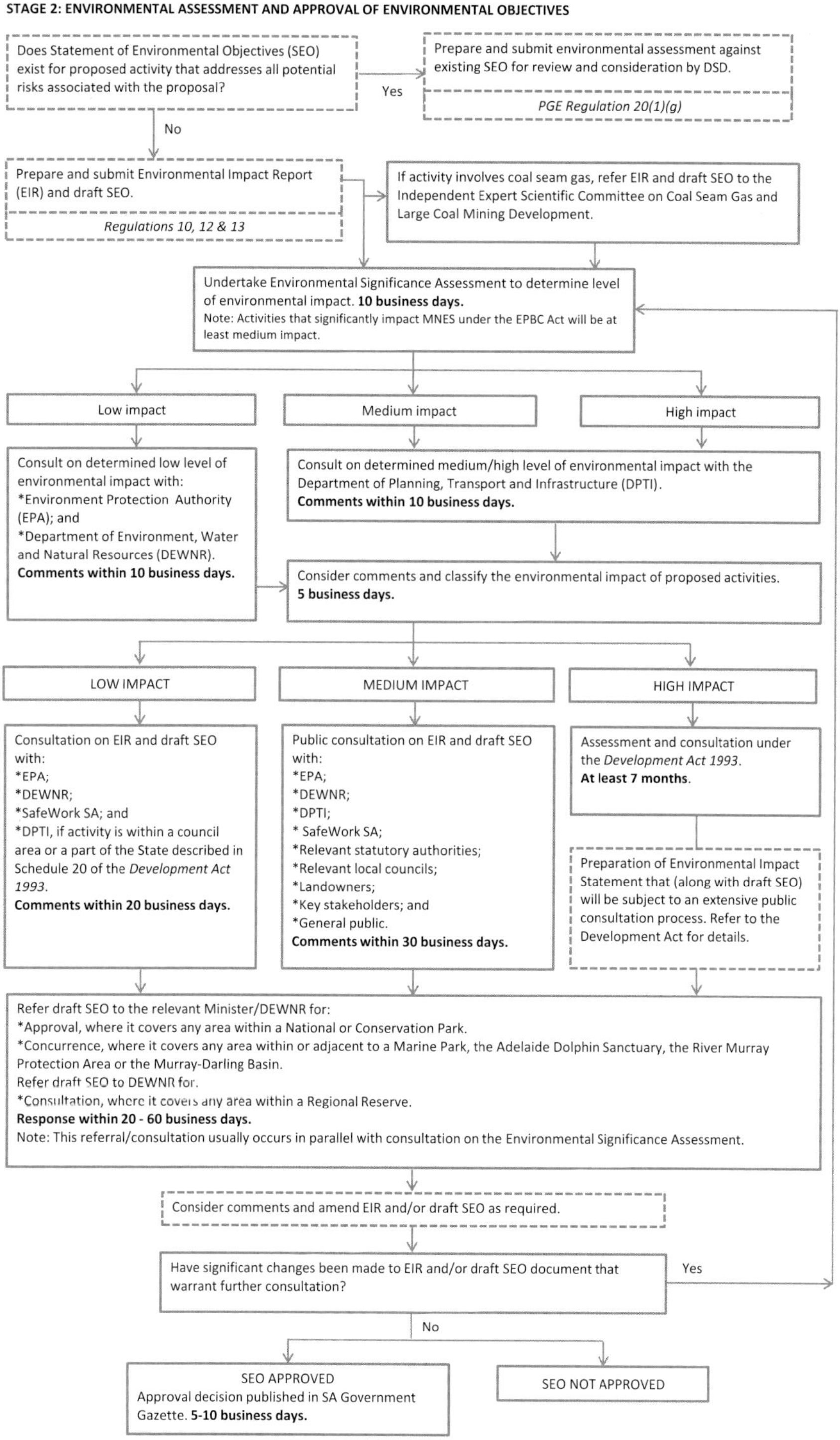

STAGE 2: ENVIRONMENTAL ASSESSMENT AND APPROVAL OF ENVIRONMENTAL OBJECTIVES

Does Statement of Environmental Objectives (SEO) exist for proposed activity that addresses all potential risks associated with the proposal?

Yes → Prepare and submit environmental assessment against existing SEO for review and consideration by DSD.

PGE Regulation 20(1)(g)

No → Prepare and submit Environmental Impact Report (EIR) and draft SEO.

Regulations 10, 12 & 13

If activity involves coal seam gas, refer EIR and draft SEO to the Independent Expert Scientific Committee on Coal Seam Gas and Large Coal Mining Development.

Undertake Environmental Significance Assessment to determine level of environmental impact. **10 business days.**
Note: Activities that significantly impact MNES under the EPBC Act will be at least medium impact.

Low impact | Medium impact | High impact

Low impact: Consult on determined low level of environmental impact with:
*Environment Protection Authority (EPA); and
*Department of Environment, Water and Natural Resources (DEWNR).
Comments within 10 business days.

Medium/High impact: Consult on determined medium/high level of environmental impact with the Department of Planning, Transport and Infrastructure (DPTI).
Comments within 10 business days.

Consider comments and classify the environmental impact of proposed activities.
5 business days.

LOW IMPACT | MEDIUM IMPACT | HIGH IMPACT

LOW IMPACT: Consultation on EIR and draft SEO with:
*EPA;
*DEWNR;
*SafeWork SA; and
*DPTI, if activity is within a council area or a part of the State described in Schedule 20 of the *Development Act 1993*.
Comments within 20 business days.

MEDIUM IMPACT: Public consultation on EIR and draft SEO with:
*EPA;
*DEWNR;
*DPTI;
* SafeWork SA;
*Relevant statutory authorities;
*Relevant local councils;
*Landowners;
*Key stakeholders; and
*General public.
Comments within 30 business days.

HIGH IMPACT: Assessment and consultation under the *Development Act 1993*.
At least 7 months.

Preparation of Environmental Impact Statement that (along with draft SEO) will be subject to an extensive public consultation process. Refer to the Development Act for details.

Refer draft SEO to the relevant Minister/DEWNR for:
*Approval, where it covers any area within a National or Conservation Park.
*Concurrence, where it covers any area within or adjacent to a Marine Park, the Adelaide Dolphin Sanctuary, the River Murray Protection Area or the Murray-Darling Basin.
Refer draft SEO to DEWNR for:
*Consultation, where it covers any area within a Regional Reserve.
Response within 20 - 60 business days.
Note: This referral/consultation usually occurs in parallel with consultation on the Environmental Significance Assessment.

Consider comments and amend EIR and/or draft SEO as required.

Have significant changes been made to EIR and/or draft SEO document that warrant further consultation?

Yes → (return to Environmental Significance Assessment)

No →

SEO APPROVED
Approval decision published in SA Government Gazette. **5-10 business days.**

SEO NOT APPROVED

Figure 18.2 Stage 2 in the licensing and approvals process for exploration, retention, production and associated activities pursuant to the South Australian *Petroleum and Geothermal Energy (PGE) Act 2000.*

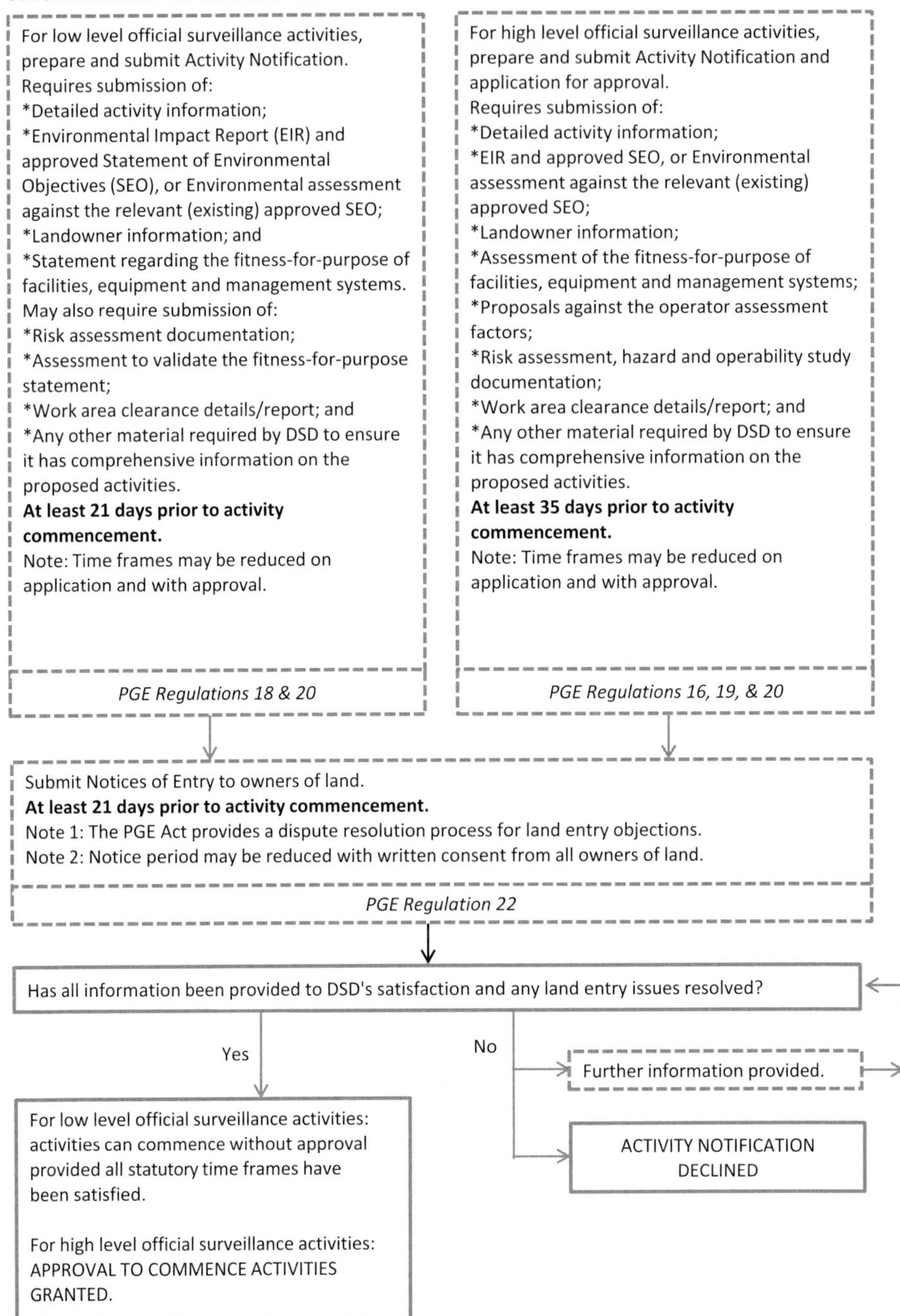

Figure 18.3 Stage 3 in the licensing and approvals process for exploration, retention, production and associated activities pursuant to the South Australian *Petroleum and Geothermal Energy Act 2000* (PGE Act). (Dashed-line boxes, items initiated by proponent or licensee; solid-line boxes, items initiated by the Department of State Development, DSD.)

At the licensing approval stage, prior to the grant of any licence, if and where applicable, a Native Title Land Access Agreement or Indigenous Land Use Agreement (ILUA) signed by all parties, the Crown, the Licensee and the relevant Native Title Claimant Group must be in place. Publicly available Native Title land access agreements (DSD 2014a), first deployed in October 2001 in South Australia, remain benchmarks for leading practice deeds that meet the requirements of the NT Act. To date Aboriginal people, the upstream petroleum industry and the South Australian Government have agreed upon conjunctive Native Title land access agreements for 53 petroleum exploration licences and two conjunctive ILUAs. Conjunctive land access agreements for petroleum have also been expeditiously agreed by current Petroleum Exploration Licences over part of the South Australian Officer Basin that coincides with lands owned by the Anangu Pitjantjatjara Yankunytjatjara (APY) and Maralinga Tjarutja (MT) peoples.

With reference to the principles for leading practice regulation, adopting the transparency principle, South Australia publishes Native Title land access agreements on the DSD's website, providing the benefits of experience for subsequent negotiations.

Prior to the granting (or refusal) of licence areas, the DSD is also required to refer some licence applications to DEWNR for comment, and in some cases approval, from the Minister for Sustainability, Environment and Conservation or the Director of National Parks and Wildlife, for regulated activities within the protected area network in South Australia. In tandem with the PGE Act, the National Parks and Wildlife Act establishes the approval regime for petroleum and geothermal energy exploration and production within the reserve system. As detailed in Figure 18.1, and in accordance with the administrative arrangement between DSD and DEWNR (DSD 2014b), licences require approval from the Minister for Sustainability, Environment and Conservation if the area falls within a National Park, a Conservation Park or the Adelaide Dolphin Sanctuary. Exploration, survey or exploration-related associated activities licences also require the Minister's approval if the licence area falls within a Marine Park, and the retention, production and production-related associated activities licences require approval within Regional Reserves. Through this process, matters of interest to DEWNR can be addressed prior to the grant of a licence in order to avoid potential land-use conflicts, which in turn gives greater certainty to the proponent with respect to the security of title. Some production and pipeline licences may also need to be referred to the Minister administering the Development Act in certain circumstances.

Stage 2: Environmental Assessment and Approval

As mentioned earlier, the grant of a PGE Act licence does not provide an automatic entitlement to land access to conduct operations. Rather, regulated activities under the PGE Act (under section 96) may not be carried out unless an approved SEO is in place, prepared on the basis of an EIR. The EIR describes the specific features of the environment where the activities will take place and identifies all potential impacts, the risks relating to the activity and the proposed risk-mitigation strategies. The SEO identifies the environmental objectives to be achieved to address the risks identified in the EIR and the criteria to be used to assess achievement of the objectives.

Examples of the information and potential impacts that the EIR and final SEO are expected to address include:

- impacts on aquifers, including pressure and contamination;
- impacts on groundwater use;
- contamination of surface water and shallow groundwater;
- soil contamination;
- impacts on native vegetation and native fauna caused by clearance required for above-ground infrastructure (e.g. track clearance, water storage ponds, flow-back storage ponds, other infrastructure etc.);
- interaction of stock or native fauna with water storage ponds;
- potential impacts of introduction or spread of pest plants and animals;
- disturbance to existing land uses (e.g. within reserves under the National Parks and Wildlife Act, pastoral land, etc.) or to local heritage features;
- air pollution and greenhouse gas emissions;
- impacts on the health and wellbeing of the local community; and
- remediation and rehabilitation requirements.

Division 3 of the PGE Act and Part 3 of the Regulations describe the information that must be provided in EIRs and SEOs.

Potential impacts on Matters of National Environmental Significance (MNES) as defined under the EPBC Act can also be addressed in the EIR and SEO where relevant.

Through the consultation requirements of the PGE Act, stakeholders, including land-holders and other government agencies, are required to be informed and consulted on the potential risks associated with proposed activities, and management strategies are to be deployed to reduce such risks to an acceptable level. Stakeholders are also provided with opportunities to raise any issues of concern that they may have prior to the commencement of regulated activities. Other agencies with the duty of care for ensuring that the objectives of the legislation they administer are met are consulted to ensure that their requirements are included within the objectives detailed in the SEO.

The DSD expects that licensees will initiate consultation with stakeholders prior to and during the development of their EIR and SEO, in order to describe their planned activities and the potential impacts, positive or otherwise, which may be experienced by the stakeholders. This is also an opportunity for the licensee to respond to any queries that their stakeholders may have and to understand concerns to ensure that they are addressed within the EIR and SEO.

Once an EIR and draft SEO have been prepared and submitted for assessment, the DSD uses the information provided in the EIR to complete an environment-significance assessment to determine the level of environmental impact of the activity. The significance assessment is conducted in accordance with publicly documented criteria (DSD 2014c); it assesses the level of certainty in the predicted impacts such as those listed above and their potential consequences related to the proposed activities and the degree to which these consequences can be managed. The environmental significance criteria enable the identification of any deficiency in stakeholder consultation during the development of the EIR and draft SEO. Where the DSD's assessment identifies such a deficiency, the determined level of environmental significance may be greater and likely to trigger more extensive stakeholder consultation by the DSD. This ensures that relevant stakeholders are provided with appropriate time for opinions to be considered and represented equitably in advance of SEO and subsequent activity approvals.

The combination of the outcomes of the significance assessment criteria lead to the determination of a level of significance for each event relating to the activity, cumulating in the determination of an overall level of environmental impact of the activity as low, medium or high. The level of environmental impact assigned to a particular activity in turn determines the consultation that the DSD undertakes, both with co-regulatory agencies on the level assigned and, more broadly, on the content of the EIR and draft SEO documents. The consultation arrangements are outlined within the PGE Act and regulations and within administrative arrangements between the DSD and its co-regulatory agencies, which are all available on the DSD website (DSD 2014b).

Where activities are assessed as being of low environmental impact, the DSD consults on its determination and the content of the EIR and draft SEO with the EPA and DEWNR. It also consults on the content of the EIR and draft SEO documents with SafeWork SA and with DPTI if the area is within a council area or an area described in Schedule 20 of the Development Act.

Where activities are assessed as of medium environmental impact, the DSD consults on the determined level with DPTI and initiates a public consultation process, inviting comments on the EIR and draft SEO from the public and also directly from the EPA, DEWNR, DPTI, SafeWork SA, the relevant statutory authorities and local councils, landowners and stakeholders. During the public consultation process the EIR and draft SEO are made available to the public through the DSD website and at its office for at least 30 business days. Members of the public are notified of the consultation process through an advertisement in the local newspaper as well as on the DSD website, and in addition directly affected stakeholders are provided with targeted correspondence from the DSD.

For activities assessed as being of high environmental impact, the DSD consults with the DPTI on this determination and, where DPTI agree with the assessment, such proposed activities are referred to DPTI for assessment and consultation under the Development Act. This requires preparation of an EIS and extensive public consultation.

For all activities within a National or Conservation Park, a Marine Park or the Adelaide Dolphin Sanctuary, the draft SEO is referred to DEWNR for approval from the Minister for Sustainability, Environment and Conservation in line with agreements within the administrative arrangement between DSD and DEWNR. For activities within the River Murray Protection Area or the Murray–Darling Basin, DSD will seek concurrence on the SEO approval with DEWNR.

Concerns raised during consultation are incorporated into the EIR and draft SEO documents as appropriate, enabling changes to be made to address the comments prior to approval by the Minister. As noted previously, all this happens well before any company can apply to undertake any on-ground activities regulated pursuant to the PGE Act.

Notwithstanding that the Minister's approval of an SEO will incorporate the input from the consultation process, any final decision which may not satisfy any legitimate stakeholder concerns is subject to a review and appeal process pursuant to Part 15 of the PGE Act and, as a last resort, by application to the District Court of South Australia.

All SEOs and associated EIRs are public documents and can be found on the DSD website (DSD 2014d).

Stage 3: Activity Notification and Application for Approval (Including Notice of Entry)

As described previously, the grant of PGE Act petroleum exploration, retention, production and pipeline licences does not provide an automatic entitlement to land access for regulated upstream petroleum operations. Once the EIR and SEO are in place, a licensee can apply for approval to undertake a specific activity that is described within those documents. Along with the activity approval application the licensee provides DSD with an activity notification which contains detailed activity information including:

- an environmental assessment of the activity against the SEO, including, where relevant, assessment as to whether the activity may have potential significant impacts on MNES;
- landowner information (including copies of notices of entry sent to landowners);
- an assessment of the fitness for purpose of the licensee management systems and any facilities or equipment to be used;
- work-area clearance details and report;
- risk assessment documentation;
- any further information or material as required by DSD to ensure that the department has comprehensive information on the proposed activities.

Where MNES are identified, referral to the Commonwealth Minister for the Environment will be made by the licensee or the DSD, for assessment and a decision as to whether the activity requires approval under the EPBC Act.

Licensees are classified as carrying out activities requiring high- or low-level official surveillance. The level of official surveillance determines the information that must be provided in the notification, the level of scrutiny that DSD applies during review of the notification and the period of notice prior to the proposed commencement of activities. The PGE Act outlines operator assessment factors that consider the licensee's policies, procedures, management systems and track record to classify the licensee's level of official surveillance. Initially licensees are classified as carrying out high-level official surveillance activities and must address the operator assessment factors within their activity notification. High-level official surveillance operators must apply for approval to undertake activities at least 35 days in advance of the proposed activity commencement date and cannot commence until approval is provided.

Operators can apply to be classified as carrying out low-level official surveillance activities through a demonstration of the operator assessment factors and, once classified as low-level, official surveillance can provide a shorter period of notice (at least 21 days) to the Minister through the DSD before commencing activities. Approval is not required; however, detailed activity information must still be submitted with the activity notification. If further information is requested the licensee cannot commence until the department has comprehensive information on the activity.

Mutual trust for compatible, sustainable, land access for upstream petroleum operations is traditionally indemnified with formal land access agreements struck between licensees, people potentially affected and enterprises. To provide an impetus for fair and sustainable land access for petroleum, geothermal energy and gas storage operations in South Australia, the PGE Act defines 'owner of land' to cover all persons who may be directly affected by regulated activities, entitling them to notices of entry and compensation. Owners of land are provided with opportunities to raise concerns prior to the commencement of regulated activities, and state regulations require operations to effectively manage risks and meet community expectations for net outcomes, otherwise the activities will not be approved. The outcome is demonstrable leverage to all persons who may be directly affected by regulated activities, not just those holding land titles but also people such as Native Title claimants, persons holding a tenement over or in relation to the land and anyone leasing potentially affected land for enterprises.

Notice of entry is provided to owners of land at least 21 days prior to the licensee's entry to the land to conduct an activity and forms part of the activity notification process. Owners of land are provided with information on the nature of the activities to be carried out, including any anticipated events and the management of their consequences to minimise risks to an acceptable level, to enable the landowner to make informed decisions on whether this would have an impact on the land.

Owners of land are entitled to object to the licensee's proposed entry by giving notice to the licensee within 14 days of the notice of entry. In this circumstance the licensee must notify the Minister that their entry is disputed, and the activity cannot be undertaken until the dispute is resolved. The licensee and the owner of the land should attempt to reach an agreement of terms under which the licensee may enter the land or, if the risks of the activity to the owner of land are too high then the licensee may choose either to modify the activity and re-issue the activity notification or to cancel the activity. In rare cases where the licensee and the landowner cannot resolve the dispute, the Minister may attempt to mediate between the parties or either party may apply to the Warden's Court for resolution. To date, disputed notices of entry have been resolved through satisfactory negotiation and have not reached the Warden's Court.

Owners of land are entitled to appropriate compensation from licensees for any losses, deprivation or reasonable costs sustained during both the process of negotiating land access and for the full period of land access, right through to the decommissioning of any facilities.

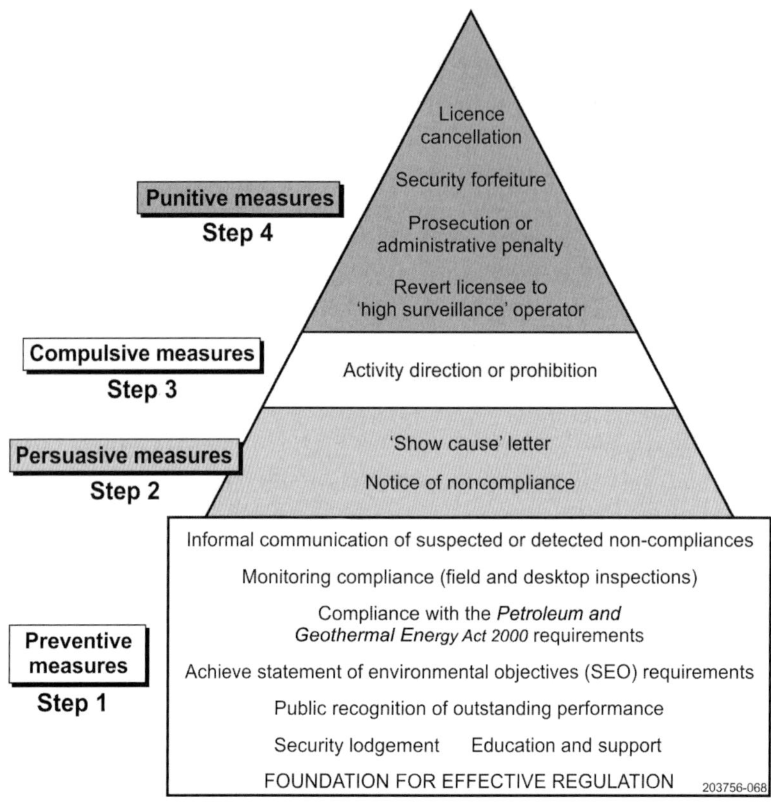

Figure 18.4 Compliance and enforcement pyramid used by the DSD for monitoring and enforcing compliance with the *Petroleum and Geothermal Energy (PGE) Act 2000.*

Compliance and Enforcement

The DSD continuously monitors licensee performance and compliance with the PGE Act. South Australia's approach to provide fair, predictable and trustworthy regulation entails a publicly available compliance policy, which is available on the DSD website (DSD 2014e). This policy adopts a compliance and enforcement pyramid (Figure 18.4) for monitoring and enforcing compliance. The pyramid details a series of steps and measures available to the DSD for facilitating, monitoring and, where necessary, enforcing compliance. The DSD aims to maintain its regulatory activities at 'Step 1: Preventive measures', shown as the base of the pyramid in Figure 18.4. In cases where industry fails to adequately and appropriately respond to detected noncompliance, 'Step 2: Persuasive measures' is instigated. Only in extreme and exceptional cases would the DSD expect to utilise Steps 3 and 4, 'Compulsive' and 'Punitive' measures, respectively, to enforce compliance and achieve acceptable environmental or administrative outcomes.

The DSD prepares a PGE Act Annual Compliance Report for the purpose of outlining:

- the compliance monitoring and surveillance activities carried out by the DSD during each year for activities regulated under the PGE Act;
- the regulatory performance of the petroleum and geothermal industries in accordance with the requirements of the PGE Act;
- all serious incidents that may have occurred from the previous year; and
- all step 2, 3 or 4 enforcement actions (see Figure 18.4) that may have been taken during the year.

The DSD's Petroleum and Geothermal Energy Act Compliance Report and Company Annual Reports are publicly available through its website (DSD 2014e).

As well as information provided through the activity notifications, the DSD regularly meets with licensees to discuss their activities and compliance, and it conducts ongoing monitoring and surveillance through both field and desktop studies. In addition, licensees are required to submit annual reports reporting on activities undertaken within each licence area during the respective licence year and on their performance and compliance with the PGE Act and the relevant environmental objectives. Company annual reports also provide information on the activities proposed for the ensuing licence year.

Where there have been instances of serious and reportable incidents as defined under section 85 of the PGE Act, licensees are required to investigate such incidents to determine the root cause and undertake corrective actions to prevent their recurrence.

In addition to risk assessments and fitness-for-purpose assessments conducted prior to the construction of facilities, licensees are required to undertake an assessment, at least every five years, of: the risks imposed by facilities on public health and safety; the environment; and the security of production or supply of natural gas. Through the provision of information as specified in the PGE Act and Regulations, the licensee must demonstrate that all risks are being systematically managed to be as low as reasonable practicable and that the integrity of facilities is being adequately maintained.

Co-Regulation and Co-Regulatory Agencies

The most highly leveraged aspect of the PGE Act is its definition of the 'environment' as the social, natural and economic environment, and its application of SEOs to set standards for environmental risk management and environmental outcomes in alignment with co-regulation. This means in effect that a breach of an SEO is a breach of the PGE Act and other cognate legislation and thus aligns objectives across government.

As described previously in this chapter, through collaboration with co-regulatory agencies and processes outlined in administrative agreements, the DSD maintains a one-window-to-government approach for the regulation of upstream petroleum, geothermal energy and pipeline activities in South Australia. Licensees have obligations under legislation other than the PGE Act and, where possible, the objectives of other legislation are captured within SEOs for activities under the PGE Act. This is achieved only by maintaining good working relationships with co-regulatory agencies and by maintaining an understanding of the requirements for PGE Act licensees under other legislation. The DSD values the expertise and assistance of its co-regulatory agencies, particularly when seeking advice during consultation on the content of EIR and SEO documents.

Concluding Remarks

Regulation for the compatible multiple use of land in Australia is undertaken with both risks and net benefits in mind. Considerable net benefits flow from the community ownership of subsurface resources when a development effectively manages risks to social, natural and economic environments. Industry must act early to engage and inform stakeholders effectively so they can make informed decisions on activities.

Trustworthy, efficient and effective regulation is fundamental to attracting investment with community support. The key ingredients of best practice regulation are frameworks that: elicit community trust and investor confidence; provide certainty; entail robust public consultation processes; are transparent; enable flexibility; are open to amendment; are efficient; are practical; and focus on outcomes. New energy development technologies will necessitate evolutionary improvement to regulatory frameworks, and best practice regulation will also continually evolve.

A one-window-to-government approach to regulation enables co-regulators to do their jobs in parallel rather than in series, one after the other. This fosters efficiency without reducing stringent standards for ecologic, social, heritage and economic outcomes.

References

Goldstein, B.A., Alexander, E., Cockshell, D., Malavazos, M. and Zabrowarny, J. 2007, The virtuous life-cycle for exploration and production (E&P): lead and lag factors, *APPEA Journal*, 47, 387–401.

Department of State Development (DSD) 2014a, Native title and ILUAs. Available from http://www.petroleum.statedevelopment.sa.gov.au/environment/native_title,_aboriginal_lands,_iluas [accessed 9 October 2015].

Department of State Development (DSD) 2014b, Administrative arrangements. Available from http://www.petroleum.statedevelopment.sa.gov.au/legislation/regulation/admin_arrangements [accessed 9 October 2015].

Department of State Development (DSD) 2014c, Environmental impact classification. Available from http://www.petroleum.statedevelopment.sa.gov.au/legislation/regulation/level [accessed 9 October 2015].

Department of State Development 2014d, SEO, EIR and ESA Reports. Available from http://www.pir.sa.gov.au/petroleum/environment/register/seo,_eir_and_esa_reports [accessed 9 October 2015].

Department of State Development 2014e, Petroleum and Geothermal Energy Act annual compliance reporting. Available from http://www.petroleum.statedevelopment.sa.gov.au/legislation/compliance/petroleum_act_annual_compliance_report [accessed 9 October 2015].

Malavazos, M. 2001, The South Australian Petroleum Act 2000: principles and philosophy of best practice regulation, *MESA Journal*, 21, 33–35.

19

Best Practice for Community Engagement: Determining Who is Affected and What is at Stake

PETA ASHWORTH[1]

Introduction

Community engagement and citizen participation have been important themes in democratic theory for some time. Early literature in the area of *community engagement* emerges from the analysis of attempts by American governments to democratise social programmes in the 1970s through *public participation* (Contandriopoulos, 2004). This coincided with the development of new social movements such as environmentalism and feminism (Dryzek *et al.*, 2003). These movements also demanded increased participation and engagement from governments on social issues.

Reed (2008) also noted the increasingly important role of engagement in his review on environmental and natural resource management decisions. He suggests this is typically due to the complexity and uncertain nature of most environmental problems, which, more often than not, tend to be multi-scale, affecting multiple actors and agencies. One might argue that these are the current characteristics of the unconventional gas industry, both in Australia and internationally.

The overriding aim of community engagement is to involve community members in decisions or policy making that is likely to affect them now or in the future (Keeney, 1998). Over the past few decades there has been a stark move away from the traditional one-way, top-down approach to an approach that is far more inclusive, where possible trying to engage as many individuals and groups, the stakeholders, who may be impacted by a new development or decision. However, this can be challenging when considering the range of stakeholders involved in a decision, particularly when you include minority groups or those more marginalised in society, who are often hard to reach. Adhering to principles of inclusivity, engaging across the broad spectrum across of stakeholders, enables those participating to proactively identify issues and find solutions collaboratively using shared learning (AccountAbility, 2008).

Coinciding with the focus on the importance of two-way engagement is an increase in the research literature, frameworks, theories, models, toolkits and standards that relate to community engagement. Applications of the literature, frameworks and theories to the

[1] I would like to acknowledge the insightful reviews from Dr Michal Moore, Dr Justine Lacey and Dr Fabien Medvecky, which really helped me to shape this chapter.

unconventional natural gas industry are explored in this chapter. As part of this, the concepts of a social licence to operate (SLO), technology assessment (TA) and responsible innovation (RI) are discussed, as all appear highly relevant. Reflecting on their similarities and differences sheds light on the practical considerations that project developers, policy makers and others need for deploying successful unconventional gas projects within communities and beyond.

Defining Engagement

For the purposes of this chapter, *engagement* is described as the developing and maintaining of relationships – an ongoing interaction between parties as opposed to a one-way conversation (Nicholls, 2003). *Community* may refer to the general public, a particular geographical area or a specific group or set of groups with particular interests with which an individual might identify (McIntyre, 1996).

Individuals and groups who have the power to affect, or be affected by, an outcome or particular issue have been described as *stakeholders* (Freeman, 1984) and these tend to be the focus of engagement activities within communities.

Analysing the range of stakeholders and understanding their ability to engage helps to determine the most suitable processes for engagement (Reed *et al.*, 2009). Within each community individuals may have diverse values, attitudes and beliefs and therefore the community may not always be homogenous. In turn, each individual citizen may be involved in a unique mix of communities, on the basis of their age, gender, interests and circumstances (Solomon, 1999; Hashagen, 2002).

Determining whom to engage from within a community thus becomes a critical component of any engagement activity and is often influenced by a combination of the factors outlined above (Reed *et al.*, 2009). In contested or emergent industries, the range of stakeholders to be considered often reaches far beyond the immediate community. It will most likely also include governments (local, state and national), community service organisations and environmental non-government organisations (ENGOs). However, the position of the relevant governments can change within any activity, depending on their levels of interaction, responsibilities and interests within a project.

Types of Participation and Engagement

Reed (2008), in his review of the changing nature of the participation that has occurred over the past few decades in environmental problems, grouped the literature into four separate typologies. These include: a typology based around the differing degrees of participation that progress across a spectrum of involvement; the second typology, on the nature of participation, is based on the direction of communication flows; the third typology is based around theoretical approaches – be they normative, focusing on process, and/or pragmatic – as a means to an end; the fourth typology is based around the objectives for

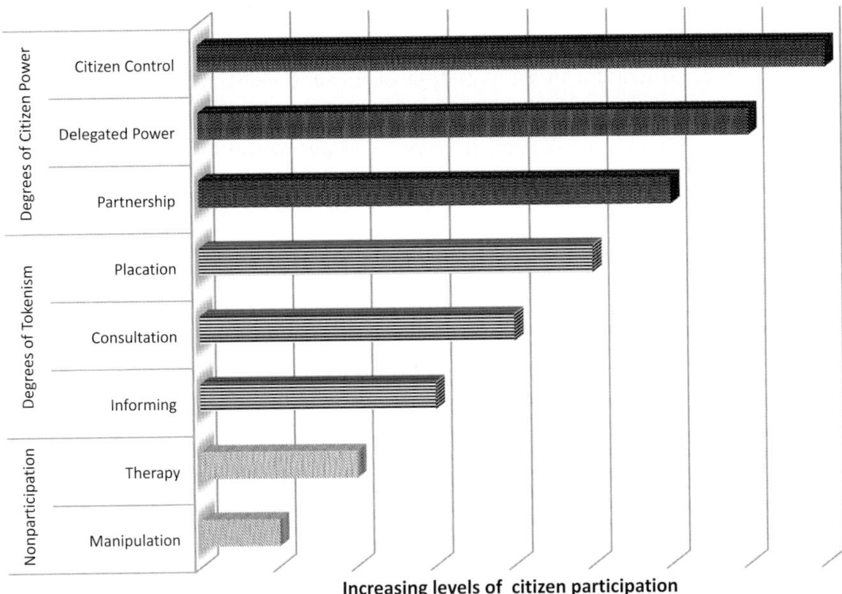

Figure 19.1 Increasing levels of citizen participation (adapted from Arnstein's ladder of citizen participation). Reprinted by permission of The American Planning Association (http://www.planning.org).

which participation is used, that is, whether it may be research driven, development driven and so forth (Reed, 2008, p. 2419).

Frameworks of Participation and Engagement

One well-known framework that suggests progression through a spectrum of participation is Arnstein's (1969) "ladder of citizen participation". The eight levels of participation are grouped into three subsections ranging from "non-participation" to "degrees of tokenism" and "degrees of citizen power". "Non-participation" focuses on manipulation of the public to gain support for policy makers through education and public relations. "Degrees of tokenism", as the name suggests, is characterised by organisations engaging because it is something they have to do but tends to be top down; the organisation maintains the balance of power. At the opposite end is active engagement, "degrees of citizen power", where those not in power, "the have-nots", are granted complete control of the process (Arnstein, 1969). The increasing levels of participation and the associated hierarchies are illustrated in Figure 19.1.

Alternatively, the International Association for Public Participation (IAP2) outlined a public participation spectrum where the level of public impact and engagement increases as you move across the spectrum (IAP2, 2014). The spectrum begins with a commitment to *inform* the public through fact sheets, websites and open houses as the lowest level of engagement, which is predominantly top down. Next along the spectrum is *consult*, where feedback from the public is sought on decisions or alternatives. This is followed by *involve* with a commitment to work with the public to ensure that their concerns and aspirations are reflected in decisions. The fourth pillar is *collaborate*, reflecting more of a partnership approach to decision-making, through to *empowerment* at the opposite end, where the public holds the final decision-making power (IAP2, 2014; ACCC, 2013).

Thomas (1998) suggested that the most appropriate programme of public participation will be influenced by what the organisation or government is hoping to achieve by the engagement activity. Ultimately this choice will be affected by a range of factors including the level of public interest, political will, time frame, available resources and the likely degree of controversy expected. All these are relevant to the current state of play with unconventional gas.

Considering Social Licence to Operate and Engagement

The term "social licence to operate" (SLO) has also emerged as one that has relevance to processes of stakeholder engagement and how companies operate within a community. The roots of the SLO concept emerged from the mining industry in response to social risk (Moffat and Zhang, 2014) and is now being applied across many other sectors including paper manufacturing (Gunningham *et al.*, 2004), alternative energy technologies (Hall *et al.*, 2013) and even humanitarian aid (Jijelava and Vanclay, 2014). The SLO has been described as the ongoing acceptance and approval from *a community of an industry or operation* (Thomson and Boutilier, 2011) or *other stakeholders who can affect its profitability* (Graafland, 2002).

The SLO represents the unwritten rules of relationships with a community and its stakeholders that go beyond compliance (Franks *et al.*, 2013). It centres on trust and expectations of how things should be done by operators within a local community and across broader society (Gunningham *et al.*, 2004). An SLO takes time to achieve but is something that can be very quickly and easily lost. Lacey *et al.* (2012) suggested that it is much easier to establish when an operation does *not* hold a social licence; this situation will be characterised by complaints from neighbours, blockades and community protests which may ultimately lead to political impositions and potentially even formal licence restrictions.

More recently, research of Moffat and Zhang (2014), using online surveys of participants living within a coal-seam gas area, quantified key elements that constitute an SLO. Their research identified the factors that predicted trust in the company, ultimately leading to the acceptance of the operation in the local area, were impacts on social infrastructure, quality of contact with the community and evidence of fair procedures when the organisation dealt with local stakeholders and community (Moffat and Zhang, 2014). These insights are helpful for companies wishing to work in different communities as they provide a focus of what factors are important to the stakeholders living and working within the community. However, these factors can apply equally to those who may oppose an operation and this is evidenced in the community engagement guidelines released by the "Lock the Gate" movement opposed to the coal-seam gas industry in Australia. The guidelines highlight many similar factors as being important for building momentum to oppose projects.

In addition, despite the concept of SLO suggesting an underlying commitment to engagement and participation in decision-making, there is no real evidence which demonstrates that to achieve an SLO there is a need to vest decision-making power in the citizens of the community. One could argue that the factors that constitute an SLO may be more centred on "degrees of tokenism", that is, the minimum requirement for a company to keep communities and broader stakeholders, including government regulators, happy. Of course, an adverse event may occur that challenges a company's SLO. This then requires a much more inclusive and empowering decision-making process to ensure that the impacted stakeholders' concerns are heard and implemented appropriately in order to maintain the SLO.

Technology Assessment and Engagement

Huijts *et al.* (2012) proposed a framework for technology acceptance that is also relevant when considering engagement. The authors suggest that when it comes to various technologies, impacted individuals will base their acceptance on:

> (1) the overall evaluation of costs, risks and benefits, (2) moral evaluations, depending on the extent to which the technology has a more positive or negative effect on the environment or society and (3) positive or negative feelings related to the technology, such as feelings of satisfaction, joy, fear or anger.
>
> *Huijts et al. (2012, p. 526)*

In addition to positive and negative effects, Huijts *et al.* (2012) suggested that attitudes and personal norms will be directly influenced by the perceived costs. These include: both non-monetary costs, such as the effort required to understand or use the technology, and also the financial costs; the perceived risks that may threaten group values, morals and conventions such as safety; and any uncertainty associated with the technology. However, the perceived benefits are the collective benefits the technology is likely to bring. In their example using carbon capture and storage (CCS), the benefits included reduced greenhouse gas emissions, energy security and other personal benefits such as access to the technology and improved environmental conditions. Subjective norms (social pressure) were also believed to influence outcomes both positively and negatively depending on individual attitudes (Ajzen, 1991).

The perceived fairness of the distribution of the costs, risks and benefits is also an important factor (Wolsink, 2005). That is, are they distributed equally across all stakeholder groups or do some stakeholders carry more risks while others benefit? With some projects, the concept of distributive fairness is often challenged because it is the local communities who carry the burden of the technology, while many of the benefits may be predominantly global. This has often been a criticism in coal mining regions where local communities feel they carry the burden of the mine (dust, noise, and other health impacts) while revenue through taxes on coal exports goes to state and federal governments. To overcome these perceptions, many companies establish community benefit funds and make investments to improve local amenities to ensure local benefits are also received for hosting the project. However, the issue of compensation for projects, either for individuals or for groups, is complex and needs to be considered carefully so as not to be seen as buying acceptance of either individuals or the broader host community (ter Mors *et al.*, 2012).

The Importance of Trust

Like SLO, Huijt *et al.*'s (2012) framework suggests that all the above factors are mediated by trust. If those responsible for the project are trusted then this is likely to enhance its acceptance; if those who oppose the technology are trusted then this is likely to decrease acceptance (Siegrist and Cvetkovich, 2000). At the same time, the former authors suggest that, on the basis of work by Earle and Siegrest (2008), trust and procedural fairness influence one another to some degree (Huijt *et al.*, 2012, p. 529). Research has demonstrated that processes that include community representatives and other interested stakeholder groups in the decision-making, allowing them a voice that is listened to and acknowledged, are considered fairer. Therefore, such processes are more likely to build trust and increase potential acceptance (Terwel *et al.*, 2010). Furthermore, deciding which stakeholders to include, their relationship to the community and the time and resources available for engagement will all strongly influence the outcomes of the engagement activity.

Trust is something that is built up over time and, like SLO, can easily be eroded. It is generally not transferable. An organisation's reputation will precede its operations, and if a positive image has been formed in other communities it can be helpful when establishing the organisation in a new area. However, if the company's reputation has been negative it is also likely to have a negative perception in the new community.

This was demonstrated in early work completed by the United States Department of Energy's (US DOE's) Regional Sequestration Partnerships, which found a number of similar factors to Huijts *et al.*'s (2012) technology assessment framework that would impact on whether a CCS project would be accepted. Trust was critical, in particular, whether community representatives felt that they could trust project owners and government to take care of their problems. Previous relationships and experiences within communities, as well as the existing reputation issues of the industry (i.e. did any legacy issues exist?), were also important factors. Often such legacy issues may have occurred a long time in the past but, if experienced within a local community or by particular stakeholders, they will heavily influence the perception of the company and its associated operations (Bradbury *et al.*, 2009).

In addition, researchers found that a lack of confidence in government, industry and science to manage the associated health, environmental, and social risks was likely to compound negative perceptions. Common questions participants raised included: "Will the process be fair? Is it transparent? Will anyone listen to us? Can we have a say in what happens? Who can I call if I have a problem?" (Bradbury *et al.*, 2009). Perceptions of fairness and transparency are really about the process of how decisions are made: how much input into the decision can a community or other stakeholders have, and is there is some flexibility in relation to the final outcome? If projects come with a predetermined outcome then perceptions of procedural fairness will more likely be low and may result in some contestation or angst amongst the stakeholders and the local communities. It is important to note that in these situations the stakeholders who oppose a decision may not necessarily live in the immediate community or impacted area; however, they may have interests in the project for a multitude of reasons and as such may object or protest the decision.

Responsible Innovation

Responsible research and innovation (RRI) or responsible innovation (RI) has recently emerged, especially in Europe, as a science policy framework that also has relevance to the unconventional gas industry. It is particularly relevant when considering the controversies that have arisen around "fracking". The idea of RI is to align technological innovation with broader social values as well as to support institutional decisions concerning the goals and trajectories of research and innovation under conditions of uncertainty, ambiguity and ignorance. Responsible innovation has been defined as

> ... a transparent, interactive process by which societal actors and innovators become mutually responsive to each other with a view to the (ethical) acceptability, sustainability and societal desirability of the innovation process and its marketable products (in order to allow a proper embedding of scientific and technological advances in our society).
>
> *von Schomberg (2013)*

It seeks to reconfigure traditional approaches to technological governance by moving the focus away from the governance of risk to the governance of innovation and by propagating a collective, future-oriented, ethic of care and stewardship. Thus RI focuses not only on the impacts of innovation but also on their purposes and processes, thereby empowering a measure of social agency in technological choices and making such choices more publicly accountable. Subjecting "fracking" from the unconventional gas world to a process of RI would include both public dialogue, extended ethical review and traditional engineering risk assessment, as outlined above (UK ESRC, 2012).

Why Bother with Engagement?

The degree to which an organisation is committed to uphold principles such as RI and engage with communities or stakeholders in its decision-making processes will determine the scope and purpose of the engagement, the strategies and techniques to be adopted and the stakeholders required to engage. If a strategic approach is being taken, this will often involve a formal risk–opportunity analysis. It requires the organisation to weigh up the benefits, or otherwise, of adopting particular approaches and what the likely costs of not engaging might be. If an organisation is interested only in informing the public about current issues or trying to gauge currents levels of public opinion on an issue then the methods chosen to do this are quite different and involve considerably less risk than those methods that fully involve the public in decision-making processes (Hashagen, 2002).

Stirling (2005: p. 219) discussed the dichotomy that exists in engagement processes used to appraise science and technology and their acceptability. He referred to one process as the analytic method, which focuses on *risk and cost–benefit analysis* and is usually aimed at the expert or influential stakeholder group. The second is described as a *transparent, all-inclusive,* participation process, which tends to focus on more marginal issues such as environment protection and social justice. Whichever the motivation for such participation might be, researchers tend to group motivations into three types of viewpoints, which have been described as: *normative*, because it is the right thing to do; *instrumental*, because it is a better way to achieve particular ends: or *substantive*, because it leads to better ends (Fiorino, 1989; Stirling, 2005).

Nevertheless there is still much debate in the literature about whether the reason for engagement should be to manage or influence the stakeholders in order to improve the ability to operate and reduce risk of failure (exercising power) or whether it should be to understand the concerns of the stakeholder groups and hence improve the social value of the organisation's business (an exercise in partnering). These two reasons are not necessarily mutually exclusive, but they do reflect very different attitudes towards stakeholders. In one, the stakeholder is seen as a potential interferer with an inadequate source of knowledge and information. In the other, the stakeholder is respected as a source of knowledge and information. Over time, the increased accountability of governments and industry has meant there is greater acknowledgement of the importance of accessing a range of views to inform decision-making and therefore a much greater tendency to collaborative learning from the range of stakeholders rather than to the expert–lay dichotomy which has existed in so many engagement activities around new technologies.

Reed *et al.* (2009) described many organisations' experiences with stakeholder engagement as a journey – often begun because a negative event has occurred and needs to be addressed. Once the issue has been successfully resolved the organisations begin to observe and understand the benefits of engagement. This usually results in a more proactive and strategic approach to stakeholder and community engagement, which ultimately leads to better partnerships across all levels. Unfortunately, however, it appears that corporate memory is limited, and industry and governments tend not to learn from previous experiences. This seems to happen often with mining and resource extraction operations and has been witnessed more recently with unconventional gas operations. The failure of some companies to engage adequately with a range of stakeholders and apply the considerations outlined above has resulted in moratoriums being put in place in some states. This also begs the question of the role of government and politicians in processes of implementation.

Common Factors Required for Best Practice Engagement

Looking across these frameworks and theories, there are a number of factors common to most, which seem essential for industries and government wishing to engage stakeholders successfully and operate within communities. These include the minimisation and management of social and environmental impacts, procedural fairness in project processes, trust in the operator, provision of timely, accurate information and the development of shared knowledge to create meaningful relationships with those in the community. Each factor and its implications for the unconventional gas industry is explored in more detail below.

Minimisation and Management of Impacts

It is unlikely that any new operation will have no negative impacts on a community (Moffat and Zhang, 2014). However, stakeholder and host community evaluations of costs, risks and benefits that a project brings to a community was confirmed by Huijts *et al.* (2012) to be important. Working with local stakeholders and associated communities from the earliest possible opportunity, in an open and transparent manner, can help to minimise such impacts and build trust in a company and its reputation. Regardless of what message a company brings about their project, individuals within the community, along with other stakeholders, will weigh up what they perceive to be the likely benefits of hosting such a project as compared to the risks and costs that it poses. Trading-off costs and benefits in this way helps individual stakeholders and their associated groups and communities to decide whether to support such a project.

Work by Zhang and Moffat (2015) confirmed, across two samples of the Australian population (the numbers N in the samples were respectively 210 and 2590), that, when key impacts and benefits were considered about the mining industry's impact, environmental impact was the major factor that led to the rejection of mining. On the flip side, employment creation and promoting mining community development were seen as the most important benefits leading to greater acceptance of mining. Given the large sample size in the study these findings are likely to hold across other resource extraction industries and therefore will be important factors for the unconventional gas industry to consider.

What also plays heavily into such evaluations is whether any legacy issues exist, that is, events that may have left a negative impression on individuals and/or the host community. Conversely, a company may already hold a negative reputation associated with their actions in other areas, which usually precedes them in their next operating area. For example, in a recent review of the land access negotiations for a three-dimensional (3D) seismic study by the South West Hub project in Western Australia (www.dmp.wa.gov.au/South_West_Hub .com.au), a key factor that stopped landholders allowing access to their property was if they had been negatively impacted in a previous Alcoa land buyback process. In order to increase its buffer zone adjacent to the Wagerup Alumina refinery, Alcoa bought additional houses and farmland in and around the Yarloop town site in the early mid 2000s. Some locals benefited substantially from the process while others fell outside the buyback zones and this resulted in some local resentment, which remains in the community today (Ashworth, 2014).

Building Trust

Work by Terwel and colleagues (2011) explored how the role of trust in stakeholders impacted the public acceptance of carbon capture and storage (CCS), a relatively new technology with some perceived risks and uncertainties. Although their work focused on CCS, the implications of their findings are also relevant for the unconventional gas industry, particularly in relation to an organisation's perceived competence and integrity. Terwel *et al.* (2011) found that the public attitude was dependent on people's trust in the project stakeholders and their perceived organisational competence and integrity, and not solely on the technology itself. This in turn impacted how individuals perceived information provided to them about the properties of the technology and whether they would rely on the positions advocated by project stakeholders. If competence-based trust was high then the individuals tended to perceive smaller risks and larger benefits of CCS, whereas low integrity-based trust caused people to run counter to these positions.

Terwel *et al.* (2011) also found that congruence between organisations' inferred motives and stakeholder communications facilitated trust. Thus, when an industrial stakeholder communicated the environmental benefits of CCS before economic benefits it would activate less trust. Conversely, it was expected that environmental non-government organisations (NGOs) would communicate first about the environmental benefits of the technology. As in the work by ter Mors *et al.* (2010), the researchers found that public trust in the environmental NGOs involved in CCS was higher than public trust in the industrial stakeholders. This was due to the inferred public-service motives of environmental NGOs, who were perceived to be involved in CCS due to their concern for the natural environment and future generations, as opposed to the industrial organisation's motives, such as profit maximisation and image building (Terwel *et al.*, 2011).

Governments and other political decision-makers were not the focus of Terwel *et al.*'s (2012) study. However, these authors suggested that mistrust of government often arises due to a lack of transparency in decision-making and public scepticism toward the "hidden agendas" of politicians. The authors suggested that governments should build mechanisms which allow them to be more open about how decisions are made, to help build trust in government decision-making across all levels of government.

A Fair Process

The processes used to resolve conflict and ensure innovation in determining possible solutions will significantly influence how equitable or just a decision will be perceived (Lacey, 2008; Wilder, 2008). This can be as simple as ensuring that the community feel that they have a voice that has been heard during the process and project organisers have responded to them accordingly. Adequate resources – both time and money – can help to ensure that a fair process has prevailed because often processes require flexibility to accommodate the needs of the more marginalised in the community and therefore can take longer than anticipated.

More recently, the establishment of community reference groups (CRGs) has been seen as a particularly useful way of engaging communities about both potential and ongoing projects (Ashworth *et al.*, 2012). In this case individuals from within the community agree to act as a liaison between the operation and their respective communities, sharing information and concerns. Such bodies usually comprise key leaders from across the community, government representatives and any others who have an interest in the project. Representation on CRGs can be achieved through a process of selection, or volunteering or a mix of both. Regardless of the particular process, what is important is that all meetings are documented and the information shared widely to others outside the CRG. Whatmore and Landström (2011) reported on the power and influence of such groups, describing a process that brought together social and natural scientists in the United Kingdom with local people affected by floods to collaboratively inform flood management policy at the highest levels, ultimately winning the support of local politicians.

Methods and techniques used to ensure a fair process can include the use of surveys, if broader community views are sought, and one-on-one interviews, focus groups and town hall meetings. However, if issues are contested then unless the process is well managed (see for example Ashworth *et al.* (2013) and their large-group-process methodology) it is unlikely that town hall meetings will be constructive for either party. In such cases, investing in more personalised and small group engagement is likely to be more appropriate.

As part of ensuring a fair process, it is important to cast the net wide to ensure that not just mainstream individuals are given a chance to have an input. Stakeholder analysis of the range of individuals and groups that exist within a community or region is one way to help ensure representativeness in an engagement process (Reed *et al.*, 2009). Building databases of the range of stakeholders that exist can be a helpful tool for industry and government agencies to draw upon when needing to engage.

Clear, Accurate and Timely Information

It is not surprising that often many individuals lack the necessary information about new projects, and therefore information provision, and dialogue around the topic, is an essential component of any community engagement activity. Where possible, if the information can be relayed from a trusted and objective source it is more likely to have credibility. Pisarski and Ashworth (2013) outlined a way to achieve "objective" information that involves the use of a diverse expert group, whose members may have with opposing views about the technology or project and who work to compile the latest and most accurate information about the project and associated technologies. This information is then presented to different groups from across the community and beyond. Using an expert who has a good understanding of the technology being discussed also helps to win trust in the information. Equally, stakeholders from across the community, or communities, can also share their views and be listened and responded to as a way of building a collaborative approach to decision-making about projects.

The above-mentioned information is best relayed in a variety of forms, but for new projects it is crucial to ensure that there are ample opportunities for face-to-face interactions, as this allows individuals to question the information being presented and to receive responses to their concerns. The literature indicates the need for early engagement as a way to ensure that local communities have the information they need to make an informed decision about a project. Early research of Ashworth *et al.* (2012) suggested that having communications experts participate as part of the technical team from the beginning of a project helps to ensure a proactive approach to early engagement. Social site characterisation is a method that has been put forward where a community is not only studied for its technical suitability but also its social suitability. Part of this process encourages establishing a baseline understanding of the community, the issues they are facing and their levels of technical knowledge about the project. Once a baseline is established it is much easier to inform ongoing engagement activities and gather the information for which stakeholders are asking. This can also help to mitigate the chance of negative reactions towards a project or at least keep concerns in the open (Ashworth *et al.*, 2010).

The need for an independent and trusted source of information for new projects is highlighted in an excerpt from the Queensland Parliament, Australia. In a particular session, an independent Member of Parliament was discussing the issue of a new proposed underground coal gasification project. The project proponents had recently held a public meeting about the project in the Member's local area. Some 80 residents attended the evening to express their concerns and hear more about the project. The Member highlighted that there was no independent voice to answer questions being raised. This is best evidenced in the excerpt below:

> There was no positive alternat[iv]e independent view put forward; it was either opposition or nothing. That caused me concern. It is my understanding that people who have concerns are not necessarily opposed to this project; they just want independent information. I ask the minister to release any independent information that she may have in regard to this particular project.
>
> *Queensland Parliament, 2010*

Shared Knowledge

The need for independent and trusted experts has been well established. Unlike a project proponent, an independent expert often has no vested interest in the project; they are purely there to answer questions from a professional and technical basis. The expert's credentials demonstrate their independence and help to build trust in what they have to say. However, this is not to say that the information from the expert is the only information required. The whole reason for engaging with communities is to establish a process for sufficiently sharing knowledge from a variety of sources to allow a public discussion to take place. Governments may also play a role in these processes, sharing information about their role as the regulator, explaining legal requirements or observing what issues may be emerging.

Without such processes, negativity amongst communities can escalate and potentially cause major harm to a project's successful deployment. This reinforces the necessity of early and proactive engagement around new or refurbished projects. The question, then, is: should this engagement be regulated and to what extent?

Policy Considerations to Enhance Best Practice Community Engagement

The frameworks outlined in this chapter assume that community engagement should take place voluntarily, beyond compliance, and companies should see it as an investment in building relationships with a community. Such a notion suggests that policy and regulations that demand community engagement may be counter to building such meaningful relationships. However, as not all companies will prioritise community needs as important, there is perhaps a minimum level of engagement that policy makers can encourage.

Mandating Engagement Through Regulations

Therefore, strengthening the regulatory component to include mandatory engagement structures may help to enhance the building of a relationship between project proponents and a community. However, the sheer act of mandating the engagement shapes what is talked about and how it will take place. For example, mandated companies may choose to focus on what is required to "tick the box" in the eyes of the regulator, rather than truly engaging and understanding stakeholder concerns or potential issues within a community. The challenge, therefore, for regulators is not only to structure regulatory regimes in such a way as to facilitate more comprehensive engagement processes but also to find a way to support proponents with a range of benefits (Gunningham *et al.*, 2004).

Despite regulation, at an early stage of a project's development, if proponents choose to go beyond compliance of the required regulation then this is likely to bring enormous benefits to them. Proponents' commitment to engage with stakeholders within a community can help to engender favour with that community, particularly, if project proponents listen and respond in a genuine manner. Responding in such a way can reduce the number of complaints and more often than not alleviate the need for local protests. Not only will proponents enjoy more positive community engagement; this type of response can also reduce the regulatory burden. If communities do not oppose projects and proponents are seen to be working collaboratively with these communities, governments are less likely to need to step in and mandate engagement activities (Ashworth and Cormick, 2011).

The importance of engagement by political authorities is also supported in Terwel *et al.*'s (2011), research on trust in CCS deployment, where they found that

> People are more likely to accept CCS policy decisions if they trust the political authorities that are responsible for making these decisions. Political authorities can establish trust by giving voice to interested parties and by communicating to the public how decisions are made.

In the Netherlands, the failure of the Dutch government to ensure a transparent and open engagement process around the deployment of a new CCS project in Barendrecht was one of the many reasons that resulted in the project being cancelled and, furthermore, the introduction of subsequent regulation to prevent any onshore storage of carbon dioxide in the near future (Brunsting *et al.*, 2011).

The Need for Regulations to Guide Operations

What is also important for projects to operate successfully is adequate regulations in place to guide how they operate. Such regulations provide greater confidence in communities, and their stakeholders, that potential risks to their health, safety and environment will be protected. Zhang and Moffat's study (2015) confirmed across a large statistical sample ($N = 2590$) that confidence in governments to hold mining industries accountable through adequate legislation played an important role in moderating the negative effect of environmental impacts on the acceptance of mining activities. This perhaps is most recently illustrated in Queensland, where early examples of environmental impacts negatively impacted overall perception of the coal seam gas industry. There were thus a number of lobby groups that openly opposed coal seam gas operations in Queensland and beyond. This had the result that a moratorium was put in place in New South Wales until further investigation could take place (James and Daniel, 2013). There are similar examples of such issues arising from the environmental impacts of the coal seam gas industry in other countries around the world, for example in the United States and in the United Kingdom (Muffson, 2012; Pidd, 2014).

On the other hand, an example of where legislation has helped build trust in the management of the coal seam gas industry was the establishment in July 2013 of the Gasfields Commission Act. The Act was established in Queensland Parliament to

> . . . manage and improve the sustainable coexistence of landholders, regional communities and the onshore gas industry in Queensland.
>
> *Queensland Parliament (2013)*

The Gasfields Commission comprises of a full-time commissioner (the Chair) and up to six part-time commissioners across areas that are impacted by the coal-seam gas industry. The gas commissioners who applied for the roles were all long-standing, well-respected members of their communities. Most have been farmers, or have relatives who are farmers, and therefore are seen to relate well to other landholders. Therefore landholders, who are potentially most impacted by the coal-seam gas industry, identify strongly with the commissioners, who tend to have the trust of the landholders and subsequently the community in which they live. A commissioner has a number of designated functions and powers and provides an intermediary role between government, industry and local communities. For example, the commissioners ensure that adequate consultation and information is available for enhancing decision-making and the co-existence of the emergent industry with local stakeholders. It is a model that has application in other states, territories, and even internationally, if it were implemented in a similar way, where local well-respected figures with some standing in the community are independently selected and appointed to commissioner roles.

Implications and Considerations for the Unconventional Gas Industry

The key points arising from the literature reviewed in this chapter suggest that for the unconventional gas industry there are a number of important considerations for engaging communities about their intended operations. Fundamental to any chance of success is the critical importance of building trust with communities in which they wish to operate. Engaging early in an open and transparent manner will help to do this and at the same time will ensure that the community is able to provide input into decisions that may directly impact them. Ensuring a staged approach to the project, where decision points are openly communicated, can also be helpful in building trust in the project process.

Being aware of potential environmental impacts, particularly around precious resources such as water, and working to minimise any of these will also help to build a more positive view of a company's operations. Applying the model of Zhang and Moffat (2015) for achieving an SLO, minimising environmental impacts and ensuring that operations are well managed to protect the environment is one of the most important undertakings the unconventional gas industry can do to enhance its credibility and standing within the community. One technical consideration that will also be helpful is to ensure that baseline monitoring is established prior to the commencement of any technical components of the project. Such monitoring is likely to be the responsibility of the proponents. However, there may be opportunities for governments to gather some of this information as part of their overall characterisation of a site, if the site has the potential for multiple uses. Past experience has shown that having baseline data available and shared openly between proponents, community and policy makers can help to provide additional assurances about the integrity of a project's operations.

Engaging in dialogue to understand community concerns and responding appropriately will also help to build a positive reputation of the company with those in the community. As every community's needs will be somewhat different, taking the time to discover key local issues, whether it be waiting times at the doctor's surgery or the need for new childcare or other local services, and finding ways to support such initiatives is likely to be met with more positive support than opposition. The establishment and use of a community reference group that has an experienced Chair to manage interactions between members is also an effective strategy for building meaningful relationships with the community, particularly, if those selected for the reference group are well respected and trusted within the community. However, establishing clear terms of reference in conjunction with the project proponent and members of the reference group is essential to ensure transparent operations of the reference group. In addition, ensuring that representatives of the group are committed to sharing information and liaising with the broader community about what is happening in the project is also helpful as part of a broader community engagement strategy.

As part of the dialogue process, having objective information in language that is easy to understand will help to ensure that the engagement process is seen to be fair. This can be done through the use of credible experts as well as building local advocates – such as those within the community liaison group – to create shared knowledge about the project. There are multiple processes that have been developed for engaging communities, and allowing adequate time and space for communities to respond and ask questions is also important for ensuring best practice engagement. It is through these activities that legacy issues can be raised and, if the process is respected by stakeholders from across the community, can start to be addressed. For some, though, such legacy issues may be so deep seated that there is almost nothing that can be done to overcome their entrenched views. Recognising when this occurs is also important, as it demonstrates respect for the impacted stakeholders and builds greater credibility for the project proponent across the community.

Policy makers also have an important role to play in providing comfort to communities about the unconventional gas industry. In particular, ensuring appropriate regulations are in place to protect and minimise the risks to the health and safety of individuals and the environment will help to build the industry's social licence to operate. Where possible, ensuring the consistency of such regulations across a nation will not only make it easier for communities to understand them but will also minimise transaction costs for companies that may be operating across state and territory borders.

In closing this chapter, it should be noted that there has been a long history of resource projects that sometimes negatively impact some stakeholders, when the projects are implemented in communities. Such negative impacts can be polarising, with different stakeholders from within a community adopting different positions towards the project. Given that many communities in regional and remote areas can be quite small, such polarisation can be particularly devastating. It is worth reflecting that although these best practice principles have been tried and tested for building project acceptance, they can only be a guide for industry.

Every community will respond differently on the basis of their culture, prior experiences, ideologies and values. A company may be diligent in their implementation of the above considerations but still not always successful in gaining acceptance and approval for their project. What becomes important in such cases is that the company responds in a positive manner towards the community and respects their concerns. Establishing a respectful and open relationship is critical for any ongoing deliberations between industry, communities and government, not only in that particular area but also beyond. If an issue does become contested then an independent mediator can be helpful in finding the best way forward.

References

ACCC (2013). AER Stakeholder Engagement Framework. Australian Competition and Consumer Commission. https://www.aer.gov.au.

AccountAbility (2008). AA1000 Stakeholder Engagement Standard 2011. http://www.accountability.org/standards/aa1000ses/index.html.

Ajzen, I. (1991). The theory of planned behavior. *Organizational Behavior and Human Decision Processes* 50, 179–211.

Arnstein, S.R. (1969). A ladder of citizen participation. *American Planning Association Journal* 35 (4), 216–224. Reprinted by permission of The American Planning Association (http://www.planning.org).

Ashworth, P. (2014). Lessons from project level community engagement. Project No. 7–0414–0227. ANLEC R&D: Canberra.

Ashworth P., Cormick, C. (2011). Enabling the social shaping of CCS technology. In I. Havercroft, R. Macrory& R.B. Stewart (eds.), *Carbon Capture and Storage: Emerging Legal and Regulatory Issues*. Hart Publishing: London.

Ashworth, P., Einsiedel, E., Howell, R., Brunsting, S., Boughen, N. Boyd, A. *et al.* (2013). Public preferences to CCS: how does it change across countries? In *Proc. 11th International Conference on Greenhouse Gas Control Technology*, 18–22 November 2012, Kyoto, Japan. Energy Procedia, Elsevier.

Ashworth, P., Bradbury, J., Wade, S., Feenstra, C.F.J., Greenberg, S., Hund, G. *et al.* (2012). What's in store: lessons from implementing CCS. *International Journal of Greenhouse Gas Contro* 9, 402–409.

Ashworth, P., Boughen, N., Mayhew, M., Millar, F. (2010). From research to action: now we have to move on CCS communication. *International Journal of Greenhouse Gas Control* 4, 426–433.

Bradbury, J., Ray, I., Peterson, T., Wade, S., Wong-Parodi, G., Feldpausch, A. (2009). the role of social factors in shaping public perceptions of CCS: results of multi-state focus group interviews in the US. *Energy Procedia* 1(1), 4665–4672.

Brunsting, S., De Best-Waldhober, M., Feenstra, C.F.J., Mikunda, T. (2011). Stakeholder participation practices and onshore CCS: lessons from the Dutch CCS case Barendrecht. *Energy Procedia* 4, 6376–6383.

Contandriopoulos, D. (2004). A sociological perspective on public participation in health care. *Social Science & Medicine* 58(2), 321–330.

Cormick, C. (2002). Australian attitudes to GM food and crops – changes in public attitudes to GM technology, *Pesticide Outlook* 13 (6), 261–264.

Desbarats, J., Upham, P., Riesch, H., Reiner, D., Brunsting, S., de Best-Waldhober, M. *et al.* (2010). Review of the public participation practices for CCS and non-CCS projects in Europe. Institute for European Environmental Policy; Brussels.

Duggan, J. (2014). China petrochemical plant may be halted after protests. *The Guardian*, 2 April 2014. http://www.guardian.co.uk/environment/2014/Apr/02/China-petrochemical-plant-may-be-halted-after-protests. Accessed 10 November 2014.

Dryzek, J.S., *et al.* (2003). *Green States and Social Movements*. Oxford University Press: New York.

Earle, T., Siegrist, M. (2008). On the relation between trust and fairness in environmental risk management. *Risk Analysis* 28, 1395–1414.

Fiorino, D. (1989). Environmental risk and democractic processes: a critical review. *Columbia Journal of Environmental Law* 14, 501–547.

Freeman, R.E. (1984). *Strategic Management: A Stakeholder Approach*. Pitman: Boston.

Franks, D.M., McNab, K., Brereton, D., Cohen, T., Weldegiorgis, F., Horberry, T. *et al.* (2013). Designing mining technology for social outcomes. Final Report of the Technology Futures Project.

Graafland, J. (2002). Profits and principles: four perspectives. *Journal of Business Ethics* 35, 293–305.

Gunningham, N., Kagan, R.A., Thornton, D. (2004). Social license and environmental protection: why businesses go beyond compliance. *Law Social Inquiry* 29 (2), 307–341.

Hall, N., Ashworth, P., Devine-Wright, P. (2013). Societal acceptance of wind farms: analysis of four common themes across Australian case studies. *Energy Policy* 58, 200–208.

Hashagen, S. (2002). Models of community engagement, Scottish Community Development Centre, pp. 1–12.

Head, B. (2007). Community engagement: participation on whose terms? *Australia Journal of Political Science* 42, 3.

Huijts, N., Molin, E., Steg, L. (2012). Psychological factors influencing sustainable energy technology acceptance: a review-based comprehensive framework. *Renewable and Sustainable Energy Reviews* 16, 525–531.

Huijts, N., Molin, E., van Wee, B. (2014). Hydrogen fuel station acceptance: a structural equation model based on the technology acceptance framework. *Journal of Environmental Psychology* 38, 153–166.

IAP2 (2014). Core values for the practice of public participation. IAP2 International Federation 2014. https://www.iap2.org.au/resources/iap2s-public-participation-spectrum.

James, M., Daniel, S. (2013). Moratorium on coal seam gas extraction in Sydney's drinking water zone. http://www.abc.net.au/news/2013–11–12/moratorium-placed-on-coal-seam-gas-extraction-in-sydney27s-dri/5087252. Accessed 14 December 2014.

Jasanoff, S. (2003). Technologies of humility: citizen participation in governing science. *Minerva* 41, 223–244.

Jijelava, D., Vanclay, F. (2014). Assessing the social licence to operate of development cooperation organizations: a case study of Mercy Corps in Samtskhe-Javakheti, Georgia. *Journal of Social Epistemology (Special Issue, A Social Licence to Operate)* 28 (3–4), 297–317.

Keeney, R. (1998). *Value-Focussed Thinking. A Path to Creative Decision Making*. Harvard University Press.

Kuppler, S. (2012). From government to governance? (Non-)Effects of deliberation on decision-making structures for nuclear waste management in Germany and Switzerland. *Journal of Integrative Environmental Sciences* 9 (2), 103–122.

Lacey, J. (2008). Utilising diversity to achieve water equity. *Rural Society* 18 (3), 244–254.

Lacey, J., Parsons, R., Moffat, K. (2012). *Exploring the Concept of a Social Licence to Operate in the Australian Minerals Industry*. CSIRO: Brisbane.

McIntyre, J. (1996). Tools for ethical thinking and caring: a reflexive approach to community development theory and practice in the pragmatic 90s. Community Quarterly.

Moffat, K., Zhang, A. (2014). The paths to social licence to operate: an integrative model explaining community acceptance of mining. *Resources Policy* 39, 61–70.

Muffson, S. (2012). In North Dakota, the gritty side of an oil boom. *The Washington Post*, 18 July 2012. http://www.washingtonpost.com/business/economy/in-north-dakota-the-gritty-side-of-an-oil-boom/2012/07/18/gJQAZk5ZuW_story.html. Accessed 10 December 2012.

Nicholls, F. (2003). When science is not enough: why public engagement is essential for carbon storage projects. Towards Zero Emissions: Strategies and Technologies for Decreasing Industrial GHG Emissions, Brisbane.

Pidd, H. (2014). Anti-fracking group stages day of action. *The Guardian*, 19 August 2014. http://www.guardian.co.uk/environment/2014/Aug/19/anti-fracking-group-stages-day-of-action. Accessed 12 November 2014.

Pisarski, A., Ashworth, P. (2013). The citizen's round table process: canvassing public opinion on energy technologies to mitigate climate change. *Journal of Climatic Change* 119 (2), 533–546.

Queensland Parliament (2013). Gasfields Commission Act 2013.

Queensland Parliament (2010). Hansard Record of Proceedings: First session of the Fifty Third Parliament, Tuesday 23 March, 2010. www.parliament.qld.gov.au/hansard/, p. 954.

Reed, M. (2008). Stakeholder participation for environmental management: a literature review. *Biological Conservation* 141, 2417–2431.

Reed, M., Graves, A., Dandy, N., Posthumus, H., Hubacek, K., Morris, J., *et al.* (2009). 'Who's in and why?' A typology of stakeholder analysis methods for natural resource management. *Journal of Environmental Management* 90, 1933–1949.

Siegrist, M., Cvetkovich, G. (2000). Perception of hazards: the role of social trust and knowledge. *Risk Analysis* 20, 713–720.

Solomon, F. (1999). Zen and the art of stakeholder involvement: the 1999 AMEEF travelling scholarship. In *Proc. Minerals Council of Australia Environmental Workshop*, Townsville, Queensland. Minerals Council of Australia.

Stern, P.C., Fineberg, H.V. (eds.) (1996). *Understanding Risk: Informing Decisions in a Democratic Society*. National Academy Press: Washington, DC.

Stirling, A. (2005). Opening up or closing down? Analysis, participation and power in the social appraisal of technology. In M. Leach, I. Scoones and B. Wynne (eds.), *Science and Citizens: Globalization and the Challenge of Engagement,* Zed Books: London, pp. 218–231.

ter Mors, E., Weenig, M., Ellemers, N., Daamen, D. (2010). Effective communication about complex environmental issues: perceived quality of information about carbon capture and storage (CCS) depends on stakeholder collaboration. *Journal of Environmental Psychology* 30 (4), 347–357.

ter Mors, E., Terwel, B.W., Daamen, D.D.L. (2012). The potential of host community compensation in facility siting. *International Journal of Greenhouse Gas Control* 11S, S130–S138.

Terwel, B., Harinck, F., Ellemers, N., Daamen, D. (2010). Voice in political decision-making: the effect of group voice on perceived trustworthiness of decision makers and subsequent acceptance of decisions. *Journal of Experimental Psychology* 16, 173–186.

Terwel, B.W., Harinck, F., Ellemers, N., Daamen, D.D.L. (2011). Going beyond the properties of CO_2 capture and storage (CCS) technology: how trust in stakeholders affects public acceptance of CCS, *International Journal of Greenhouse Gas Control* 5, 181–188.

Terwel, B.W., ter Mors, E., Daamen, D.D.L (2012). It's not only about safety: beliefs and attitudes of 811 local residents regarding a CCS project in Barendrecht. *International Journal of Greenhouse Gas Control* 9, 41–51.

Thomas, I. (1998). *Environmental Impact Assessment in Australia*, 2nd edn. The Federation Press: Sydney, Chapter 4.

Thomson, I., Boutilier, R.G. (2011). The social licence to operate. In Darling, P. (ed.), *SME Mining Engineering Handbook. Society for Mining, Metallurgy, and Exploration.* SME: Colorado, pp. 673–690.

UK ESRC (2012). Fracking and public dialogue. Economic and Social Research Council. http://www.esrc.ac.uk/news-and-events/features-casestudies/features/20493/carousel-fracking-and-public-dialogue.aspx.

von Schomberg R. (2013). A vision of responsible research and innovation. In R. Owen., J. Bessant and M. Heintz (eds.), *Responsible Innovation*. John Wiley: London, pp. 51–74.

Whatmore, S.J., Landström, C. (2011). Flood apprentices: an exercise in making things public. *Economy and Society* 40 (4), 582–610.

Wilder, M. (2008). Equity and water in Mexico's changing institutional landscape. In J. Whiteley, H. Ingram and R. Perry (eds.), *Water, Place and Equity*. Cambridge: MIT Press, pp. 95–116.

Wolsink, M. (2007). Wind power implementation: The nature of public attitudes: Equity and fairness instead of 'backyard motives'. *Renewable and Sustainable Energy Reviews* 11 (6), 1188–1207.

Zhang, A., Moffat, K. (2015). A balancing act: the role of benefits, impacts and confidence in governance in predicting acceptance of mining in Australia. *Resources Policy* 44, 25–34.

20

Managing the Impact of Coal Seam Gas Water Extraction in the Surat Basin

RANDALL COX

Introduction

Water has always been extracted to some extent in the process of producing petroleum and gas products. Conventional production operations involve pumping petroleum and gas that has accumulated in natural geologic traps in porous geological formations. Because groundwater is unavoidably extracted in the process of production it is difficult to regulate the production of the petroleum product and the extraction of water separately. However, the volume of water associated with conventional petroleum and gas operations tends to be relatively small, and the locations of operations tend to be remote from major water users. As a result, in Queensland a statutory authorisation exists to take groundwater in the process of producing petroleum and gas products.

Coal seam gas (CSG) operations are fundamentally different. They involve pumping water from coal formations to reduce the water pressure and thereby release gas that is adsorbed on coal particles. A CSG well initially produces water but, as the water pressure in the formation falls, gas starts to desorb and flow to the well. Progressively the gas content increases and the water content decreases. The amount of water produced is relatively large. Water bores producing water from the coal formations near the development area are likely to be affected by the pressure reductions, and water bores producing water from overlying and underlying formations can also be affected if water can flow between the formations.

In the Surat Basin the amount of water produced by CSG operations is large by comparison with conventional operations. There are existing water users extracting water from the target coal formation. There are four CSG operations and pressure impacts will extend laterally causing overlapping impacts between the four operations. Also, because of hydraulic connectivity between the coal formation and overlying and underlying aquifers, there will be some water pressure reductions in those aquifers.

Because of the scale of planned CSG development in the Surat Basin, in late 2010 the Queensland Government introduced additional regulatory arrangements for the management of water resources in areas of petroleum and gas development. Some complementary regulatory arrangements relate to the standard of construction of CSG wells needed to avoid damage to groundwater resources, to the treatment and beneficial use of groundwater after it is extracted and to the disposal of brine. Those topics are not the primary focus of this chapter, however.

The regulatory arrangements introduced in 2010 also deal with the management of the impacts of CSG water extraction on groundwater resources and associated water supply bores and springs. That part of the framework and the progress that has been made in implementing arrangements under the framework compromise are the topics of this paper. A particular focus is the way in which the cumulative impacts from multiple CSG operations in an area are managed.

Groundwater Management Framework

Overview

The CSG industry in the Surat Basin is being implemented in the presence of an existing agricultural industry dependent on the use of groundwater. The decision to establish the CSG industry was in effect a decision to manage the coal measures in the area as a hydrocarbon resource rather than as a groundwater resource. The groundwater management framework provides a means of managing the impacts caused by CSG water extraction and ensuring that CSG operators meet the costs of that activity. The framework is set out in Chapter 3 of the *Water Act 2000* (Qld). The following principles are the basis for the regulatory framework.

Make Good: It is not possible to establish a CSG industry without a major lowering of water pressure in the target coal formation. This lowering is likely to affect water supply bores tapping the coal formation and to some extent bores tapping overlying or underlying aquifers. Therefore operators are required to make good any impairment of private bore supply. To the greatest extent possible, these arrangements should be implemented before the CSG operations impact on the bore.

Baseline Bore Data: In order to make good the impairment of private bore supplies, operators are required to collect baseline data about water bores that could be affected.

Underground Water Impact Reporting: Even where knowledge about the groundwater flow system is sufficient to make reasonable predictions about long-term impacts, depressurisation will cause major new stresses to the groundwater flow system. Therefore operators are required to monitor pressure responses in the system as the industry develops and maintain a groundwater flow model to predict long-term pressure impacts. They are required to prepare and periodically update an underground water impact report (UWIR) setting out the results of those assessments and planned management arrangements.

Cumulative Management: Where the impacts from adjacent operations overlap, a cumulative approach to management should apply to ensure that the total impact is assessed and integrated impact management arrangements are implemented.

Make Good Impairment of Supply (Water Act, Chapter 3, Part 5)

Under the framework a petroleum tenure holder has a responsibility to 'make good' impairment of supply from a water bore that results from CSG water extraction. The supply from a water bore is considered to be impaired if a decline in water level has caused the quantity or quality of water to deteriorate to the extent that the supply is no longer suitable for the required purpose. The water supply can be made good by any measures agreed by the parties. It may involve deepening a water supply bore, replacing the bore or establishing a new water supply from some other source.

The intention is that for any bore the make-good actions are triggered in advance of any impact occurring. When a water pressure reduction of more than a threshold amount is predicted to occur within three years then the tenure holder is required to carry out a detailed bore assessment and enter into a make-good agreement with the bore owner. However, if at any time the regulator considers that a bore supply has been impaired then the regulator can also direct the tenure holder to conduct a bore assessment, which in turn triggers a make-good agreement.

Because of the nature of groundwater flow, the maximum impact on a water bore may occur after CSG water extraction ceases. Therefore, before exiting a tenure, a tenure holder is required to enter into a make-good agreement with the owner of any bore predicted to be impacted by more than the threshold amount at any time in the future. It can be expected that predictions about long-term impact will have become progressively more accurate through progressive revisions of the assessments before the end of the tenure.

Arrangements for underground water impact reporting and the collection of baseline data about private water bores provide the basis for giving effect to the obligation to make good impairment of supply.

Underground Water Impact Reporting (Water Act, Chapter 3, Part 2)

Tenure holders are required to prepare a UWIR every three years during the life of a project except in areas where the cumulative management framework has been triggered, as detailed in a later section. The report includes: predictions of the impact of CSG water extraction on groundwater pressures in the target coal formation as well as overlying and underlying aquifers over the life of the project; a water monitoring strategy; and a spring impact management strategy.

The UWIR has to be updated every three years to incorporate new knowledge. When it is approved, implementation of the water monitoring strategy and of the spring impact management strategy becomes a legal obligation for the tenure holder.

Baseline Bore Data (Water Act, Chapter 3, Part 3)

Tenure holders are required to collect baseline information about private water bores on a tenure before production commences. They are required to submit a baseline assessment plan to the regulator setting out the timing for the collection of the data. The regulator publishes guidelines specifying the data that are to be collected.

The impacts of CSG development on a petroleum tenure can extend beyond the boundary of the tenure. The extent of future impacts that will occur beyond the boundaries of a tenure are identified when a UWIR is prepared and then updated every three years. Arrangements for the collecting of baseline data for bores in affected areas outside the petroleum tenure are set out in the UWIR.

Cumulative Management Framework (Water Act, Chapter 3, Part 2; Chapter 3A)

Water pressure reductions from CSG water extraction propagate horizontally and therefore the impacts from the individual operations can overlap. As a result, more than one operator can contribute to a reduction in the water level in a water supply bore. Under the regulatory framework, where these circumstances exist a cumulative management area (CMA) can be established. The cumulative management framework is described in a later section.

Roles and Responsibilities

The Queensland Department of Environment and Heritage Protection (EHP) is the regulator. Tenure holders are accountable to EHP for their responsibilities under the framework. The EHP discharges this responsibility along with the regulation of a range of environmental matters in relation to the CSG projects. It establishes CMAs where appropriate.

The CSG Compliance Unit within the Department of Natural Resources and Mines (DNRM) carries out both engagement and compliance activities for DNRM and, by arrangement, for EHP. It is the point of contact for landholders on a wide range of issues related to CSG development, including issues with water bores. If there is a disagreement or breakdown in the make-good arrangements then the Compliance Unit supports individual bore owners with regard to technical aspects.

The Queensland GasFields Commission comprises seven commissioners with a wide range of backgrounds. The Commission facilitates the successful coexistence of the onshore gas industry with other industries by bringing stakeholders together to identify and facilitate the resolution of issues. The Commission provides advice to Ministers.

The Office of Groundwater Impact Assessment (OGIA) carries out cumulative assessment for CMAs using regional groundwater flow modelling. Its role is central in the implementation of the cumulative management framework. It develops integrated management arrangements and assigns responsibility to individual petroleum tenure holders for implementation of individual parts of those integrated arrangements. Further details are provided in a later section.

Cumulative Management Framework

Overview

When a CMA is established, OGIA becomes responsible for the preparation of a UWIR and for updating the UWIR every three years to incorporate new knowledge about the system. This obligation displaces the responsibilities of individual tenure holders in the CMA to prepare individual reports. The UWIR prepared by OGIA must include the following:

- Predictions of the cumulative impacts from whole-of-life water extractions by all petroleum and gas operations in the CMA on water pressures in all aquifers in the area;
- Specification of integrated regional management arrangements, including a water monitoring strategy and spring impact management strategy;
- Assignment of responsibilities for implementing separate parts of the management strategies to individual petroleum tenure holders. For example, individual petroleum tenure holders are assigned responsibility for implementing separate parts of the regional monitoring network, and rules are prescribed by which a single tenure holder becomes responsible for any necessary make-good actions for a water bore.

While the assessment obligations of tenure holders in a CMA are displaced to be carried out by OGIA, the tenure holders pay a levy to meet the cost of OGIA in carrying out those activities.

Cumulative Impact Assessment

In a CMA it is more efficient and more effective for the assessment of future impacts to be carried out centrally. If an individual tenure holder were to attempt to assess the impact caused by the tenure holder's individual operation, it would be necessary to construct a model of the groundwater system beyond the boundaries of the tenure. This would involve the collection of hydrogeological data from neighbouring tenure holders. This could pose difficulties as a tenure holder might resist providing data to a neighbour if the data were considered to include information considered to relate to commercial interests. Also, assessments carried out by neighbouring tenure holders could be based on different understandings of the regional groundwater flow system, which would lead to confusion.

Therefore the cumulative management framework provides that OGIA carry out the assessment. It has the power to collect information from tenure holders in order to carry out those responsibilities. Tenure holders can provide technical information to OGIA as an independent body for the purposes of groundwater assessment without compromising commercial interests.

Integrated Management Strategies

In a CMA, monitoring objectives will be best met through a single integrated water monitoring network. If an individual tenure holder were to design a monitoring network for an individual operation, it would be necessary for the network to include the construction of monitoring works on adjacent tenures. Where multiple neighbouring operations exist, this would lead to duplication and land access difficulties. The same issues apply in relation to the monitoring of springs.

Therefore the cumulative management framework provides that OGIA include in the UWIR a regional water monitoring strategy and spring impact management strategy.

Assignment of Responsibilities to Tenure Holders

While water and spring monitoring strategies are designed centrally by OGIA and specified in the UWIR, the responsibility for implementation rests with individual tenure holders. The UWIR assigns responsibilities for individual parts of the regional strategies to individual tenure holders.

Similar issues apply in relation to make-good actions. Where there are multiple adjacent operations, in the absence of an independent body to assign responsibilities it would be difficult to administer the requirement that tenure holders make good the impairment of individual bore supplies. The reason is that a bore supply could be impacted by more than one CSG operation. For example, a bore supply could be located on a tenure held by one operator but not yet developed and be affected by operations on an adjacent tenure held by another tenure holder. To avoid these issues and provide clarity for both tenure holders and bore owners, the UWIR can include rules for the assignment of responsibilities for make-good action. While the rights of a bore owner with regard to the making good of impairment of supply remain defined by the *Water Act 2000* (Qld) with linkages to predictions of impacts in the UWIR, the rules in the UWIR are the means by which a single tenure holder is identified as the responsible tenure holder.

OGIA

The Office of Groundwater Impact Assessment (OGIA) is an independent entity housed within the Queensland Department of Natural Resources and Mines. The manager of OGIA must carry out the functions of the OGIA without interference by any person. It has powers to obtain data from tenure holders in order to carry out its functions.

A levy on tenure holders funds OGIA; the levy is established under the provisions of the *Water Act 2000* (Qld), with the structure of the levy set out in the *Water Regulation 2004*. It consults with an advisory body comprising CSG and non-CSG interests on planned annual expenditure, after which the responsible Minister of the Queensland Government can approve an annual budget. It then raises the levy progressively during the year. Funds raised can only be used for the operations of OGIA. Unspent levy funds are returned to tenure holders by way of a credit on levy charges in the following year.

The Office of Groundwater Impact Assessment has existed since 1 January 2013. The functions currently administered by OGIA were first established in late 2010 and were administered by the then Queensland Water Commission.

Surat Underground Water Impact Report

The Process of Preparing the Surat UWIR

The Surat CMA was established in early 2011. It covers an area that extends far enough from the development area in the Surat and Southern Bowen Basins to include the area of possible impacts. When the CMA was established the preparation of the Surat UWIR commenced.

A draft Surat UWIR was released for public consultation in mid 2012. The draft UWIR was adjusted on the basis of submissions received and a final Surat UWIR was submitted to the regulator and approved in December 2012.

Hydrogeology of the Area

The Surat CMA covers part of three geologic basins: the northern Surat Basin, the southern Bowen Basin and the western Clarence–Moreton Basin. The Bowen Basin is the deepest and oldest and runs north–south through the centre of the region. Overlying this is the Surat Basin, which covers most of the central and southern parts of the CMA. The Clarence–Moreton Basin interfingers with the Surat Basin across the Kumbarilla Ridge to the east. Overlying these basins are areas of unconsolidated younger alluvial sediments and volcanics. The most important of these is the Condamine Alluvium, which is used extensively for irrigation. Figure 20.1 shows the distribution of the basins in the area.

The Great Artesian Basin (GAB) is a groundwater system comprising all or parts of several geologic basins. Within the Surat CMA, the GAB includes the Surat Basin, the equivalent formations in the Clarence–Moreton Basin and the upper parts of the Bowen Basin. The GAB is a multilayered system, comprising the major part of the groundwater flow system shown in Figure 20.2. The target coal formation in the Surat Basin is the Walloon Coal Measures. Major aquifers of the GAB include the Hutton Sandstone and the Springbok Sandstone, which underlie and overlie the Walloon Coal Measures. The Bandanna Formation is the target coal formation in the Bowen Basin. The coal formations comprise many thin lenses of coal in a non-carboniferous matrix.

Within the Surat CMA some 87 000 Ml per annum are extracted from the GAB and deeper sediments, mostly for stock supply but also for other uses. A further 127 000 Ml per annum are extracted from shallow alluvial and volcanic aquifers, mostly for irrigation purposes. In total there are some 20 000 private water supply bores in the CMA.

Water extraction as part of conventional petroleum and gas operations commenced in 1962. Coal seam gas production from the Bandanna Formation commenced in the Bowen Basin in 1993 and from the Walloon Coal Measures in the Surat Basin in 2003. Water extraction associated with conventional operations is steady at some 2000 Ml per annum. Water extraction associated with CSG operations was about 18 000 Ml per annum when the Surat UWIR was prepared in 2011 and will rise substantially as the industry expands.

Connectivity of Formations

The coal formations supporting CSG production are part of a multilayered groundwater flow system. Water pressures in the coal formations are substantially reduced in the process of producing CSG. As the industry develops these pressure impacts will be transmitted to

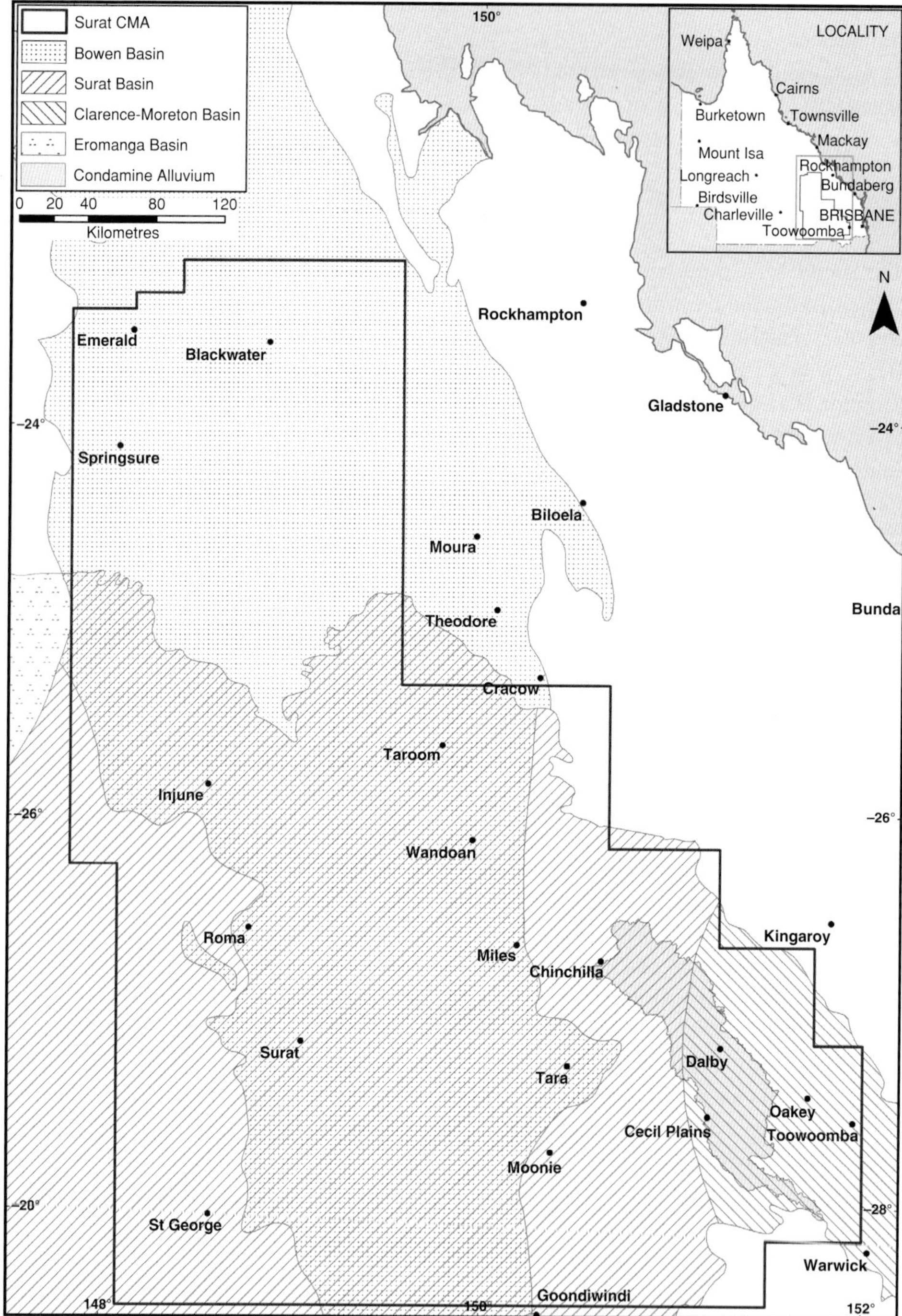

Figure 20.1 Geologic basins in the Surat cumulative management area.

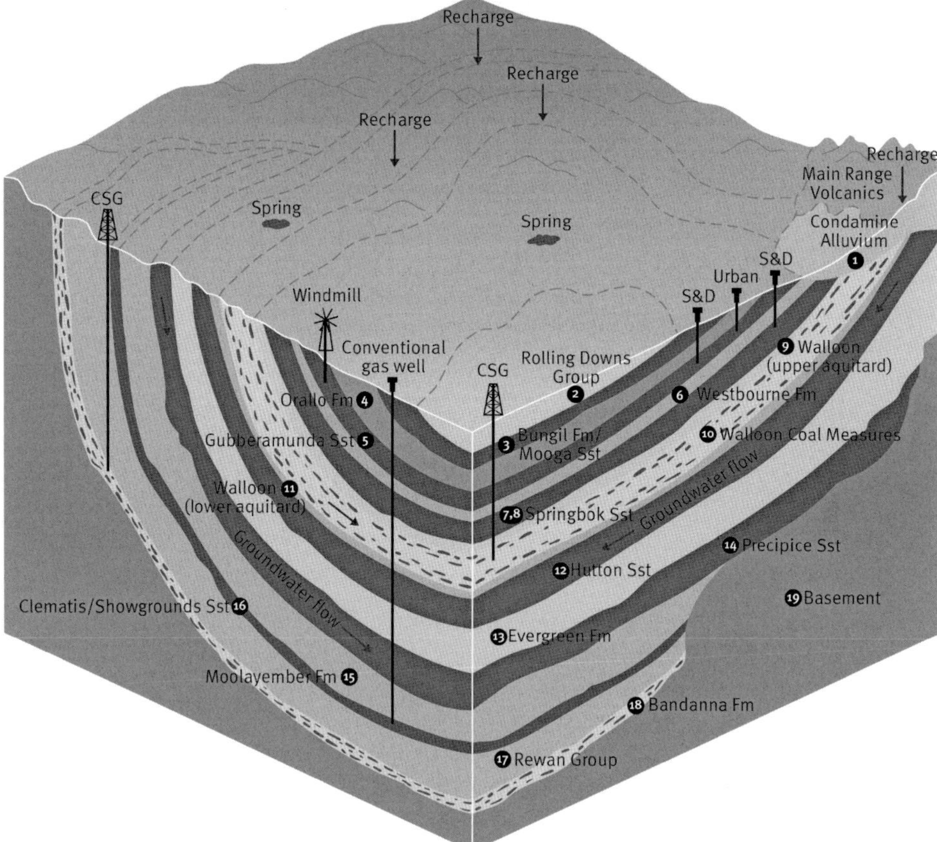

Figure 20.2 Conceptual representation of the groundwater flow system.

some extent to overlying and underlying aquifers, where there is hydraulic interconnectivity between the formations. Appropriate representation of the interconnectivity between the formations was a priority in the development of the regional groundwater flow model. Important areas of interconnectivity within the CMA are as follows.

The Condamine Alluvium overlies the eastern margin of the area of planned development. Basal alluvial clays and the upper weathered horizon of the underlying Walloon Coal Measures together comprise a transition zone that functions as an aquitard of variable thickness. This area of interconnectivity is of particular significance because of the economic importance of the Condamine Alluvium.

The upper and lower non-coal-bearing parts of the Walloon Coal Measures form aquitard layers separating the gas-bearing horizons of the Walloon Coal Measures from the overlying and underlying aquifers. The lower aquitard is generally thicker and of a more consistent thickness than the upper aquitard.

For the most part the Bandanna Formation is isolated from aquifers by thick low-permeability formations. However, there is a relatively small area north of Roma where the Precipice Sandstone, the basal aquifer of the Surat Basin, rests unconformably on the Bandanna Formation.

Predicted Water Pressure Impacts

A regional groundwater flow model was constructed covering an area of 550 km × 600 km, with 19 layers and a grid size of 1.5 km. It was calibrated against existing data, and uncertainty analysis techniques were used to improve confidence in predictions of water pressure impacts made using the model. The model was externally reviewed and found to be fit for purpose. Industry development plans were obtained from each tenure holder in the area in order to produce a cumulative development profile to be used for assessment purposes. The industry development plans will change over time and the model is used to review annually predictions of impacts resulting from those changes.

Representation of Affected Areas

The UWIR contains a map of the extent of the immediately affected areas, in accordance with regulatory requirements. The immediately affected area for an aquifer is the area where the water pressure in the aquifer is predicted to fall by more than a threshold value within three years, as a result of petroleum and gas operations. The threshold values are 5 m for sandstone aquifers and 2 m for alluvial aquifers. The maps are of legal significance, because on approval of the UWIR the responsible tenure holder is required to carry out a detailed bore assessment of the existing water supply bores sourcing water from an aquifer in its immediately affected area and to enter into agreements with the owners of the bores about any measures necessary to make good impairment of supply. Although there is an underlying general requirement that impairment of supply be made good irrespective of the amount of pressure reduction that has been caused by the impairment, a predicted pressure reduction of the threshold value within three years is a trigger for proactive action.

In the UWIR, immediately affected areas of any significant extent exist only for the coal formations, as impacts in overlying and underlying aquifers will tend to occur later in time. On the basis of records existing at the time, 85 bores were identified in the immediately affected areas.

The UWIR also contains maps of the extent of the long-term affected areas. The long-term affected area for an aquifer is the area where the water pressure is predicted to fall by more than 5 m at any time in the future as a result of CSG operations. On the basis of records existing at the time, 529 bores were identified in the long-term affected areas. The UWIR also includes maps showing the distribution patterns of expected impacts for each aquifer. The long-term impact is summarised in Figure 20.3 and is described in the following sections.

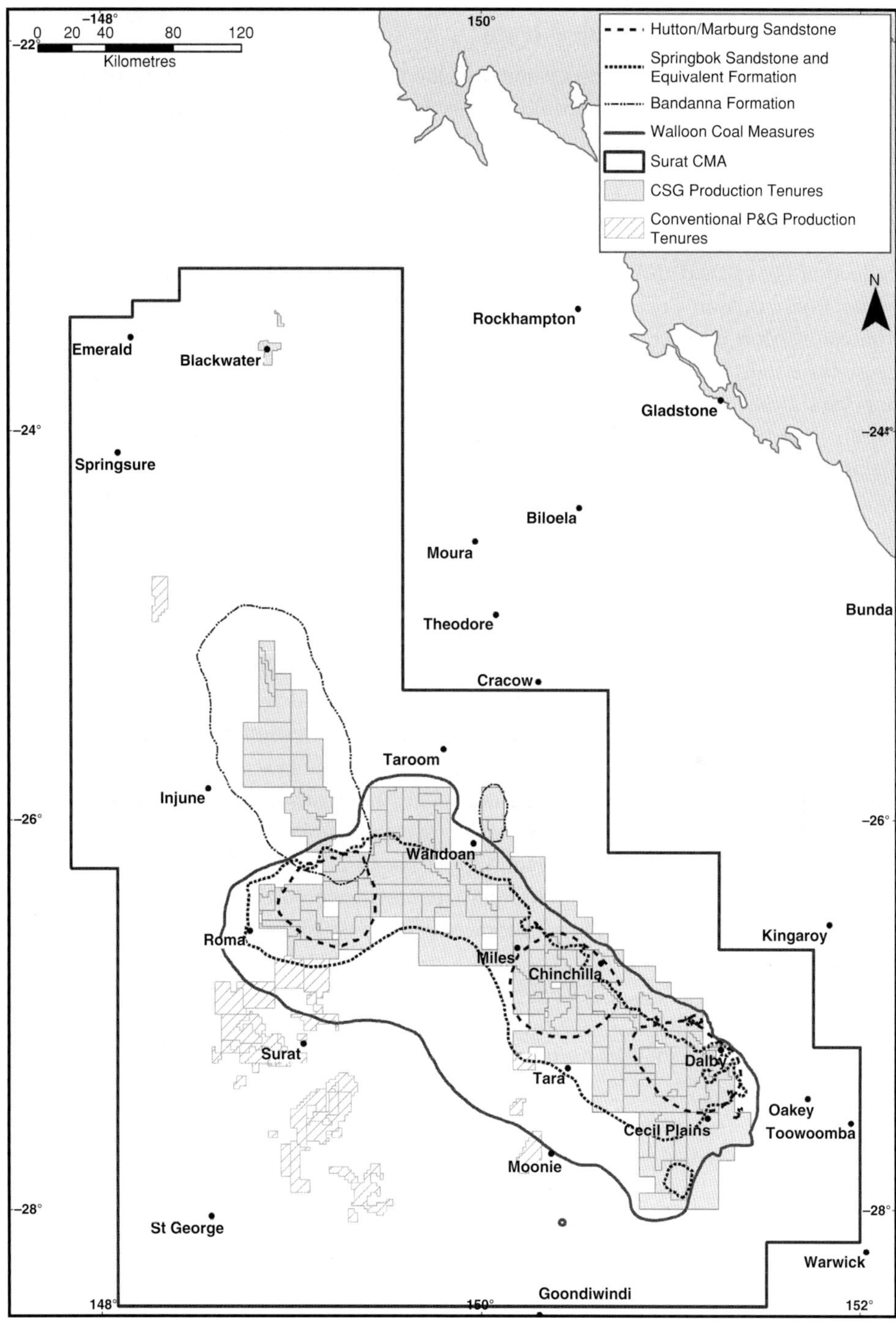

Figure 20.3 Areas where impacts on aquifers will exceed 5 m in the long term.

Long-Term Impact in Individual Aquifers

The Walloon Coal Measures present the CSG target formation in the Surat Basin. For most of the impacted area, the long-term impact is expected to be less than 150 m. Within the production area, the magnitude of impact reflects the depth of the top of the coal formation, because CSG operators advised that operational practice for CSG production is to lower the pressure in the coal seams to approximately 35 to 40 m above the top of the uppermost coal seam. As a result, in the more westerly areas, where the coal formation is deep, the pressure reduction is expected to be up to 700 m. There are 400 private water bores that source water from Walloon Coal Measures within the predicted long-term affected area for this formation. Most of these are located to the east, where the formation is relatively shallow and impacts are smaller. Half the affected bores are expected to experience an impact of less than 21 m.

The Bandanna Formation is the CSG target formation in the Bowen Basin. For most of the impacted area the long-term impact is expected to be less than 200 m. As for the Walloon Coal Measures, the pressure reduction will be greater in areas where the coal formation is deep. There are no private water supply bores that source water from the Bandanna Formation within the predicted long-term affected area for this formation.

The Springbok Sandstone overlies the Walloon Coal Measures. For the most part the aquifer is separated from the productive coal seams by an upper low-permeability layer of the Walloon Coal Measures, although this layer is thin or absent in some areas. Over most of the affected area the maximum impact is expected to be less than 20 m. There are 104 bores that source water from the Springbok Sandstone within the predicted long-term affected area for this formation.

The Hutton Sandstone underlies the Walloon Coal Measures. It is separated from the productive coal seams by a low-permeability layer of the Walloon Coal Measures. Over most of the affected area the maximum impact is expected to be less than 5 m, although there are small areas where maximum impacts may reach 18 m. There are 23 bores that source water from the Hutton Sandstone within the predicted long-term affected area for this formation.

The Precipice Sandstone is a basal aquifer of the Surat Basin that has some connectivity with the Bandanna Formation of the underlying Bowen Basin in areas where the intervening units were eroded away before the Precipice Sandstone was deposited, leaving the Precipice Sandstone in contact with the Bandanna Formation. Around the contact area the long-term maximum impact is expected to reach 10 m. There are no private bores that source water from the Precipice Sandstone within the predicted long-term affected area for this formation.

The Gubberamunda Sandstone and Mooga Sandstone are shallow aquifers that are not well connected to the coal formations. Generally, impacts of less than 3 m are predicted and only small areas will be affected.

The Condamine Alluvium partially overlies the eastern margin of the area of planned CSG development. There is no long-term affected area for the Condamine Alluvium as the maximum water level impact is expected to be smaller than the 2 m threshold that applies for unconsolidated aquifers. The maximum impact is expected to be 1.2 m in a small area on the western edge of the alluvium. Over most of the area the maximum impact is expected to be approximately 0.5 m.

Affected Springs

There are 71 spring complexes in the CMA. Of these, small water-pressure impacts are predicted in the source aquifers of five complexes. At these sites tenure holders are required to assess the local hydrologic setting to improve understanding of the risk to springs and identify options for the mitigation of impacts.

Implementation of the Surat UWIR

Changes to Predictions of Impacts

As previously mentioned, the cumulative management framework is designed to ensure that make-good arrangements are in place before water supplies are impaired by any pressure reductions resulting from CSG water extraction. Tenure holders have to carry out a bore assessment and enter into make-good agreements with the owners of bores that are predicted to have an impact of more than the trigger amount within three years.

The groundwater flow model is updated every three years, and this will change predicted impacts to some extent. However, the predicted impacts are largely dependent on the scope and timing of the planned development of the CSG industry. The predicted impacts set out in the Surat UWIR are based on the cumulative industry development profile planned at the time the UWIR was prepared. Many factors can change development planning over time, however. As a result, OGIA updates the industry development profile every year and uses the current groundwater flow model to assess whether there is any significant change to predicted impacts. The results are set out in the OGIA annual reports.

The first annual report was published in December 2013. It showed that, although the planned long-term development footprint had not changed significantly, the CSG industry had developed more slowly than was planned at the time the UWIR was prepared. As a result there were some bores that had had make-good measures implemented a little earlier than necessary, given the changed timing of impacts. In one area development had been brought forward in time, triggering the need for earlier make-good action for three bores. The 2014 report showed a continuation of the same trend.

Implementation of Monitoring

Tenure holders are progressively implementing the water monitoring requirements set out in the UWIR. The monitoring network will be reviewed when the UWIR is updated having regard to the emerging data and improved understanding of the groundwater flow system.

Research Activities

Extensive knowledge has been built over many years about the regional groundwater flow system and this was used as the basis for the development of the UWIR. However, the groundwater system is complex and knowledge about it continues to improve. The new understanding will be incorporated in the preparation of subsequent revisions of the UWIR. Focus areas for OGIA research activities are as follows.

Hydrogeological Conceptualisation

The recharge and discharge mechanisms for aquifers and the way water moves through them are considerations in the study of all groundwater flow systems. However inter-aquifer flow is a particular focus in the CMA because there will be a major lowering of water pressure in the Walloon Coal Measures and this will propagate to some extent into the overlying and underlying aquifers. The OGIA will continue its analysis of geological and geophysical hydrochemical data into the future to improve understanding of connectivity, so that it can be optimally represented in the progressive redevelopment of groundwater regional flow modelling.

Condamine Connectivity

While the potential for inter-aquifer flow is important across the region, as mentioned above, it is particularly important in the footprint of the Condamine Alluvium. The Condamine Alluvium is an important aquifer because it provides irrigation water for a major agricultural area. The alluvium overlies the Walloon Coal Measures in places. Although the leakage of water from the alluvium that will result from the lowering of water pressures in the Walloon Coal Measures is expected to be relatively small, it is important that the potential impacts are understood as clearly as possible. Therefore, OGIA will continue to carry out connectivity research at a more intense level in this area.

Groundwater Flow Modelling

Groundwater flow modelling is a highly developed discipline. However, the CSG development in the CMA poses specific challenges. The movement of water in the coal measures is complex and water behaves differently where gas is also moving in a formation. The OGIA will continue to develop improved approaches to groundwater flow modelling and apply them in building future generations of the regional groundwater flow model.

Spring Function and Monitoring Methodology

There are springs in the CMA that support high-value groundwater dependent ecosystems. A change in water pressure in the source aquifer at the location of a spring is a primary consideration in assessing the risk to a spring. However, the nature of spring function and how a spring is connected to the source aquifer will affect the susceptibility of a spring to a change in water pressure in the source aquifer. The OGIA will continue to investigate spring function at key sites and to develop new ways to monitor the flow of water to springs, so that understanding of the risk to springs continues to improve.

Community Engagement

The Office of Groundwater Impact Assessment carries out formal community consultation when the UWIR is updated. Between updates of the UWIR, stakeholder engagement is carried out in a range of ways. It supports the Queensland GasFields Commission in community engagement activities. Accordingly, it provides briefings for the Community Leaders Committees convened by the Queensland GasFields Commission throughout the region. It also responds to requests from local government and industry groups for briefings. It has strong links to the petroleum tenure holders and works closely with the tenure holders to facilitate data exchange and to collaborate on research projects.

Future Directions – Broadening the Regulatory Framework

The regulatory framework for managing the impact of water extraction described in this chapter applies only to the taking of water by petroleum tenure holders. As mentioned previously, the framework has been developed because a decision to develop a CSG industry is effectively a decision to manage the target coal formation as a petroleum resource rather than as a groundwater resource. Water is pumped from the coal at a rate deliberately intended to lower water pressures and at a rate that could not and need not be sustained in the long term. The framework is directed at the management of the impacts of that water extraction.

Other water extraction activities, including water extraction to support mining development, are managed under an allocation system that seeks to share a water resource between users by directly or indirectly limiting the volume that can be taken under the water entitlement system. However, in the case of water extraction to support mining operations, similar considerations apply as for CSG operations. For example, if a major open-cut mine is constructed that intersects a groundwater resource then groundwater will flow into the mine. A decision to approve the construction of a mine is effectively a decision to approve the taking of the water and ensure its safe operation and needs to be considered at the time of the approval of the mine.

In late 2014 the Queensland Government introduced before the Parliament changes to legislation in order to better align the regulation of the taking of water for mining purposes with the taking of water for petroleum and gas production. The intention was to provide greater certainty for bore owners about the protection of water supplies from mining activity, greater certainty for miners about the right to take the water that is unavoidably taken as part of a mining operation and to ensure that any taking of groundwater that is not an unavoidable consequence of the mining or petroleum activity is managed under the water entitlement system, as is the case for all other water users. Under the proposed changes the regulatory framework described in this chapter would progressively apply to those mining activities that significantly affect ground resources.

Conclusion

The CSG industry is a major extractor of water in the Surat Basin. The extraction will result in a major lowering of water pressures in the target coal formations, and there will be some impact on overlying and underlying formations because of connectivity between formations. These impacts are unavoidable if the industry is to develop.

Queensland has introduced a regulatory framework to deal with these issues. The framework ensures that impairment of water bore supplies is made good by petroleum tenure holders and that this action is taken before significant impact occurs.

The water pressure impacts will occur well into the future and therefore decisions about water management need to be based on the predictions of future impacts. While there is sufficient information about the groundwater system to enable sound predictions to be made, the regulatory framework provides for the predictions to be progressively refined and management arrangements adjusted accordingly. Importantly, the framework includes arrangements for the assessment and management of the cumulative impacts of multiple adjacent CSG operations.

In the Surat Basin the framework is being found to be an important factor in achieving the successful coexistence of the CSG industry with other industries and community interests in the area.

Reference

Queensland Water Commission (QWC) (2012). Underground Water Impact Report: Surat Cumulative Management Area, Queensland Government, Brisbane.

21

Whole-of-Landscape Assessment and Planning in the Management of Unconventional Gas Exploration and Production in Australia

JOHN WILLIAMS, ANN MILLIGAN AND TIM STUBBS

Introduction

In principle, searching for and producing unconventional gas in Australia can be seen as just one more land use to be imposed on this continent's very old landscapes. However, that is no reason for complacency. Gas exploration and production entails risks to land, water and air, as has been highlighted by experiences overseas and in early gas exploration activities in eastern Australia. These have naturally led to anxiety in Australian communities, especially in regions overlying basins prospective for unconventional gas.

It is a defensible proposition that the only development activities that should be acceptable in a region are those that allow the landscape to maintain its function indefinitely. It would be folly to secure one natural resource while putting at risk renewable long-term resource use.

In this chapter, we reiterate that the adoption of a knowledge-based long-term regional strategic land-use planning approach for the regulation of unconventional gas activities should help avoid perverse outcomes. Ideally, *all* productive developments in a landscape should be assessed for their use of resources and the likelihood that they will have impacts on the local environment, ecology, water resources and people. In fact, a number of such initiatives are now in progress and, while they are based on the need to limit issues arising from unconventional gas, it seems that other landscape developments are being included in the assessments. We draw attention to some of these initiatives and also recall a practical example from 2012: with it, the Namoi Catchment Management Authority showed the way to evaluate the impacts of mining in their landscape which was already highly valued for its food and fibre production and its conservation of unique environmental assets.

A Case for Whole-of-Landscape Assessment and Planning

The terms *whole-of-landscape assessment* or *landscape-scale assessment* are used to indicate strategic and bioregional approaches for assessing proposed landscape developments (Australian Government, 2015). These approaches differ strongly from the species-by-species protection or project-by-project approaches that continue to be the most common ways of assessing the probable impacts of unconventional gas exploration and development in Australia. By contrast, the Environmental Impact Statement (EIS) approach cannot deal effectively with the cumulative impacts of multiple developments over time or space, and often leads to (Dovers, 2002) 'death by 1000 cuts'!

There is an extensive literature on how to manage landscapes as integrated systems – how to use knowledge of landscape processes to work out, upfront, where urban development, mining or agriculture can safely operate without compromise to water resources, biodiversity, other land uses or landscape environmental function. While a systematic application of such approaches has proved difficult (Ashton, 1998; Bellamy *et al.*, 2002) there is emerging evidence in South Africa and Australia that such approaches are worthwhile. Government, industry and the community should be able to use them to manage the increasing complexity of future land-use decisions (Ashton, 1998; Curtis and Lefroy, 2010; Curtis *et al.*, 2014; McKenzie, 2013; Roux *et al.*, 2006; Wilde, 2013).

Whole-of-landscape assessment is best seen as an essential component of robust strategic regional land-use planning. It is a process that can help society understand how the geochemistry, hydrology and ecology of a region interact to generate the landscape's biophysical functionality. It links readily to land-use planning for an individual catchment (or watershed), by using principles of integrated catchment (watershed) management (ICM) (Ashton, 1998) to model a mosaic of appropriate land uses given the underlying capacity of natural systems to support a desired set of values.

The need for a '*whole-of-system analysis*' and '*cumulative risk assessment*' for regions prospective for unconventional gas was highlighted by Williams *et al.* (2012). They recommended that the approach used for assessing unconventional gas developments (and any other developments) should be, first, to understand the regional landscape's capacity and then to determine whether there is capacity for the development without crossing landscape limits which impact on long-term landscape functionality. Second, they recommended that the current development approval processes should be updated to approve new developments only on the basis of landscape limits and the expected cumulative impacts of the existing and proposed developments.

These recommendations were consistent with and supported by concerns from scientists, economists, engineers (Poisel, 2012; Randall, 2012) and the community (ANEDO, 2013) about the potential impacts of unconventional gas exploration and production (e.g. Moran and Vink, 2010; National Water Commission, and CSIRO, cited in Williams *et al.*, 2012, Chapter 3).

Key Issues to Consider in Whole-of-Landscape Assessment and Planning for Unconventional Gas Exploration and Development

All land uses – not just gas operations – have cumulative impacts on the functioning of the landscape and ecosystems, and it is vital to see those impacts as a whole. This is exactly what can be done using a whole-of-landscape assessment approach for land-use planning. In relation to gas operations, however, there are specific potential impacts which need to be considered, separately and together. Such issues in Australia can differ from those associated with gas activities overseas, partly because Australia has a small population occupying a huge and very ancient landscape. The summary presented below is based on Williams *et al.* (2012). Although action is now being taken to foresee or manage the occurrence of some of or all these issues, it is sensible to keep the whole range of issues in mind as a context, particularly if other countries wish to draw on Australian experience.

Land Surface Impacts on Biodiversity

Unlike the production of conventional oil and gas, the production of coal seam gas and shale gas involves drilling many wells a few tens of metres deep and relatively close together. That imposes a large 'footprint' on the land surface, which, by its scale and nature, cuts across landscape and biological habitat (see Figure 21.1). Some original coal seam gas developments in Australia have an average of 1.1 well pads and 1.6 kilometres of road per square kilometre of land (Eco Logical Australia, 2012). The density of the wells, particularly for coal seam gas, facilitates depressurising the coal seam and driving the desorption of methane from the coal matrix (Freij-Ayoub, 2012). Across a gas field, thousands of wells potentially need to be connected by pipelines and roads. Fortunately, multidirectional drilling technology is now helping to reduce those numbers: the fewer the well pads, gas-gathering systems, road and firebreaks, the smaller the intrusion on habitat and other land uses.

Many land uses, not just unconventional gas projects, require vegetation to be cleared, thus introducing invasive species, including weeds, bringing in extra traffic and noise, interrupting long-established traditional or heritage practices and rights and fragmenting habitat patches.

A number of scientific studies (Ries *et al.*, 2004; Saunders *et al.*, 1991) have confirmed that native fauna and flora are negatively affected when native vegetation areas across a landscape are fragmented. Evidence from coal seam gas developments to date indicates that severe effects are possible, particularly in landscapes that have already been extensively cleared for other productive uses. Fragmentation of remaining blocks of native vegetation is yet to be dealt with adequately in the policy and regulatory environments of Australian governments, either state or federal. The clearing of native vegetation needs to be managed in a way that balances the needs of and the permissions available to landholders, gas and mining companies and other developers, without threatening biodiversity. A whole-of-landscape approach for assessment and planning would provide a powerful long-term mechanism to deal with this very important issue in Australia.

Land Surface Impacts on Food and Fibre Production

Food and fibre production is perceived to be at risk from the cumulative fragmentation involved in developing an unconventional gas field. There are fears of loss or contamination of strategic agricultural land and its vital water resources. To date, the cumulative

Figure 21.1 An aerial view of roads and other infrastructure in a coal seam gas field near Dalby State Forest, southern Queensland. The image is 6.8 km wide. *Source:* Eco Logical Australia (2012).

fragmentation of productive land has not received enough attention in overall landscape assessment and planning (ANEDO, 2013).

Unconventional gas production generally compromises the landscape for productive agricultural and pastoralist activities (which now potentially include carbon sequestration) and is a vexed issue in relation to irrigated agriculture; however, the irrigation use of treated 'produced water' from coal seams has shown promise in short-term trials. Extensive grazing is one form of food and fibre production that may be better than others at co-existing with gas production (Williams *et al.*, 2012, p. 35).

A balanced co-existence of gas production and the various forms of agriculture and forestry should be possible with careful management (Williams *et al.*, 2012). For such a co-existence to extend over the long term, research in Australia suggests that a whole-of-landscape assessment will be an essential component of robust strategic regional land-use planning. To evolve a mosaic of appropriate land uses, given the underlying capacity of natural systems to support a desired set of values, will require mechanisms such as integrated catchment (watershed) management (ICM) to be implemented within a regional governance framework. This has been attempted recently in New South Wales (NSW) (NRC, 2010, 2012a, 2012b; Wilde, 2013).

Impacts on Communities

New developments can be seen as positive if they bring jobs and population into an area to boost the use of local schools and businesses and to help address regional underemployment (e.g. Rolfe *et al.*, 2011). On the negative side, state governments have been perceived to discount the quality of the surface soils owned by landholders, when committing the land beneath the topsoil to mining and energy companies. In Australia, landholders' rights relate only to the topsoil; ownership of all beneath is retained by the Crown. Through this double ownership, landholders generally have no legal standing when trying to prevent mining companies from entering their 'freehold' land. Compensation may not be sufficient to overcome landholder resistance, which is an expression also of perceptions of negative impacts on amenity and inadequate benefits for their neighbours and their communities.

Building trust is key to securing a 'social licence to operate' for any major resource project, including unconventional gas operations, and it is important to have a transparent approach in the collection and dissemination of reliable data (NSW CSE, 2013). Communities are more likely to accept information as credible if it comes from a source perceived to be truly independent (Lacey *et al.*, 2012; Lloyd *et al.*, 2013). Involving local people and landholders in the collection and understanding of environmental monitoring data has also been shown to increase trust (see Chapter 19 of this volume).

Social research suggests that there are better opportunities for the unconventional gas industry if it makes a direct financial return to the communities most affected by its operations, improves communication and collaboration with stakeholders and invests in infrastructure. These approaches facilitate ongoing access and strengthen the social licence to operate (Petkova *et al.*, 2009). The challenge for the industry is to articulate an agenda that balances its own commercial needs with broader expectations about contributions to the development of affected communities and regions.

The situation is recognised in some initiatives. In Western Australia, for example, it was recently reported that landholders and oil and gas companies have reached an agreement on the principles of land protection and access (WA DMP, 2015b).

In the NSW Namoi River catchment, past experience (Eco Logical Australia, 2012) has shown that a balanced co-existence of unconventional gas development and various forms of agriculture should be possible, with careful management supported by bioregional planning and cumulative risk assessment. Indeed, the Namoi region is to be the case study area for a new Land-Use Conflict Taskforce Project (Foley, 2015). The Namoi catchment is currently used for grazing, broad-scale cropping, horticulture and forestry and for growing irrigated cotton and wheat; more than half the Namoi River basin is covered by native vegetation (Welsh *et al.*, 2014).

For the co-existence of communities and new land-use developments, including gas operations, it is necessary to have open dialogue, respect and transparency. It will also be important that the community has confidence not only that unconventional gas operations and effects are being effectively monitored but also that concerns will be identified and remediated, or operations stopped, before a serious problem arises. In the case of shale gas, many of the most prospective areas are subject to Native Title or are Designated Aboriginal Lands and it is important to ensure that traditional owners are aware of the nature and scale and possible impacts of shale gas developments from the start (Hawke, 2014). The industry also has the potential (Trigger *et al.*, 2014) to help address the aspirations of Aboriginal people to build greater economic self-sufficiency.

Potential Impacts on Surface Water Resources and Aquatic Ecosystems

The management of unconventional gas to minimise its impact on groundwater has received widespread attention. Cronshaw and Grafton (2016, Chapter 15 in this volume) and Cox (2016, Chapter 20 in this volume) deal with these matters at length. Here we direct attention to impacts on inland surface waters and their ecosystems.

Water resources are affected in different ways by coal seam gas extraction and shale gas extraction. In Australia, for seam gas production, groundwater is generally extracted from the coal seam or overlying geological strata in order to release the pressure holding the gas in the coal. The water produced in this process is usually of low quality and large volume, and its storage and/or disposal at the ground surface present significant costs and management challenges. Shale gas and tight gas are usually found in strata a kilometre or more below ground. To extract these gases, water is pumped into the well to produce pressure to fracture ('frac') the rock and release the gas. In arid or semi-arid areas or in otherwise dry conditions, access to sufficient water for fracking is not necessarily easy. It is likely to require a water licence, possibly adding to existing competition for licences for urban, agricultural or other industrial uses.

Surface waters (rivers and streams) in Australia have naturally low (and variable) discharge and flow regimes. Extracting large volumes of water from them leads to severe ecological impacts – this is one reason why extractions are licensed and regulated in populated regions. Similarly, these surface waterbodies cannot be used as receiving zones for water that has been pumped to the ground surface, whether it is groundwater extracted to depressurise a coal seam or water recovered after deep fracking. Such produced or recovered water is often brackish, saline or contaminated with substances such as metals and radionuclides, which can be toxic to plants, animals and humans (Batley and Kookana, 2012; Moran and Vink, 2010; NWC, 2011, 2012; QWC, 2012; Vink *et al.*, 2008). Contaminated produced water needs careful storage and transport or treatment. It cannot be spilt or leaked into crops, native vegetation or surface waters. Even after treatment of the water, its disposal into natural streams can affect stream ecosystems if it is not matched to stream temperature and natural flow regimes (Levick *et al.*, 2008; Smythe-McGuiness *et al.*, 2012), which can vary from no-flow to flood. Disposal, even of clean water, is not simply a matter of emptying it into the nearest stream because, depending on the volume, timing and quality, such actions can have negative effects on the ecological health and biodiversity of that stream.

The water itself can be treated, by reverse osmosis for example, and in some cases treated water is pumped to water storages for use in agricultural production. It is still not clear how the loads of salts resulting from that treatment are being dealt with. Salt management remains a significant challenge for industry (Khan and Kordek, 2014).

The recharging of drained coal seams is a further matter which has not received a great deal of attention. Once coal seam gas production has ceased the seams need to be recharged with water, by for example re-injection, but the source of the recharge water is not clear. The original 'produced water' is likely to have been disposed of long before.

The report of the Chief Scientist (NSW CSE, 2014) observes: 'There is a need to understand better the nature of risk of pollution or other potential short- or long-term environmental damage from [coal seam gas] and related operations, and the capacity and cost of mitigation and/or remediation and whether there are adequate financial mechanisms in place to deal with these issues.' Thanks to the increasing amount of work in Australia becoming available on the subject of coal seam gas production in relation to the protection of water resources, it is now clear that the potential impacts of coal seam gas on water resources are significant, require very careful attention and merit being the focus of much public concern, governance, regulation and monitoring, as covered elsewhere in this volume.

Legacy Issues

An issue for the future will be how to maintain the integrity of wells that are no longer in use. By 'integrity' we mean the complete separation and sealing-off of the well structure from the strata and aquifers it intersects.

The gas in a coal field is finite, and when it runs out there will be thousands of wells to be decommissioned. Unless the abandoned wells retain their integrity into the future, connections between strata, kilometres long, which may include confined aquifers and water-bearing materials with very different chemistries, could lead to currently unforeseen problems. There is also some risk of fire from leaking hydrocarbon gases. Practicable and assessable decommissioning, regulation and governance of defunct gas wells must be given careful attention. These wells will be in the landscape for ever.

This matter is also touched on by the NSW Chief Scientist's report (NSW CSE, 2014, p. 10): 'Legacy issues, including better understanding of inappropriately abandoned wells, need attention.'

Steps Towards Whole-of-Landscape Assessment and Planning

There are indications that Australian governments are exploring assessment procedures that shift away from individual project-by-project approvals, which generally fail to take into account the long-term cumulative impact of multiple developments.

First steps are being taken towards a more strategic and long-term approach to guiding the development and sustainable use of natural resources and managing the collective impacts of developments on the environment and community wellbeing. Implicit in this is that non-statutory whole-of-landscape assessment and landscape planning informs and appropriately binds subsequent statutory planning concerned with the legally enforceable regulation and management of changes to land use and development.

By 2012, there was some acceptance of the need for 'comprehensive scientific catchment-based analyses of the ecology, hydrology and geology of an area – that is, comprehensive bioregional assessment' (quoting Williams *et al.*, 2012, p. 43). In New South Wales and Queensland the state governments had each set up strategic regional land-use planning to look at land-use conflicts and the Queensland Government had passed the *Strategic Cropping Land Act 2011* (since replaced by the *Regional Planning Interests Act 2014*).

The Australian Government had responded to community concerns about coal seam gas and coal mining by forming the Independent Expert Scientific Committee (IESC) in 2012 under the *Environmental Protection and Biodiversity Conservation Act 1999*, to examine water-related impacts from coal seam gas and large-scale coal mining. Other inquiries were under way or completed.

The establishment of the IESC was an important step forward; the committee provides advice on coal-mining and associated developments that could affect water resources (IESC, 2015a). Of critical importance will be the development of a formal nexus between the work of the IESC and the state regulatory processes for land-use planning, particularly in NSW and Queensland, the states with most coal seam gas operations. The exchange and procurement of relevant information, and the capacity to deliver that information in regional strategic planning, will be of paramount importance.

At the time of writing, Australian state and federal governments and the mining industry are increasingly announcing initiatives in joint and cumulative assessments of landscapes, and potential impacts on them from developments, as discussed below. Guidelines are being produced for industry and for assessment, and there is a stronger focus on data gathering and compilation. Known risks are being noted and that there is an intention to manage them.

Some Ways Forward

Bioregional Assessments

The Australian Government through the IESC has begun a program of bioregional assessments, to understand 'the potential impacts of coal seam gas and large coal mining developments on water resources and water-related assets' (AGBAP, 2015). Six bioregions in eastern and central Australia are currently being assessed; all have significant coal and gas deposits (Figure 21.2).

Bioregional assessment studies are involving scientists from a range of bodies, including federal government departments and statutory research institutions. They are guided by a published set of methods (Barrett *et al.*, 2013) and are intended to develop 'multi-layered records of the natural environment in specific bioregions'. The studies are looking at the 'ecology, hydrology, geology and hydrogeology of bioregions' so as to explore potential effects, both immediate and cumulative, at the ground surface and below ground, and they are not confined to individual sites (IESC, 2015b). Local communities are involved in this as well, contributing expertise in natural resources management, catchment management and traditional Indigenous knowledge.

The assessments are progressively producing and publishing data, asset databases, descriptions, registers of water-dependent assets, geographic information, models, diagrams, analyses and assessments of risks and potential likelihoods and impacts from coal-related operations relevant to water resources, but they are not at present considering the risks and impacts to economic or social assets.

While state government agencies may cooperate in this information gathering, the bioregional assessments are independent of the state-based regional land-use planning processes by which statutory land use is determined.

The Bioregional Assessment Programme intends its products to be 'a single authoritative source for all interested parties to refer to when considering the water-related impacts of potential coal seam gas and coal mining developments' (AGBAP, 2015). That includes the IESC itself, when it is advising 'government regulators on the water-related impacts of specific development proposals'. This appears to be an exercise not just in identifying situations, issues and potential issues but also in keeping the public informed.

The assessments aim to understand the scale of the impacts of coal seam gas developments on other surface and groundwater resources, with confidence limits, and to propose how to monitor and review and make further studies towards ensuring that impacts on water resources are minimal.

Targets that could be affected in a bioregion and assessed by this programme are, for instance, important wetlands or vegetation or fauna species that could be affected by changed water quality or availability. Availability can include 'changes in baseline variables, flow regimes, hydraulic salinity' and other impacts could come from eco-toxicity or effects on human health or from a reduction in water quality by contaminants (Barrett *et al.*, 2013).

After compiling what is known of the occupants, landscape and geology of a bioregion (its context), the intention is that the programme will produce for each region: a modelled analysis of the hydrogeology in relation to coal seams and the surface ecology; an analysis of the risks, taking account of 'uncertainties from models and data'; and a 'synthesis of outcomes' which the IESC can use when providing advice on developments in that region. Although there may be insufficient data to make a 'quantitative analysis of impacts', it is expected that the programme will provide a 'defensible baseline statement as to the current

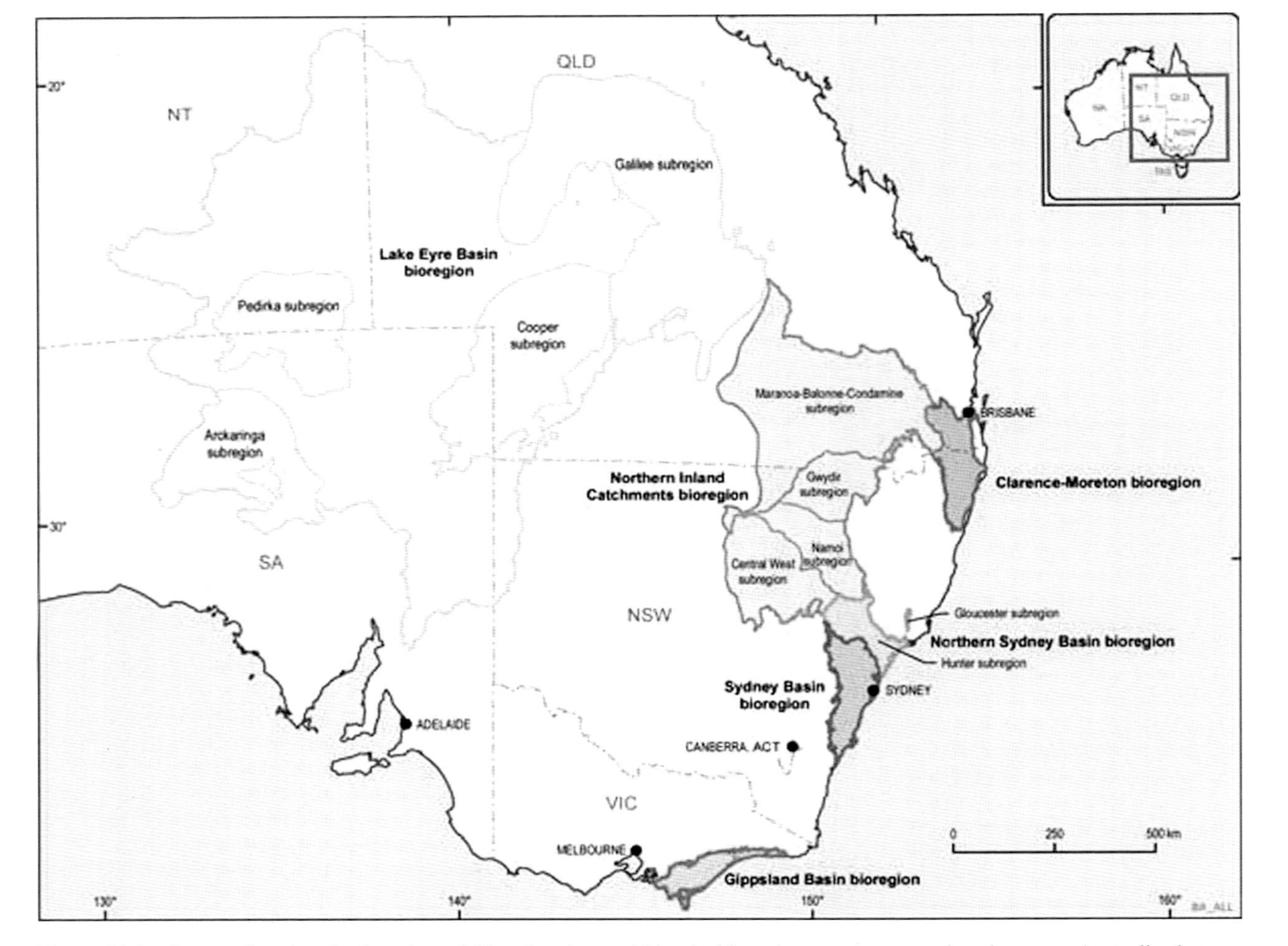

Figure 21.2 A map showing the location of 13 subregions within six bioregions across central and eastern Australia that are being assessed by the Bioregional Assessment Programme. *See* http://www.bioregionalassessments.gov.au/. © Commonwealth of Australia.

state of scientific knowledge on the impacts of coal seam gas and coal mining development on water resources within a bioregion and its subregions' (Barrett *et al.*, 2013).

A limitation, within the context of this present chapter on unconventional gas, is that the Bioregional Assessment Programme focuses only on 'coal and coal seam gas resource developments' and, therefore, does not consider other unconventional gas systems; nor is the programme set up to consider whole-of-landscape assessment and planning. Nevertheless, it is an important step forward in the evolution of the process in Australia.

Multiple Land Use Framework

The Standing Council on Energy and Resources (SCER), now known as the Council of Australian Governments (COAG) Energy Council, has produced a Multiple Land Use Framework (MLUF). It is intended 'to enable government, community and industry to effectively and efficiently meet land access and use challenges, expectations and opportunities' (SCER, 2013a), and distinguishes between multiple land uses (at the one time) and sequential land uses.

This framework has been developed to help solve problems of land access and conflicting land uses, particularly related to mining and energy resources, within existing state policies and regulations. However, 'the underlying concept can extend to all sectors, interests and values including but not limited to agriculture, minerals and energy resources, environmental, heritage, cultural, tourism, infrastructure, community and forestry. The framework supports the ability of local and regional communities and governments to maximise land use in a flexible, environmentally sustainable, manner over time' (SCER, 2013a).

The framework document claims that: 'The MLUF is a fresh approach to changes in land use. By reducing tensions that can arise between stakeholders, we achieve a better economic, social and environmental outcome that leads to sustainable outcomes for future generations' (SCER, 2013a).

In the separate 'background' document (SCER, 2013b), the MLUF (Figure 21.3) is summarised as: 'four desired outcomes; eight principles to guide land access and land use decisions; and nine components to consider in planning, preparing and assessing land access and land use decisions.' It is expected to improve: recognition of the possible needs and benefits of land uses; informed discussions and confidence among communities and landholders; and land-use choices based on the land's potential and the consequences of uses (SCER, 2013b). The document provides case studies from across the world, as well as details of the research leading to the MLUF itself, and the resulting principles.

The state and territories resource agencies will be the drivers for designing the implementation model. It appears that the Australian Petroleum Production and Exploration Association Ltd (APPEA) strongly supports the MLUF (Byers, 2014).

In principle, the MLUF could provide the framework for implementation of whole-of-landscape assessment and planning, as we have outlined. It is perhaps rather early to see whether it might evolve in that manner.

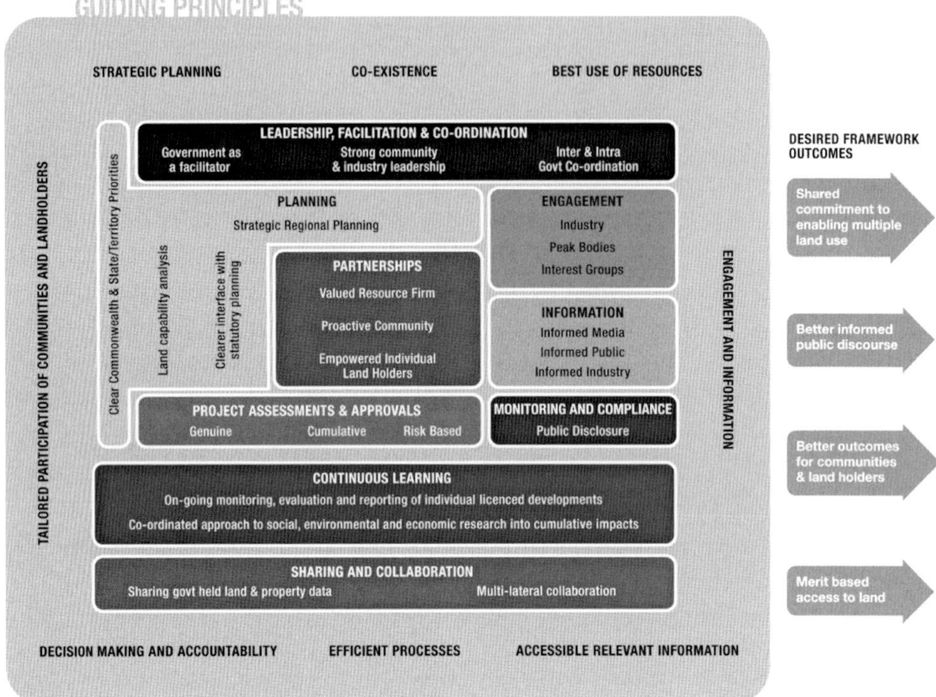

Figure 21.3 The concept diagram for the Multiple Land Use Framework developed by The Standing Council on Energy and Resources (SCER), now known as the COAG Energy Council. *Source:* SCER (2013b, p. 32).

Integrated Catchment Management Leading to Integrated Development Management

How can whole-of-landscape assessment and planning evolve further, as a means for sustainably incorporating unconventional gas development into the Australian landscape?

The answer is that regional governance will need to be adopted as a means of implementation. Assessment and land-use planning must be *owned* by the communities that live in the landscape.

Experience, internationally and in Australia, provides evidence that communities where there are effective and harmonious interactions between science, society, industry and government are more likely to achieve the ideals of integrated catchment management (Roux *et al.*, 2006). An outcome of that is 'integrated development management', occurring in socially acceptable ways that promote political stability and human wellbeing (Ashton *et al.*, 2006; NRC, 2010, 2012a, 2012b; Roux *et al.*, 2006; Wilde, 2013). The success of the above-mentioned interactions depends on effective and trustworthy interpersonal relationships between individuals and institutions in which knowledge and experiences is shared in a unified learning system (Curtis *et al.*, 2014). This allows all participants to move beyond traditional roles of commander-controller, knowledge provider and knowledge consumer, to a true partnership where interdependencies are recognised (McKenzie, 2013) and all parties can negotiate feasible, desirable and acceptable outcomes (Roux *et al.*, 2006).

The challenge in Australia now is to take the strong conceptual basis of ICM and build institutions, governance and social processes that pioneer a new era in managing the environment and development, leading to integrated development management (Ashton, 1998). Regionally based institutions are able to access community support, local knowledge and relevant scientific expertise.

The advantages of regional management as a concept were recognised by T. A. Lang, a famous water engineer in Queensland, in 1944. He said (cited in Powell, 1993, by Wentworth Group, 2014): 'there are sound reasons for adopting a regional basis, rather than a political one, when planning the development and management of natural resources'.

As the Wentworth Group has said more recently, 'Each region has unique features that define it and create a shared identity and sense of purpose amongst its occupants. These interests transcend political boundaries . . . The benefit of a regional model is that it operates at a scale large enough to manage the pressures on our landscapes, yet it is small enough to use local knowledge to tailor solutions to suit those landscapes. It produces better results for taxpayers, as well as supporting economic opportunities and social benefits that a healthy landscape provides to many rural, coastal, and urban communities' (Wentworth Group, 2014).

Australian governments have long debated the value of incorporating regional concepts into the planning process. In recognition of this, over the past decade Australia has moved to a regional model for natural resource management. Integrated management of these complex biophysical, economic and social issues, at the regional scale at which these landscape processes function, is an essential precondition to a healthy and productive environment, regardless of whether people live in urban, coastal or rural regions. There are 54 regional natural resource management bodies currently working across Australia. They involve landholders, citizens, industry and agencies of governments in their work to achieve healthy and productive landscapes that support local communities and those industries that depend on a healthy environment (NRC, 2010; Wentworth Group, 2014; Williams, 2012).

The regional model (Figure 21.4) has proven to be a cost effective means of bringing science, local communities and governments together, so that people become directly involved in coordinating on-ground investments in their landscapes.

Regional planning should have the additional benefit of being able to coordinate actions across government agencies, industry, philanthropic investors and non-government organisations. To achieve these outcomes will not be easy. At the heart of this task ahead is the need to build the capacity for and delivery of strategic regional land-use planning which is executed collectively and owned, respected and implemented by all sectors of the community, industry and government.

Regional Catchment Action Plans (NRC, 2012a, 2012b; Wilde, 2013; Williams, 2012) are strategic regional plans for improving the health, productivity and resilience of our landscapes. These were pioneered in NSW over a period of over ten years with Catchment

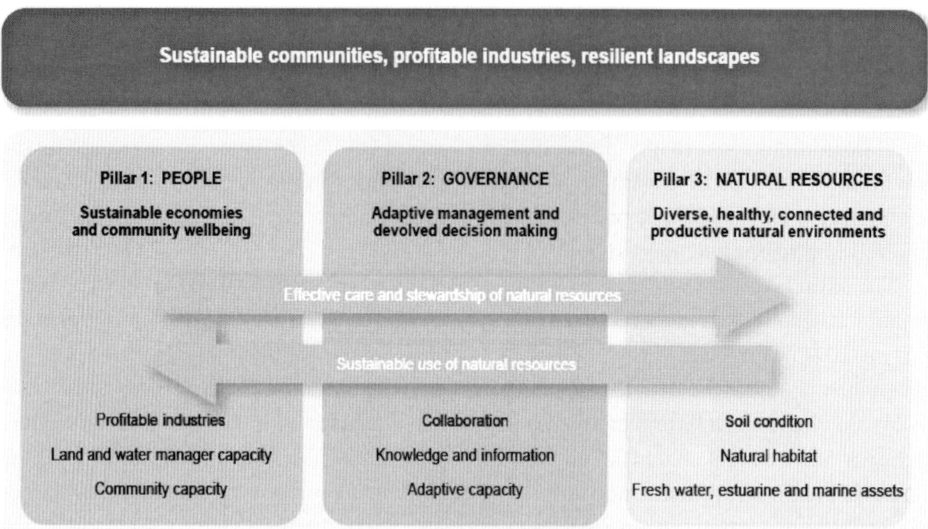

Figure 21.4 Southern Rivers Catchment Action Plan framework. *Source:* Wilde (2013, p. 5).

Management Authorities (NRC, 2010) which were responsible for their creation and development along with community, industry, agencies of state governments and the Australian Government. The Catchment Action Plans were regularly audited by the NSW Natural Resources Commission (NRC, 2012a, 2012b). These plans identified within a spatial and temporal framework what the community and governments value about these landscapes and explain what needs to be done to ensure the long-term, sustainable, management of a region's natural resources. This is perhaps the most tractable non-statutory planning framework that has been implemented so far in Australia. It could evolve to incorporate whole-of-landscape assessment and planning (Wilde, 2013) on the basis of a realistic whole-of-system framework. Such a framework should enable stakeholders to foresee some of, if not all, the interactions likely from multiple land uses and their effects on the natural functioning of landscapes.

As acknowledged by, e.g., the NSW Chief Scientist and Engineer (2013, 2014), stakeholders need to understand what a regional landscape is capable of sustaining, so that they can model its limits to development without losing function. Social and legislative approval systems need to relate closely to landscape limits that are based on well-designed and active in-progress monitoring which supports both the assessment of risks to landscape limits and appropriate regulation and compliance arrangements. Good data about the ecology, hydrology, geology and human aspects of each region need to be collected (e.g. Minerals Council of Australia, 2015) and combined with suitable and vigilant regulations and governance.

The framework for whole-of-landscape assessment and planning as incorporated into regional-catchment action plans would appear, then, to meet much of what has been learned in the pursuit of better incorporation of unconventional gas production into the sustainable use of Australian landscape. The components are available. The challenge will be to evolve the policy and governance to pull it all together.

Practical Cumulative Risk Assessment

As well as the MLUF and the Bioregional Assessments Programme initiatives already mentioned, there are a number of other examples where cumulative impacts are being noted and planned for, around Australia, at a government level, in the minerals industry itself and in a ground-breaking instance at regional level in 2011.

Examples

In Western Australia the State Government has recently announced a regulatory framework for shale and tight gas which is intended to protect Western Australian waters, environment and public health (WA DMP, 2015a).

In NSW, the Chief Scientist's report in late 2014 (NSW CSE, 2014) recommended (Recommendation 8): 'That Government move towards a target and outcome-focused regulatory system . . . [including]: regularly reviewed environmental impact and safety targets . . . ; appropriate and proportionate penalties for non-compliance; automatic monitoring processes that can provide data (sent to and held in the openly accessible Whole-of-Environment Data Repository) which will help detect cumulative impacts at project, regional and sedimentary basin scales which can be used to inform the targets and the planning process'. The report also warned that there would need to be 'significantly more data' collected and that there was 'a need to understand better how the different resources and their development regimes interact'. The NSW State Government accepted all the recommendations (Clayton Utz, 2014).

Also in NSW, in October 2015, the Government set up a Land Use Conflict Taskforce, 'to develop options for resolving land-use conflict in targeted hot spot areas throughout NSW'. The group is to focus on 'agricultural, community and mining interests especially in view of mining and petroleum developments currently progressing through the planning assessment system'. Further, it is to 'develop, assess and present options to relevant Ministers' and is to report to the departments of the State Premier and to the departments of planning, environment, industry, primary industries and heritage (Foley, 2015).

Other changes to strengthen the governance and regulation of unconventional gas and multiple land use in NSW include the Regional Strategic Land Use Policy, which the Government created as 'a system to protect strategic agricultural land and water resources, and create jobs and investment for regional communities by providing certainty for landholders and mining companies' (NSWG, 2015).

In November 2015 changes were proposed for the NSW Government's mining and petroleum laws in the near future. If passed, they would be likely to alter aspects of the administration of mining and petroleum titles in New South Wales, including compliance and land-access arrangements for exploration titles (Warwick Giblin, OzEnvironmental Pty Ltd, personal communication, 20 November 2015).

In the minerals industry itself, the Minerals Council of Australia published the *Cumulative Environmental Assessment Industry Guide* (MCA, 2015). It is intended to 'assist individual proponents/companies in conducting well-designed, leading practice cumulative environmental impact assessments', and also to be 'useful to regulatory bodies and other interested stakeholders in understanding cumulative impact assessments'. The guide was prepared by two environmental consulting companies, one being Eco Logical Australia; its chapters outline when and how to make a successful cumulative impact assessment that is fit for purpose and provides case studies and summaries of relevant legal cases.

At a regional level, the natural resources management body responsible for the Namoi River catchment took action in 2011 (Williams *et al.*, 2012). Working with the consulting group, Eco Logical Australia, the Namoi Catchment Management Authority (CMA) led the way into cumulative management frameworks. The team demonstrated a holistic and strategic assessment of the cumulative impacts of coal seam gas and mining in that region, based on a framework they had developed for assessing cumulative risks that mining posed to their natural resource assets (Eco Logical Australia, 2011; Williams *et al.*, 2012, p. 90).

The Namoi catchment area has large coal reserves. Thus, the region is under pressure to allow additional coal mining as well as the production of coal seam gas. The CMA could see that, in the long term, the region's mineral resources could be a threat to the natural resource assets of the catchment but could also potentially bring the region substantial economic benefits.

The CMA decided to do a risk assessment of the natural resources impacts of individual mineral development projects in the catchment and also to examine the impacts that *could* result from a number of mining developments occurring at the same time. They had the data they needed about the natural resources of the region, having completed the regional catchment planning process and developed the Namoi Catchment Action Plan to identify objectives and targets for the catchment.

The strategic vision the CMA had for that catchment provided the framework for the risk assessment. Bioregional assessments and the MLUF are potentially providing similar frameworks now for other landscapes in relation to mining and coal seam gas development.

As Williams *et al.* (2012) described, the CMA staff identified the natural resource asset classes and the datasets available for each class. These formed the basic building blocks for the risk assessment process (Table 21.1).

The framework that the CMA developed (Figure 21.5) can assess both the risks associated with an individual project and the cumulative risks of any new project or projects when added to the existing pressures on the natural resources (Williams *et al.*, 2012, p. 91).

Using this framework and a GIS modelling tool, the CMA produced a cumulative risk statement on the individual and cumulative impacts associated with any real or hypothetical mining scenario and listed the potential outputs that the tool would produce for a hypothetical scenario of four new mines in the catchment.

Table 21.1 *Natural resource assets suitable for risk assessment in the Namoi catchment*

Theme	Asset	No. datasets (Appendix I)
Land (Agriculture)	Land use	20
	Soil type	17
Biodiversity	Threatened species	15
	Viable populations (connectivity)	5
	Vegetation condition (intactness)	21
	Vegetation extent (cover)	6
	Vegetation type[a]	8
Water	Groundwater	15
	Surface water	18

Source: Table redrawn from Eco Logical Australia (2011, p. 5).
[a] Includes wetlands and groundwater-dependent ecosystems

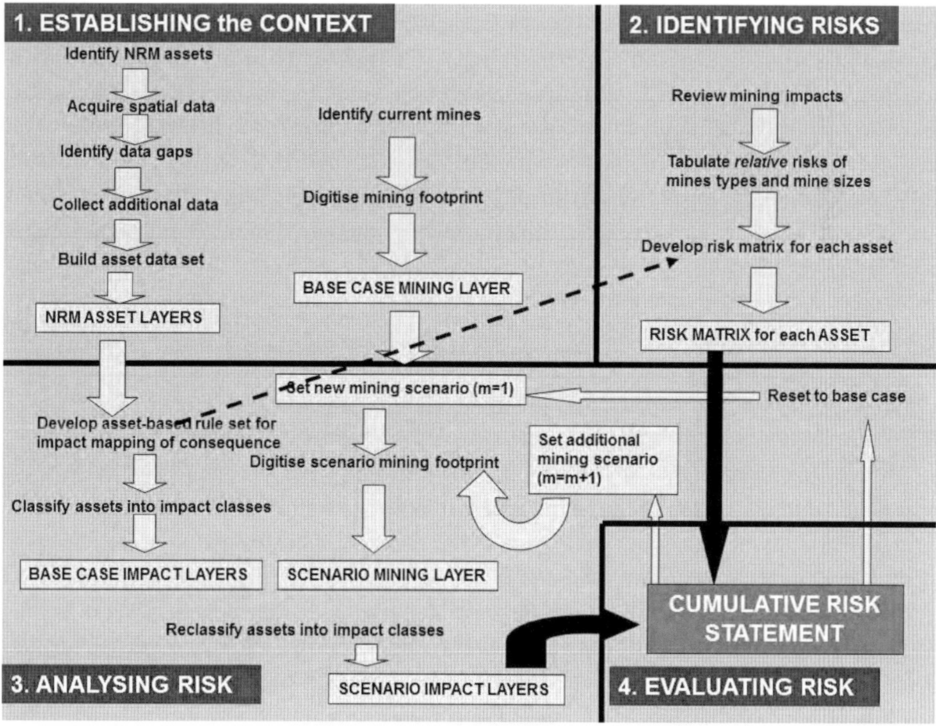

Figure 21.5 Namoi CMA proposed framework for cumulative risk assessment. *Source:* Williams *et al.* (2012, p. 91).

The kinds of tool developed by the ecological consultants for the Namoi CMA and for the Minerals Council of Australia (MCA, 2015) should enable mining and coal seam gas developers to run a range of scenarios to determine how best to structure their operation to minimise, or remove completely, any negative impacts on the natural resource assets of a landscape or region.

Conclusion

This chapter has outlined the need for, and progress being made towards, cumulative and whole-of-landscape assessment of the risks associated with new developments in landscapes in Australia, particularly in relation to the unconventional gas potential.

It is evident that the need to model and plan for the possible impacts of multiple developments is recognised at the highest levels in the Australian governments and the mining industry. Information and analyses, legislation, frameworks and regulatory mechanisms are coming into being and should help policy, governance, the community and the industry to manage those and other important issues. Cumulative risk analysis and whole-of-landscape frameworks are becoming recognised as a desirable first tool for strategic environmental assessment before unconventional gas fields are developed.

Concurrently with the development of unconventional gas in Australia, state and federal governments and the minerals industry have sought solutions to environmental assessment and built frameworks for bioregional assessment, multiple land use and cumulative impact assessment. These types of framework seek to provide a unifying means for state agencies to move towards more comprehensive landscape assessment. In addition, there have been some pioneering developments in both Victoria (GBCMA, 2013) and New South Wales (NRC, 2012a, 2012b) in regional and catchment action planning which have, in principle, built on whole-of-landscape assessment. This pioneering approach, piloted and implemented over a ten-year period in NSW, has been governed and delivered by statutory regional catchment management authorities. The development of capacity within these regional bodies to execute whole-of-landscape assessment and planning is well documented and other regions should be able to replicate it. Many components necessary to implement whole-of-landscape assessment and planning within a regional governance mechanism have been shown to be feasible. However, it is vital not to assume all will fall into place.

Whole-of-landscape assessment and planning which incorporates cumulative risk assessment and which is delivered through a regional authority with a basis of governance that brings continuity, transparency and community, and also industry and government ownership, to the assessment and land-use planning process will require a commitment to significant and ongoing reform. It has the potential to provide the understanding of cumulative impact and an evaluation of community values against landscape assets over the long term. Success will depend on whether regulators and communities choose to use the knowledge derived from whole-of-landscape assessment and how it is valued against the opinions of vested interests.

Whatever the approach used for cumulative impact assessment and whole-of-landscape planning, it is essential that it be supported by spatially adequate explicit ecological, hydrological and geological data, which are not yet available. The geological stratigraphy, hydrogeology and landscape ecology to support the natural resources information will require ongoing investment in research and programmes to document the biophysical characteristics of the landscape for some years yet. Investment in acquiring, storing and retrieving this knowledge and information using modern digital technology should be seen as part of the way in which a society builds capacity to plan its future, so that it can live more sustainably within the boundaries and functional thresholds of its landscapes. The modelling of groundwater and surface water resources and likely impacts on them must be sophisticated, transparent, based on clear assumptions and independent of industry pressures in order to achieve best practice decision-making.

Regulations and governance, in our view, are most effective when they follow and are informed by a knowledge of landscape functional processes. The tools and the social process involved in whole-of-landscape assessment and planning have the potential to manage risks and minimise the ways in which unconventional gas developments affect the functionality of our landscapes – that is, the integrity of the hydrological, geochemical and ecological processes on which humans depend.

We consider that multiple-use mineral- and food-producing basins are possible, provided that we remember the essential long-term need to retain landscape functionality. For this to be achieved the whole-of-landscape assessment and planning must inform, support and be coupled to an enabling regulatory environment.

References and Further Reading

Ashton, P.J. (1998). Integrated catchment management: balancing resource utilization and conservation. Division of Water, Environment & Forestry Technology, CSIR, Pretoria, South Africa. http://awiru.co.za/pdf/astonpeter.pdf [accessed 1 December 2015].

Ashton, P.J., Turton, A.R. and Roux, D.J. (2006). Exploring the government, society, and science interfaces in integrated water resource management in South Africa. *Journal of Contemporary Water Research & Education*, **135**, 28–35. http://onlinelibrary.wiley.com/doi/10.1111/j.1936–704X.2006.mp135001004.x/abstract [accessed 2 December 2015].

Australian Government (2015). Chapter 10: Strategic assessments and bioregional planning. Department of the Environment, Water, Heritage and the Arts. https://www.environment.gov.au/system/files/resources/5d70283b-3777–442e-b395-b0a22ba1b273/files/10-assessments.pdf [accessed 28 November 2015].

Australian Government Bioregional Assessment Programme (AGBAP) (2015). Bioregional assessments [webpage]. An inter-agency government partnership between the Department of the Environment, Geoscience Australia, the Bureau of Meteorology and the CSIRO. Available from: http://www.bioregionalassessments.gov.au/ [accessed 5 December 2015].

Australian Network of Environmental Defenders Offices (ANEDO) (2013). Submission on the Draft National Harmonised Regulatory Framework for Coal Seam Gas 2012. http://www.scer.gov.au/files/2013/03/ANEnviromentalDefendersOffice.pdf [accessed 27 November 2015].

Barrett, D.J., Couch, C.A., Metcalfe, D.J. *et al.* (2013). Methodology for bioregional assessments of the impacts of coal seam gas and coal mining development on water resources. Report to the Interim Independent Expert Scientific Committee on Coal Seam Gas and Coal Mining. http://www.iesc.environment.gov.au/publications/methodology-bioregional-assessments-impacts-coal-seam-gas-and-coal-mining-development-water/ [accessed 27 November 2015].

Batley, G.E. and Kookana, R.S. (2012). Environmental issues associated with coal seam gas recovery: managing the fracking boom. *Environmental Chemistry*, **9** (5), 425–428.

Bellamy, J., Ross, H., Ewing, S. and Meppem, T. (2002). *Integrated catchment management: learning from the Australian experience for the Murray–Darling Basin*. Final report, January 2002. CSIRO Sustainable Ecosystems, Canberra, Australia. http://www.mdba.gov.au/sites/default/files/archived/mdbc-NRMreports/2234_ICM_Learning_from_the_Aust_exp_MDB.pdf [accessed 2 December 2015].

Byers, D. (2014). Letter to the Agricultural Competitiveness Taskforce, Department of the Prime Minister and Cabinet re: Agriculture Competitiveness Green Paper. Australian Petroleum Production and Exploration Association Limited (APPEA), 16 December 2014. http://agwhitepaper.agriculture.gov.au/GP%20Submissions%20for%20publication/GP304%20Australian%20Petroleum%20Production%20and%20Exploration%20Association.pdf [accessed 5 December 2015].

Clayton Utz (2014). NSW opens the door to more CSG projects, but more details to come [webpage]. Available from: http://www.claytonutz.com/publications/news/201411/14/nsw_opens_the_door_to_more_csg_projects_but_more_details_to_come.page [accessed 27 November 2015].

Cox, R. (2016). Managing the impact of coal seam gas water extraction in the Surat Basin. Chapter 20 in this volume.

Cronshaw, I. and Grafton, R.Q. (2016). Risk and opportunities of unconventional natural gas: Australia and the United States. Chapter 5 in this volume.

Curtis, A. and Lefroy, E.C. (2010). Beyond threat- and asset-based approaches to natural resource management in Australia. *Australasian Journal of Environmental Management*, **17** (3), 134–141.

Curtis, A., Ross, H., Marshall, G. R. *et al.* (2014). The great experiment with devolved NRM governance: lessons from community engagement in Australia and New Zealand since the 1980s. *Australasian Journal of Environmental Management*, **21** (2), 175–199. http://dx.doi.org/10.1080/14486563.2014.935747 [accessed 1 December 2015].

Dovers, S. (2002). Too deep a SEA? Strategic environmental assessment in the era of sustainability. In *Strategic Environmental Assessment in Australasia*, eds. S. Marsden and S. Dovers. Sydney: The Federation Press, Chapter 2, pp. 24–46.

Eco Logical Australia (2011). Proposed framework for assessing the cumulative risk of mining on natural resource assets in the Namoi Catchment. Project 11COFNRM-0006 prepared for Namoi CMA. September 2011. http://education.nwlls.com/client/multimedia/namoi__risk_assessment_final_v5_14sept11.pdf [accessed 27 November 2015].

Eco Logical Australia (2012). Shale gas development in Australia: potential impacts and risks to ecological systems. Final report prepared for the Australian Council of Learned Academies (ACOLA). January 2013.

Foley, M. (2015). Coal conflict strategy is being developed by NSW government, *The Land*, 20 November 2015. http://www.theland.com.au/story/3506923/coal-conflicts-focus-of-new-state-strategy/ [accessed 27 November 2015].

Freij-Ayoub, R. (2012). Opportunities and challenges to coal bed methane production in Australia. *Journal of Petroleum Science and Engineering*, **88–89**, 1–4.

Goulburn Broken Catchment Management Authority (GBCMA) (2013). Goulburn Broken regional catchment strategy 2013–2019. State of Victoria, Goulburn Broken Catchment Management Authority, Shepparton, Victoria, Australia. http://www.gbcma.vic.gov.au/downloads/RegionalCatchmentStrategy/GBCMA_RCS_2013–19.pdf [accessed 5 December 2015].

Hawke, A. (2014). Report of the independent inquiry into hydraulic fracturing in the Northern Territory. Hydraulic Fracturing Inquiry, Northern Territory Government, Darwin, Northern Territory, Australia. 28 November 2014. http://www.hydraulicfracturinginquiry.nt.gov.au/docs/report-inquiry-into-hydraulic-fracturing-nt.pdf [accessed 5 December 2015].

Independent Expert Scientific Committee on Coal Seam Gas and Large Coal Mining Development (IESC) (2015a). Webpage available from: http://www.iesc.environment.gov.au [accessed 27 November 2015].

Independent Expert Scientific Committee on Coal Seam Gas and Large Coal Mining Development (IESC) (2015b). Advice on bioregional assessments [webpage]. Available from: http://www.iesc.environment.gov.au/bioregional-assessments [accessed 27 November 2015].

Khan, S. and Kordek, G. (2014). Coal seam gas: produced water and solids. Report prepared for the Office of the NSW Chief Scientist and Engineer (OCSE). School of Civil and Environmental Engineering, The University of New South Wales, Sydney, Australia. http://www.chiefscientist.nsw.gov.au/__data/assets/pdf_file/0017/44081/OCSE-Final-Report-Stuart-Khan-Final-28-May-2014.pdf [accessed 5 December 2015].

Lacey, J., Parsons, R. and Moffat, K. (2012). *Exploring the concept of a Social Licence to Operate in the Australian minerals industry: results from interviews with industry representatives.* EP125553. Brisbane: CSIRO.

Levick, L., Fonseca, J., Goodrich, D. *et al.* (2008). The ecological and hydrological significance of ephemeral and intermittent streams in the arid and semi-arid American southwest. US Environmental Protection Agency and USDA/ARS Southwest Watershed Research Center, EPA/600/R-08/134, ARS/233046.

Lloyd, D.J., Luke, H. and Boyd, W.E. (2013). Community perspectives of natural resource extraction: coal-seam gas mining and social identity in Eastern Australia. *Coolabah*, **10**, 144–164. http://epubs.scu.edu.au/cgi/viewcontent.cgi?article=2476&context=esm_pubs [accessed 1 December 2015].

McKenzie, F. (2013). Pathways to collaborative action: transforming agricultural, land and food systems. *Ecoagriculture Discussion Paper No. 10.* Washington, DC: EcoAgriculture Partners.

Minerals Council of Australia (MCA) (2015). Cumulative Environmental Impact Assessment Industry Guide. An industry guide prepared by the Minerals Council of Australia. July 2015. http://www.minerals.org.au/file_upload/files/reports/Cumulative_Environmental_Impact_Assessment_Industry_Guide_FINAL.pdf [accessed 27 November 2015].

Moran, C. and Vink, S. (2010). *Assessment of impacts of the proposed coal seam gas operations on surface and groundwater systems in the Murray–Darling Basin.* Centre for Water in the Minerals Industry, Sustainable Minerals Institute, The University of Queensland.

National Water Commission (NWC) (2011). Position Statement – coal seam gas and water. Australian Government National Water Commission, Canberra, Australia. December 2010. http://www.nwc.gov.au/__data/assets/pdf_file/0003/9723/Coal_Seam_Gas.pdf [accessed 9 December 2015].

National Water Commission (NWC) (2012). Coal seam gas update – June 2012. Australian Government National Water Commission, Canberra, Australia. http://www.nwc.gov.au/nwi/position-statements/coal-seam-gas [accessed 9 December 2015].

Natural Resources Commission (NRC) (2010). Progress towards healthy resilient landscapes: implementing the standard, targets and catchment action plans. Progress report, December 2010. NSW Government Natural Resources Commission, Sydney, Australia. http://www.nrc.nsw.gov.au/_literature_108159/2010_report_-_Full_report [accessed 1 December 2015].

Natural Resources Commission (NRC) (2012a). Standard for Quality Natural Resource Management. May 2012. NSW Government Natural Resources Commission, Sydney, Australia. http://www.nrc.nsw.gov.au/_literature_177537/Standard%20for%20Quality%20NRM [accessed 1 December 2015].

Natural Resources Commission (NRC) (2012b). Framework for assessing and recommending upgraded catchment action plans. Version 2, June 2012. NSW Government Natural Resources Commission, Sydney, Australia. http://www.nrc.nsw.gov.au/LiteratureRetrieve.aspx?ID=105039&A=SearchResult&SearchID=60953024&ObjectID=105039&ObjectType=6 [accessed 1 December 2015].

New South Wales Government (NSWG) (2015). Strategic regional land use [webpage]. Available from: http://www.nsw.gov.au/initiative/strategic-regional-land-use [accessed 27 November 2015].

New South Wales Government Chief Scientist and Engineer (NSW CSE) (2013). Initial Report on the Independent Review of Coal Seam Gas Activities in NSW. http://www.chiefscientist.nsw.gov.au/__data/assets/pdf_file/0016/31246/130730_1046_CSE-CSG-July-report.pdf [accessed 27 November 2015].

New South Wales Government Chief Scientist and Engineer (NSW CSE) (2014). Final Report of the Independent Review of Coal Seam Gas Activities in NSW. http://www.chiefscientist.nsw.gov.au/__data/assets/pdf_file/0005/56912/140930-CSG-Final-Report.pdf [accessed 27 November 2015].

Petkova, V., Lockie, S., Rolfe, J. and Ivanova, G. (2009). Mining developments and social impacts on communities: Bowen Basin case studies. *Rural Society*, **19** (3), 211–228.

Poisel, T. (2012). Coal seam gas exploration and production in New South Wales: the case for better strategic planning and more stringent regulation. *Environmental and Planning Law Journal*, **29** (2), 129–151.

Powell, J.M. (1993). *The emergence of bioregionalism in the Murray–Darling Basin.* Murray–Darling Basin Commission, Canberra, Australia.

Queensland Water Commission (QWC) (2012). Underground Water Impact Report for the Surat Cumulative Management Area. Office of Groundwater Impact Assessment (consultation draft by Queensland Water Commission, May 2012), Queensland Government Department of Natural Resources and Mines, Brisbane. http://dnrm.qld.gov.au/__data/assets/pdf_file/0016/31327/underground-water-impact-report.pdf [accessed 27 November 2015].

Randall, A. (2012). Coal seam gas – toward a risk management framework for a novel intervention. *Environmental and Planning Law Journal*, **29** (2), 152–162.

Ries, L., Fletcher Jr., R.J., Battin, J. and Sisk, T.D. (2004). Ecological responses to habitat edges: mechanisms, models, and variability explained. *Annual Review of Ecology, Evolution, and Systematics*, **35**, 491–522.

Rolfe, J., Gregg, D., Ivanova, G., Lawrence, R. and Rynne, D. (2011). The economic contribution of the resources sector by regional areas in Queensland. *Economic Analysis and Policy*, **41** (1), 15–36.

Roux, D.J., Rogers, K.H., Biggs, H.C., Ashton, P.J. and Sergeant, A. (2006). Bridging the science–management divide: moving from unidirectional knowledge transfer to knowledge interfacing and sharing. *Ecology and Society*, **11** (1), 4. Online at http://www.ecologyandsociety.org/vol11/iss1/art4 [accessed 30 November 2015].

Saunders, D.A., Hobbs, R.J. and Margules, C.R. (1991). Biological consequences of ecosystem fragmentation: a review. *Conservation Biology*, **5** (1), 18–32.

Smythe-McGuiness, Y., Lobegeiger, J., Marshall, J. *et al.* (2012). *Macroinvertebrate Responses to Altered Low-Flow Hydrology in Queensland Rivers*. National Water Commission, Canberra.

Standing Council on Energy and Resources (SCER) (2013a). Multiple land use framework. Standing Council on Energy and Resources, Canberra. December 2013. http://www.scer.gov.au/files/2013/12/Endorsed-MLUF.pdf [accessed 5 December 2015].

Standing Council on Energy and Resources (SCER) (2013b). Multiple land use framework background document. Standing Council on Energy and Resources, Canberra. December 2013. http://www.scer.gov.au/files/2013/12/Land-Access-MLUF-and-CSG-Background-Document.pdf [accessed 5 December 2015].

Trigger, D., Keenan, J., de Rijke, K. and Rifkin, W. (2014). Aboriginal engagement and agreement-making with a rapidly developing resource industry: coal seam gas development in Australia. *The Extractive Industries and Society*, **1** (2), 176–188.

Vink, S., Kunz, N., Barrett, D. and Moran, C. (2008). *Scoping study: Groundwater impacts of coal seam gas development – assessment and monitoring*. The University of Queensland Centre for Water in the Minerals Industry, Sustainable Minerals Institute, St Lucia, Queensland, Australia.

Welsh, W., Hodgkinson, J., Strand, J. *et al.* (2014). *Context statement for the Namoi subregion: Product 1.1 from the Northern Inland Catchments Bioregional Assessment*. Department of the Environment, Bureau of Meteorology, CSIRO and Geoscience Australia, Australia.

Wentworth Group of Concerned Scientists (2014). Blueprint for a healthy environment and a productive economy. November 2014. http://wentworthgroup.org/wp-content/uploads/2014/11/Blueprint-for-a-Healthy-Environment-and-a-Productive-Economy-November-2014.pdf [accessed 30 November 2015].

Western Australia Department of Mines and Petroleum (WA DMP) (2015a). *Guide to the Regulatory Framework for Shale and Tight Gas in Western Australia: A Whole-of-Government Approach*. 2015 edition. Government of Western Australia, Department of Mines and Petroleum, Perth. http://www.dmp.wa.gov.au/Documents/Petroleum/WEB_Shale_and_Tight_Gas_Framework.pdf [accessed 27 November 2015].

Western Australia Department of Mines and Petroleum (WA DMP) (2015b). New gas agreement for agriculture sector [webpage]. Tuesday, 3 November 2015. Government of Western Australia, Department of Mines and Petroleum, Perth. http://www.dmp.wa.gov.au/News/New-gas-agreement-for-16635.aspx [accessed 27 November 2015].

Wilde, B. (2013). Strategic planning on the coast: the benefits of applying systems and resilience approaches. In *Proc. 22nd NSW Coastal Conference 2013, Port Macquarie*. NSW Government Natural Resources Commission, Sydney, Australia. http://www.coastalconference.com/2013/papers2013/Bryce%20Wilde.pdf [accessed 30 November 2015].

Williams, J. (2012). Catchment management – setting the scene: an overview of catchment management models in Australia. *Water*, April 2012, 1–5.

Williams, J., Stubbs, T. and Milligan, A. (2012). An analysis of coal seam gas production and natural resource management in Australia. A report prepared for the Australian Council of Environmental Deans and Directors by John Williams Scientific Services Pty Ltd, Canberra, Australia. http://acedd.org.au/wp-content/uploads/2013/05/CSG-Analysis-Report.pdf [accessed 27 November 2015].

22

Unconventional Energy in British Columbia: A Post-*Tsilhqot'in* View

WILLIAM NIKOLAKIS

Introduction

The landmark decision by the Supreme Court of Canada (SCC) in *Tsilhqot'in Nation vs. British Columbia*[1] ("*Tsilhqot'in*"), confirmed for the first time in Canada's legal history a claim for Aboriginal title. The *Tsilhqot'in* decision arrived at a time in British Columbia (BC) of significant attention to unconventional energy. In 2013, the BC Liberal Government was re-elected on a platform that included an LNG strategy. This strategy aims to spur upstream and downstream development of the province's gas resources, to capitalise on Asia's 'gas boom' (Carr, 2012). Given the glacial pace of the modern treaty-making process, many of BC's First Nations will be looking at ways to obtain title – either through the courts or through negotiation – with implications for energy development in the province, and more broadly across Canada. In this chapter it is argued that the language of consent in *Tsilhqot'in* may help further encourage efforts to legitimate 'free, prior and informed consent' (FPIC), which is shaping how business is done globally.

For those First Nations with a strong claim for Aboriginal title, and the resources to pursue a claim in the courts, the decision in *Tsilhqot'in* confirms that the Crown must obtain their consent before infringing on their title (including those with pending claims). If the Crown does not obtain the consent of these First Nations, and the proposal is rejected, then the Crown must justify any infringement on title. The Crown must show that it has consulted these title-holders in good faith; mining and forestry projects may provide that justification. But, if the Crown chooses to move forward on the basis of a justification path, it will hand considerable power to the courts to decide what activities can occur on title lands. This scenario will lead to additional costs and delays for projects, imposing a considerable burden on First Nations, firms, the Crown, the economy and the courts. Achieving consent is likely the most efficient and effective way forward for the Crown and for gas producers.

It must be acknowledged that the *Tsilhqot'in* case included more than 300 days of court time, cost millions in legal fees over three decades and required volumes of evidence to support the Tsilhqot'in Nation's claim for title. Many First Nations, for myriad reasons, simply will not have the resources or evidence to support a title claim of this magnitude.

[1] *Tsilhqot'in Nation vs. British Columbia*, 2014 2 SCR 257.

Some First Nations may not want to pursue a claim for title, but they may draw on political and economic resources to enhance their jurisdiction in their traditional territories. The absence of Aboriginal title will not preclude these First Nations from engaging with the Crown or firms on unconventional energy projects – there are examples of First Nations without title establishing partnerships and joint ventures with LNG producers and pipeline operators across BC.

This chapter illustrates how the expansion of Aboriginal title marks an important point between First Nations, the Crown and unconventional energy producers; the language of consent from the *Tsilhqot'in* decision is part of a broader set of forces which are legitimating FPIC globally and which will reshape how extraction firms do business with Indigenous peoples.

Background and Context

The BC Government's LNG Strategy is focused on developing shale gas deposits in its northeast – and this includes the expansion of production, pipelines and liquefaction and purification facilities to meet growing demand in Asian markets (BC Government, 2014a). The strategy documents the fact that investment has grown considerably in the sector, from \$1.8 billion in 2000 to over \$7.1 billion in 2010. The Government's ambition was to build the province's first LNG-export facility by 2015 and to add another three by 2020 (BC Government, 2014a). The province recognised that it could not meet this aspiration without partnerships with First Nations in northwestern BC and the interior – where the pipelines and export facilities are to be located.

The focus of most of BC's upstream gas activity is in the northeast of the province. In 2012, 2 billion cubic feet of shale gas, or 25% of Canadian production, was derived from the region (Chong and Simikian, 2014). Gas production is the region's biggest economic driver and it is expected to grow; as of November 2013, there were estimated shale gas reserves of 2.9 trillion cubic feet in the area (Jeakins, 2014). The largest shale-gas plays in BC are the Horn River Basin, Cordova Embayment, the Montney and the Liard Basin, which, combined, account for the majority of the northeast's gas production.[2] First Nations in this region signed a treaty in 1899 with the Crown, entitled Treaty 8.[3] Treaty-8 nations have had the longest dealings with gas producers. However, with the provincial government heralding a stronger push for pipelines and production facilities, more First Nations will be engaged by the sector.

Some First Nations have already agreed to pipelines and production facilities in their territory (see for example TransCanada, 2015; Haisla First Nation, 2015). While others are opposed such as the Lax Kw'alaams on BC's northwest coast, who recently rejected a proposal by Pacific NorthWest LNG; the latter had sought the First Nation's consent to

[2] Note that since the 1970s there has been an offshore oil and gas moratorium in BC.

[3] Treaty 8 covers 840 000 km^2, and includes parts of BC, the Northwest Territories, Alberta and Saskatchewan. The Treaty affirms hunting and fishing rights for First Nations off reserve land. The Crown can take up land for the purposes of settlement or industry after consultation with relevant First Nations.

construct a pipeline and terminal in their territory (CBC, 2015). There is diversity in the choices being made by First Nations, but central to this decision are trade-offs between livelihoods and socio-ecological risk. There are also concerns about equity and distributive justice: are First Nations being offered a fair share of the benefits flowing from LNG and will they be true economic partners in LNG projects?

After *Tsilhqot'in*, the legal landscape re-emphasises that gas proponents and the Crown must engage with First Nations,[4] to a standard that may be more comprehensive than what is being practised. But *Tsilhqot'in* leaves more questions than answers. Should the Crown be making efforts to obtain consent? What is consent and who should it be obtained from? Should proponents be making efforts to ensure that any information they convey to First Nations is sufficient to meet the standard of informed consent? The answers to these questions can be negotiated between First Nations, the Crown and firms; or they can be hammered out in the courts, which is costly, time intensive and has uncertain results. Either way, these questions must be answered.

For First Nations who have a strong claim for title to their territory – that is, along with sufficient proof and evidence of their occupation they also have enough resources to support this claim through the courts – the *Tsilhqot'in* decision affirms a space for them to play a more active role in development in their territories. They will have the power to provide or withhold their consent to a project. However, if consent is withheld then the Crown can infringe on title. But such and infringement must be justified, or the public benefit must outweigh the harm to Aboriginal interests. The substantive issues to be decided by courts in this justification process will be distilled down to whether jobs and income outweigh the infringement of economic, cultural, spiritual, religious, social and ecological values held by title holders. This deliberation will inevitably be from an Anglo-Canadian perspective, though the court in *Tsilhqot'in* emphasised the importance of the Aboriginal perspective in reaching decisions of this nature in the courts.

Consent may become legally, if not practically, necessary for any energy company doing business in BC. More broadly, consent has attracted considerable attention at a global level and there is a variety of mechanisms legitimating consent as a standard. These include, for example, non-state market driven (NSMD) actors (such as certification schemes, NGOs and socially responsible investment (SRI) investors) and at the level of international law the United Nations Declaration on the Rights of Indigenous Peoples (UNDRIP), which provides for a consent regime of Free Prior and Informed Consent (FPIC).[5] International law is also influencing the decisions of domestic and tribal courts and has given effect to the UNDRIP and FPIC through common law (Carpenter and Riley, 2014). Looking ahead, consent will loom larger at a global level, as both an objective and a process, and will potentially be a standard practice for unconventional energy producers to obtain social licence.

[4] Note that *Tsilhqot'in* related to an authorisation of a timber licence; there is no LNG extraction or facilities planned for Tsilhqot'in territory.

[5] UNDRIP (2008), available at: http://www.un.org/esa/socdev/unpfii/documents/DRIPS_en.pdf.

What are Aboriginal Rights and Title?

Aboriginal rights and title are inherent rights of Aboriginal peoples in Canada, which confirm the pre-existing relationship of Aboriginal peoples to their land. Aboriginal title is an Aboriginal right: these rights are collective in nature and flow from an Aboriginal peoples' continued use and occupation of certain areas prior to British sovereignty (Slattery, 1987). The Canadian Constitution, 1982, affirms Aboriginal rights but there is little guidance in the Constitution on what these rights are. Aboriginal rights asserted by Aboriginal groups include practices that may or may not be recognised by the Crown. Hence, the courts have played an important role in defining these rights. Slattery (2007) segments Aboriginal rights into specific and generic rights. Generic rights include rights to land (Aboriginal title), subsistence rights (harvesting), the right to self-governance and the right to practise culture and language (among other things). Specific rights include those agreed to in treaties or by the courts for a specific group, including for example, the right to fish[6] or, for other groups, the right to sell fish into the market place.[7]

While these rights are recognised and affirmed under the Constitution they are not absolute and can be extinguished.[8] The *Sparrow* decision set out that legislation can infringe Aboriginal rights protected under the Constitution if it meets a two-step justification analysis, that is (1) the legislation must further a 'compelling and substantial" purpose and (2) the Crown must address the 'priority' of the infringed Aboriginal interest in a way that is consistent with the Crown's fiduciary obligation – that is, the Crown affects this right only as much as is necessary.[9] In *R. v. Van der Peet*, the SCC set out a test on what constitutes a valid Aboriginal right.[10] The test to determine the existence of an Aboriginal right asks whether the practice is central to a distinctive culture that has continued since contact and, if this is proven, then these rights are protected under s. 35 (1) of the Constitution. This test has evolved in order to address criticisms that it was freezing Aboriginal rights in time (Slattery, 2007).

Aboriginal title exists independently of other Aboriginal rights.[11] For example, a group does not have to prove Aboriginal title to exercise other Aboriginal rights (such as hunting or berry harvesting). Aboriginal title recognises that Aboriginal customary land tenure has survived the assertion of British sovereignty. Aboriginal title includes elements of Canadian, French and Aboriginal law – it is *sui generis*, that is, it is unique in law. To obtain a declaration of Aboriginal title through the courts, Aboriginal claimants must objectively prove that their occupation of the land was continuous since the time the Crown asserted sovereignty,[12] as well as exclusive and sufficient to ground title (Woodward *et al.*, 2008). Aboriginal title is tangible and capable of being registered in a land titles registry but, even after such a declaration of title, the Crown maintains underlying or radical title to land,[13]

[6] *R. v. Sparrow* [1990] 1 S.C.R. 1075 ("*Sparrow*").
[7] *Ahousaht Indian Band and Nation v. Canada (Attorney General)*, 2009 BCSC 1494 ("*Ahousaht Fishing Rights Case*"). However, this right to sell fish commercially is highly circumscribed.
[8] *Sparrow*. [9] *Sparrow* at pp. 1113–1119. [10] *R. v. Van der Peet* [1996] 2 S.C.R. 507.
[11] *R. v. Adams* [1996] 3 S.C.R. 101. [12] In BC this is deemed to be 1846.
[13] This is not unusual; the Crown has the ability to expropriate land from private land owners for fair compensation.

which is subject to Aboriginal land interests where they are established. The Crown has a special fiduciary relationship with First Nations and is the only entity that can alienate Aboriginal title land, which action must be for the benefit of the affected First Nations. Any failure in this fiduciary duty can result in a claim for damages against the Crown.[14]

Aboriginal title has both a positive and a negative component.[15] The positive dimension is that Aboriginal title gives the title holders a right to exclusive use and occupation of the title land, including traditional and economic uses. The negative dimension is that any uses of title land must be consistent with the collective nature of the group's title. That means that title lands cannot be used for a purpose that "deprives future generations of the control and benefit of the land".[16] Hence, there is a sustainability component to activities on title land.

The *Tsilhqot'in* Decision

The Tsilhqot'in Nation comprise six First Nations groups, with 3285 members, who share a common culture and history in central BC (BC Government, 2014b). In 2014, the SCC awarded the Tsilhqot'in Nation title to 1900 square kilometres of land in their traditional territories. The litigation commenced in 1983, when the BC Government allocated a forest licence to a company in Tsilhqot'in territory. The Tsilhqot'in opposed this and sought a declaration banning logging in the area. In 2002, after negotiations between the Crown and Tsilhqot'in broke down, the Tsilhqot'in included a claim for Aboriginal title before the court. The trial hearing of the title claim lasted over five years, with the trial judge finding enough evidence to award title over the land claimed. However, the claim was rejected owing to a procedural matter.[17] This trial decision was appealed by the provincial government because of the expansive definition taken by the trial judge on what is sufficient occupation to ground title.

The BC Court of Appeal (BCCA) rejected the claim for Aboriginal title on the basis of evidence that the Tsilhqot'in were "semi-nomadic".[18] The BCCA took a narrow approach to Aboriginal title, what is termed the "postage stamp" view. This postage stamp perspective requires regular and constant occupation of an area by Aboriginal claimants to ground title, for example areas must be permanent village sites (Woodward *et al.*, 2008). The BCCA concluded that the Tsilhqotin's claim for Aboriginal title was too broad and that in certain areas there was insufficient occupation. However, certain hunting and fishing rights were recognised. The Tsilhqot'in were granted leave to appeal to the SCC. The key issue to be decided was whether the test for determining Aboriginal title should adopt the postage stamp perspective or a broader approach like that determined by the trial judge?

[14] *Guerin v. R.* [1984] 2 S.C.R. 335. [15] *Tsilhqot'in* at para. 15. [16] *Tsilhqot'in* at para. 15.

[17] *Tsilhqot'in Nation v. British Columbia*, 2007 BCSC 1700. The Court in this decision held that the way in which the case was pleaded by the Tsilhqot'in, for an all-or-nothing situation for title to their territory, meant that the claim could not be awarded. However, the Court also recognised that the Tsilhqot'in have a right to obtain a moderate livelihood from trading furs in the area claimed; that the BC Forest Act does not apply to Aboriginal title lands; and that BC infringed the Tsilhqot'in's Aboriginal rights and title without justification.

[18] *William v. British Columbia*, 2012 BCCA 285.

Chief Justice McLachlin, writing for a unanimous SCC, took a broader perspective on title, concluding that Aboriginal title flows from an occupation of the land that is a regular and exclusive use of the land, from at least 1846, when the Crown asserted sovereignty to BC. To meet this standard, building on the 1997 decision of *Delgamuukw*,[19] claimants must objectively prove that this occupation is (1) sufficient, (2) continuous (if present occupation is relied upon) and (3) exclusive. In making this determination, the court must look at both the Aboriginal and common law perspectives. Courts should also examine the claimant groups' characteristics, for example, the number of members, the way of life (nomadic or sedentary etc.) and the characteristics of the land-base (size, topography and sufficient availability of resources to support people).[20] Although "[n]ot every use or nomadic passage will be sufficient to ground title",[21] regular use for fishing, hunting or resource gathering could be sufficient to prove Aboriginal title. This view is far more expansive than the traditional perspective of occupation, which was narrow and focused on sites which were intensively and continuously occupied, such as village sites.

First Nations groups who obtain a declaration of title, either through agreement or by court order, will have a right to exclusive occupation over the lands on which title is exercised. Subject to certain conditions, the title-holders can decide how they want to use the land and its resources, including the right to generate revenues. The Court summarised the nature of title at para. 88 in *Tsilhqot'in*:

"Aboriginal title confers on the group that holds it the exclusive right to decide how the land is used and the right to benefit from those uses . . . [However] the uses must be consistent with the group nature of the interest and the enjoyment of the land by future generations."

The Court emphasised that this right to enjoy the economic fruits from the title land is not just as passive actors:

" . . . this is not merely a right of first refusal with respect to Crown land management or usage plans. Rather, it is the right to proactively use and manage the land."[22]

However, the Crown can infringe on title, but it must demonstrate that it is achieving a broader public good. It can do this by satisfying a three-part justification test:

"To justify overriding the Aboriginal title-holding group's wishes on the basis of the broader public good, the government must show: (1) that it has discharged its procedural duty to consult and accommodate, (2) that its actions were backed by a compelling and substantial objective; and (3) that its actions are consistent with the Crown's fiduciary obligation to the group," *Sparrow*.[23]

To satisfy the first part of the infringement test the Crown must demonstrate that it consulted in good faith with the title holders and accommodated their interests.

[19] The test in *Delgamuukw* required claimants to prove that their occupation of lands prior to the assertion of Crown sovereignty was: sufficient, continuous and exclusive.
[20] *Tsilhqot'in* at para. 37. [21] *Tsilhqot'in* at para. 44. [22] *Tsilhqot'in* at para. 94. [23] *Tsilhqot'in* at para. 77.

The second part of the infringement test requires the Crown to have a compelling and substantial objective, which may include economic activities that generate income and jobs. However, the economic and social interests flowing from development to the public must be weighed and balanced against the Aboriginal interests. The kinds of project that may infringe on title include mining and forestry projects.[24] However, the Court ruled that granting rights to third parties to harvest timber on Tsilhqot'in land was a serious infringement that will not lightly be justified, imposing a higher threshold for activities relating to third parties.[25]

The third part of the test requires the government action to be consistent with the fiduciary duty owed by the Crown to First Nations. This duty includes recognition that Aboriginal title is collective in nature and should be preserved for future generations. It must be considered against the theme emphasised by the Court, that: "incursions on Aboriginal title cannot be justified if they would substantially deprive future generations of the benefit of the land".[26] Hence unsustainable activities for which the long-term ecological costs outweigh the short-term economic benefits may not be sufficient grounds to infringe upon title. The fiduciary duty also requires that government action relating to title land be rationally connected to a goal, and the government must not go further than is necessary to achieve that goal. The benefits generated from the activity must outweigh the adverse impact to the Aboriginal interest.[27] The governing ethos in dealings between the Crown and First Nations is not one of competing interests but of reconciliation.[28]

To avoid claims of invalid infringement or a breach of duty to consult, the SCC concluded that Governments and individuals proposing to use or exploit land, whether before or after a declaration of Aboriginal title, can avoid a charge of infringement or failure to adequately consult by obtaining the consent of the interested Aboriginal group.[29] Importantly, the SCC decided that any First Nations with a title claim pending must have their title preserved, and the Crown should also obtain their consent before development is approved. If title is established, the Crown will have to re-assess past conduct to determine whether it has validly infringed on title. For example, legislation affecting title land may be rendered invalid if it unjustifiably impacts title.[30] The message from Canada's top court is that the Crown must seek consent from title holders or those who may have a strong claim for title. If the Crown does not obtain consent and moves forward with an infringement of title, there is significant risk. For instance, damages could be payable to First Nations for any losses, and any permit authorisations such as gas tenures may be invalidated – leading to project closures and delays.

It is likely that the SCC made it clear that consent is necessary because this both procedurally and substantively protects Aboriginal title and brings far more certainty to the development process. Consent reshapes the paradigm for doing business in BC and other parts of Canada. First Nations can be viewed as true partners for development, and their values recognised and integrated in development in their territories.

[24] *Tsilhqot'in* at para. 83. [25] *Tsilhqot'in* at para. 120. [26] *Tsilhqot'in* at para. 86. [27] *Tsilhqot'in* at para. 87.
[28] *Tsilhqot'in* at para. 17. [29] *Tsilhqot'in* at para. 97. [30] *Tsilhqot'in* at para. 92.

Consent: The New Mandate?

The language of consent is not new – earlier treaties between colonial governments and First Nations required the consent of First Nations to surrender their lands in exchange for payments, land and rations (Asch and Macklem, 1991).[31] The Royal Proclamation, 1763, recognized that Aboriginal title existed in certain areas of North America and it set out guidelines that land could only be ceded to the Crown through treaty; the Crown could then sell this land to settlers. This Royal Proclamation was followed in parts of Canada but not in BC. The Crown in BC only started negotiating modern treaties in 1993, and progress on concluding these has been less than overwhelming, to date.[32]

As the expression of Aboriginal title has grown in jurisprudence, the concept of consent has taken on a renewed importance. How it will develop is not yet known, but it is likely to have important implications for resource developers across Canada and perhaps even beyond, to common law countries such Australia and New Zealand. The decision in *Tsilhqot'in* shows a significant shift in jurisprudence – only 50 years earlier in BC the doctrines of *terra incognita* and *terra nullius*, or 'no man's land', were affirmed in *R. v. White and Bob*.[33] These doctrines offered little in the way of consent from Indigenous peoples – in fact they enabled the Crown to unilaterally extinguish Aboriginal rights or title without consent.[34] The concept of consent is on a continuum of rights that are developed from interactions between Indigenous peoples, governments and industry. This continuum of rights follows a pattern of transgressions against Indigenous peoples by government and industry that were followed by measures to redress these transgressions – facilitated by the courts and NSMD actors. These response measures typically lead to enhanced rights for Indigenous peoples. This pattern is not uniform, and there may be instances where rights regress. However, the continuum is helpful in evaluating and benchmarking the performance of jurisdictions on Indigenous rights against peers.

The expansion of Aboriginal property rights in BC demonstrates this continuum in practice. In recent times the language of consent appeared in *Delgamuukw*, where the muscles of Aboriginal title were first flexed by the SCC. The SCC held in *Delgamuukw* that where a group has Aboriginal title, there may be instances that "require the full consent of an aboriginal nation, particularly when provinces enact hunting and fishing regulations in relation to aboriginal lands".[35] Consent expanded in the *Haida* decision, where the Haida argued that a transfer of forest licences in their territories to third parties was invalid without their consent. The court in *Haida* set out a spectrum for the duty to consult First Nations by the Crown (and not by third parties), ranging from a minimum duty to notify and listen to Aboriginal interests in weaker claims, to consultation and an

[31] The language of consent was also present in documents authorising the annexing of Australia (Neate, 1989), where colonial administrators were ordered to obtain the consent of native peoples before annexing land or resources.

[32] The process in BC has been slower than land claims processes in Yukon, Northwest Territories, northern Quebec and Nunavut. There were some treaties signed on Vancouver Island, the Douglas Treaties between 1850–1854 and Treaty 8 in 1899; however, in northern BC the modern treaty process in place for over 20 years has only concluded three treaties, with the Nisga'a, Maa-nuulth and Tsawassen First Nations.

[33] (1965), 52 D.L.R (Zd) 481.

[34] *Delgamuukw v. British Columbia*, 1993 CanLII 4516 (BCCA) at para. 180.

[35] *Delgamuukw v. British Columbia* [1997] 3 S.C.R. 1010 ('*Delgamuukw*') at para. 168.

accommodation[36] of Aboriginal interests where there is a strong claim for title.[37] Referencing *Delgamuukw*, the Court held that on serious issues affecting title the "full consent of [the] Aboriginal nation" may be required.[38] In *Haida* the court suggested that "Aboriginal consent spoken of in *Delgamuukw* is appropriate only in cases of established rights, and then by no means in every case.[39] Ten years after *Haida,* in *Tsilhqot'in* we see a fuller expression by Canada's top court on when consent must be obtained by the Crown, that is, consent must be obtained by the Crown, before decisions are made that infringe upon title, from both title holders and also from those whose title claim is pending. However, the SCC did not elaborate a process for obtaining consent, so there is little guidance in practice. Unless a process can be agreed upon we can expect to see the characteristics of the consent process hammered out in the courts.

While we see the courts adopting a requirement for consent, it is not required in all circumstances, nor will all First Nations have a strong enough claim for title that necessitates the Crown to obtain their consent. However, while the Crown may not need to obtain consent, firms may voluntarily adopt consent as a business practice. Typically, consent is not supported by extractive firms; for example, the International Council on Mining and Metals (ICMM 2010) contended that Indigenous consent for resource projects involves a moral dilemma that provides Indigenous peoples with a veto power that undermines state sovereignty. The fact that consent could undermine state sovereignty has meant that few countries have formally adopted the FPIC principle as a binding rule or law (Lyster, 2011). This position was confirmed by Canada, who challenged a statement by the World Council on Indigenous Peoples Outcome Document on FPIC.[40] So, while consent has the potential to create a more equitable approach to development, at its current stage it is contested. It is important to note that even in *Tsilhqot'in* there are no limits imposed on the Crown's sovereignty, so there is no veto power given to the title holders.

It is important to acknowledge that the courts are not alone in helping to further consent along its continuum. There is a range of NSMD governance mechanisms that aim to promote those firms and activities that reduce or mitigate negative impacts from globalisation and development (Cashore, 2002). The NSMD mechanisms include diverse actors, such as NGOs and civic actors, along with public and private actors, who can work in alliance to promote, define and select preferred goods and practices through diverse mechanisms. These mechanisms can include the certification of goods and processes, corporate social responsibility (CSR) measures and public reporting on impacts and market campaigns to either boycott or promote the activities of firms (Guay *et al.*, 2004). It is within this dynamic that NSMD governance mechanisms, such as forest certification and free, prior and informed

[36] Accommodation means that the Crown should take steps to avoid irreparable harm or minimise the effects of infringement until a claim is resolved: *Haida* at para. 47.

[37] *Haida Nation v. British Columbia (Minister of Forests)* [2004] 3 S.C.R. 511 ("*Haida*").

[38] *Haida* at para. 24. [39] *Haida* at para. 48.

[40] Canada challenged the call for FPIC, arguing that it offered a veto to Indigenous Peoples, and fettered the Crown's sovereignty: http://www.canadainternational.gc.ca/prmny-mponu/canada_un-canada_onu/statements-declarations/other-autres/2014–09–22_WCIPD-PADD.aspx. However, recently elected Canadian Prime Minister Justin Trudeau stated he would implement the UNDRIP.

consent (FPIC), become legitimated through the practices of firms and actors (Bernstein and Cashore, 2007). As Nikolakis *et al.* (2014) documented, legitimation can occur through the shareholder advocacy approaches pursued by some SRI mutual or pension funds; this has been effective in encouraging change among firms regarding consent and has included pressuring Chevron to develop an Indigenous People's policy and GoldCorp to adopt FPIC for its Marlin mine in Guatemala.

According to FPIC there is a different resource development paradigm, where Indigenous peoples are active participants in the development process. However, the gains on consent made by the courts, at international law and through the activities of NSMD actors can be contested by governments and firms, which slows its course.

Consent in Practice: Free, Prior and Informed Consent?

The SCC did not provide guidance on how consent is to be obtained by the Crown. However, the concept of FPIC, which is not explicitly mentioned in the *Tsilhqot'in* decision, offers a framework that represents both an objective and a procedure for dealings between Indigenous peoples and external parties (i.e. government and industry). The procedural aspect of FPIC is for: "Free" consent, or consent obtained without duress or oppression; "Prior" consent, or consent given before the decision is made; and "Informed", or educated consent. This value-laden framework has sometimes created more questions than answers, for example, who should the consent be obtained from and how should this consent be obtained (i.e. by referendum or another approach?). In *Saramaka People v. Suriname*[41] the Inter American Court on Human Rights offered that FPIC should be given effect through a specific Indigenous group's rights and customs. This includes determining who should give consent and how they should provide it; in this decision the court reasoned that the community should determine whether it be through customary approaches, such as through chiefs, or through modern plebiscite approaches. This decision by the Inter American Court on Human Rights made clear that a one-size-fits-all approach will not work for FPIC, for what is legitimate to one community may not be legitimate to another. In practice this means that governments and industry will have to work closely with groups to define how FPIC is to be given expression through the groups' traditional institutions, customs and values.

There are statutory regimes from an early vintage where informed consent is a requirement,[42] but this idea of FPIC was only formally given expression in the UNDRIP in 2007. The UNDRIP has been recognised and signed by countries like Australia and Canada, but not implemented as domestic law. Among many things the UNDRIP calls for

[41] This 2007 judgment is available at: www.forestpeoples.org/sites/fpp/files/publication/2010/09/surinameiachrsaramakajudgmentnov07eng.pdf.

[42] For example, the *Aboriginal Land Rights Act (ALRA) (Northern Territory) 1976* (Commonwealth), in Australia, provides for informed consent in section 19 (5) (a), stating that traditional Aboriginal owners must understand the nature and purpose of any dealings with their land and consent to any changes. Also, in the Philippines, the *Indigenous People's Rights Act 1997* requires the government to ensure FPIC prior to any decisions that will affect Indigenous people's lands and resources.

good faith and FPIC in administrative decisions and legislation which affect Indigenous peoples' rights (Article 19) or for developments which affect Indigenous people's land and territories to exploit or utilise water, minerals or other resources (Article 32). It is important to note that the UNDRIP is not binding on state parties until it has been ratified and implemented in domestic jurisdictions. While advances like the UNDRIP occur at the international stage, Tsotsie (2007) described the gap between advancements in international law and their implementation, in particular a lack thereof in the domestic legal frameworks of state parties. Carpenter and Riley (2014) observed, however, that elements of international human rights instruments, such as the UNDRIP, are being adopted into common law through the decisions of local and tribal courts.

The UNDRIP is also important in setting expectations for how resource development should proceed on traditional lands (Carifio, 2005). Indeed, there have been a number of recent cases, especially in Latin America, where resource development has been affected, or in some cases abandoned, where Indigenous rights (with reference to FPIC) have not been respected. In March 2011, the Colombian Supreme Court suspended three industrial projects on the basis of noncompliance with the Indigenous FPIC right. In its verdict, the Court pointed out that neither the Colombian government nor the companies carried out an adequate consultation process with the Indigenous communities affected by the projects (INDH, 2011). Talisman Energy took the decision to leave Peru after several years of protests against its oil projects on Indigenous lands; in this case there was significant national and international pressure to adopt FPIC. Talisman started to work in the area in 2004 with the consent of some families, but they never undertook a broader FPIC process, which subsequently led to conflict and criticism from the broader collective (Talisman Energy, 2012).

While there are concerns that FPIC establishes a veto power on development for Indigenous groups, in practice it requires a comprehensive process to include Indigenous peoples in development. The requirement that Indigenous peoples be informed in coming to their decision creates an onus on government and industry to educate Indigenous peoples on the project impacts in ways that encourage true understanding – this places Indigenous peoples in a more participatory role in the development process (Carifio, 2005). The concept of FPIC also requires some independent standard to decide whether the decision considered by Indigenous groups in fact constitutes free, prior, informed and true consent. This standard would need to be objective, independent and binding for it to be credible, and it might need to be reviewed by the courts.

There are arguments that focus on the potential uncertainty that consent could create for companies wanting to develop Aboriginal land. Indeed, in the *Haida* decision in Canada the SCC emphasised that the duty to consult "does not give Aboriginal groups a veto over what can be done with land".[43] Conversely, delays from protest, conflict and litigation can add considerable uncertainty to projects where the consent of Indigenous peoples has not been obtained, potentially adding billions in costs (Buxton and Wilson, 2013).

[43] *Haida* at para. 48.

While there is still a tremendous amount of work to be done in implementing FPIC globally, the *Tsilhqot'in* decision reshapes the paradigm, at least in BC, by highlighting the importance of consent when dealing with lands claimed by Aboriginal groups. Court decisions like that in *Tsilhqot'in* may have a broader impact through jurisprudence in common law jurisdictions and, combined with NSMD governance and international law, may further encourage consent along its continuum.

First Nations Legal Orders on Title Lands

An important question answered by the Court in the *Tsilhqot'in* decision was whether the *Forest Act*,[44] a provincial statute that regulates the ownership, management and use of forests, applies to Aboriginal title lands. The Court concluded that Aboriginal title land does not fall within the definition of Crown land in the *Land Act*.[45] That means that the *Forest Act*, which provides for the allocation of timber on Crown land for harvest, no longer applies to trees on land subject to a declaration of Aboriginal title but only to Crown lands.[46]

Lands under claim remain as Crown land until a declaration of title is ordered by the courts; then these lands reverts to being title lands.[47] The provincial legislature could amend the *Forest Act* to include within its ambit forests on Aboriginal title lands; however, if the legislation imposes unreasonable limitations and undue hardship or constrains the title holders from exercising their rights in preferred ways (i.e. the most efficient method) then the legislation may be an infringement of the title holders constitutional rights.[48] The issuance of timber harvest licences on title lands to third parties will constitute an infringement unless the Crown obtains the consent of the title holders or the Crown justifies its infringement.[49]

So, provincial regulations that have a general application, such as the *Forest Act*, can apply to the exercise of Aboriginal rights and Aboriginal title lands so long as they do not infringe upon constitutionally protected Aboriginal rights; in the event that there is such an infringement, this can be agreed to by the title holders or the infringement may be justified by the Crown. The aim is to strike a balance between preserving Aboriginal rights and achieving broader regulatory objectives, such as managing forests.[50] In *Tsilhqot'in*, the licences granted under the *Forest Act* were inconsistent with the duties owed to the Tsilhqot'in people, hence, the authorisations were invalid.

The approach mapped out by the Court in *Tsilhqot'in* reflects aspects of the UNDRIP, in particular Article 25, which calls for states to support the right of Indigenous peoples to maintain and strengthen their distinctive relationships with land and resources, and Article 32 (2), which calls for FPIC in relation to the development of natural resources in the territories of Indigenous peoples, including the exploitation of water. While the *Tsilhqot'in* decision emphasises that Aboriginal title enables title-holders the ability to use resources for economic benefit, their ownership of water was not decided. To date there have been no SCC cases in Canada which have affirmed inherent Aboriginal rights to water, although

[44] R.S.B.C. 1995, c. 157. [45] R.S.B.C. 1996, c. 245. [46] *Tsilhqot'in* at para. 115. [47] *Tsilhqot'in* at para. 116.
[48] *Tsilhqot'in* at para. 151. [49] *Tsilhqot'in* at para. 124. [50] *Tsilhqot'in* at para. 105.

lower courts have suggested their existence (Phare, 2009). The Crown in BC has vested rights to the ownership and management of water through legislation in the *Water Act*, to be replaced by the *Water Sustainability Act* in 2015.[51] Going forward, there may be a space for title-holders to create their own laws and regulations to manage resources on title lands. Indeed, the Tsilhqot'in moved to affirm the Nemiah Declaration after obtaining title; the Declaration is a law that governs resource use on Tsilhqot'in lands.[52] This declaration calls for a ban on commercial logging and mining on title lands.

As Indigenous groups are increasingly provided with greater scope for governance and economic rights in their territories, such as through Aboriginal title, it will be necessary to rebuild Indigenous legal orders to support these governing institutions and to rebuild an engaged citizenry (Napoleon, 2007). Developing rules on how to use natural resources will be necessary to generate economic outcomes for First Nations that ensure title is preserved for future generations. The challenge will be to match these Indigenous laws with competing values and non-Indigenous laws on the land base; this will require focused efforts to reduce conflict.

It may also follow from the decision in *Tsilhqot'in* that where other statutes, such as the incoming *Water Sustainability Act,*[53] refer to Crown land and give powers to authorise activities on land that becomes title land, the statute will no longer apply. So it is likely that section 24 of the *Water Sustainability Act,* which authorises permits to construct, maintain or operate works on Crown land, does not apply to Aboriginal title lands. Any approvals on land that then becomes title land would require the consent of title holders or any infringement to be justified. This has considerable implications for oil and gas activity on lands that may be subject to title claim. The activities of the BC Oil and Gas Commission, as set out under the *Oil and Gas Activities Act*,[54] such as permitting a master licence to cut timber or to construct works, would be restricted on Aboriginal title land (which is protected by s. 35 (1) of the *Constitution Act 1982*). Perhaps it is within this vacuum that Indigenous laws will emerge and be recognised as legitimate.

If a title-group withholds consent to the approval of works on their title lands then the Crown can attempt to justify an infringement. If this justification is challenged by title holders then the Court will be the arbiter of this decision, which creates considerable uncertainty for all parties involved. In doing this, the Crown must demonstrate that it has met its procedural duty to consult title-holders and that the activity to be authorised is supported by a compelling and substantial objective. The SCC referenced that it is likely that natural resource projects will be justified in infringing upon title.[55] But, the Crown has a fiduciary obligation to the title holders which requires the Crown to act in ways that respect and protect the Aboriginal title interests for future generations. At para. 86 the Court ruled that:

> ... incursions on Aboriginal title cannot be justified if they would substantially deprive future generations of the benefit of the land.

[51] Section 5 of the *Water Sustainability Act.*
[52] Affirmation of the Nemiah Declaration, enacted 19 March 2015. http://www.tsilhqotin.ca/PDFs/Nemiah_Declaration.pdf.
[53] Bill 18, 2014. [54] S.B.C. 2008, c. 36. [55] *Tsilhqot'in* at paras. 83–84.

Following this reasoning then it becomes less clear whether unconventional energy, with its externalities such as land clearing, seismic testing and water quality issues, will be consistent with the Crown's duty to preserve title interests for future generations. This point is likely to attract considerable judicial attention as unconventional energy expands across BC and the rest of Canada.

Concluding Insights: Development Post-*Tsilhqot'in*

Now that Aboriginal title has been affirmed in BC, there will no doubt be more awards of title. The *Tsilhqot'in* decision will also have a ripple effect across Canada and in common law jurisdictions abroad. For unconventional gas developers this will mean there is an additional landlord, one who comes to the bargaining table with a different set of values. On title lands we may also see different sets of rules developed in relation to land and resource management that may be above those required by mainstream regulations.

What is likely to have a fundamental impact in Canada and abroad is the language of consent that has developed in Canadian jurisprudence. However, it would be misguided to suggest that consent has reached its zenith in *Tsilhqot'in*. For most First Nations without title, consultation continues (which may range from simple notification of impending development to consultation and accommodation of interests). But it must also be acknowledged that consent has come a long way towards being a legitimate practice, and the courts are not alone in furthering consent. Non-state market driven (NSMD) governance actors and international law are also actively furthering consent, which is moving along a continuum towards FPIC – with transgressions by the state and companies against Indigenous groups followed up with restorative measures, typically in the form of rights legitimating consent (Nikolakis *et al.*, 2014). There was no guidance by the SCC in *Tsilhqot'in* on how consent should be achieved, but it is likely that consent processes will be context driven, reflecting local customs and practices, and will replicate the FPIC framework.

While the Crown can justify an infringement on title if consent cannot be obtained, it is likely that achieving consent offers more certainty than leaving it to the courts to decide whether there is valid justification. The SCC emphasised in *Tsilhqot'in* that incursions on Aboriginal title will not be justified if this substantially deprives future generations of the benefit of the land. Unconventional energy projects on Aboriginal title land, or land where title may be pending, will require the consent of title holders. Even then, the uses to which title land can be put must reflect the collective nature of title and be consistent with the preserving of interests for future generations, which means that any project on title land must actively seek to preserve these interests even if consent has been given.

Companies working in areas where there are active land claims in Canada would be wise to actively engage First Nations and work to an FPIC standard, and some have done so. The recent proposal put forward by Pacific NorthWest LNG, seeking the consent of the Lax Kw'alaams, demonstrates this in practice; however, this example also shows that First Nations have non-financial priorities that can trump financial concerns. The potential risk to government and proponents from not obtaining consent from existing and potential title holders can include damages and a cancellation of permits to operate (either quashing or suspending them). It is clear that the landscape has changed in BC and First Nations will play a more active role as participants in development.

Canadian jurisprudence provides for a reconciliation of Aboriginal interests with the broader public good. Reconciliation is likely to be best achieved through the development of partnerships between Indigenous groups and proponents based on true consent rather than through the exercise of uneven and unilateral power, which leads to conflict and inequity. Through negotiation and the achievement of FPIC we may begin to observe a shift to a more equitable and sustainable business paradigm.

References

Asch, M., and Macklem, P. (1991). Aboriginal rights and Canadian sovereignty: an essay on *R. v. Sparrow. Alberta Law Review* 29, 498.

Bernstein, S., and Cashore, B. (2007). Can non-state global governance be legitimate? An analytical framework. *Regulation & Governance* 1 (4), 347–371.

BC Government (2014a). British Columbia's natural gas strategy: fuelling BC's economy for the next decade and beyond. Ministry of Energy and Mines. Available at: http://www.gov.bc.ca/ener/popt/down/natural_gas_strategy.pdf (accessed on 11 June 2015).

BC Government (2014b). Profile Tsilhqot'in National Government. Available at: http://www2.gov.bc.ca/gov/topic.page?id=1072F40CADFA4966B65DB15C3877AB31 (accessed on 13 November 2014).

Buxton, A., and Wilson, E. (2013). *FPIC and the Extractive Industries, A Guide To Applying The Spirit Of Free, Prior And Informed Consent In Industrial Projects*. London, UK: International Institute for Environment and Development.

Canadian Broadcasting Corporation (CBC) (2015). Lax Kw'alaams Band reject $1B LNG deal near Prince Rupert, 13 May 2015. Available at: www.cbc.ca/news/canada/british-columbia/lax-kw-alaams-band-reject-1b-lng-deal-near-prince-rupert-1.3072293 (accessed on 11 June 2015).

Carr, S. (2012). Fuelling our economy: British Columbia's LNG strategies. BC Ministry of Energy, Mines and Natural Gas, BC Government, Victoria, BC. Available at: http://www.cleanenergybc.org/media/Plenary_2-Steve_Carr-MoE.pdf.

Carifio, J. (2005). Indigenous peoples' right to free, prior, informed consent: reflections on concepts and practice. *Arizona Journal of International & Comparative Law* 22 (1), 19–39.

Cashore, B. (2002). Legitimacy and the privatization of environmental governance: How non–state market–driven (NSMD) governance systems gain rule–making authority. *Governance* 15 (4), 503–529.

Chong, J. and Simikian, M. (2014). Shale gas in Canada: resource potential, current production and economic implications. Economics, Resources and International Affairs Division, Government of Canada, Ottawa. Available at: http://www.parl.gc.ca/Content/LOP/ResearchPublications/2014–08-e.htm#a9.

Guay, T., Doh, J.P., and Sinclair, G. (2004). Non-governmental organizations, shareholder activism, and socially responsible investments: ethical, strategic, and governance implications. *Journal of Business Ethics* 52 (1), 125–139.

Haisla First Nation (2015). Economic development projects. Available at: http://haisla.ca/economic-development/projects/ (accessed on 11 June 2015).

INDH (National Institute of Human Rights of Chile) (2011). Histórico fallo de la Corte Constitucional de Colombia a favor de los derechos territoriales, a la consulta previa y la autonomía. Available at: http://www.indh.cl/historico-fallo-de-la-corte-constitucional-de-colombia-a-favor-de-los-derechos-territoriales-a-la-consulta-previa-y-la-autonomia.

International Council on Mining and Metals (ICMM) (2010). *Good Practice Guide. Indigenous Peoples and Mining*. London, UK.

Jeakins, P. (2014). *BC's story: Effectively regulating natural gas and oil*. Presentation to a Yukon Select Standing Committee. Victoria, BC: BC Oil and Gas Commission.

Lyster, R. (2011). REDD+, transparency, participation and resource rights: the role of law. *Environmental Science and Policy* 14 (2), 118–126.

Napoleon, V. (2007). Thinking about Indigenous legal orders. Research Paper for the National Centre for First Nations Governance. Available at: http://fngovernance.org/ncfng_research/val_napoleon.pdf.

Neate, G. (1989). *Aboriginal Land Rights Law in the Northern Territory*. Chippendale, NSW: Alternative Publishing Company Co-operative.

Nikolakis, W., Nelson, H., and Cohen, D. (2014). Are indigenous peoples important to sustainable development? Evidence from socially responsible investment mutual funds in North America, *Organization & Environment* 27 (4), 368–382.

Phare, M.A.S. (2009). *Denying the Source: The Crisis of First Nations Water Rights*. Victoria: Rocky Mountain Books Ltd.

Slattery, B. (2007). A taxonomy of Aboriginal rights. In H. Foster, H. Raven and J. Webber (eds.) *Let Right Be Done: Aboriginal Title, the Calder Case, and the Future of Indigenous Rights*. Vancouver: UBC Press, pp. 111–128.

Slattery, B. (1987). Understanding aboriginal rights. *Canadian Bar Review* 66, 727–783.

Talisman Energy (2012). Talisman announces decision to exit Peru. Available at: http://www.talisman-energy.com/operations/latin_america/peru/peru_news/talisman-announces-decision-to-exit-peru.html.

TransCanada (2015). Prince Rupert Gas Transmission announces project agreement with Gitanyow First Nation. Available at http://transcanada.mwnewsroom.com/Files/33/33e08bf3-b367-441d-870d-c6f4bbcf5719.pdf (accessed on 11 June 2015).

Tsosie, R.A. (2007). Indigenous people and environmental justice: the impact of climate change. *University of Colorado Law Review*, 78, 1625.

Woodward, J., Hutchings, P., and Baker, L.A. (2008). *Rejection of the "postage stamp" approach to Aboriginal title: the Tsilhqot'in Nation decision.* Prepared for the Continuing Legal Education Society of British Columbia, 12 January 2008. Victoria: BC, Canada.

23

Fugitive Emissions from Coal Seam Gas Production

STUART DAY[1]

Introduction

Carbon dioxide emissions from natural gas combustion are generally lower than from other fossil fuels. For example, when used for electricity generation, CO_2 emissions from a gas-fired power station are about half those from a conventional coal-fired plant (Jaramillo *et al.*, 2007). However, when assessing the overall greenhouse impact of natural gas it is important to account for fugitive emissions of methane and other greenhouse gases released during production, processing and distribution operations. Because of methane's high global warming potential, even modest levels of fugitive emissions have the potential to reduce the greenhouse benefit of gas over other fuels (Alvarez *et al.*, 2011; Wigley, 2011; McJeon *et al.*, 2014).

As a consequence of the increasing worldwide production of natural gas the level of fugitive emissions from gas production has come under scrutiny over the past few years, especially those from the unconventional gas industry. During 2011 a group of researchers from Cornell University in the United States estimated that fugitive emissions from shale gas production in the US at that time were much higher than previously estimated, accounting for as much as 7.9% of production (Howarth *et al.*, 2011). Although controversial, several subsequent studies of the US unconventional gas industry also found levels of fugitive emissions consistent with the Howarth *et al.* study (e.g. Pétron *et al.*, 2012; Karion *et al.*, 2013; Caulton *et al.*, 2014). However, the results have been varied indicating that there still remains a considerable level of uncertainty.

The United States is by far the leading producer of unconventional gas and most research into fugitive emissions from gas production has been focussed on the US industry. However, there are a number of other countries with significant reserves of onshore gas that are in varying degrees of development. In Australia, the coal seam gas (CSG) industry has developed rapidly over the last decade and now supplies a significant proportion of domestic gas, especially in Queensland where CSG comprises almost 90% of production (DNRM, 2015). The Australian CSG industry is set to expand even further; three liquefied natural gas plants were due to come on line in late 2014. But, like the unconventional

[1] Email: stuart.day@csiro.au, Telephone +612 4960 6052.

gas industry more generally, there has been relatively little publicly available information on the level of fugitive emission from the Australian CSG industry. Various comparisons have been made with the high emissions reported for some of the US shale and tight gas operations (Grudnoff, 2012); however, given that there are some major differences between the production of CSG and other forms of unconventional gas, differences in geological setting as well as differences in the practices and regulation of the Australian and the US industries, this comparison may not be valid.

In this chapter we consider recent advances in the understanding of fugitive emissions from the unconventional gas industry in general but particularly in relation to CSG production in Australia.

Sources of Fugitive Emissions from CSG Production

In both conventional and unconventional gas operations, fugitive emissions arise from unintentional releases such as from leaking equipment and system upsets (i.e. accidental releases) as well as from intentional releases of gas. Intentional emissions include those from venting and flaring, the normal operation of pneumatic devices and some maintenance activities where gas must be released to allow work to be performed on items of equipment. Emissions can occur over the entire gas supply chain from the well and field production facilities, to gas processing and through to transmission and distribution at the point of sale. In the case of Australian CSG, a significant proportion of production is used to produce liquefied natural gas (LNG), the production of which may also contribute to fugitive emissions.

Much of the downstream infrastructure for processing and distributing gas is common to both conventional and unconventional sources, so it is reasonable to expect that fugitive emissions would be similar in each case. However, because unconventional gas requires different production methods and many more wells than conventional reservoirs, it is likely that there are differences in the fugitive emissions from conventional and unconventional sources. Moreover, the production methods applied to shale and tight gas reservoirs are quite different from those used for CSG, and this may also influence fugitive emissions.

Shale and tight gas occur in source strata with low permeability and require horizontal drilling and hydraulic fracture stimulation for economic extraction (Cipolla *et al.*, 2012). Most of the gas in these reservoirs is stored within the pores in a compressed form (i.e. as free gas) although some may be adsorbed in organic material in shale source rocks. After the hydraulic fracture treatment is complete the pressure within the seam causes the injection fluid to flow back to the surface, and any gas entrained in the flowback water can potentially be released as fugitive emissions.

Coal seam gas, however, is largely stored as adsorbed gas within the microporous structure of coal, with relatively little free gas (Saghafi, 2010; Moore, 2012). Consequently, an appreciable gas flow does not begin until the seam pressure is reduced by the removal of seam water and most CSG wells therefore require a downhole dewatering pump to sustain gas flow. The production profile of CSG wells is usually characterised by an initial period of high water flow that gradually decreases with increasing gas production (Moore, 2012). Hydraulic fracturing stimulation may be used on CSG wells to improve production rates but at present only about 8% of Australian wells have required this treatment (NSW Chief Scientist and Engineer, 2014).

The drilling and completion of production wells is a major activity in developing a CSG field and therefore provides the potential for fugitive releases of methane. Although there is relatively little free gas available in the coal for fugitive release, permeable and porous strata above or below the target coal seams may contain free gas that could be released during drilling. Gas wells are constructed with a steel liner that is cemented into the surrounding rock strata, which normally ensures that all the gas produced reaches the surface via the cased well bore. The cement plays an important role in determining the flow integrity of the well, preventing flow up the outside of the casing. If the casing is not properly installed or subsequently degrades in service, some gas may migrate to other strata or to the surface (Erno and Schimtz, 1996; Darrah *et al.*, 2014).

It has also been proposed that subsurface emissions of a much larger scale are possible. Tait *et al.* (2013) proposed that drilling and hydraulic fracturing operations associated with CSG production in Queensland were resulting in the formation of cracks in the overlying strata that provided a further emission route. While this is speculative at present, methane and other hydrocarbon emissions have long been associated with oil and gas production fields, and elevated soil gas and atmospheric methane levels are often indicative of underlying gas reserves (Klusman, 1993). There is some evidence that oil and gas extraction can actually reduce these natural emissions as the reservoir is depleted (Duffy *et al.*, 2007).

Fugitive emissions may also originate from surface infrastructure, which includes the individual well pads and the gas- and water-gathering pipeline networks that connect the wells to processing facilities. Coal seam gas well pads usually comprise a well head attached to the casing and often a water pump, a separator to remove water from the produced gas, piping to connect to the gas and water gathering lines along with a range of components such as pressure regulators, valves, flanges and control devices. If the well has a water pump then an engine, which is usually fuelled by gas from the well, is also located on the pad to power the pump.

Methane emissions may originate from leaking equipment both on the well pad and elsewhere throughout the field-production infrastructure. Other common sources include pneumatic control devices that are operated by the gas pressure within the gas lines, and some vents installed within the systems. In the CSG industry, it is common for high-point vents to be installed in the water-gathering system, which allow methane to vent periodically from the network. The large produced-water holding ponds and treatment facilities associated with CSG operations are also potential sources of methane. Uncombusted fuel in the engine exhaust from pumps and compressors is also a source of methane emissions.

Some pre-production operations including exploration and pilot-phase testing may also result in fugitive emissions. Generally these operations are undertaken in areas where there is no infrastructure, so any gas produced cannot be used commercially. Venting this gas would produce substantial fugitive emissions but Australian practice is usually to flare or utilise the gas to power production equipment. Since methane has a global warming potential 25 times that of CO_2, combusting excess methane significantly reduces greenhouse emissions as compared with venting.

Finally, while they are not necessarily from CSG directly, it is worth noting that methane emissions may occur throughout urban gas-reticulation systems. Recent studies in the US have shown that in some cases large amounts of gas may be lost through ageing pipeline systems (Phillips *et al.*, 2013).

Estimating Fugitive Emissions

In very broad terms, fugitive emissions are estimated by either bottom-up or top-down approaches. Bottom-up involves measuring or estimating emissions from individual parts of a process and then calculating total emissions by aggregating the individual components. This approach is often used to prepare national inventories where data supplied by industry are compiled by government regulators such as the Australian Government's Department of the Environment (Department of the Environment, 2015). One advantage of the bottom-up approach is that it may provide information on specific items of equipment or processes, which is useful in identifying and mitigating emission sources. However, there are frequently a large number of processes and components that must be accounted for to properly characterise the industry. Consequently there is a risk that some emissions will not be counted.

An alternative approach is to use top-down methods, which generally cover much larger areas than bottom-up methods and are often based on atmospheric transport methods. Much of the most recent research into fugitive emissions from the US unconventional gas industry, for instance, has used top-down atmospheric methods (Karion *et al.*, 2013; Miller *et al.*, 2013; Caulton *et al.*, 2014).

Although in principle there should be agreement between bottom-up and top-down methods, in practice this is often not the case and their results may differ by as much as a factor of two or more, with the bottom-up methods yielding the lower estimate (Nisbet and Weiss, 2010; Brandt *et al.*, 2014). Some reasons for the discrepancy include uncertainty regarding the emission factors often used in bottom-up methods, incomplete activity data, or compilation of data with some emission sources missing.

Conversely, top-down methods tend to capture all sources of methane including those not associated with the industry under investigation; hence the methodology requires complex data analysis and interpretation to remove the contribution from these sources. If the latter are not properly accounted for, an overestimation of the relevant emissions may occur.

Many of the methods used for estimating greenhouse emissions from the oil and gas industries, including fugitive emissions, are detailed in the American Petroleum Institute's Compendium of Greenhouse Gas Emissions Methodologies for the Oil and Natural Gas Industries (API, 2009). Typically, fugitive emissions are estimated using information on a particular activity and an associated emission factor. The activity may be a process or even a particular item of equipment, while the emission factor is an averaged emission rate determined for the activity. Emissions are then calculated by multiplying the activity data by the emission factor:

$$Emission\ Rate = Activity \times Emission\ Factor$$

While this approach is relatively simple to apply, it may introduce significant uncertainties because often emission factors are based on limited measurement data for the activity. For example, some of the uncertainty estimates provided in the API Compendium for emission factors are more than +/− 250%. In some cases the activity data may also be a significant source of uncertainty, which further adds to the uncertainty of the estimate.

Directly measuring emissions is potentially more accurate but while some sources such as vents may be relatively straightforward to measure, others present more of a challenge. As well as the technical difficulties of measuring emissions from diffuse sources, the size and geographical distribution of production facilities often mean that it is not practical to measure all sources using bottom-up methods.

Before considering methods for measuring fugitive emissions directly, it is important to distinguish between measuring the methane concentration and measuring the emission rate or flux from a source. Accurate measurement of methane concentration is essential for plant maintenance and safety reasons because it indicates the presence of a leak. Most gas facilities therefore have leak detection and repair regimes to locate and repair equipment leaks. Generally, leak detection involves operators inspecting equipment using handheld gas detectors or infrared imaging cameras to locate leaks, which are then repaired as necessary. Standard methods such as USEPA Method 21 are available to ensure that leak testing is performed to appropriate standards.

Obviously leak detection and repair is important in minimising fugitive emissions but the methane concentration by itself tells us little about the emission rate. In a constrained system such as a pipe (e.g. a vent), the emission rate is easily determined by multiplying the methane concentration in the gas stream by the volumetric flow rate through the pipe. Emissions from diffuse sources such as leaks or drilling operations, where the flow rate cannot be readily determined, are more difficult to quantify. In these cases, although elevated methane concentrations may be detected near the source, the concentration varies with factors unconnected with the leak rate such as distance from the source and the prevailing ambient conditions (i.e. the wind velocity, time of day, height of the boundary layer etc.).

For measuring emission rates from individual items of equipment there are several options available. One method involves sealing the component in a flexible enclosure (i.e. bagging it) and passing a stream of air or other carrier gas through the enclosure at a known rate. The concentration in the diluted air stream is measured using a suitable gas analyser and the leak rate is calculated by multiplying the concentration by the dilutent gas flow rate. Although this procedure is accurate, and is the basis of a standard method (USEPA, 1995), it is costly and very time consuming, so alternative methods are often used. One such method is the 'high-flow' technique developed by the USEPA and Gas Research Institute during the early 1990s (Kirchgessner *et al.*, 1997). In this system, methane from a leak is entrained in an air flow from a high volume air pump. The instrument measures the volumetric flow rate and the methane concentration of the mixed air stream to yield the leak rate.

Another alternative is to use the concentration data alone to infer the emission rate rather than directly measure it. Here, the methane concentration measured during screening tests is related to the emission rate by correlation with previously measured emission inventories for each component. However, this approach is subject to very high uncertainties because of the generally limited data on which the correlations are based (Broomfield and Donovan, 2012). The inherent variability of concentration measurements, due to variability in ambient conditions, also renders this method very inaccurate.

Emissions of gas migrating to the surface outside well casings can be measured using surface flux chambers. Various designs for flux chambers are used (Denmead, 2008) although they all operate by enclosing an area of soil with a chamber placed on the ground surface to capture gas emitted from the latter. In one system, the methane concentration in the chamber is measured over a period of time. Since there is no exchange of air with the outside atmosphere, the concentration of gas increases within the chamber as gas flows from the soil into the chamber during the course of the experiment. The emission flux is a function of the rate of concentration increase within the chamber and of the volume and area enclosed by the chamber. Flux chambers have been used previously for measuring well casing emissions (Erno and Schimtz, 1996; Day *et al.*, 2014) but the method has also been applied to mapping emissions from other gas emission sources. For instance, flux chambers were used in a detailed study of methane emissions from CSG infrastructure in Colorado (LTE, 2007) and from abandoned oil and gas wells in New York State and Pennsylvania (Etiope *et al.*, 2013). The method is capable of providing detailed contour maps of emission sources but, because the area covered by a chamber is typically <1 m^2 and because of factors such as soil heterogeneity and microbial variability, many measurements are required to fully characterise a given site.

Because of the difficulty in measuring emissions from every item of equipment across the entire industry, many researchers have adopted methods based on atmospheric dispersion. One of the advantages of this approach is that it can be applied at a range of scales, from individual well pads (Allen *et al.*, 2013; Day *et al.*, 2014) through to entire production regions (Karion *et al.*, 2013; Caulton *et al.*, 2014).

Atmospheric dispersion models are often used to calculate ground level concentrations of pollutants downwind of emission sources (Holmes and Morawska, 2006); however, the models can also be used in reverse to estimate source strength. In the reverse mode ground level downwind concentration, wind speed, atmospheric stability and a number of other parameters are used as input to the model to calculate the emission flux. This approach was demonstrated by Hirst *et al.* (2004) to quantify ethane emissions in an oil and gas field at a distance several kilometres downwind of the source and by Loh *et al.* (2009), who measured methane emissions within about 30 m of the source. Because concentration data are usually measured only at ground level, the vertical extent of the plume must be estimated using predictions from atmospheric dispersion methods. Although properly set up experiments can yield accurate emission fluxes, in some cases estimation of the vertical component introduces a significant source of uncertainty to the result.

One way of reducing this uncertainty is to determine the vertical profile of the plume by measuring the methane concentration within the plume at different heights. This can be achieved on large scales using an aircraft equipped with a suitable analyser to fly traverses across the plume in order to measure methane concentrations over the entire transect. The emission flux is calculated by multiplying the integrated methane profile by the average wind speed at the time of the traverse. This method was recently used to estimate emissions from the Marcellus shale region in Pennsylvania (Caulton *et al.*, 2014). In Australia, airborne traverses were used to estimate methane emissions from metropolitan areas of Sydney and Brisbane (Carras *et al.*, 1991).

This method can also be used at a smaller scale. In one example, a vehicle-based system designed for estimating emissions from natural gas wells and other local infrastructure was successfully demonstrated (Tsai *et al.*, 2012). However, in this case, since the system was ground based, it was not suitable for plumes more than about 5 m high at the traverse location.

An alternative method involves using a tracer gas that is released at a known rate from the same location as the methane source. Provided that the tracer is not reactive and is subject to the same dispersion behaviour as the target methane source, the emission rate may be calculated by multiplying the tracer release rate by the ratio of the methane concentration enhancement (i.e. the measured methane minus the background level) and the tracer enhancement (Leuning *et al.*, 2008; Allen *et al.*, 2013).

Other atmospheric methods include a variation on the inverse method described above. In this case methane concentrations are measured at a series of fixed monitoring stations positioned around the source. The concentration and local meteorological data are then combined using a suitable inverse model to calculate both the location of the sources within the field and the emission fluxes (Humphries *et al.*, 2012; Luhar *et al.*, 2014).

Some atmospheric techniques such as inverse modelling using fixed monitoring stations can potentially yield continuous monitoring, which is important since fugitive emission rates and locations can change over time. Most other methods for measuring fugitive emissions are periodic, making detecting temporal changes more difficult.

One of the principal disadvantages of wide-area atmospheric methods is that emissions from sources other than that under investigation may be detected. For instance, methane from natural or agricultural sources will be mixed with methane emissions from CSG activities. This complicates interpretation of the data, and careful analysis is required to ensure that emissions from only the CSG sources are quantified. Knowledge of the area of surveillance may help to identify some of the non-CSG sources, but other methods may also be required. In a study undertaken in the Denver–Julesburg gas field in Colorado, for example, Pétron *et al.* (2012) used the elevated levels of heavier hydrocarbons associated with the natural gas to identify spurious methane sources effectively. However, Australian coal seam gas generally does not have significant levels of hydrocarbon marker compounds, so other techniques may be needed for identifying emissions from Australian CSG fields. One option could be to measure the isotopic composition of the methane to discriminate emissions from CSG sources from other methane sources such as landfills, coal mines etc. (Zazzeri *et al.*, 2015). However, sophisticated analyses of the fugitive gases, generally at trace levels, are then necessary and this adds a level of complication to the procedure.

Fugitive Emissions Research

Estimates of methane leakage from the gas industry have been made since at least the early 1970s (Kirchgessner *et al.*, 1997). Early estimates of global loss rates ranged from about 7 to 70 Tg CH_4 per year[1], which corresponded to about 1% to 10% of total gas production. However, these estimates tended to rely on data derived from unaccounted-for-gas, which is the difference between the amounts of gas sold through the distribution network and that supplied to customers. While some of this gas is in fact lost due to leakage there are numerous other unrelated factors (e.g. measurement uncertainty, gas used to power compressors etc., accounting errors) that severely limit its use as a method for estimating fugitive emissions (Haydell, 2001).

The first truly comprehensive study of fugitive emissions from the gas industry was performed during the 1990s by the USEPA (Kirchgessner *et al.*, 1997). This investigation examined leakages and intentional releases of methane from all points in the US gas supply chain, from wells through to the point of sale. Detailed measurements were made of leaks from individual items of equipment at processing facilities, pumping and regulation stations, pipelines and meters throughout the US. This bottom-up study involved more than 200 000 individual measurements at 33 sites across the US. Despite the scope of the study, it also illustrated the difficulty of bottom-up studies as the measurements represented only a small sample of the total industry.

The results of the USEPA study showed that gas production accounted for about a quarter of emissions with the remainder associated with processing, transport, storage and distribution. On the basis of 1992 gas production data, fugitive emissions from the gas industry as a whole were estimated to represent about 1.4% of the total production, although the uncertainty of the estimate was considered to be more than 30%.

When the USEPA study was conducted, US gas production was overwhelmingly from conventional sources. Since then the unconventional gas industry has grown rapidly and now accounts for around one third of gas production and is expected to continue to grow strongly in the coming decades (EIA, 2014). However, until very recently, little additional research on emissions from the unconventional gas industry had been undertaken since the original USEPA study.

In 2011, Howarth *et al.* (2011) published a paper that reported estimates of fugitive emissions of methane from shale gas production in the US. The authors compiled data from various sources such as the USEPA and industry for several shale gas fields in the US. The data were used to estimate emissions from the production chain: well completion, venting and equipment leaks, processing losses and transport and distribution losses. They concluded that emissions from shale gas well completions were much higher than for conventional gas production, suggesting that 1.9% of total gas production is released during the water flowback and 'drill-out' stages, in comparison with their estimate of 0.01% for conventional wells. The higher emissions from shale gas wells were mainly attributed to methane entrained in the water which flows back from the well during the first few days to weeks after the wells are hydraulically fractured. Because of the higher contribution of the wellhead emissions, Howarth *et al.* (2011) estimated total fugitive emissions from the shale gas industry to be between 3.6% and 7.9% of total production over the life of the wells.

The Howarth *et al.* (2011) study assumed that all gas produced during flowback was vented to the atmosphere but critics of this study argued that gas capture and flaring are more typical of industry practice than venting (Cathles *et al.*, 2012; Cathles, 2012). Since the Howarth *et al.* (2011) study was published, however, various other studies based on atmospheric methods or satellite observations have consistently yielded emission estimates for US gas fields that are similar to or even higher than the Howarth *et al.* estimate. In the first of these, Pétron *et al.* (2012) reported methane emission estimates based on measurements made in the Denver Julesburg Basin in Colorado, where natural gas and condensate are produced from largely tight gas formations. Pétron *et al.* found strongly correlated atmospheric methane and other hydrocarbon concentrations, which they attributed to venting and leaking of natural gas and flashing from condensate storage tanks. Using a top-down approach, methane emissions were estimated to correspond to between 2.3% and 7.7% of total production, which is consistent with the estimates of Howarth *et al.* (2011). Cathles (2012) subsequently noted that condensate released from storage tanks is rich in propane relative to methane whereas the gas venting from wells contains a lower proportion of propane, which suggests that using atmospheric ratios of propane to methane may overestimate the methane flux.

Despite this criticism, other studies have found similar levels of emissions. Karion *et al.* (2013) conducted airborne surveys of the Uintah oil and gas field in Utah to estimate emissions from the region. They flew downwind of the field while measuring the horizontal and vertical methane concentration profiles within the resultant plume but, unlike the Pétron *et al.* (2012) study, used a mass-balance approach to calculate the methane flux directly and thus did not need to use marker hydrocarbon compounds. Their estimates of methane emissions from the gas field were equivalent to 8.9% $\pm$ 2.7% of gas production.

A similar airborne study was made in the Marcellus shale region of Pennsylvania (Caulton *et al.*, 2014). The study region covering about 2800 km^2 included more than 3400 shale gas wells along with some coal mines that would also be expected to be sources of methane. Emissions from gas production in the region varied significantly ranging from 2.8% to 17.3% of production, the upper limit corresponding to the largest emission rate so far reported.

Although the estimated emissions were very high, the authors of this study noted that many wells in the area did not exhibit elevated downwind methane concentrations, suggesting that leakage from these wells was relatively low. A large proportion of the emissions were thought to originate from only a small number of wells; it was estimated that as much as 30% of the emissions were derived from only 40 wells, or about 1% of the total in the study area. This is consistent with the observation made by Brandt *et al.* (2014), who suggested that a small proportion of large-emission sources were producing most of the emissions.

As well as airborne measurements, satellite observations have been used recently to estimate emissions from gas production regions across the United States (Schneising *et al.*, 2014; Kort *et al.*, 2014). In the study by Schneising *et al.*, methane concentrations in the atmosphere were measured using the SCIAMACHY sensor on the ENVISAT satellite and combined with prevailing wind data to estimate emissions from the Bakken and Eagle Ford fields in the United States. Like other top-down studies, emissions were found to be large, with leakage rates corresponding to production losses of 10.1% ± 7.3% and 9.1% ± 6.2%.

Kort *et al.* (2014) also used SCIAMACHY data to examine emissions within the Four Corners region of the US, located where the states of Arizona, New Mexico, Utah and Colorado meet. The data showed that this region has the largest methane anomaly in the US, with emissions estimated to be equivalent to 10% of the entire US inventory for natural gas systems.

A potential disadvantage of satellite-based observations is that the spatial resolution is coarse; in the case of the SCIAMACHY data the resolution is around 0.5° × 0.5°. However, because the system has a regular orbit (35 days in the case of the SCIAMACHY sensor) it is possible to examine long-term variations in emissions, which is generally not possible using aircraft or other periodic methods.

Two other recent studies have re-examined measurement data from a range of sources to derive emissions from natural gas operations in the United States (Miller *et al.*, 2013; Brandt *et al.*, 2014). Miller *et al.* (2013) determined the spatial distribution of methane emissions throughout the United States using atmospheric methane data from airborne and fixed monitoring stations. This study considered all sources of anthropogenic methane emissions, including fugitive emissions from oil and gas production. The results indicated that bottom-up inventories of US emissions were underestimating emissions by a factor ~1.5 to 1.7. However, they also suggested that in the southern central region of the US methane emissions were about 2.7 times higher than current inventories, which they attributed to an underestimation of fugitive emissions from fossil fuel extraction.

Brandt *et al.* (2014) reviewed the technical literature relating to natural gas emissions in the US and Canada from the past 20 years. Their analysis confirmed the discrepancy between top-down and bottom-up methods, with top-down estimates averaging about 1.5 times higher. A number of reasons for the under-reporting of bottom-up methods were suggested, including unrepresentative sampling, increases in the prevalence of hydraulic fracturing in recent years and errors with emission factors. One of the key findings of this study was that many investigations seem to point to a relatively small number of 'super emitters' that account for most emissions. A follow-on from this point was that it seems unlikely that some recent regional studies that report high methane emissions are representative of the gas industry overall. It was further suggested that hydraulic fracturing is not likely to be a dominant contributor to fugitive emissions from the natural gas industry and that other sources such as abandoned oil and gas wells and natural seeps may contribute to emissions.

In one of the few recent bottom-up studies reported, Allen *et al.* (2013) performed a very comprehensive study of emissions from individual gas production facilities across the US. This study was similar in approach to that conducted by the USEPA during the early 1990s (Kirchgessner *et al.*, 1997) except that it was concerned only with unconventional gas, although not CSG. Measurements were made at various stages of gas production operations at 190 sites across the United States and included well completions (specifically the period of flowback after hydraulic fracturing), gas-well unloading (i.e. the removal of liquids from well bores), well workovers and surface infrastructure on well pads.

This work found that emissions from the well completions and liquid unloadings were much lower than the current estimates used by the USEPA for compiling the US national greenhouse gas inventory. Emissions due to leaking equipment were found to be generally consistent with current estimates while emissions from pneumatic devices were somewhat higher than current estimates.

Overall, methane emission rates measured during this study represented approximately 0.42% of the total gas production. This is significantly lower than the top-down estimates. However, it was acknowledged that the results only included well measurements; downstream emissions, which contribute to total emissions, were not included in this estimate.

In Australia, there have been several studies that considered fugitive emissions from the CSG industry, although far fewer than those conducted in the US. Several approaches for examining emissions have been used. Maher *et al.* (2014) made a series of measurements within CSG fields in Surat Basin in Queensland and near Casino in NSW. They made mobile surveys of ambient methane and CO_2 concentrations using a vehicle fitted with a cavity ring-down spectrometer. Measurements were made during winter and early morning when atmospheric mixing is at its lowest, in order to yield the highest trace-gas concentrations.

Elevated methane and CO_2 concentrations were found in both the surveyed regions, especially in Queensland. These higher levels were attributed to CSG activities although the authors noted that there are a number of other significant sources of methane in the region including intensive agriculture.

Although this study provides a useful mapping of methane concentrations within the two production regions, the concentration data alone is not a reliable indicator of the magnitude of the emissions. Further work is required to identify the sources producing the elevated methane levels in the region and to estimate their emission rates.

Similar surveys were made by Iverach *et al.* (2014) of a Queensland gas field. These workers also discussed some of the complexities of identifying methane sources based on ambient concentration data alone.

To date, only one Australian study has attempted to measure the rates of fugitive emissions from CSG activities (Day *et al.*, 2014). In that study emission rates from equipment leaks and other methane sources on well pads were measured at 43 sites throughout Queensland and NSW. Well pad emissions were found to be derived from equipment leaks, vents, the operation of pneumatic equipment and exhaust from the engines used to power dewatering pumps. The mean emission rate of all these sources for all wells was $3.2\,g\ min^{-1}$, which is equivalent to about 0.02% of the average gas production from these wells. This is very much less than emission rates measured in the US. While it is likely that there are real differences between emissions from the US and the Australian unconventional gas industries, it is much too early to suggest that this single study is indicative of the scale of fugitive emissions from CSG production since it represents only a very small component of the Australian CSG industry. In fact the 43 wells examined represent less than 1% of the total number of wells currently operating in Australia. Moreover other parts of the production chain were not considered and these are likely to contribute to overall emissions. Indeed, the highest methane concentration levels reported by Maher *et al.* (2014) were apparently made in the vicinity of large gas processing plants, suggesting that these may be significant emission sources. Emissions from the large network of pipelines, water treatment plants and gathering lines associated with the CSG industry have yet to be examined in detail.

Mitigation of Fugitive Emissions

Minimising fugitive emissions from gas production is an important aspect of environmental management within the industry. However, effective mitigation strategies rely on a detailed understanding of the emission routes and the relative magnitude of the emissions.

An area of considerable controversy regarding emissions from unconventional gas has been the use of hydraulic fracturing. As discussed previously, some workers have produced estimates of emissions during the flowback period of shale and tight gas operations that are extremely high (e.g. Howarth *et al.*, 2011) while others have argued that emissions are much lower as a result of reduced-emission completion practices. Reduced-emission completions involve either capturing gas produced during flowback or, where this is not practical, flaring. To illustrate the effectiveness of these practices, O'Sullivan and Paltsev (2012) calculated the maximum gas that could potentially be released during a hydraulic fracturing operation and then compared this with the amount that would be released after modern industry methods are applied. On the basis of their analysis, the average potential emission from the gas fields considered was about 284 t CH_4 per event, which, assuming that the average flowback duration is nine days, equates to 31.5 t CH_4 per day. However, O'Sullivan and Paltsev noted that the US unconventional gas industry did not normally vent all the gas produced during flowback; instead they suggested that current field practice is better represented by 70% capture, 15% flaring and only 15% venting. Using this assumption, emissions during flowback were estimated to be only 7.5 t CH_4 per day.

The field measurements of emissions made by Allen *et al.* (2013) yielded even lower results. Although the Allen *et al.* data showed a fairly wide range of emissions, ranging from less than 0.01 t CH_4 to more than 17 t CH_4 per event, the average value was 1.7 t CH_4 per event, which equates to an average emission rate of 0.8 t CH_4 per day per well. It is quite clear, therefore, that reduced-emission completions are an effective strategy for reducing the emissions from shale and tight gas hydraulic fracturing operations.

The situation for CSG, however, is less clear. First, only a relatively small proportion of Australian CSG wells are subject to hydraulic fracturing stimulation (NSW Chief Scientist and Engineer, 2014), although this may increase in the future as less permeable seams are targeted for production. Second, even for CSG wells that are hydraulically fractured, it is unlikely that the gas flow characteristics will be the same as those for other unconventional gas wells, owing to the fact that coal seam gas generally only starts to flow in significant quantities once the pressure is reduced, i.e. the water is removed. However, in the absence of measured data it is difficult to predict the true level of emissions from Australian CSG well completion activities.

Apart for reduced-emission completions, there may be other opportunities for mitigating emissions. Pneumatic equipment powered by natural gas often vents quantities of methane during normal operation, which depending on the design, may be either on a continuous basis or intermittently while it operates. Estimates of the emission rates vary but the results of the study undertaken by Allen *et al.* (2013) confirmed that emissions from these devices represent a substantial proportion of fugitive emissions from gas production. In the United States, the USEPA Gas Star programme encourages operators to replace or retrofit gas-operated pneumatics with 'low bleed' systems, and the results from the programme have reportedly yielded savings of the order of 36.4 bcf ($\sim 1.0 \times 10^9$ m^3). Replacing gas-powered pneumatic systems with air-actuated devices has the potential to eliminate these emissions altogether, and a number of Australian CSG companies are now installing these devices throughout their production facilities (Hardisty *et al.*, 2012). Recent emission measurements made on Australian CSG wells seem to confirm the reduction of emissions associated with these low emission systems (Day *et al.*, 2014).

The proper maintenance of equipment also has the potential to reduce emissions. For example, during 2010 the Queensland Department of Employment, Economic Development and Innovation (DEEDI) initiated an audit of the CSG industry in Queensland in which 2719 CSG wells were inspected for leakage (DEEDI 2011a). Of these, five were found to have methane concentrations in the immediate vicinity higher than the lower explosive limit (LEL, i.e. >5% or 50 000 ppm) and a further 29 had methane levels between 10% and 100% of the LEL. These wells were subsequently repaired. It was also reported that numerous much lower-level leaks were detected (i.e. below 10% LEL with most below 1% LEL), although the number was not stated. Leaks that required repair were mainly from valves, flanges and other connections but there were also several instances where gas was leaking from around the well casing. The report noted that the methodology used by the companies performing the inspections was inconsistent, so that it was not possible to compare some datasets. This led to the development of a code of practice to ensure consistent inspection and reporting across the Queensland industry (DEEDI 2011b).

Conclusions

Minimising fugitive emissions from natural gas production is critical to the sustainability of the industry, but there continues to be controversy about the magnitude of emissions, especially those from unconventional gas production. Recent research in the United States is leading to a better understanding of emissions, suggesting that in some parts of the industry emissions are high although this may not be the case across the whole industry. Indications are that much of the emission is confined to a relatively few large-emission sources. Presumably, large-emission sources will be the easiest to detect, so a key priority must be to identify these sources and develop appropriate mitigation strategies.

In Australia, there is less information available on emissions from the CSG industry but an initial study suggests that, on well pads at least, emissions may be relatively low. However, it must be remembered that less than 1% of the total number of wells currently operating in Australia have been examined. With production set to expand substantially over the next few years, it will be important to monitor carefully emissions from the industry. In addition, the current measurements only relate to well pads. Emissions from well completions, which are one of the largest sources of emissions from other forms of unconventional gas, are poorly defined. Other parts of the Australian CSG production chain are also subject to high uncertainties because emissions are generally estimated rather than measured. Downstream infrastructure such as water treatment and gas processing facilities, pipelines and LNG plants are all potential sources of fugitive emissions that will contribute to the overall greenhouse footprint of the CSG industry. There is also a need to determine background emissions prior to large-scale gas developments, so that the effect of gas extraction can be reliably quantified.

The proper characterisation of fugitive emissions from the Australian CSG industry is likely to involve a combination of bottom-up and top-down methods. Top-down methods in particular offer the prospect of measuring emissions on a continuous basis across the whole industry, but in fact many Australian CSG fields are co-located with large-methane-emission sources such as coal mines and agricultural activities. Hence the application of these methods to Australian CSG operations will require careful design and interpretation to ensure accurate results.

References

Allen, D.T. Torres, V.M., Thomas, J., Sullivan, D.W., Harrison, M., Hendler, A., *et al.* (2013). Measurements of methane emissions at natural gas production sites in the United States. *Proceedings of the National Academy of Science* **110**, 18023–18024.

Alverez, A.A., Pacala, S.W., Winebrake, J.J., Chameides, W.L., Hamburg, S.P. (2012). Greater focus needed on methane leakage from natural gas infrastructure. *Proceedings of the National Academy of Science* **109**, 6435–6440.

API (2009). *Compendium of Greenhouse Gas Emissions Methodologies for the Oil and Natural Gas Industry*. American Petroleum Institute, Washington DC.

Broomfield, M., Donovan, B. (2012). Monitoring and control of fugitive methane from unconventional gas operations. United Kingdom Environment Agency, Bristol. http://cdn.environment-agency.gov.uk/scho0812buwk-e-e.pdf.

Brandt, A.R., Heath, G.A., Kort, E.A., O'Sullivan, F., Pétron, G., *et al.* (2014). Methane leaks from North American natural gas systems. *Science* **343**, 733–735.

Carras, J.N., Thomson, C.T., Williams, D.J. (1991). Measurement of methane fluxes in urban plumes. Seventh Joint Conference on Applications of Air Pollution Meteorology with AWMA, Jan. 14-18, 1991, New Orleans, La. Boston, Mass.: American Meteorological Society.

Cathles, L. (2012). Assessing the greenhouse impact of natural gas. *Geochemisty, Geophysics, Geosystems* **13**, Q06013.

Cathles, L., Brown, L., Team, M., Hunter, A. (2012). A commentary on 'The greenhouse-gas footprint of natural gas in shale formations' by R.W. Howarth, R. Santoro and Anthony Ingraffea. *Climatic Change* **113**, 525–535.

Caulton, D.R., Shepson, P.B., Santoro, R.L., Sparks, J.P., Howarth, R.H., Ingraffea, A.R. *et al.* (2014). Toward a better understanding and quantification of methane emissions from shale gas development. *Proceedings of the National Academy of Science* **111**, 6237–6242.

Cipolla, C., Lewis, R., Maxwell, S., Mack, M. (2012). Appraising unconventional resource plays: separating reservoir quality from completion effectiveness. IPTC 14677. In *Proc. International Petroleum Technology Conference, Bangkok, Thailand.*

Darrah T.H., Vengosh, A., Jackson, R.B., Warner, N.R., Poreda, R.J. (2014). Noble gases identify the mechanisms of fugitive gas contamination in drinking-water wells overlying the Marcellus and Barnett Shales. *Proceedings of the National Academy of Science Early Edition*. www.pnas.org/cgi/doi/10.1073/pnas.1322107111.

Day, S., Dell'Amico, M., Fry, R., Javanmard Tousi, H. (2014). Field measurements of fugitive emissions from equipment and well casings in Australian coal seam gas production facilities. 41 pp. CSIRO, Australia.

DEEDI (2011a). Coal seam gas well head safety program. Final Report. Department of Employment, Economic Development and Innovation, Queensland. http://mines.industry.qld.gov.au/assets/petroleum-pdf/Coal-Seam-Gas-Well-Head-Safety-Program-Inspection-Report-2011.pdf.

DEEDI (2011b). Code of practice for coal seam gas well head emissions detection and reporting. Department of Employment, Economic Development and Innovation, Queensland. http://mines.industry.qld.gov.au/assets/petroleum-pdf/code_practice_well_leak_class.pdf.

Denmead, O.T. (2008). Approaches to measuring fluxes of methane and nitrous oxide between landscapes and the atmosphere. *Plant Soil* **309**, 5–24.

Department of the Environment (2015). National Inventory Report 2013. Commonwealth of Australia, Canberra. http://www.environment.gov.au/climate-change/greenhouse-gas-measurement/publications/national-inventory-report-2013.

DNRM (2015). Queensland's petroleum and coal seam gas 2013–14. Department of Natural Resources and Mines, Queensland. (https://www.dnrm.qld.gov.au/__data/assets/pdf_file/0020/238124/petroleum.pdf).

Duffy, M., Kinnaman, F.S., Valentine, D.L., Keller, E.A., Clark, J.F. (2007). Gaseous emission rates from natural petroleum seeps in the Upper Ojai Valley, California. *Environmental Geosciences* **14**, 197–207.

EIA (2014). Annual energy outlook 2014. US Energy Information Administration, Washington DC. http://www.eia.gov/forecasts/aeo/pdf/0383%282014%29.pdf.

Erno, B., Schmitz, R. (1996). Measurements of soil gas migration around oil and gas wells in the Lloydminster area. *Journal of Canadian Petroleum Technology* **35**, 37–46.

Etiope, G., Drobniak, A., Schimmelmann, A. (2013). Natural seepage of shale gas and the origin of the 'eternal flames' in the Norther Appalacian Basin, USA. *Marine and Petroleum Geology* **43**, 178–186.

Grudnoff, M. (2012). Measuring fugitive emissions: is coal seam gas a viable bridging fuel? Policy Brief No 41, The Australia Institute, Canberra. https://www.tai.org.au/index.php?q=node%2F19&pubid=1032&act=display.

Hardisty, P.E., Clark, T.S., Hynes, R.G. (2012). Life cycle greenhouse gas emissions from electricity generation: a comparative analysis of Australian energy sources. *Energies* **5**, 872–897.

Haydell, M. (2001). Unaccounted-for gas. In *Proc. American School of Gas Measurement Technology*, pp. 148–153. http://www.asgmt.com/default/papers/asgmt2002/docs/21.pdf.

Hirst, B., Gibson, G., Gillespie, S., Archibald, I., Podlaha, O., Skeldon, K.D., *et al.* (2004). Oil and gas prospecting by ultra-sensitive optical gas detection with inverse gas dispersion modelling. *Geophysical Research Letter* **31**, L12115.

Holmes N. S., Morawskw, L. (2006). A review of dispersion modelling and its application to the dispersion of particles: an overview of different dispersion models available. *Atmospheric Environment* **40**, 5902–5928.

Howarth, R., Santoro, R., Ingraffea, A. (2011). Methane and the greenhouse-gas footprint of natural gas from shale formations. *Climatic Change* **106**, 679–690.

Humphries, R., Jenkins, C., Leuning, R., Zegelin, S., Griffith, D., Caldow, C. *et al.* (2012). Atmospheric tomography: a Bayesian inversion technique for determining the rate and location of fugitive emissions. *Environmental Science and Technology* **46**, 739–1746.

Iverach, C., Lowry, D., France, J., Fisher, R., Nisbet, E., Baker, A. *et al.* (2014). The complexities of continuous air monitoring in attributing methane to sources of production. In *Proc. Australian Earth Sciences Convention, Newcastle, Australia* July 7–10. Geological Society of Australia.

Jaramillo, P., Griffin, W.M., Matthews, H.S. (2007). Comparative life-cycle air emissions of coal, domestic natural gas, LNG, and SNG for electricity generation. *Environmental Science and Technology* **41**, 6290–6296.

Karion, A., Sweeney, C., Pétron, G., Frost, G., Hardesty, R.M., Kofler, J. *et al.* (2013). Methane emissions estimate from airborne measurements over a western United States natural gas field. *Geophysical Research Letters* **40**, 4393–4397.

Kirchgessner, D.A., Lott, R.A., Cowgill, R.M., Harrison, M.R., Shires, T.M. (1997). Estimate of methane emissions from the US natural gas industry. *Chemosphere* **35**, 1365–1390.

Klusman, R.W. (1993). *Soil Gas and Related Methods for Natural Resource Exploration.* Wiley, Chichester, UK.

Kort, E.A., Frankenberg, C., Costigan, K.R., Lindenmaier, R., Dubey, M.K., Wunch, D. (2014). Four corners: the largest US methane anomaly viewed from space. *Geophysical Research Letters* **41**, 6898–6903, doi:10.1002/2014GL061503.

Leuning, R., Etheridge, D., Luhar, A., Dunse, B. (2008). Atmospheric monitoring and verification technologies for CO_2 geosequestration. *International Journal of Greenhouse Gas Control* **79**, 14.

Loh, Z., Leuning, R., Zegelin, S.J., Etheridge, D.M., Bai, M., Naylor, T., Griffith, D. (2009). Testing Lagrangian atmospheric dispersion modelling to monitor CO_2 and CH_4 leakage from Geosequestration. *Atmospheric Environment* **43** (16), 2602–2611.

LTE (2007). Phase II Raton Basin gas seep investigation Las Animas and Huerfano Counties, Colorado. Project No. 1925, Oil and Gas Conservation Response Fund. (http://cogcc.state.co.us/documents/library/AreaReports/RatonBasin/PhaseII/Phase%2011%20Seep%20Investigation%20Final%20Report.pdf, accessed 4 August 2106).

Luhar, A.K., Etheridge, D.M., Leuning, R., Loh, Z.M., Jenkins, C.R., Yee, E. (2014). Locating and quantifying greenhouse gas emissions at a geological CO_2 storage site using atmospheric modeling and measurements. *Journal of Geophysical Research: Atmospheres* **119**, doi:10.1002/2014JD021880.

McJeon, H., Edmonds J., Bauer, N., Clarke, L., Fisher, B., Flannery, B.P. *et al.* (2014). Limited impact on decadal-scale climate change from increased use of natural gas. *Nature* **514**, 482.

Miller, S.M., Wofsya, S.C., Michalak, A.M., Kort, E.A., Andrews, A.E., Biraude, S.C. *et al.* (2013). Anthropogenic emissions of methane in the United States. *Proceedings of the National Academy of Science* **110**, 20018–20022.

Moore, T.A. (2012). Coalbed methane: a review. *International Journal of Coal Geology* **101**, 36–81.

Nisbet, E., Weiss, R. (2010). Top-down versus bottom-up. *Science* **328**, 1241.

NSW Chief Scientist and Engineer (2014). Independent Review of Coal Seam Gas Activities in NSW Information Paper: Fracture stimulation activities. NSW Government, Australia. http://www.chiefscientist.nsw.gov.au/__data/assets/pdf_file/0008/56924/140930-Final-Fracture-Stimulation.pdf.

O'Sullivan, F., Paltsev, S. (2012). Shale gas production: potential versus actual greenhouse gas emissions. *Environmental Research Letters* **7**, 044030.

Pétron, G., Frost, G., Miller, B.R., Hirsch, A.I., Montzka, S.A., Karion, A. *et al.* (2012). Hydrocarbon emissions characterization in the Colorado Front Range: a pilot study. *Journal of Geophysical Research – Atmospheres* **117**. D04304.

Phillips, N.G., Ackley, R., Crosson, E.R., Down, A., Hutyra, L., Brondfield, M. *et al.* (2013). Mapping urban pipeline leaks: methane leaks across Boston. *Environmental Pollution* **173**, 1–4.

Saghafi, A. (2010). Potential for ECBM and CO_2 storage in mixed gas Australian coals. *International Journal of Coal Geology* **82**, 240–251.

Schneising, O., Burrows, J.P., Dickerson, R.R., Buchwitz, M., Reuter, M., Bovensmann H. (2014). Remote sensing of fugitive methane emissions from oil and gas production in North American tight geologic formations. *Earth's Future* **2**, 548–558, doi:10.1002/2014EF000265.

Tait, D.R., Santos, I. Maher, D.T., Cyronak, T.J., Davis, R.J. (2013). Enrichment of radon and carbon dioxide in the open atmosphere of an Australian coal seam gas field. *Environmental Science and Technology* **47**, 3099–3104.

Tsai, T., Rella, C., Crosson, E. (2013). Quantification of fugitive methane emissions with spatially correlated measurements collected with novel plume camera. *Geophysical Research Abstracts* **15**, EGU2013-11020. EGU General Assembly 2013.

USEPA (1995). *Protocol for Equipment Leak Emission Rates*. US Environmental Protection Agency, Washington DC.

Wigley, T. (2011). Coal to gas: the influence of methane leakage. *Climatic Change* **108**, 601–608.

Zazzeri., G., Lowry, D., Fisher, R.E., France, J.L., Lanoisellé, M., Nisbet. E.G. (2015). Plume mapping and isotopic characterisation of anthropogenic methane sources. *Atmospheric Environment* **110**, 151–162.

Appendix: Units

Table A.1 *Units of measurement and abbreviations*

Units		Metric prefixes			Other abbreviations	
J	joule	k	kilo	10^3 (thousand)	bcm	billion cubic metres
l	litre	M	mega	10^6 (million)	m^3	cubic metre
t	tonne	G	giga	10^9 (billion)	ft^3	cubic feet
g	gram	T	tera	10^{12} (trillion)	bbl	barrel
W	watt	P	peta	10^{15}	Mtoe	million tonnes of oil equivalent
W h	watt hour	E	exa	10^{18}	Gcal	gigacalorie
		b	billion	10^9	Mbtu	million British thermal units

Source: adapted from BREE 2014, Guide to the Australian Energy Statistics

Table A.2 *Conversion between units of volume*[a]

	gal US	gal UK	bbl	ft^3	l	m^3
For one unit	Multiply by					
US gallon (gal)	1	0.8327	0.02381	0.1337	3.785	0.0038
UK gallon (gal)	1.201	1	0.02859	0.1605	4.546	0.0045
barrel (bbl)	42.0	34.97	1	5.615	159.0	0.159
cubic foot (ft^3)	7.48	6.229	0.1781	1	28.3	0.0283
litre (l)	0.2642	0.220	0.0063	0.0353	1	0.001
cubic metre (m^3)	264.2	220.0	6.289	35.3147	1000	1

Source: IEA 2005, Energy Statistics Manual.

[a] To find the conversion from a unit A on the left, move across that row until you reach the column headed by the desired new unit, B. The column entry that you have reached gives the number F by which B (the column heading) must be multiplied to obtain A (the unit at the left of the row in question). Thus A = FB. For example, 1 US gallon = 0.1337 × 1 ft^3.

Table A.3 *Conversion between units of energy (cf. footnote to Table A.2)*

	PJ	Gcal	Mtoe	Mbtu	GW h
For one unit	Multiply by				
petajoule (PJ)	1	238 800	0.02388	947 800	277.8
gigacalorie (Gcal)	4.1868×10^{-6}	1	10^{-7}	3.968	1.163×10^{-3}
Mtoe	41.868	10^{7}	1	3.968×10^{7}	11 630
million btu (Mbtu)	1.0551×10^{-6}	0.252	2.52×10^{-8}	1	2.931×10^{-4}
gigawatt hour (GW h)	0.0036	860	8.6×10^{-5}	3412	1

Source: IEA 2005, Energy Statistics Manual.

Table A.4 *Conversion rates (cf. footnote to Table A.2)*

	Mcm	bcm	Tcm	Mcf	bcf	Mt LNG	GJ	TJ	PJ	Mbtu
For one unit	Multiply by									
Mcm	1	10^{-3}	10^{-6}	35.31	3.53×10^{-2}	7.35×10^{-4}	38 800	38.80	3.88×10^{-2}	36 775
bcm	10^{3}	1	10^{-3}	35 313	0.03531	0.735	3.88×10^{7}	38 800	38.80	3.68×10^{7}
Tcm	10^{6}	10^{3}	1	3.53×10^{7}	35 313	735	3.88×10^{10}	3.88×10^{7}	38 800	3.68×10^{10}
Mcf	0.028	2.83×10^{-5}	2.83×10^{-8}	1	10^{-3}	2.08×10^{-5}	1099	1	1.1×10^{-3}	1041
bcf	28.32	0.028	2.83×10^{-5}	10^{3}	1	0.021	1.1×10^{6}	1099	1.099	1.04×10^{6}
Mt LNG	1361	1.361	1.36×10^{-3}	48 045	48.04	1	5.28×10^{7}	52 787	52.79	4.93×10^{7}
GJ	2.58×10^{-5}	2.58×10^{-8}	2.58×10^{-11}	9.1×10^{-4}	9.1×10^{-7}	1.89×10^{-8}	1	10^{-3}	10^{-6}	0.948
TJ	0.026	2.58×10^{-5}	2.58×10^{-8}	0.910	9.1×10^{-4}	1.89×10^{-5}	10^{3}	1	10^{-3}	948
PJ	25.77	0.026	2.58×10^{-5}	910	0.910	0.019	10^{6}	10^{3}	1	9.48×10^{5}
Mbtu	2.72×10^{-5}	2.72×10^{-8}	2.72×10^{-11}	9.6×10^{-4}	9.6×10^{-7}	2.03×10^{-8}	1.055	1.06×10^{-3}	1.06×10^{-6}	1

Glossary and Acronyms

3D Seismic: Geophysical method that allows three-dimensional images of the subsurface to be constructed by measurement of the reflection of seismic energy, i.e. waves of elastic energy travelling through rock, comparable to sound waves in air. The seismic energy is created by a controlled seismic energy source (explosives or mechanical vibrations) and detected by an array or receivers.

AGL: A large and long-standing Australian gas and power distributor, which has more recently moved upstream, developing and owning gas production and power generation including renewables.

ANH: Colombian National Hydrocarbons Agency

ANLA: Colombian National Authority of Environmental Licenses

bcm: Volumetric measure defined as a billion cubic metres, widely used in European gas industry.

Beneficial use (of produced water): Water extracted from strata from which unconventional gas is produced can under circumstances be used for other purposes, with appropriate treatment and regulatory oversight and approvals.

BG Group: A British multinational oil and gas company with operations in more than 25 countries. It was born out of the privatised British Gas.

BTEX: A group of volatile monocyclic hydrocarbons, including benzene, toluene, ethylbenzene and xylene.

Casing: A pipe that is lowered into a well and cemented in place to maintain the integrity of the well.

CBM: Coal bed methane, generally called coal seam gas (CSG) in Australia.

CMA: Cumulative management area

CNOOC: China National Offshore Oil Corporation, one of the largest Government owned oil and gas companies.

CSG: Coal seam gas

DEC: New York Department of Environmental Conservation.

DEP: Department of Environmental Protection (Pennsylvania).

Dewatering: The process of removing water from coal beds. The resultant reduction in hydraulic pressure enables gas to flow and methane to be produced economically.

DNRM: Queensland Department of Natural Resources and Mines.

EHP: The Queensland Department of Environment and Heritage Protection

EIA: United States Energy Information Administration

EIS: Environmental Impact Statement.

Eminent domain: The ability to resume private land for utilities such as roads, railways or other utilities.

ENSO: El Niño Southern Oscillation

EPA: United States Environmental Protection Agency.

GAB: The Great Artesian Basin

Henry Hub: A physical location on the Louisiana–Texas border where gas is most actively traded in the United States, effectively providing a much quoted reference price for North American gas.

Horizontal drilling: Advanced drilling technique that enable an oil or gas well to dramatically increase its intersection with hydrocarbon-bearing strata.

Hydraulic fracturing: A technique to enhance or create fracture paths for fluid flow within a rock mass by opening existing fractures or the creating new fractures by the application of high-pressure fluid on the rock.

IESC: Australian Federal Independent Expert Scientific Committee.

Injection: Delivery of fluid into a reservoir via a well.

LNG: Liquefied natural gas – methane cooled to the point where it liquefies, sharply reducing its volume and thus increasing its energy density, making transport economic over long distances and over seaborne routes.

Mbtu: Million British thermal units, widely used in North America as a basis for pricing, and also in the LNG industry.

MESD: Colombian Ministry of Environment and Sustainable Development

Million tonnes LNG: The common unit for LNG-importing countries (Japan and Korea).

MME: Colombian Ministry of Mines and Energy

Natural gas: Methane CH_4 (with or without impurities such as nitrogen). Natural gas is often classified as either biogenic (of biological origin) or thermogenic (of thermal or heat origin).

NICNAS: Australian National Industrial Chemicals Notification and Assessment Scheme.

NORM: Naturally occurring radioactive material

OGIA: The Queensland Office of Groundwater Impact Assessment.

PEL: Petroleum Exploration Licence (NSW).

Permeability: The ability of a fluid to flow through geological formations, in this case of methane through host rocks such as shale or coal.

Petajoule (PJ): One petajoule = 10^{15} joules or 1 million GJ. The heat energy content of about 37 000 tonnes of black coal or 29 million litres of petroleum.

Porosity: The open pore space within a rock, usually filled with fluid.

Proppant: Fine-grained material that is injected with fracturing fluids to hold fractures open during a hydraulic fracturing treatment.

Reservoir: A bed of rock containing oil or natural gas.

Shale: A fine-grained sedimentary rock that may contain oil or natural gas, which may not be extractable naturally.

Shale gas: Methane derived from low permeability sedimentary formations.

Spudding: The initiation of well drilling after the drill rig has been established on-site.

Stimulation: Any of a range of techniques to increase the natural permeability of the rock. These techniques include hydraulic fracturing, chemical stimulation (including the injection of acids) and temperature cycling.

TPES: Total primary energy supply, namely all inputs into the energy sector, including inputs for electricity generation.

Tubing: A small-diameter pipe that is run inside the well casing to serve as a conduit for the passage of oil and gas to the surface. The tubing can be a permanent or temporary part of the borehole.

TW h: A terawatt hour is a measure of energy equivalent to 10^{12} watts of power sustained for one hour, equal to 3.6×10^{12} kilojoules. It is a convenient unit for measuring electricity production and use at the national level.

UCG: Unconventional gas, referring chiefly to methane produced from unconventional sources and including, notably, shale gas, tight gas and coal bed methane.

UWIR: Underground water impact report

Well: Industry term for holes drilled into the earth for the purpose of gathering data or the injection or production of fluids.

Index